Legumes under Environmental Stress

Legumes under Environmental Stress

Yield, Improvement and Adaptations

EDITED BY

Mohamed Mahgoub Azooz

South Valley University, Egypt

Parvaiz Ahmad

S.P. College, Srinagar, J&K, India

WILEY Blackwell

Registered Office
John Wiley & Sons, Ltd, The Atrium, Southern Gate, Chichester, West Sussex, PO19 8SQ, UK

Editorial Offices
9600 Garsington Road, Oxford, OX4 2DQ, UK
The Atrium, Southern Gate, Chichester, West Sussex, PO19 8SQ, UK
111 River Street, Hoboken, NJ 07030-5774, USA

For details of our global editorial offices, for customer services and for information about how to apply for permission to reuse the copyright material in this book please see our website at www.wiley.com/wiley-blackwell.

Library of Congress Cataloging-in-Publication Data

Azooz, M. M.
Legumes under environmental stress : yield, improvement and adaptations / Mohamed Mahgoub Azooz, Parvaiz Ahmad.
pages cm
Includes bibliographical references and index.
ISBN 978-1-118-91708-4 (cloth)
1. Legumes–Effect of stress on. 2. Legumes–Yields. 3. Legumes–Adaptation. I. Ahmad, Parvaiz. II. Title.
SB177.L45A96 2015
633.3–dc23
2014025842

A catalogue record for this book is available from the British Library.

Wiley also publishes its books in a variety of electronic formats. Some content that appears in print may not be available in electronic books.

Set in 8.5/12pt Meridien by SPi Publisher Services, Pondicherry, India

1 2015

Contents

Contents

List of contributors

Mohammad Abass Ahanger
School of Studies in Botany
Jiwaji University
Gwalior
India

Fakiha Afzal
Atta-ur-Rahman School of Applied Biosciences
National University of Sciences and Technology (NUST)
Islamabad
Pakistan

Parvaiz Ahmad
Department of Botany
S.P. College
Srinagar
Jammu and Kashmir
India

Shakeel Ahmad
Department of Agronomy
Bahauddin Zakariya University
Multan
Pakistan

Md. Mahabub Alam
Laboratory of Plant Stress Responses
Department of Applied Biological Science
Kagawa University
Kagawa
Japan

Mohammad Ali
Institute of Biotechnology
Bahauddin Zakariya University
Multan
Pakistan

Saroj Arora
Department of Botanical and Environmental Sciences
Guru Nanak Dev University
Amritsar
India

Sadia Arshad
Atta-ur-Rahman School of Applied Biosciences
National University of Sciences and Technology (NUST)
Islamabad
Pakistan

Mohamed Mahgoub Azooz
Department of Botany
Faculty of Science
South Valley University
Qena
Egypt

Mustafeez Mujtaba Babar
Atta-ur-Rahman School of Applied Biosciences
National University of Sciences and Technology (NUST)
Islamabad
Pakistan

P. Osman Basha
Department of Genetics and Genomics
Yogi Vemana University
Vemanapuram
India

Fozia Bashir
Department of Botany
Krishna College of Science and Information Technology
Bijnor
Uttar Pradesh
India

Renu Bhardwaj
Department of Botanical and Environmental
Sciences
Guru Nanak Dev University
Amritsar
India

Syeda Fatma Hasan Bukhari
Atta-ur-Rahman School of Applied Biosciences
National University of Sciences and Technology (NUST)
Islamabad
Pakistan

Viola Devasirvatham
Faculty of Agriculture and Environment
Plant Breeding Institute
The University of Sydney
Cobbitty
NSW
Australia

Genoveva Devinar
Laboratorio de Fisiología Vegetal
Departamento de Ciencias Naturales
Facultad de Ciencias Exactas
Físico-Químicas y Naturales
Universidad Nacional de Río Cuarto
Río Cuarto
Argentina

Dilfuza Egamberdieva
Department of Microbiology and Biotechnology
Faculty of Biology and Soil Science
National University of Uzbekistan
Tashkent
Uzbekistan

Masayuki Fujita
Laboratory of Plant Stress Responses
Department of Applied Biological Science
Kagawa University
Kagawa
Japan

Pooran M. Gaur
International Crops Research Institute
for the Semi-Arid Tropics
Patancheru
Telangana
India

Vandana Gautam
Department of Botanical and Environmental Sciences
Guru Nanak Dev University
Amritsar
India

Subramaniam Gopalakrishnan
International Crops Research Institute for the Semi-Arid
Tropics (ICRISAT)
Patancheru
Andhra Pradesh
India

Neha Handa
Department of Botanical and Environmental Sciences
Guru Nanak Dev University
Amritsar
India

Abdul Hannan
Office of Assistant Land Reclamation Officer – Factory area
near Madni Masjid
Sargodha
Directorate of Land Reclamation
Irrigation Department
Punjab
Pakistan

Mirza Hasanuzzaman
Department of Agronomy
Faculty of Agriculture
Sher-e-Bangla Agricultural University
Dhaka-1207
Bangladesh

Syed Sarfraz Hussain
Australian Centre for Plant Functional
Genomics (ACPFG)
University of Adelaide
Glen Osmond
Australia

Ayesha Ihsan
Atta-ur-Rahman School of Applied Biosciences
National University of Sciences and
Technology (NUST)
Islamabad
Pakistan

Sumira Jan
Center for Research and Development (CORD)
University of Kashmir
Srinagar
India

Harpreet Kaur
Department of Botanical and Environmental Sciences
Guru Nanak Dev University
Amritsar
India

Tajinder Kaur
Department of Botanical and Environmental
Sciences
Guru Nanak Dev University
Amritsar
India

Alvina Gul Kazi
Atta-ur-Rahman School of Applied Biosciences
National University of Sciences and Technology (NUST)
Islamabad
Pakistan

Ammarah Khan
Atta-ur-Rahman School of Applied Biosciences
National University of Sciences
and Technology (NUST)
Islamabad
Pakistan

P.S. Sha Valli Khan
Department of Botany
Yogi Vemana University
Vemanapuram
India

Saifullah Khan
Atta-ur-Rahman School of Applied Biosciences
National University of Sciences
and Technology (NUST)
Islamabad
Pakistan

Tahira Khan
Atta-ur-Rahman School of Applied Biosciences
National University of Sciences
and Technology (NUST)
Islamabad
Pakistan

Sukhmeen Kaur Kohli
Department of Botanical and Environmental
Sciences
Guru Nanak Dev University
Amritsar
India

Vinod Kumar
Department of Botanical and Environmental Sciences
Guru Nanak Dev University
Amritsar
India

Arafat Abdel Hamed Abdel Latef
Department of Botany
Faculty of Science
South Valley University
Qena
Egypt

Analia Llanes
Laboratorio de Fisiología Vegetal
Departamento de Ciencias Naturales
Facultad de Ciencias Exactas
Físico-Químicas y Naturales
Universidad Nacional de Río Cuarto
Río Cuarto
Argentina

Virginia Luna
Laboratorio de Fisiología Vegetal
Departamento de Ciencias Naturales
Facultad de Ciencias Exactas
Físico-Químicas y Naturales
Universidad Nacional de Río Cuarto
Río Cuarto
Argentina

Haythem Mhadhbi
Laboratory of Legumes
Centre of Biotechnology of Borj Cedria
Hammam lif
Tunisia

Photini V. Mylona
Agricultural Research Center of Northern Greece
Thermi
Greece

Kamrun Nahar
Laboratory of Plant Stress Responses
Department of Applied Biological Science
Kagawa University
Kagawa
Japan

Alexios N. Polidoros
Department of Genetics and Plant Breeding
School of Agriculture
Aristotle University of Thessaloniki
Greece

Saiema Rasool
Forest Biotech Laboratory
Department of Forest Management
Faculty of Forestry
Universiti Putra Malaysia
Serdang
Selangor
Malaysia

Hammad Raza
Atta-ur-Rahman School of Applied Biosciences
National University of Sciences and Technology (NUST)
Islamabad
Pakistan

Abdul Gaffar Sagoo
MLL Land Reclamation Research Station, Bhakkar, Punjab,
Pakistan

Anket Sharma
Department of Botanical and Environmental Sciences
Guru Nanak Dev University
Amritsar
India

Ram Sharma
International Center for Agricultural Research in the Dry Areas (ICARDA)
Tashkent
Uzbekistan

Bujun Shi
Australian Centre for Plant Functional Genomics (ACPFG)
University of Adelaide
Glen Osmond
Australia

Vyacheslav Shurigin
Department of Microbiology and Biotechnology
Faculty of Biology and Soil Science
National University of Uzbekistan
Tashkent
Uzbekistan

Ravinder Singh
Department of Botanical and Environmental Sciences
Guru Nanak Dev University
Amritsar
India

Daniel K.Y. Tan
Faculty of Agriculture and Environment
Plant Breeding Institute
The University of Sydney
Cobbitty
NSW
Australia

Ashwani Kumar Thukral
Department of Botanical and Environmental Sciences
Guru Nanak Dev University
Amritsar
India

Richard M. Trethowan
Faculty of Agriculture and Environment
Plant Breeding Institute
The University of Sydney
Cobbitty
NSW
Australia

Muhammad Waqas
Directorate General of Agricultural Extension
Office of Agricultural Officer
Kahna Nau
Lahore
Pakistan

Najam-us-Sahar Sadaf Zaidi
Atta-ur-Rahman School of Applied Biosciences
National University of Sciences and Technology (NUST)
Islamabad
Pakistan

Preface

Legumes represent the most utilized plant family with 20,000 species and are among the most important crops worldwide, having major impacts on agriculture, the environment, and human/animal nutrition and health. Legumes rank third behind cereals and oilseeds in world production, accounting for 27% of the world's primary crop production. Grain legumes constitute an important dietary constituent for humans and animals and these alone contribute 33% of the dietary protein nitrogen (N) needs of humans besides being a source of income and livestock feed. These perfectly match the requirements of small-scale, low-income farmers in the developing countries where they accounted for 61.3 million hectares in 2002, compared to 8.5 million hectares in developed countries. The primary dietary legumes are common beans, pea, chickpea, broad bean, pigeon pea, cowpea and lentil. Legumes are also major sources for vegetable oil, with soybean and peanut providing more than 35% of the world's processed vegetable oil.

Legumes are often exposed to environmental stresses (biotic and abiotic) that decrease productivity throughout the world. Abiotic stresses (salt, drought, temperature, UV, nutrient deficiency) alone are responsible for more than 50% yield reductions of some major crops. Abiotic stress causes osmotic and oxidative stress within the plant. The oxidative stress is caused by the generation of reactive oxygen species (ROS). These ROS react with biomolecules like proteins, nucleic acids, membrane lipids, etc. and hamper their normal functioning in the cell. Plants respond to these stresses through synthesis of metabolites and antioxidant enzymes that enhance tolerance mechanisms in plants under stress. Biotechnology approaches are also used for the improvement of legume crops under environmental stresses. The present volume comprises 17 chapters that provide detailed information on legumes, yields, mitigation strategies for different abiotic stresses, and new approaches in alleviating environmental stress in legumes.

Chapter 1 gives an overview of legumes and breeding under abiotic stress. Different abiotic stresses, and breeding of cool and warm season food legumes, are well documented. Chapter 2 discusses the effect of salt stress on leguminous crops. Omics approaches for understanding salt stress responses in legumes are also mentioned. Chapter 3 throws light on the effect of different abiotic stresses on legumes. The effects of nutrient deficiency, methods to control nutrient deficiency and the role of these nutrients in alleviating abiotic stress are well explained. Chapter 4 is about chickpea, its role and responses under abiotic and biotic stress. The role of omics in investigating chickpea under abiotic stress, and breeding of chickpea to resist biotic stress are also explained in detail. Chapter 5 deals with the effect of temperature stress on chickpea at different growth stages. Chapters 6 and 7 explain the effects of pesticides on legumes. Chapter 8 highlights the symbiotic association between legumes and rhizobia under abiotic stress, the nodulation process, and the effect of osmotic stress on nodule integrity and functioning. Chapter 9 deals with microbial strategies for improving legume production under hostile environments. Chapter 10 discusses the role of abscisic acid (ABA) in legumes under abiotic stress. ABA regulation of leaf expansion, and nodulation under abiotic stress is well documented. Chapter 11 describes the exogenous application of phytoprotectants in legumes to combat environmental stress. The roles of phytohormones, compatible solutes, nutrients and trace elements are nicely explained. Chapter 12 is about genetic and molecular responses of legumes under changing environments. Chapter 13 deals with omics approaches and abiotic stress tolerance in legumes. Transcriptomics, proteomics, genomics, metabolomics, transgenomics, functional genomics and phenomics of different legumes under stress are well documented. Chapter 14 discusses microRNA (miRNA)-mediated regulatory functions under abiotic stresses in legumes. Here the authors discuss miRNA identification, functional diversity, and expression profiling under abiotic stress in legumes. Chapter 15 deals with biotechnological approaches to overcome biotic and abiotic constraints in legumes. Chapter 16 explains gene pyramiding and omics approaches to enhance stress tolerance in leguminous plants. Finally, Chapter 17 discusses how to combat phosphorus deficiency on alkaline calcareous soils by using an adsorption isotherm technique for legume crops in arid environments.

This volume is a comprehensive account of current knowledge about the physiological responses and adaptability of legumes to salt, temperature and other environmental stresses. We have tried our best to ensure the accuracy of the information in this volume, however, there is a possibility that some errors remain, for which we seek readers' indulgence and feedback. We are also very grateful to the authors for their valuable contributions. Gratitude is also due to the publisher, John Wiley & Sons, Ltd., particularly Gudrun Walter (Editorial Director, Natural Sciences), Fiona Seymour (Senior Project Editor), Laura Bell (Assistant Editor), Renee Last, and all the other staff members of Wiley, who were directly or indirectly associated with us in this project for their constant help, valuable suggestions, and efforts in bringing out the timely publication of this volume.

M.M. Azooz
Parvaiz Ahmad

About the editors

Prof. Dr Mohamed Mahgoub Azooz

Prof. Dr Mohamed Mahgoub Azooz is a Professor of Plant Physiology in the Department of Botany, Faculty of Science, South Valley University, Qena, Egypt. He received his BSc in 1984 and MSc in 1990 from Assiut University, Egypt; and a PhD in 1997 from South Valley University, Qena, Egypt, in collaboration with Tübingen University, Germany. He has been seconded to King Faisal University, Saudi Arabia, to act as full-time faculty member (from 2005 to present) in the Department of Biological Sciences, College of Science. Prof. Azooz has published more than 90 research articles in peer-reviewed journals and contributed 13 chapters in internationally published books. He has also authored three books and edited five volumes with international publishers. He has supervised and approved many MSc students and PhD scholars, and is a member of the editorial boards and reviewer of many international journals. Prof. Azooz has participated in varying capacities as principal investigator, co-investigator, consultant and collaborator in more than 20 research projects. He has been awarded scientific publication prizes 2012, 2013 and 2014 from both King Faisal University and South Valley University for his scientific articles published in international scientific journals with high impact factors. Prof. Azooz has been included in Marquis *Who's Who in the World* (28th edition, 2011), and in the 'Top 100 Scientists' list published by the International Biographical Centre (Cambridge, UK, 2011). His current research is on the physiological, biochemical and molecular responses of plants against various biotic and abiotic stresses and their mechanisms of tolerance, with the main focus on medicinal and economic plants, aiming to improve their growth and productivity using some growth stimulators.

Dr Parvaiz Ahmad

Dr Parvaiz is Senior Assistant Professor in the Department of Botany at Sri Pratap College, Srinagar, Jammu and Kashmir, India. In 2000 he completed his postgraduate studies in botany at Jamia Hamdard, New Delhi, India. After receiving a Doctorate from the Indian Institute of Technology (IIT), Delhi, India, he joined the International Centre for Genetic Engineering and Biotechnology, New Delhi, in 2007. His main research areas are stress physiology and molecular biology. He has published more than 35 research papers in peer-reviewed journals and 29 book chapters. He is also an editor of 13 volumes (one with Studium Press Pvt. India Ltd., New Delhi, India; nine with Springer USA; and three with Elsevier USA). He is a recipient of the Junior Research Fellowship and Senior Research Fellowship from the Council of Scientific and Industrial Research (CSIR), New Delhi, India. In 2007 Dr Parvaiz was awarded the Young Scientist Award under a Fast Track scheme by the Indian Department of Science and Technology (DST). Dr Parvaiz is actively engaged in studying the molecular and physiobiochemical responses of different agricultural and horticultural plants under environmental stress.

CHAPTER 1

Legumes and breeding under abiotic stress: An overview

Arafat Abdel Hamed Abdel Latef[1] and Parvaiz Ahmad[2]

[1] *Department of Botany, Faculty of Science, South Valley University, Qena, Egypt*

[2] *Department of Botany, S.P. College, Srinagar, Jammu and Kashmir, India*

1.1 Introduction

The present world population of 7.2 billion is expected to reach 9.6 billion by the middle of the 21st century due to the high growth rate, particularly in developing countries. There is a need to produce about 70% more food to feed this excessive population (Varshney & Roorkiwal, 2013).

Legumes belong to the family Fabaceae/Leguminosae (with about 700 genera and 18,000 species). Legume crops can be divided into two groups according to their ability to grow in different seasons, namely cool season food legumes and warm or tropical season food legumes (Miller *et al.*, 2002; Toker & Yadav, 2010). The cool season food legumes include broad bean (*Vicia faba*), lentil (*Lens culinaris*), lupins (*Lupinus* spp.), dry pea (*Pisum sativum*), chickpea (*Cicer arietinum*), grass pea (*Lathyrus sativus*) and common vetch (*Vicia sativa*) crops (FAOSTAT 2009; Andrews & Hodge, 2010). These are among the world's oldest cultivated plants (Materne *et al.*, 2011). Dry pea, chickpea, broad bean and lentil are the four major cool season grain legume crops produced for human consumption. They are grown on all continents except Antarctica. Lupin species – e.g. *Lupinus albus* (white lupin) and *Lupinus luteus* (yellow lupin) – and vetches – in particular, common vetch – are important for animal feed (Andrews & Hodge, 2010). On the other hand, the warm season food legumes include pigeon pea (*Cajanus cajan*), cowpea (*Vigna unguiculata*), soybean (*Glycine max* L.), mung bean (*Vigna radiata* var. *radiata*) and urd bean (*Vigna mungo*) crops, which are mainly grown in hot and humid climatic conditions. Warm season food legumes are popular in different parts of world; for example, pigeon pea is mainly grown in India and African countries, cowpea and soybean are important crops in the USA, while mung bean and urd bean are important crops in Southeast Asian countries, especially in the Indian subcontinent (Singh *et al.*, 2011).

Legumes rank third after cereals and oilseeds in world production and have major effects on the environment, agriculture, and animal and human nutrition and health (Graham & Vance, 2003; Dita *et al.*, 2006; Mantri *et al.*, 2013). Legumes are a primary source of amino acids and provide around one-third (20–40%) of all dietary protein (Zhu *et al.*, 2005; Kudapa *et al.*, 2013). Legumes produce secondary metabolic compounds that can protect the plant against pathogens and pests (Kudapa *et al.*, 2013).

Legumes are second to cereals in providing food for humans worldwide (Kamal *et al.*, 2003; Ashraf *et al.*, 2010; Kudapa *et al.*, 2013). In comparison with cereal grains, legume seeds are rich in protein, and thus are a source of nutritionally rich food (Ahlawat *et al.*, 2007; Ashraf *et al.*, 2010; Kudapa *et al.*, 2013). Grain legumes such as chickpea, pigeon pea, cowpea, dry pea, lentil, mung bean, urd bean, bean (*Phaseolus vulgaris* L.), broad bean and grass pea are the main source of dietary protein for vegetarians, and are an integral part of the daily diet in several forms worldwide. In addition, grain legumes, predominantly peanut (*Arachis hypogaea* L.) and soybean are also major sources for vegetable oil, providing more than 35% of the world's processed vegetable oil (Sharma *et al.*, 2010).

Legumes under Environmental Stress: Yield, Improvement and Adaptations, First Edition. Edited by Mohamed Mahgoub Azooz and Parvaiz Ahmad.

Legumes play an important role in diet and they are often referred to as 'poor man's meat'. Legumes are an important source of protein, oil, fibre and micronutrients, and play a vital role in cropping cycles due to their ability to fix atmospheric nitrogen (El-Enany *et al.*, 2013; Mantri *et al.*, 2013).

Under conducive environmental conditions, legumes develop symbiotic associations with arbuscular mycorrhizal (AM) fungi, leading to the formation of sites of phosphorus nutrient exchange called arbuscules (Parniske, 2008; Mantri *et al.*, 2013).

Biological fixation of nitrogen (N) is considered more ecofriendly than industrial N fixation because the NH_3 produced in the former process is readily assimilated into organic forms by the plant (Valentine *et al.*, 2011). Biological nitrogen fixation (BNF) in legume nodules occurs with differentiated forms of rhizobia, termed bacteroids, within specialized structures called symbiosomes, inside the host plant cells (Arrese-Igor *et al.*, 2011). Thus, these symbiotic associations have strongly driven the investigation and application of biotechnology tools for legumes (Dita *et al.*, 2006).

It is estimated that crops grown on 90% of arable lands experience one or more environmental stresses. Abiotic stress causes more than 50% of crop loss worldwide (Rasool *et al.*, 2013; Rodziewicz *et al.*, 2014). 'Abiotic stress' is a broad term that includes multiple stresses (drought, waterlogging, salinity, heat, chilling and mineral toxicities) and negatively affects the adaptability and yield of legumes. Application of biotechnology tools to legume crops can help in solving or reducing the problems resulting from abiotic stress.

This chapter aims to review the main abiotic stresses that have a negative impact on the production of some important food legumes. It also summarizes the selection criteria and available genetic resources for stress resistance under abiotic stress conditions.

1.2 Legumes under abiotic stress

1.2.1 Legumes under drought

Drought is a type of water stress that is imposed due to lack of rainfall and/or inadequate irrigation. About 60% of all crop production suffers from drought conditions (Grant, 2012; Naeem *et al.*, 2013). For legumes, drought stress has adverse effects on total biomass, pod number, seed number, seed weight and quality, and seed yield per plant (Toker *et al.*, 2007b; Charlson *et al.*, 2009; Khan *et al.*, 2010; Toker & Mutlu, 2011; Impa *et al.*, 2012; Hasanuzzaman *et al.*, 2013; Pagano, 2014). Drought alone resulted in about a 40% reduction in soybean yield (Valentine *et al.*, 2011). Faba bean and pea are known to be drought-sensitive, whereas lentil and chickpea are known as drought-resistant genera (Toker & Yadav, 2010). Singh *et al.* (1999) arranged warm season food legumes in increasing order of drought tolerance: soybean < blackgram < greengram < groundnut < Bambara nut < lablab < cowpea. Sinclair and Serraj (1995) reported that legumes such as faba (broad) bean, pea and chickpea export amides (principally asparagine and glutamine) in the nodule xylem are generally more tolerant to drought stress than cowpea, soybean and pigeon pea, which export ureides (allantoin and allantoic acid).

The symbiotic nitrogen fixation (SNF) rate in legume plants rapidly decreased under drought stress due to (i) the accumulation of ureides in both nodules and shoots (Vadez *et al.*, 2000; Charlson *et al.*, 2009), (ii) decline in shoot N demand, (iii) lower xylem translocation rate due to a decreased transpiration rate, and (iv) decline of metabolic enzyme activity (Valentine *et al.*, 2011). Several reports have indicated that drought stress led to inhibition in nodule initiation, nodule growth and development as well as nodule functions (Vadez *et al.*, 2000; Streeter, 2003; Valentine *et al.*, 2011). The decrease in SNF under drought conditions was associated with the reduction of photosynthesis rate in legumes (Ladrera *et al.*, 2007; Valentine *et al.* 2011).

In many nodules of legumes, water stress resulted in stimulation of sucrose and total sugars (González *et al.*, 1995, 1998; Ramos *et al.*, 1999; Streeter, 2003; Gálvez *et al.*, 2005; Valentine *et al,*. 2011). This was consistent with a study on pea mutants, which showed that sucrose synthase (SS) is essential for normal nodule development and function (Craig *et al.*, 1999; Gordon *et al.*, 1999).

Drought stress induces oxidative damage in legumes and this has a harmful effect on nodule performance and BNF (Arrese-Igor *et al.*, 2011). Some reports suggest that nodules having an increment in enzymatic antioxidant defence can display a higher tolerance to drought/salt stress in common bean (Sassi *et al.*, 2008) and chickpea (Kaur *et al.*, 2009). In addition to this, Verdoy *et al.* (2006) reported improved resistance to drought stress in *Medicago truncatula* by overexpression of Δ-pyrroline-5-carbolyate synthetase resulting in accumulation of high proline levels.

Generally, the mechanisms of drought tolerance include (i) escape, (ii) avoidance, or (iii) resistance (Ishitani *et al.*, 2011; Toker & Mutlu, 2011; Impa *et al.* 2012; Rapparini & Peñuelas, 2014). There are several screening and selection techniques for drought tolerance in food legumes; however, few techniques have been successful under field conditions (Toker & Mutlu, 2011):

1 Line source sprinkler irrigation systems (Saxena *et al.*, 1993).
2 Root trait characteristics (root length, root density, root biomass, root length density; Serraj *et al.*, 2004) and the 'root-box pin board' method (Singh & Matsui, 2002).
3 Delayed sowing strategy (Singh *et al.*, 1997).
4 Comparison of lines under non-stressed and stress conditions by defined formula (Silim & Saxena, 1993; Toker & Cagirgan 1998).
5 Rain-out shelter tunnels (Abdelmula *et al.*, 1999; Amede *et al.*, 1999; Link *et al.*, 1999).

[Methods 1–5 can be useful for large-scale screening, but are labour and time consuming (Toker & Mutlu, 2011).]

6 Delayed canopy wilting (DCW) in soybean (Charlson *et al.*, 2009).
7 Delayed leaf senescence (DLS) trait in cowpea (Hall *et al.*, 2002).
8 Leaf pubescence density (LPD) in soybean for drought-prone environments (Du *et al.*, 2009).
9 Recovery ability after wilting (RAW) in chickpea (Toker *et al.*, 2007b).
10 The use of carbon isotope discrimination ($\Delta^{13}C$) in screening. This is described for some food legumes (Stoddard *et al.*, 2006; Khan *et al.* 2010), but it incurs high costs per sample (Toker & Mutlu, 2011).

1.2.2 Legumes under salinity

Salinity is a major abiotic stress limiting germination, plant vigour and yield of agricultural crops especially in arid and semi-arid regions (Munns & Tester, 2008; Abdel Latef & Chaoxing, 2011; Aggarwal *et al.*, 2012; Ahmad & Prasad 2012a, 2012b; Porcel *et al.*, 2012; Kapoor *et al.*, 2013; Abdel Latef & Chaoxing, 2014). Approximately 20% of irrigated land worldwide currently is affected by salinity, particularly in arid and desert lands, which comprise 25% of the total land area of our planet (Yeo, 1999; Rasool *et al.*, 2013). High salinity affects plants in several ways: water stress, ion toxicity, nutritional disorders, oxidative stress, alteration of metabolic processes, membrane disorganization, reduction of cell division and expansion, and genotoxicity (Hasegawa *et al.*, 2000, Munns, 2002; Zhu, 2007; Shanker & Venkateswarlu, 2011; Gürsoy *et al.*, 2012; Djanaguiraman & Prasad, 2013). Together, these effects reduce plant growth, development and survival (Rasool *et al.*, 2013; Hameed *et al.*, 2014).

Food legumes are relatively salt sensitive compared with cereal crops, thus farmers do not consider growing food legumes in salinized soils (Saxena *et al.*, 1993; Toker & Mutlu, 2011; Egamberdieva & Lugtenberg, 2014). The sensitivity in legumes may be due to salt affecting bacterial activity and nitrogen fixation (Materne *et al.*, 2007; Toker *et al.*, 2007a; Toker & Mutlu, 2011; Egamberdieva & Lugtenberg, 2014). Salt stress led to reduction in shoot growth of soybean, chickpea, pea, faba bean and mung bean plants (Elsheikh & Wood, 1990, 1995; Delgado *et al.*, 1994; Hussain *et al.*, 2011; Saha *et al.*, 2010; Rasool *et al.*, 2013).

The response of BNF in contrasting tolerance lines of *Medicago ciliaris* to salt stress did not show a clear trend in relation to nodule carbohydrate metabolism (Ben-Sala *et al.*, 2009). Nodules of common bean (Sassi *et al.*, 2008) and chickpea (Kaur *et al.*, 2009) display a higher tolerance to osmotic/salt stress due to increased enzymatic antioxidant defence (Arrese-Igor *et al.*, 2011).

Salinity stress significantly decreased the activities of nitrogenase and phosphate enzymes (acid and alkaline) in faba bean (Rabie *et al.*, 2005; Hussain *et al.*, 2011). The effect of salinity stress on growth and some metabolic activities of mung bean was investigated by Saha *et al.* (2010). They concluded that salinity stress suppressed the early growth of mung bean seedlings. Salinity also damaged the photosynthetic machinery by causing reduced chlorophyll content, and also induced the accumulation of proline, malondialdehyde (MDA) and H_2O_2 in roots and leaves of mung bean plants. Furthermore, salinity stress caused increments in the activity of superoxide dismutase (SOD), catechol peroxidase (CPX) and catalase (CAT) in root and leaves of mung bean plants. Recently, Rasool *et al.* (2013) reported that tolerance of chickpea genotypes (SKUA-06 and SKUA-07) to salinity seems to be related to the efficiency of the enzymatic antioxidants SOD, CAT, ascorbate peroxidase (APX) and glutathione reductase (GR) against accumulation of reactive oxygen species (ROS), which would maintain the redox homeostasis and integrity of cellular components.

Several criteria have been used in screening for salinity tolerance including germination, radicle length, dry weight production, shoot length, cell survival, plant biomass, nodulation, number of pods, grain yield and K^+/Na^+ ratio (Toker *et al.*, 2007a; Flowers *et al.*, 2009; Toker & Mutlu, 2011).

1.2.3 Legumes under waterlogging

Waterlogging occurs when water enters the soil faster than it can drain away under gravity. Waterlogging is a major abiotic factor causing losses in food legumes (Toker & Mutlu, 2011; Ashraf, 2012; El-Enany *et al.*, 2013). It negatively affected germination, seedling emergence, root and shoot growth, and plant density by up to 80%, besides causing seedling diseases (Toker & Mutlu, 2011).

When mung bean plants were subjected to waterlogging stress, the activities of various enzymatic antioxidants such as SOD, CAT, APX and GR decreased markedly (Ahmed *et al.*, 2002). These authors also stated that oxidative damage was not directly involved in the impairment of photosynthetic machinery of plants under waterlogged conditions. In contrast, increase in the activities of different enzymatic antioxidants – SOD, CAT, peroxidase (POD) and APX – was recorded in pigeon pea genotypes when subjected to varying degrees of waterlogging stress (Kumutha *et al.*, 2009).

El-Enany *et al.* (2013) carried out a pot experiment with three replicates of 75% and 50% water deficit (WD) and one-fold field capacity waterlogging (WL) on cowpea (*Vigna sinensis*) plants. The data revealed that both stresses significantly decreased the fresh and dry weights of roots and shoots, number of nodules per plant and nitrogenase activity. Antioxidant metabolites like phenolic compounds, ascorbic acids, proline, MDA and H_2O_2 were significantly increased under WD and WL. The activities of certain antioxidant enzymes (SOD, CAT and APX) under both stresses were determined (El-Enany *et al.*, 2013).

Waterlogging reduces the endogenous levels of nutrients in different parts of the plant (Ashraf *et al.*, 2010, 2012). Oxygen deficiency in the root zone causes a marked decline in the selectivity of K^+/Na^+ uptake and impedes the transport of K^+ to the shoots (Ashraf *et al.*, 2012). When *Medicago sativa* was subjected to flooding stress, a marked reduction in leaf and root nutrient composition (P, K, Ca, Mg, B, Cu and Zn) was recorded in plants (Smethurst *et al.*, 2005). Among cool season food legumes, faba bean is more tolerant to waterlogging than lentil, pea and chickpea (Siddique, 2000).

Many management practices used to reduce the effects of waterlogging involve paddock selection, sowing time, seeding rate and drainage (Toker & Mutlu, 2011). Genetic variation in waterlogging tolerance in food legumes deserves attention (Toker *et al.*, 2007a; Toker & Mutlu, 2011).

1.2.4 Legumes under temperature extremes

Temperature is one of the major factors affecting the yield and quality of legumes (Christophe *et al.*, 2011). Heat stress often is defined as high temperatures that cause *irreversible* damage to plant function or development after a certain period of exposure (Bhattacharya & Vijaylaxmi 2010; Hasanuzzaman *et al.*, 2013). Plants can be damaged in different ways by either high day or high night temperatures, and by either high air or high soil temperatures. Also, crop species and cultivars differ in their sensitivity to high temperatures. High temperature may negatively affect photosynthesis, respiration, water relations and membrane stability, and also modulate levels of hormones and primary and secondary metabolites. Furthermore, throughout plant ontogeny, enhanced expression of a variety of heat-shock proteins, other stress-related proteins, and production of ROS constitute major plant responses to heat stress (Bhattacharya & Vijaylaxmi, 2010; Hasanuzzaman *et al.*, 2013).

Cool-season annual species are more sensitive to hot weather than warm-season annuals (Hall, 2001). Heat stress affected nitrate assimilation in legumes by lowering synthesis of ureides and decreasing levels and activities of nitrate reductase and glutamate synthase (Hungria & Vargas, 2000; Christophe *et al.*, 2011). In nodules, heat stress may either affect nitrogenase activity leading to decreased N_2 fixation efficiency or accelerate nodule senescence resulting in reduced nodule longevity (Bordeleau & Prévost, 1994; Hungria & Vargas, 2000; Christophe *et al.*, 2011).

Chilling stress is usually limited to plants that are native to or growing in tropical or subtropical regions of the world. Plants vary greatly in their sensitivity to chilling stress. Chill-sensitive plants have been defined as plants that are killed or injured by temperatures up to 15–20 °C above the freezing point of the tissues (Bhattacharya & Vijaylaxmi, 2010; Hasanuzzaman *et al.*, 2013). Chill-resistant plants can grow at temperatures near 0 °C (Bhattacharya & Vijaylaxmi, 2010; Hasanuzzaman *et al.*, 2013).

Oktem *et al.* (2008) investigated the effects of cold and drought stress on growth parameters and antioxidant responses in shoots and roots of lentil seedlings subjected to drought and cold (4 °C) stress for 5 days. They reported that the length and fresh weight of shoots decreased significantly under both stress conditions, contrary to the increase in these growth parameters for roots under the same conditions. The increase in proline levels was more pronounced under cold stress in shoots and roots. The oxidative damage resulted in increase of malondialdehyde (MDA), and hydrogen peroxide (H_2O_2) was markedly higher in shoots under cold stress. Both stress conditions caused a significant increase in malondialdehyde levels in root tissues. SOD activity was differentially altered in shoot and root tissues under drought and cold stress. The CAT activity was higher in roots under drought stress, whereas APX activity increased in root tissues under cold stress (Oktem *et al.*, 2008; Bhattacharya & Vijaylaxmi, 2010).

1.2.5 Legumes under soil acidity

Approximately 40% of the world's arable land is considered to be acidic (Valentine *et al.*, 2011). In natural ecosystems, soil acidity determines the availability of mineral nutrients such as phosphorus (P) and also determines the level and severity of phytotoxic elements such as aluminium (Al), manganese (Mn) and iron (Fe) (Muthukumar *et al.*, 2014). Al ions present in acidic soils cause (i) induction of ROS and lipid peroxidation, which limit crop yield in these soils (Yamamoto *et al.*, 2002; Muthukumar *et al.*, 2014); and (ii) extrinsic toxicity through calcium (Ca) and magnesium (Mg) deficiency (Kinraide *et al.*, 2005; Muthukumar *et al.*, 2014).

Three possible groups of mechanisms appear to operate in plants that can tolerate acidic conditions (Muthukumar *et al.*, 2014). These include the following:

1. Exclusion of toxic ions such as Al and Mn from the root apex.
2. Tolerance to toxic levels of Al and Mn through detoxification in the plant symplasm.
3. Enhanced efficiency in the uptake of limiting nutrients from acid soils (Kochian *et al.*, 2004; Bhalerao & Prabhu, 2013).

Soil acidity is a major factor affecting the growth and yield of legumes in many of the world's agricultural systems, due to the effect of phosphorus (P) deficiency and Al toxicity on SNF; it might be considered as a main limitation on the contribution of legumes to the global N cycle (Valentine *et al.*, 2011).

1.2.6 Legumes under nutrient deficiency

In agricultural soils, deficiency of some elements negatively affects nitrogen fixation in legumes and consequently reduces their yield. Toker and Mutlu (2011) reported that in chickpea, N and P deficiencies resulted in yield losses of 790,000 and 653,000 t/year, respectively, worldwide. In most legume-growing soils, N and P are at either low or medium levels, whereas potassium (K) is usually sufficiently available to support growth, although it can be deficient in some soils (Srinivasarao *et al.*, 2003). Ca and Mg are generally deficient in acid soils (pH < 5.5).

Sulphur (S) deficiency has been reported on light-textured soils in India, and the application of S at 20 kg/ha is recommended for these soil types (Srinivasarao *et al.*, 2003). S deficiency is also seen in calcareous soils with a pH of 8.0 or higher (Toker *et al.*, 2011). Iron (Fe) deficiency has been recorded in many legume crops such as chickpea, lentil, lupin, pea, bean and soybean (Erskine *et al.*, 1993; Toker *et al.*, 2010).

Al toxicity induced reduction of SNF due to (i) the inhibition of rhizobial growth in the soil; (ii) the retardation of nodulation; and (iii) the possible alteration in organic acid metabolism (Valentine *et al.*, 2011). Boron (B) toxicity or deficiency induced suppression of normal growth in pea or faba bean (Dwivedi *et al.*, 1992; Poulain & Almohammad, 1995).

Toker and Mutlu (2011) reported that in legume species the relative sensitivity to zinc (Zn) deficiency is high for common bean relative to soybean (Alloway, 2009). Lentil, chickpea and pea were found to be more sensitive to Zn deficiency than oilseeds and cereals (Tiwari & Dwivedi, 1990). Differential Zn efficiency was reported among navy bean genotypes (Jolley & Brown, 1991; Moraghan & Grafton, 1999). Zn deficiency caused delay in pod maturity in bean (Blaylock, 1995).

1.3 Breeding of cool season food legumes

In 1970s, the breeding of legumes started with the establishment of the International Centre for Agricultural Research in Dry Areas (ICARDA) in Syria and the International Crops Research Institute for

Semi-Arid Tropics (ICRISAT) in India, supported by the Consultative Group in International Agricultural Research (CGIAR), as well as through strengthening of the agricultural research systems of different conditions (Materne *et al.*, 2011). Genetic resources for use in cool season food legume breeding are maintained at ICARDA, ICRISAT and also by other national programmes, particularly in the USA, Canada, Australia, India and a number of other important repositories (Materne *et al.*, 2011).

1.4 Breeding of cool season food legumes under abiotic stress

Legumes are considered more sensitive than cereals to a wide range of abiotic stresses (Dita *et al.*, 2006). Materne *et al.* (2011) reported that matching a crop's phenology to an environment, including the avoidance of drought and heat, is a key part of improving adaptation and increasing crop yields, and has been a major global focus in breeding for local and broad adaptation of all the cool season food legumes (Materne & Siddique, 2009; Khan *et al.*, 2010). One of the major achievements of ICARDA's collaborative lentil research is broadening the narrow genetic base of lentil in South Asia through introgression of genes from ICARDA germplasm (Materne *et al.*, 2011).

In short growing season climates (both winter- and spring-sown crops), selection for earlier flowering genotypes has been an important trait for avoidance of late season abiotic stress (e.g. terminal drought and high temperatures) (Materne *et al.*, 2011). In chickpea, whilst the *Kabuli* type is generally considered more drought sensitive than *Desi* types (Leport *et al.*, 2006), ICRISAT developed an extra-short-duration *Kabuli* variety (ICCV 2), which improved yields and expanded production. Since the release of this cultivar, even earlier-maturing germplasm has been developed and combined with a double-podding trait (Ahmad *et al.*, 2005).

Salinity is attracting greater attention from researchers and breeding programmes internationally (Materne *et al.*, 2011). Lentil cultivars with improved tolerance to NaCl have been released already in Australia (Materne & Siddique, 2009). Flowers *et al.* (2010) give a comprehensive overview of studies conducted to explore genetic variation to salt sensitivity in chickpea. Greater efforts have also been focused on quantifying thresholds, and it was recently reported that subsoil chloride (Cl) concentration was the most effective indicator of reduced grain yields rather than salinity, and that growing chickpea on soils with Cl > 600 mg/kg should be avoided due to high yield losses (Dang *et al.*, 2010). Similarly, faba bean has been reported to be more sensitive to Cl than Na, and genetic variation for tolerance to the individual ions was observed (Tavakkoli *et al.*, 2010). Screening methodologies range from pot-based to field methods. More recently, attention has been focused on improving genetic knowledge that could provide molecular markers for salt tolerance in the near future (Varshney *et al.*, 2009).

Cold tolerance has been an important trait for improvement in crop adaptation in many countries (Materne *et al.*, 2011). For chickpea, chilling temperatures at the reproductive phase often result in pod abortion. Clarke *et al.* (2004) successfully used pollen selection methods to develop and release two cultivars that produce pods under lower temperatures than other cultivars. In the USA and Turkey, large yield increases have been achieved by sowing lentil in winter rather than spring, using genotypes tolerant to cold temperatures during winter (Materne & McNeil, 2007). Similarly, very high tolerance of seedlings to cold temperatures has been identified in faba bean (Link *et al.*, 2010). In Australia, lentil breeding lines with improved tolerance to boron have been developed that could improve yields by up to 91% in the target region, based on controlled environment experiments (Hobson *et al.*, 2006). Whilst genetic variation has been identified in chickpea (Hobson *et al.*, 2009), only limited research in this crop has been undertaken. Genetic variation has been identified in field pea (Redden *et al.*, 2005), and the overall level of tolerance of this crop is greater than in lentil and chickpea (Materne *et al.*, 2011).

1.5 Breeding of warm season food legumes

Singh *et al.* (2011) reported that efforts focused on the breeding of warm season food legumes have been made in different international centres supported by CGIAR. Among these centres, ICRISAT has focused research on pigeon pea, and the International Institute of Tropical Agriculture (IITA) has a global mandate for cowpea improvement. The Asian Vegetable Research and

Development Centre (AVRDC) was established for the improvement of mung bean worldwide. Besides, the US Department of Agriculture (USDA) has focused research activities on soybean. The Indian Institute of Pulses Research, Kanpur, a leading centre of the Indian Council of Agriculture Research, and other Agriculture Universities in India are also involved in genetic improvements in warm season legume crops, including pigeon pea, mung bean and urd bean. These national and international centres are involved in collection, evaluation and sharing of germplasm, and also undertake breeding programmes for genetic improvement. The international centres also distribute the segregating populations and inbred lines to partner countries for selection and release as varieties, resulting in stimulation of breeding internationally. Hall *et al.* (1997) and Singh *et al.* (1997) have described cowpea breeding programmes in different regions of the world. The bean/cowpea CRSP (Cowpea Collaborative Research Program) is also catalysing and supporting research on cowpea improvement in the USA, Cameroon and Senegal. Significant research on various aspects of cowpea improvement is also being carried out in Brazil, Nigeria, Burkina Faso, Senegal, Mali and India and, to a lesser extent, in a number of other countries. These efforts have led to the development of different types of cowpea cultivar, including *Vigna unguiculata*, *Vigna biflora* (or catjang) and *Vigna sesquipedalis* (Hall *et al.*, 1997).

1.6 Breeding of warm season food legumes under abiotic stress

Singh *et al.* (2011) summarized some important target traits in breeding programmes for improving the warm season food legumes against abiotic stress, as detailed below.

1.6.1 Short duration and photo-thermal insensitivity

These are important traits in soybean, mung bean and urd bean, because the development of short-duration and photo-thermally insensitive genotypes creates plants suitable for different cropping systems, and also avoids terminal drought (Singh *et al.*, 2011). In cowpea, photosensitive cultivars not only flower early but also become extremely dwarf in habit when day length is under 12.5 h (Ishiyaku & Singh, 2001), and a complete association of photosensitivity has been observed with dwarfing, which is controlled by a single recessive gene (Ishiyaku & Singh, 2001). In urd bean, earliness and photo-thermosensitivity are recessive traits and are controlled by major genes (Sinha, 1988). Thus selection of genotypes with early vigour holds tremendous importance in breeding programmes. As a result, some of the very popular early varieties, such as Narendra Urd 1, KU 300, Sarla, Vamban and Urd 3, have been developed in India for commercial cultivation. Since urd bean is also cultivated in the spring/summer season, Pant U 19, T 9, KM 1 and TMV 1 have been developed as photo-thermoinsensitive varieties (Gupta & Kumar, 2006).

1.6.2 Leaf pubescence density

Suitability for soybean cultivation is improved by this trait in drought-prone areas, as it reduces leaf temperature and water loss by transpiration and enhances photosynthesis and vegetative vigour (Du *et al.*, 2009). Two additive genes control this trait in soybean (Pfeiffer & Pilcher, 2006). This is also an important trait of mung bean and urd bean; some lines of mung bean developed at AVRDC, e.g. V 2013, V 1281, V 3372, VC 1163D, VC 2750A, VC 2754A and VC 2768A, can withstand moisture stress (Tickoo *et al.*, 2006), including long spells of rainfall causing flooding.

1.6.3 Seed dormancy

Reduced seed dormancy is found in mung bean, resulting in preharvest sprouting during the maturity phase in the monsoon (*kharif*) season. Therefore the identification of lines with tolerance to preharvest sprouting is highly desirable, both in this crop (Tickoo *et al.*, 2006) and in urd bean.

1.6.4 Deep root system

Many desert plants have been reported not to have a deep root system, whereas deep rooting becomes more common in less extreme dry areas (Vadez *et al.*, 2007). In fact, the importance of any aspect of rooting pattern (depth, depth distribution, root length density, etc.) is totally relative to the distribution and amounts of water or nutrients in the soil profile. For example, increased root depth/root volume is useful only where there is significant water available to exploit by increasing the soil volume explored by the roots (Vadez *et al.*, 2007).

Kashiwagi *et al.* (2006) showed the importance of roots for seed yield under terminal drought conditions in chickpea. It has been reported that chickpea was able to allocate more roots to the deeper soil layers under conditions of stress than other legumes (Benjamin & Nielsen, 2005), or than more sensitive genotypes (Kashiwagi *et al.*, 2006). However, this was so only when the phenology of the genotype was well suited to the test environment. For example, the chickpea genotypes K1189 and ICC898 had adequate root length density (RLD) compared to ICC4958 (Vadez *et al.*, 2007).

It is assumed that pigeon pea is deep-rooted and that confers drought tolerance because the crop is usually grown on deep soils and completes its life cycle on residual moisture (Vadez *et al.*, 2007).

1.7 Biotechnology approaches

Plant biotechnology offers new ideas and techniques applicable to agriculture. It uses the conceptual framework and technical approaches of plant tissue culture and molecular biology to develop commercial processes and products (Sharma & Lavanya, 2002). These techniques enable the selection of successful genotypes, better isolation and cloning of favourable traits, and the creation of transgenic crops of importance to agriculture. This ability has moved agriculture from a resource-based to a science-based industry (Sharma & Ortiz, 2000; Sharma & Lavanya, 2002).

Legumes can face the threat posed by abiotic stress through several genetic improvement strategies, from classical breeding to more direct physiological genetic approaches.

1.7.1 MicroRNAs

MicroRNAs (miRNAs) play major roles in plant growth and development (Jones-Rhoades *et al.*, 2006; Khraiwesh *et al.*, 2012; Mantri *et al.*, 2013) as well as in regulating the synthesis of polypeptides from different mRNAs including those that act as transcription factors (Mantri *et al.*, 2013). They help plants to thrive under abiotic stresses such as drought, salinity and high/low temperature by regulating the expression of thousands of genes (Mantri *et al.*, 2013).

Next Generation Sequencing (NGS) technology has greatly accelerated the discovery and characterization of miRNAs in a range of diverse plant species (Sunkar & Jagadeeswaran, 2008; Zhao *et al.*, 2010; Chen *et al.*, 2011; Mantri *et al.*, 2013). The first isolation for miRNAs was in *Arabidopsis thaliana* (Park *et al.*, 2002; Reinhart & Bartel, 2002). Since then they have been isolated from a wide range of species via genetic screening (Lee *et al.*, 1993; Wightman *et al.*, 1993), direct cloning after isolation of small RNAs (Fu *et al.*, 2005; Lu *et al.*, 2005) and computational prediction strategies (Wang *et al.*, 2005; Jones-Rhoades *et al.*, 2006).

Mantri *et al.* (2013) in their review reported that miRNAs play an important role in drought tolerance. These include conserved miRNAs such as miR164, miR169, miR171, miR396, miR398, miR399, miR408 and miR2118 (Liu *et al.*, 2008). Their expression patterns vary with legumes species. For example, miR169 was downregulated in *M. truncatula* (Trindade *et al.*, 2010) but upregulated in common bean (in response to abscisic acid treatment) (Zhao *et al.*, 2009). In *M. truncatula*, miR398a,b and miR408 were strongly upregulated in shoots and roots under drought stress (Trindade *et al.*, 2010). The miR398 and miR408 repress the *COX5b*, *CSD1* and plantacyanin genes (Trindade *et al.*, 2010). Under drought and ABA treatments, Arenas-Huertero *et al.* (2009) identified a number of novel legume miRNAs in *Phaseolus vulgaris*. Among them pvu-miRS1, pvu miR1514a, miR159.2, pvu-miR2118 and pvu-miR2119 accumulated upon drought and ABA treatments. Novel miRNAs may target regulatory elements for cellular processes that may be unique to legumes (Arenas-Huertero *et al.*, 2009). Forty-four drought-associated miRNAs (30 were upregulated in drought conditions and 14 were downregulated) were identified in cowpea under drought stress (Barrera-Figueroa *et al.*, 2011). Wang *et al.* (2011) in their study on *M. truncatula* subjected to drought stress, identified 22 members of four miRNA families that were upregulated and 10 members of six miRNA families that were downregulated. Among the 29 new miRNAs/new members of known miRNA families, eight miRNAs were responsive to drought stress of which four each were upregulated and downregulated. The drought-responsive miRNAs were found to be involved in diverse cellular processes including development, transcription, protein degradation, detoxification, nutrient status and cross-adaptation (Mantri *et al.*, 2013).

Mantri *et al.* (2013) in their review reported that legumes express a variety of miRNAs in response to salt stress. In cowpea under salinity stress, Paul *et al.* (2011) identified 18 conserved miRNAs belonging to 16 miRNA families. Fifteen miRNAs were predicted and identified as transcription factors. Seven of these predicted miRNAs (vun-miR156a, vun-miR159b, vun-miR160a, vun-miR162a, vun-miR168a, vun-miR169b and vun-miR408) were experimentally validated in the root tissues and found to be upregulated during salt stress as revealed by qRT-PCR (Paul *et al.*, 2011). Dong *et al.* (2013) studied the dynamic regulation of miRNA in functioning soybean mature nodules under salt stress. The authors identified 110 known miRNAs belonging to 61 miRNA families, and 128 novel miRNAs belonging to 64 miRNA families. Among them, 104 miRNAs were dramatically differentially expressed (>2-fold or detected only in one library) during salt stress. The miR159bc, miR169c and miR319a,b were highly downregulated and gly_1, gly_3, miR171p and miR4416d were highly upregulated by salt (Dong *et al.*, 2013; Mantri *et al.*, 2013). Recently, Nageshbabu and Jyothi (2013) analysed the expression of nine different miRNAs in *Phaseolus vulgaris* seedlings in response to 0.4 M NaCl and drought stress. They reported that miR395 was most sensitive to both stresses and was upregulated under both the stressors. Further, miR396 and miR172 were upregulated after exposure to both the stresses (Mantri *et al.*, 2013; Nageshbabu & Jyothi, 2013).

Wang and Long (2010) by using RT-PCR showed miRNAs associated with cold tolerance in pea (*Pisum sativum*), and the level of their expressions increased after the cold treatment.

MicroRNAs are important signalling and regulatory factors in P deficiency stress (Mantri *et al.*, 2013). Under P starvation stress, miR399 was induced in common bean and *M. truncatula* (Valdes-Lopez & Hernandez, 2008). This miRNA plays a key role in maintaining Pi (inorganic phosphate) homeostasis in *Arabidopsis* and is induced under P deficiency causing repression of the ubiquitin conjugating enzyme UBC24, a repressor of phosphate transporters (Chiou *et al.*, 2006; Mantri *et al.*, 2013).

Zeng *et al.* (2010) identified 57 miRNAs under P deficiency in soybean. Also Sha *et al.* (2012), by using deep sequencing of soybean root and shoot libraries constructed under P stress, identified 60 known and conserved responsive miRNAs, belonging to 35 families. Also, 16 novel predicted miRNAs were identified. In a larger study, 167 miRNAs, belonging to 35 families, were identified via differential expression in response to P deficiency in white lupin; 17, 9 and 10 were found to be upregulated, while 7, 6 and 12 were downregulated in roots, stems and leaves, respectively (Zhu *et al.*, 2010). Recently, Xu *et al.* (2013) showed that 25 miRNAs were induced and 11 mRNAs were repressed under P deficiency in soybean.

Sulphur deficiency induced the suppression of mRNA395 in legume species (Szittya *et al.*, 2008; Kawashima *et al.* 2009). This mRNA regulates ATP sulphurylase (APS4) and a sulphate transporter (AST68) when maintaining S homeostasis during S deficiency (Mantri *et al.*, 2013).

Zeng *et al.* (2012) identified 30 stress-responsive miRNAs in Al-treated and non-treated roots. Of these, 10 were conserved miRNAs that belonged to seven families, 13 were unconserved and seven were novel. In soybean, miR396, miR390 and miR1510a-p5 were upregulated; miR156, miR164 and miR169 were downregulated; and miR1510a was non-responsive to Al (Zeng *et al.*, 2012). Chen *et al.* (2012) identified several *M. truncatula* miRNA (miR160, miR319, miR396, miR1507 miR1510a and miR390) as down-regulated and other two (miR166 and miR171) not responsive to Al treatment. Using a computational approach, Zhou *et al.* (2008) identified 26 new miRNA candidates including miR160, miR166, miR319, miR393 and miR398 that were responsive to mercury, cadmium and aluminium stresses. Their differential expressions were subsequently assessed in various *M. truncatula* organs and tissues (Mantri *et al.*, 2013).

1.7.2 Molecular marker-assisted breeding

Molecular markers are DNA regions tightly linked to agronomic traits in crops, identified by using genetic and genomic analysis. They can facilitate breeding strategies for crop improvement. However, the use of molecular markers in breeding programmes needs preliminary studies to identify and validate potential markers (Dita *et al.*, 2006).

Several molecular marker-related techniques, such as restriction fragment length polymorphism (RFLP), amplified fragment length polymorphism (AFLP), randomly amplified polymorphic DNA (RAPD), simple sequence repeats (SSR) and derivatives, have been reported for abiotic stresses (Kassem *et al.*, 2004; Lee *et al.*, 2004). This has enhanced knowledge of the genetic control of specific resistance and/or tolerance in many legumes by giving information on the number, chromosomal location and individual or interactive effects of the different quantitative trait loci (QTLs) involved (Dita *et al.*, 2006).

These technologies have identified specific molecular markers that may be used in breeding programmes via marker-assisted selection (MAS) to improve stress tolerance (Dita *et al.,* 2006). In legumes, the practical application of MAS for the genetic enhancement of resistance or tolerance to stress has been limited. Schenider *et al.* (1997) reported that MAS may be useful to select drought-tolerant common bean.

1.7.3 Gene pyramiding assisted by MAS

Pyramiding different resistance or tolerance traits into a genotype helps plant breeders to achieve resistance to abiotic stress. In legumes there are numerous examples of introgression and pyramiding of favorable alleles and QTLs. However, MAS has been used to help in gene pyramiding to overcome stresses in only a few cases (Dita *et al.,* 2006). Nevertheless, Schneider *et al.* (1997) indicated that MAS may be useful to select drought-tolerant common bean.

There are some exceptions where MAS has facilitated breeding efforts in several legume crops to combat important biotic stressors (Dita *et al.,* 2006). For example, MAS was successfully used for the breeding of soybean resistant to cyst nematode (Diers, 2004), of pinto bean resistant to common bacterial blight (Mutlu *et al.,* 2005) and of narrow-leaved lupin (*Lupinus angustifolius* L.) resistant to phomopsis stem blight (Yang *et al.,* 2002) and anthracnose (Yang *et al.,* 2004). Moreover, when resistance is conferred by single genes and/or easily overcome by new pathogen races, the gene pyramiding strategy facilitated by MAS can be an efficient method (Dita *et al.,* 2006).

Legume cultivars having appropriate combinations of resistance and/or tolerance to biotic and abiotic stresses, achieved through gene pyramiding, could provide durable resistance, and MAS can be a valuable tool to guide and identify the pyramiding of these genes (Dita *et al.,* 2006).

The generation of markers based on genes with altered expression patterns in response to stresses could result in more effective and targeted MAS. Some of these genes may be good candidates for future MAS studies in legumes (Dita *et al.,* 2006).

1.7.4 Somaclonal variation and *in vitro* mutagenesis

Tissue culture provides a big range of genetic variation in plants, which can be incorporated in plant breeding programmes (Jain, 2001; Dita *et al.,* 2006). Somaclonal variation, including callus cultivation and somatic embryogenesis, has the ability to generate genetic variation (Larkin & Scowcroft, 1981; Dita *et al.,* 2006). The ability to produce agronomically useful somaclones via organogenesis and somatic embryogenesis has been reported in pea (Griga *et al.,* 1995) and pigeon pea (Chintapalli *et al.,* 1997).

Indeed, combining mutagenesis techniques with MAS through TILLING (see Section 1.7.9) will make mutagenesis more suitable for legume enhancement. The main problem with these techniques is the high quantity of individuals required to find the desired trait. Nevertheless, by using *in vitro* selection systems this disadvantage can be reduced (Dita *et al.,* 2006).

1.7.5 *In vitro* selection

In vitro selection is one of the important classical breeding methods (Svabova & Lebeda 2005) and has been used for both biotic and abiotic stresses. In legumes, *in vitro* selection was applied to alfalfa (*Medicago sativa*) for selection of resistance to *Colletotrichum trifolii* (Cucuzza & Kao, 1986), *Fusarium oxysporum* (Cvikrova *et al.,* 1992) and *Verticillium albo-atrum* (Koike & Nanbu, 1997). However, no resistant lines were reported in these studies (Dita *et al.,* 2006).

Putative stress-resistant lines derived from both conventional breeding and transgenic approaches could be screened using *in vitro* selection. This is suitable for some abiotic stresses, where appropriate screening methods are unavailable or have low efficiency. Somaclonal variation and *in vitro* mutagenesis followed by *in vitro* selection offer an alternative way for breeding (Dita *et al.,* 2006).

1.7.6 Transcriptomics

Repression of genes or transcriptional activation is an important tool in the control of stress responses in plants (Chen *et al.,* 2002; Dita *et al.,* 2006). Thus, identification of differentially expressed genes is particularly important to understand stress responses in plants. To achieve this objective, tools such as microarrays (Schena *et al.,* 1995), serial analysis of gene expression (SAGE) (Velculescu *et al.,* 1995), suppression subtractive hybridization library (Diatchenko *et al.,* 1996), and quantitative measurement of transcription factor (TF) expression have been developed in addition to older techniques such as Northern blotting (Dita *et al.,* 2006).

In legumes, transcriptomic techniques are useful ways of breeding to combat environmental stresses. Jain *et al.* (2001) reported 43 drought-responsive mRNA transcripts differentially expressed in peanut under water stress. Pnueli *et al.* (2002) applied suppression subtractive hybridization screening (SSHS) in *Retama raetam*, a C_3 drought-tolerant legume. The authors revealed that dormancy, key to the survival of many species in arid environments, was followed by accumulation of transcripts encoding PR-10-like protein, a low-temperature-inducible dehydrin and a WRKY transcription factor. Umezawa *et al.* (2002), by using a modified c-DNA-AFLP technique in soybean, revealed 140 differentially expressed cDNA fragments by comparing control and iso-osmotic treated plants. Some of the responsive genes encoded ion transporters, transcription factors (TFs) and redox enzymes (Dita *et al.*, 2006). From the *Arabidopsis* database, Ishitani *et al.* (2004) selected 100–200 genes, and revealed that at least three *DREB*-like genes might be key transcriptional regulators of drought and/or cold resistance in common bean (Dita *et al.*, 2006).

1.7.7 Proteomics

A proteomics approach is used to investigate the pathways of biochemical activities and the different responses of plants to stress (Aghaei & Komatsu, 2013). Plant stress proteomics has the ability to identify possible candidate genes that can be used for the genetic enhancement of plants against stresses (Cushman & Bohnert, 2000; Ngara, 2009; Rodziewicz *et al.*, 2014).

In legumes, proteomic techniques have been applied to cowpea, pea and lupin for identification of proteins involved in responses to different abiotic stresses (Fecht-Christoffers *et al.*, 2003; Repetto *et al.*, 2003; Kav *et al.*, 2004; Pinheiro *et al.*, 2005; Cheng *et al.*, 2010).

Proteomic analysis of *M. truncatula* under drought stress reported that the decline in SS is one of the most observable changes in plant function in root nodules (Larrainzar *et al.*, 2007). However, a plant system approach including the proteome and metabolome responses of *M. truncatula* nodules to drought revealed that the decline in SS was not correlated with a decrease in malate concentration (Larrainzar *et al.*, 2009), in contrast to studies of nodules of grain legumes (González *et al.*, 2001; Gálvez *et al.*, 2005). This relatively contrasting behaviour of *Medicago* species to other legumes is intriguing (Arrese-Igor *et al.*, 2011). One theoretical explanation is that in pasture legumes such as *Medicago*, grazing has produced a strong evolutionary selective pressure on continued nodular SS activity during drought stress (Arrese-Igor *et al.*, 2011).

Proteome analysis of *M. truncatula* nodules provided good evidence of drought stress affecting the enzyme Met synthetase (Larrainzar *et al.*, 2007). Depressed Met availability had a major effect on both protein synthesis and sulphur metabolism in nodules (Arrese-Igor *et al.*, 2011).

Aghaei *et al.* (2009) and Sobhanian *et al.* (2010) studied the proteome of soybean under salt stress by using different tissues. They identified a 50S ribosome protein that was downregulated in leaves. Alam *et al.* (2010) studied the proteome analysis of soybean root under water stress. They indicated that two key enzymes involved in sugar metabolism, UDP-glucose pyrophosphorylase and 2,3-biophosphoglycerate-independent phosphoglycerate mutase, were downregulated under drought stress.

Cheng *et al.* (2010) reported 40 proteins (25 upregulated and 15 downregulated) in soybean seeds exposed to cold stress (4 °C). These proteins are involved in cell growth/division, storage, cellular defences, energy protein synthesis, transcription and transport. Zhu *et al.* (2006) reported that the activation of HSP70 in transgenic lines by its upstream gene *HsfA1* improved soybean tolerance under high temperature stress.

1.7.8 Transgenomics

Transgenic technology is one of the many tools available for modern plant improvement programmes (Jewell *et al.*, 2010). The use of transgenic approaches, or 'transgenomics', helps in understanding the mechanisms governing stress tolerance, providing good ways for the genetic enhancement of field crops, thereby alleviating some of the major constraints to crop productivity in developing countries (Sharma & Ortiz 2000; Reddy *et al.*, 2012). Transgenic plants or their germplasm can be used as sources of new cultivars or as new sources of variation in breeding programmes (Jewell *et al.*, 2010).

When plants are subjected to abiotic stresses, a number of genes are turned on causing increased levels of several osmolytes and proteins that may be responsible for conferring a certain degree of protection from these stresses. Thus, it may be necessary to transfer several potentially useful genes into the same plant in order to obtain a high degree of tolerance to drought or salt stress (Reddy *et al.*, 2012).

There are several transgenic technologies for improving stress tolerance involving the expression of functional genes (Reddy *et al.*, 2012), including those encoding enzymes required for the biosynthesis of osmoprotectants or late embryogenesis proteins, detoxification enzymes and modification of membrane lipids (Ishizaki-Nishizawa *et al.*, 1996; McKersie *et al.*, 1996; Xu *et al.*, 1996; Hayashi *et al.*, 1997; Bhatnagar-Mathur *et al.*, 2008; Reddy *et al.*, 2012; Rodziewicz *et al.*, 2014).

Osmotic adjustment (osmotic regulation) is one of the useful mechanisms for improving abiotic stress tolerance, especially if osmoregulatory genes can be triggered in response to drought, salinity and high temperature (Reddy *et al.*, 2012). Sharmila *et al.* (2009) showed that H_2O_2 produced by a prokaryotic osmoregulatory choline oxidase gene (*codA*) as a by-product during synthesis of glycine-betaine is responsible for building a stronger antioxidant system in chloroplasts of transgenic chickpea plants. Similarly, at ICRISAT, the P5CSF129A gene encoding the mutagenized D1-pyrroline-5-carboxylate synthetase (P5CS) for the overproduction of proline was introduced in chickpea. In several of these transgenic events the accumulation of proline in leaves increased significantly when the plants were exposed to water stress, along with a decrease in free radicals as measured by a decrease in the MDA levels, a lipid peroxidation product (Reddy *et al.*, 2012).

To date, genetic transformation has been reported in all the major legume crops such as *Vigna* spp., *C. arietinum, C. cajan, Phaseolus* spp., *Lupinus* spp., *Vicia* spp., *P. sativum*, soybean, groundnut, pigeon pea and chickpea (Sharma & Lavanya, 2002; Reddy *et al.*, 2012).

1.7.9 Targeting induced local lesions in genomes (TILLING)

Recently, Kudapa *et al.* (2013) in their review reported that validation of genes through genetic transformation, RNAi or virus-induced gene silencing (VIGS) is a time-consuming process in legumes, mainly due to lack of efficient transformation systems in legumes. This situation has promoted the application of TILLING to study gene function. In TILLING, candidate genes are screened across a mutant population (with point mutations), and line(s) with the mutation for the target gene are identified (McCallum *et al.*, 2000). If the identified line exhibits the expected phenotype for the candidate gene, the function of the candidate gene is supported. The TILLING approach could be preferred over RNAi for irreversibly inhibiting or eliminating the target genes in commercial crop plants since it avoids genetic transformation and increases stability of the phenotype (Barkley & Wang, 2008).

TILLING populations have been developed for several legumes. For example, in the model legumes *Medicago* (12,000 M2 plants; Rogers *et al.*, 2009) and *Lotus* (4904 M2 lines; Perry *et al.*, 2009) mutant populations were developed for use in reverse genetics. In the case of crop legumes, over 3000 M3 lines were developed in common bean and evaluated with root nodulation tests by Porch *et al.* (2009). In peanut a TILLING population of 10,000 lines has been established, and a subset of this population investigated for allergenicity (Tadege *et al.*, 2009). In chickpea, a TILLING population of ~3500 lines has been developed and is being used to identify candidate genes for drought tolerance (M. Thudi, personal communication). The use of NGS technologies for TILLING may increase the application of TILLING in crop legumes (Kudapa *et al.*, 2013).

EcoTILLING is a variant of TILLING, except that its objective is to discover naturally occurring polymorphisms as opposed to experimentally induced mutations (Kudapa *et al.*, 2013). Single nucleotide polymorphisms (SNPs), small insertions and deletions, and variations in microsatellite repeat number can be efficiently detected using the EcoTILLING technique (Kudapa *et al.*, 2013). For example, in legumes this method has been used to develop molecular markers for cyst nematode candidate resistance genes in soybean (Liu *et al.*, 2012). In mungbean it has been proven to be a valuable method for detecting polymorphisms in a collection that was previously shown to have limited diversity (Barkley & Wang, 2008).

1.8 Conclusions and future prospects

Food legumes are affected by abiotic stresses like salinity, water stress (drought and waterlogging), extreme temperatures (heat and cold) and nutrient deficiency, which ultimately lead to huge economic losses globally. Like other plant species, the breeding process in food legumes consists of four stages: (i) creating variations with hybridizations and induced mutations; (ii) selection in early generations; (iii) evaluation of selected lines; and (iv) release of varieties (Toker & Mutlu, 2011).

The biotechnological approaches of resistance breeding have provided several improved varieties of food legumes with tolerance to abiotic stresses. There is no substitute for these approaches, and they will continue to be the mainstay in the future. However, efforts are needed to improve the effectiveness of these approaches by further refining screening methods for resistance to stresses and identifying new sources of resistance genes in both cultivated and wild species. There is a need to use diverse sources of resistance in breeding programmes and to develop cultivars with tolerance to multiple stress factors.

Mutagenesis facilitates an increase in genetic variability for resistance to abiotic stresses in food legumes. Transgenic legumes provide a great chance, but genes can flow from transgenics to wild relatives, leading to environmental pollution when transgenics are grown in the areas where wild relatives exist.

Modern techniques including all the 'omics', such as proteomics, genomics, transcriptomics and metabolomics, will be helpful to study legume responses to abiotic stresses. However, successful application of 'omics' to abiotic constraints needs knowledge of stress responses at the molecular level, which includes gene expression to protein or metabolite and its phenotypic effects. Therefore, research dealing with other techniques such as MAS or even classical breeding will be able to take advantage of the results obtained from these 'omics' technologies.

Based on the above-mentioned information we can conclude that the support of biotechnology approaches to conventional breeding methods would lead to advancement in the development of improved cultivars of food legumes with tolerance to abiotic stresses.

References

Abdel Latef AA, Chaoxing H (2011) Effect of arbuscular mycorrhizal fungi on growth, mineral nutrition, antioxidant enzymes activity and fruit yield of tomato grown under salinity stress. *Sci Hort* 127: 228–233.

Abdel Latef AA, Chaoxing H (2014) Does inoculation with *Glomus mosseae* improve salt tolerance in pepper plants? *J Plant Growth Regul* doi: 10.1007/s00344-014-9414-4.

Abdelmula AA, Link W, von Kittlitz E, Stelling D (1999) Heterosis and inheritance of drought tolerance in faba bean, *Vicia faba* L. *Plant Breeding* 118: 485–490.

Aggarwal A, Kadian N, Karishma Neetu, Tanwar A, Gupta KK (2012) Arbuscular mycorrhizal symbiosis and alleviation of salinity stress. *J Appl Nat Sci* 4: 144–155.

Aghaei K, Komatsu S. (2013) Crop and medicinal plants proteomics in response to salt stress. *Frontiers Plant Sci* 8: 1–9.

Aghaei K, Ehsanpour AA, Shah, AH, Komatsu, S. (2009) Proteome analysis of soybean hypocotyl and root under salt stress. *Amino Acids* 36: 91–98.

Ahlawat IPS, Gangaiah B, Zahid MA (2007) Nutrient management in chickpea. In: Yadav SS (ed.), *Chickpea Breeding and Management*. CAB International, Wallingford, pp. 213–232.

Ahmad F, Gaur PM, Croser JS (2005) Chickpea (*Cicer arietinum* L.). In: Singh RJ, Jauhar PP (eds), *Genetic Resources, Chromosome Engineering, and Crop Improvement: Grain Legumes*. CRC Press, Boca Raton, FL, pp. 187–217.

Ahmad P, Prasad MNV (2012a) *Environmental Adaptations and Stress Tolerance in Plants in the Era of Climate Change*. Springer, New York.

Ahmad P, Prasad MNV (2012b) Abiotic Stress Responses in Plants: Metabolism, Productivity and Sustainability. Springer, New York.

Ahmed S, Nawata E, Hosokawa M, Domae Y, Sakuratani T (2002) Alterations in photosynthesis and some antioxidant enzymatic activities of mungbean subjected to waterlogging. *Plant Sci* 163: 117–123.

Alam I, Sharmin SA, Kim KH, Yang JK, Choi MS, Lee BH (2010) Proteome analysis of soybean roots subjected to short-term drought stress. *Plant Soil* 333: 491–505.

Alloway BJ (2009) Soil factors associated with zinc deficiency in crops and humans. *Environ Geochem Health* 31: 537–548.

Amede T, von Kittlitz E, Schubert S (1999) Differential drought responses of faba bean (*Vicia faba* L.) inbred lines. *J Agron Crop Sci* 183: 35–45.

Andrews M, Hodge S (2010) Climate change, a challenge for cool season grain legume crop production. In: Yadav SS, McNeil DL, Redden R, Patil SA (eds), *Climate Change and Management of Cool Season Grain Legume Crops*. Springer, Dordrecht, pp. 1–10.

Arenas-Huertero C, Perez B, Rabanal F, *et al.* (2009) Conserved and novel miRNAs in the legume *Phaseolus vulgaris* in response to stress. *Plant Mol Biol* 70: 385–401.

Arrese-Igor C, Gordon C, González EM, Marino D, Ladrera R, Larrainzer E, Gil-Quintana E (2011) Physiological response of legume nodules to drought. *Plant Stress* 5 (special issue 1): 24–31.

Ashraf MA (2012) Waterlogging stress in plants: A review. *Afr J Agric Res* 7: 1976–1981.

Ashraf MY, Ashraf M, Arshad M (2010) Major nutrients supply in legume crops under stress environments. In: Yadav SS, McNeil DL, Redden R, Patil SA (eds), *Climate Change and Management of Cool Season Grain Legume Crops*. Springer, Dordrecht, pp. 155–170.

Barkley NA, Wang ML (2008) Application of TILLING and EcoTILLING as reverse genetic approaches to elucidate the function of genes in plants and animals. *Curr Genom* 9:212 –226

Barrera-Figueroa BE, Gao L, Diop NN, *et al.* (2011) Identification and comparative analysis of drought-associated microRNAs in two cowpea genotypes. *BMC Plant Biol* 11: 127.

Benjamin JG, Nielsen DC (2006) Water deficit effects on root distribution of soybean, field pea and chickpea. *Field Crop Res* 97: 248–253.

Ben-Salah I, Albacete A, Martínez-Andújar C, *et al.* (2009) Response of nitrogen fixation in relation to nodule carbohydrate metabolism in *Medicago ciliaris* lines subjected to salt stress. *J Plant Physiol* 166: 477–488.

Bhalerao SA, Prabhu DV (2013) Aluminium toxicity in plants: a review. *J Appl Chem* 2: 447–474.

Bhatnagar-Mathur P, Vadez V, Sharma KK (2008) Transgenic approaches for abiotic stress tolerance in plants: retrospect and prospects. *Plant Cell Rep* 27: 411–424.

Bhattacharya A, Vijaylaxmi (2010) Physiological responses of grain legumes to stress environments. In: Yadav SS (ed.), *Chickpea Breeding and Management.* CAB International, Wallingford, pp. 35–86.

Blaylock AD (1995) Navy bean yield and maturity response to nitrogen and zinc. *J Plant Nutr* 18: 163–178.

Bordeleau LM, Prevost D (1994) Nodulation and nitrogen fixation in extreme environments. *Plant Soil* 161: 115–125.

Charlson DV, Bhatnagar S, King CA, *et al.* (2009) Polygenic inheritance of canopy wilting in soybean [Glycine max (L.) Merr.]. *Theoretic Appl Gen* 119: 587–594.

Chen L, Zhang Y, Ren Y, Xu J, Zhang Z, Wang Y (2011) Genome-wide identification of cold-responsive and new microRNAs in *Populus tomentosa* by high-throughput sequencing. *Biochem Biophys Res Commun* 417: 892–896.

Chen L, Wang T, Zhao M, Tian Q, Zhang WH (2012) Identification of aluminum responsive microRNAs in Medicago truncatula by genome-wide high-throughput sequencing. *Planta* 235: 375–386.

Chen WQ, Provart NJ, Glazebrook J, *et al.* (2002) Expression profile matrix of Arabidopsis transcription factor genes suggests their putative functions in response to environmental stresses. *Plant Cell* 14: 559–574.

Cheng L, Gao X, Li S, *et al.* (2010) Proteomic analysis of soybean [*Glycine max* (L.) Meer.] seeds during imbibition at chilling temperature. *Mol Breed* 26: 1–17.

Chintapalli PL, Moss JP, Sharma KK, Bhalla JK (1997) *In vitro* culture provides additional variation for pigeonpea *Cajanus cajan* (L) Millsp crop improvement. *In Vitro Cell Dev Biol-Plant* 33: 30–37.

Chiou TJ, Aung K, Lin SI, Wu CC, Chiang SF, Su CL (2006) Regulation of phosphate homeostasis by microRNA in Arabidopsis. *Plant Cell* 18: 412–421.

Christophe S, Jean-Christophe A, Annabelle L, Alain O, Marion P, Anne-Sophie V (2011) Plant N fluxes and modulation by nitrogen, heat and water stresses: A review. Based on comparison of legumes and non legume plants. In: Shanker AK, Venkateswarlu B (eds), *Abiotic Stress in Plants – Mechanisms and Adaptations.* InTech, pp. 79–119.

Clarke HJ, Khan TN, Siddique KHM (2004) Pollen selection for chilling tolerance at hybridisation leads to improved chickpea cultivars. *Euphytica* 139: 65–74.

Craig J, Barratt P, Tatge H, *et al.* (1999) Mutations at the *rug4* locus alter the carbon and nitrogen metabolism of pea plants through an effect on sucrose synthase. *Plant J* 17: 353–362.

Cucuzza JD, Kao J (1986) *In vitro* assay of excised cotyledons of alfalfa (*Medicago sativa*) to screen for resistance to *Colletotrichum trifolii*. *Plant Dis* 70: 111–115.

Cushman JC, Bohnert HJ (2000) Genomic approaches to plant stress tolerance. *Current Opin Plant Biol* 3: 117–124.

Cvikrova M, Binarova P, Eder J, Nedelnik J (1992) Accumulation of phenolic-acids in filtrate-treated alfalfa cell-cultures derived from genotypes with different susceptibility to *Fusarium oxysporum*. *J Plant Physiol* 140: 21–27.

Dang YP, Dalal RC, Buck SR, *et al.* (2010) Diagnosis, extent, impacts, and management of subsoil constraints in the northern grains cropping region of Australia. *Aust J Soil Res* 48: 105–119.

Delgado MJ, Ligero F, Lluch C (1994) Effects of salt stress on growth and nitrogen fixation by pea, faba-bean, common bean and soybean plants. *Soil Biol Biochem* 26: 371–376.

Diatchenko L, Lau YFC, Campbell AP, *et al.* (1996) Suppression subtractive hybridization: A method for generating differentially regulated or tissue specific cDNA probes and libraries. *Proc Natl Acad Sci USA* 93: 6025–6030.

Diers B (2004) Soybean genetic improvement through conventional and molecular based strategies. In: *5th European Conference on Grain Legumes, Djion, France, 7–11 June, 2004.* AEP, pp. 147–148.

Dita MA, Rispail N, Prats E, Rubiales D, Singh KB (2006) Biotechnology approaches to overcome biotic and abiotic stress constraints in legumes. *Euphytica* 147: 1–24.

Djanaguiraman M, Prasad PVV (2013) Effects of salinity on ion transport, water relations and oxidative damage. In: Ahmad P, Azooz MM, Prasad MNV (eds), *Ecophysiology and Responses of Plants under Salt Stress.* Springer Science+Business Media, pp. 89–114.

Dong Z, Shi L, Wang Y, *et al.* (2013) Identification and dynamic regulation of microRNAs involved in salt stress responses in functional soybean nodules by high-throughput sequencing. *Int J Mol Sci* 14: 2717–2738.

Du WJ, Fu SX, Yu DY (2009) Genetic analysis for the leaf pubescence density and water status traits in soybean [*Glycine max* (L.) Merr.]. *Plant Breeding* 128: 259–265.

Dwivedi BS, Ram M, Singh BP, Das M, Prasad RN (1992) Effect of liming on boron nutrition of pea (*Pisum sativum* L.) and corn (*Zea mays* L.) grown in sequence in an acid alfisol. *Fertil Res* 31: 257–262.

Egamberdieva D, Lugtenberg B (2014) Use of plant growth-promoting rhizobacteria to alleviate salinity stress in plants. In. Miransari M (ed.), *Use of Microbes for the Alleviation of Soil Stresses.* Springer Science+Business Media, New York, pp. 73–96.

El-Enany AE, Al-Anazi AD, Dief N, Al-Taisan WA (2013) Role of antioxidant enzymes in amelioration of water deficit and waterlogging stresses on *Vigna sinensis* plants. *J Biol Earth Sci* 3: B144–B153.

Elsheikh EA, Wood EM (1990) Effect of salinity on growth, nodulation and nitrogen yield of chickpea (*Cicer arietinum* L.). *J Exp Bot* 41: 1263–1269.

Elsheikh EA, Wood EM (1995) Nodulation and N fixation by soybean inoculated with salt tolerant *Rhizobia* or salt sensitive *Bradyrhizobia* in saline soil. *Soil Biol Biochem* 27: 657–661.

Erskine W, Saxena NP, Saxena MC (1993) Iron-deficiency in lentil – yield loss and geographic distribution in a germplasm collection. *Plant Soil* 151: 249–254.

FAOSTAT (2009) Food and Agriculture Organization of the United Nations Rome. Available at: http://faostat3.fao.org/faostat-gateway/go/to/home/E [accessed 15 June 2014].

Fecht-Christoffers MM, Braun HP, Lemaitre-Guillier C, VanDorsselaer A, Horst AJ (2003) Effect of manganese toxicity on the proteome of the leaf apoplast in cowpea. *Plant Physiol* 133: 1935–1946.

Flowers TJ, Gaur PM, Gowda CLL, *et al.* (2009) Salt sensitivity in chickpea. *Plant Cell Environ* 33: 490–509.

Flowers TJ, Gaur PM, Gowda CLL, *et al.* (2010) Salt sensitivity in chickpea. *Plant Cell Environ* 33: 490–509.

Fu H, Tie Y, Xu C, *et al.* (2005) Identification of human fetal liver miRNAs by a novel method. *FEBS Lett* 579: 3849–3854.

Gálvez L, González EM, Arrese-Igor C (2005) Evidence for carbon flux shortage and strong carbon/nitrogen interactions in pea nodules at early stages of water stress. *J Exp Bot* 56: 2551–2561.

González EM, Gordon AJ, Hames CL, Arrese-Igor C (1995) The role of sucrose synthase in the response of soybean nodules to drought. *J Exp Bot* 46: 1515–1523.

González EM, Aparicio-Tejo PM, Gordon AJ, Minchin FR, Royuela M, Arrese-Igor C (1998) Water-deficit effects on carbon and nitrogen metabolism of pea nodules. *J Exp Bot* 49: 1705–1714.

González EM, Gálvez L, Royuela M, Aparicio-Tejo PM, Arrese-Igor C (2001) Insights into the regulation of nitrogen fixation in pea nodules: lessons from drought, abscisic acid and increased photoassimilate availability. *Agronomie* 21: 607–613.

Gordon AJ, Minchin FR, James CL, Komina O (1999) Sucrose synthase in legume nodules is essential for nitrogen fixation. *Plant Physiol* 120: 867–877.

Graham, PH, Vance CP (2003) Legumes: importance and constraints to greater use. *Plant Physiol* 131: 872–877.

Grant OM (2012) Understanding and exploiting the impact of drought stress on plant physiology. In: Ahmad P, Prasad MNV (eds), *Abiotic Stress Responses in Plants: Metabolism, Productivity and Sustainability*. Springer Science+Business Media, pp. 89–104.

Griga M, Stejskal J, Beber K (1995) Analysis of tissue culture derived variation in pea (*Pisum sativum* L.). Preliminary results. *Euphytica* 85: 335–339.

Gupta S, Kumar S (2006) Urd bean breeding. In: Ali M, Kumar S (eds), *Advances in Mung Bean and Urd Bean*. Indian Institute of Pulses Research, Kanpur, pp. 149–168.

Gürsoy M, Balkan A, Ulukan H (2012) Ecophysiological responses to stresses in slants: A general approach. *Pak J Biol Sci* 15: 506–516.

Hall AE (2001) *Crop Responses to Environment*. CRC Press, Boca Raton, FL.

Hall AE, Singh BB, Ehlers JD (1997) Cowpea breeding. *Plant Breeding Rev* 15: 215–274.

Hall AE, Ismail AM, Ehlers JD, *et al.* (2002) Breeding cowpea for tolerance to temperature extremes and adaptation to drought. In: Fatokun CA, Tarawali SA, Singh BB, Kormawa PM, Tamò M (eds), *Challenges and Opportunities for Enhancing Sustainable Cowpea Production*. Institute of Tropical Agriculture (IITA), Ibadan, Nigeria, pp. 14–21.

Hameed A, Dilfuza E, Abd-Allah EF, Hashem A, Kumar A, Ahmad P (2014) Salinity stress and arbuscular mycorrhizal symbiosis in plants. In: Miransari M (ed.), *Use of Microbes for the Alleviation of Soil Stresses*. Springer Science+Business Media New York pp. 139–159.

Hasanuzzaman M, Gill SS, Fujita M (2013) Physiological role of nitric oxide in plants grown under adverse environmental conditions. In: Tuteja N, Gill SS (eds), *Plant Acclimation to Environmental Stress*. Springer Science+Business Media, New York, pp. 269–322.

Hasegawa PM, Bressan RA, Zhu JK, Bohnert HJ (2000). Plant cellular and molecular responses to high salinity. *Ann Rev Plant Physiol Plant Mol Biol* 51: 463–499.

Hayashi H, Mustardy L, Deshnium P, Ida M, Murata N (1997) Transformation of *Arabidopsis thaliana* with the codA gene for choline oxidase: accumulation of glycine betaine and enhanced tolerance to salt and cold stress. *Plant J* 12: 133–142.

Hobson KB, Armstrong RD, Nicolas M, Connor DJ, Materne M. (2006) Response of lentil (*Lens culinaris*) germplasm to high concentrations of soil boron. *Euphytica* 151: 371–382.

Hobson K, Materne M, Noy D (2009) Evaluation of chickpea (*Cicer arietinum*) germplasm for tolerance to high soil boron. In: *Proceedings of the 14th Australasian Plant Breeding Conference and 11th SABRAO Congress, 10–14 August 2009I*. CSIRO, Cairns, Australia.

Hungria M, Vargas MAT (2000) Environmental factors affecting N_2 fixation in grain legumes in the tropics, with an emphasis on Brazil. *Field Crops Res* 65: 151–164.

Hussain N, Sarwar G, Schmeisky H, Al-Rawahy S, Ahmad M (2011) Salinity and drought management in legume crops. In: Yadav SS, McNeil DL, Redden R, Patil SA (eds), *Climate Change and Management of Cool Season Grain Legume Crops*. Springer, Dordrecht, pp: 171–192.

Impa SM, Nadaradjan S, Jagadish SVK (2012) Drought stress induced reactive oxygen species and anti-oxidants in plants. In: Ahmad P, Prasad MNV (eds), *Abiotic Stress Responses in Plants: Metabolism, Productivity and Sustainability*. Springer Science+Business Media, pp. 131–147.

Ishitani M, Rao I, Wenzl P, Beebe B, Tohme J (2004) Integration of genomics approach with traditional breeding towards improving abiotic stress adaptation: Drought and aluminum toxicity as case studies. *Field Crop Res* 90: 35–45.

Ishitani M, Rane J, Bebee V, Sankaran M, Blair M, Rao IM (2011) Molecular breeding approach in managing abiotic stresses. In: Pratap A, Kumar J (eds), *Biology and Breeding of Food Legumes*. CAB International, pp. 276–295.

Ishiyaku MF, Singh BB (2001) Inheritance of short day-induced dwarfing in photosensitive cowpea. *Afri Crop Sci J* 9: 1–8.

Ishizaki-Nishizawa O, Fujii T, Azuma M, Sekiguchi K, Murata N, Ohtani T, Toguri T (1996) Low-temperature resistance of higher plants is significantly enhanced by a nonspecific cyanobacterial desaturase. *Nat Biotechnol* 14: 1003–1006.

Jain SM (2001) Tissue culture-derived variation in crop improvement. *Euphytica* 118: 153–166.

Jain AK, Basha SM, Holbrook CC (2001) Identification of drought-responsive transcripts in peanut (*Arachis hypogaea* L.). *Elect J Biotechnol* 4(2).

Jewell MC, Campbell BC, Godwin ID (2010) Transgenic plants for abiotic stress resistance. In: Kole C, Michler CH, Abbott AG, Hall TC (eds), *Transgenic Crop Plants*. Springer-Verlag, Berlin, Heidelberg, pp. 76–132.

Jolley VD, Brown JC (1991) Factors in iron-stress response mechanism enhanced by Zn deficiency in Sanilac, but not Saginaw navy bean. *J Plant Nutr* 14: 257–265.

Jones-Rhoades MW, Bartel DP, Bartel B (2006) MicroRNAS and their regulatory roles in plants. *Annu Rev Plant Biol* 57: 19–53.

Kamal A, Qureshi MS, Ashraf MY, Hussain (2003) Salinity induced changes in some growth and physio-chemical aspects of two soybean [*Glycine max* (L.) Merr.] genotypes. *Pak J Bot* 35: 93–97.

Kapoor R, Evelin H, Mathur P, Giri B (2013) Arbuscular mycorrhiza: Approaches for abiotic stress tolerance in crop plants for sustainable agriculture. In: Tuteja N, Gill SS (eds), *Plant Acclimation to Environmental Stress*. Springer Science+Business Media, pp. 359–401.

Kashiwagi J, Krishnamurthy L, Crouch JH, Serraj R (2006). Variability of root characteristics and their contribution to seed yield in chickpea (*Cicer arietinum* L.) under terminal drought stress. *Field Crop Res* 95: 171–181.

Kassem MA, Meksem K, Kang CH, *et al.* (2004) Loci underlying resistance to manganese toxicity mapped in a soybean recombinant inbred line population of 'Essex'×'Forrest'. *Plant Soil* 260: 197–204.

Kaur H, Gupta AK, Kaur N, Sandhu JS (2009) Differential response of the antioxidant system in wild and cultivated genotypes of chickpea. *Plant Growth Regul* 57: 109–114.

Kav NNV, Srivastava S, Goonewardene L, Blade SF (2004) Proteome-level changes in the roots of *Pisum sativum* in response to salinity. *Ann Appl Biol* 145: 217–230.

Kawashima CG, Yoshimoto N, Maruyama-Nakashita A, *et al.* (2009) Sulphur starvation induces the expression of microRNA-395 and one of its target genes but in different cell types. *Plant J* 57: 313–321.

Khan HR, Paull JG, Siddique, KHM, Stoddard FL (2010) Faba bean breeding for drought-affected environments: A physiological and agronomic perspective. *Field Crops Res* 115: 279–286.

Khraiwesh B, Zhu JK, Zhu J (2012) Role of miRNAs and siRNAs in biotic and abiotic stress responses of plants. *Biochim Biophys Acta* 1819: 137–148.

Kinraide TB, Parker DR, Zobel RW (2005) Organic acid secretion as a mechanism of aluminum resistance: a model incorporating the root cortex, epidermis, and the external unstirred layer. *J Exp Bot* 56: 1853–65.

Kochian L, Hoekenga O, Piñeros M (2004) How do crop plants tolerate acid soils? Mechanisms of aluminum tolerance and phosphorous efficiency. *Annu Rev Plant Biol* 55: 459–493.

Koike M, Nanbu K (1997) Phenylalanine ammonia-lyase activity in alfalfa suspension cultures treated with conidia and elicitors of *Verticillium albo-atrum*. *Biol Plant* 39: 349–353.

Kudapa H, Ramalingam A, Nayakoti S, *et al.* (2013) Functional genomics to study stress responses in crop legumes: progress and prospects. *Funct Plant Biol* 40: 1221–1233.

Kumutha D, Ezhilmathi K, Sairam RK, *et al.* (2009) Waterlogging induced oxidative stress and antioxidant activity in pigeonpea genotypes. *Biol Plant* 53: 75–84.

Ladrera R, Marino D, Larrainzar E, González EM, Arrese-Igor C (2007) Reduced carbon availability to bacteroids and elevated ureides in nodules, but not in shoots, are involved in nitrogen fixation response to early drought in soybean. *Plant Physiol* 145: 539–546.

Larkin PJ, Scowcroft WR (1981) Somaclonal variation – a novel source of variability from cell-cultures for plant improvement. *Theor Appl Genet* 60: 197–214.

Larrainzer E, Wienkoop S, Ladrera R, Weckwerth W, Arrese-Igor C, González EM (2007) *Medicago truncatula* root nodule proteome analysis reveals differential plant and bacteroid responses to drought stress. *Plant Physiol* 144: 1495–1507.

Larrainzer E, Wienkoop S, Scherling C, *et al.* (2009) Carbon metabolism and bacteroid functioning are involved in the regulation of nitrogen fixation in *Medicago truncatula* under drought and recovery. *Mol Plant Microbe In* 22: 1565–1576.

Lee GJ, Boerma HR, Villagarcia MR, *et al.* (2004) A major QTL conditioning salt tolerance in S-100 soybean and descendant cultivars. *Theor Appl Genet* 109: 1610–1619.

Lee RC, Feinbaum RL, Ambros V (1993) The *C. elegans* heterochronic gene lin-4 encodes small RNAs with antisense complementarity to lin-14. *Cell* 75: 843–854.

Leport L, Turner NC, Davies SL, Siddique KHM (2006) Variation in pod production and abortion among chickpea cultivars under terminal drought. *Eur J Agron* 24: 236–246.

Link W, Abdelmula AA, von Kittlitz E, Bruns S, Riemer H, Stelling D (1999) Genotypic variation for drought tolerance in *Vicia faba*. *Plant Breeding* 118: 477–483.

Link W, Balko C, Stoddard FL (2010) Winter hardiness in faba bean: Physiology and breeding. *Field Crops Res* 115: 287–296.

Liu HH, Tian X, Li YJ, Wu CA, Zheng CC (2008) Microarray-based analysis of stress-regulated microRNAs in Arabidopsis thaliana. *RNA* 14: 836–843.

Liu S, Kandoth PK, Warren SD, *et al.* (2012) A soybean cyst nematode resistance gene points to a new mechanism of plant resistance to pathogens. *Nature* 492: 256–260.

Lu C, Tej SS, Luo S, Haudenschild CD, Meyers BC, Green PJ (2005) Elucidation of the small RNA component of the transcriptome. *Science* 309: 1567–1569.

Mantri N, Basker N, Ford R, Pang E, Pardeshi V (2013) The role of miRNAs in legumes with a focus on abiotic stress response. *The Plant Genome* doi: 10.3835/plantgenome2013.05.0013.

Materne M, McNeil DL (2007) Breeding methods and achievements. In: Yadav SS, McNeil DL, Stevenson PC (eds), *Lentil: An Ancient Crop for Modern Times*. Springer, Dordrecht, pp. 241–253.

Materne M, Siddique KHM (2009) Agroecology and crop adaptation. In: Erskine W, Muehlbauer F, Sarker A, Sharma B (eds), *The Lentil: Botany, Production and Uses*. CAB International, Wallingford, UK, pp. 47–63.

Materne M, McNeil D, Hobson K, Ford R (2007) Abiotic stresses. In: Yadav SS, McNeil D, Stevenson PC (eds), *Lentil: An Ancient Crop for Modern Times*. Springer, Dordrecht, pp. 315–329.

Materne M, Leonforte A, Hobson K, Paull J, Gnanasambandam A (2011) Breeding for improvement of cool season food legumes. In: Pratap A, Kumar J (eds), *Biology and Breeding of Food Legumes*. CAB International, pp. 49–62.

McCallum CM, Comai L, Greene EA, Henikoff S (2000) Targeting induced local lesions in genomes (TILLING) for plant functional genomics. *Plant Physiol* 123: 439–442.

McKersie BD, Bowley SR, Harjanto E, Leprince O (1996) Water-deficit tolerance and field performance of transgenic alfalfa over-expressing superoxide dismutase. *Plant Physiol* 111: 1177–1181.

Miller PR, McConkey BG, Clyton GW, *et al.* (2002) Pulse crop adaptation in Northern Great Plains. *Agron J* 94: 261–272.

Moraghan JT, Grafton K (1999) Seed zinc concentration and the zinc efficiency trait in navy bean. *Soil Sci Soci America J* 63: 918–922.

Munns R (2002) Comparative physiology of salt and water stress. *Plant Cell Environ* 25: 239–250.

Munns R, Tester M (2008) Mechanisms of salinity tolerance. *Ann Rev Plant Biol* 59: 651–681.

Muthukumar T, Priyadharsini P, Uma E, Jaison S, Pandey RR (2014) Role of arbuscular mycorrhizal fungi in alleviation of acidity stress on plant growth. In: Miransari M (ed.), *Use of Microbes for the Alleviation of Soil Stresses*. Springer Science + Business Media, New York, pp. 43–71.

Mutlu N, Miklas P, Reiser J, Coyne D (2005) Backcross breeding for improved resistance to common bacterial blight in pinto bean (*Phaseolus vulgaris* L.). *Plant Breed* 124: 282–287.

Naeem M, Nasir Khan M, Masroor M, Khan MMA, Moinuddin (2013) Adverse effects of abiotic stresses on medicinal and aromatic plants and their alleviation by calcium. In: Tuteja N, Gill SS (eds), *Plant Acclimation to Environmental Stress*. Springer Science + Business Media, New York, pp. 101–146.

Nageshbabu R, Jyothi MN (2013) Profile of small interfering RNAs from French bean *Phaseolus vulgaris* under abiotic stress conditions. *Int J Pharma Bio Sci* 4: 176–185.

Ngara R (2009) A proteome analysis of drought and salt stress responsive proteins of different sorghum varieties. PhD Thesis, University of the Western Cape.

Oktem HA, Eyidoan F, Demirba D, *et al.* (2008) Antioxidant responses of lentil to cold and drought stress. *J Plant Biochem Biotechnol* 17: 15–21.

Pagano MC (2014) Drought stress and mycorrhizal plants. In: Miransari M (ed.), Use of Microbes for the Alleviation of Soil Stresses. Springer Science + Business Media, New York, pp. 97–110.

Park W, Li J, Song R, Messing J, Chen X (2002) CARPEL FACTORY, a Dicer homolog, and HEN1, a novel protein, act in microRNA metabolism in Arabidopsis thaliana. *Curr Biol* 12: 1484–1495.

Parniske M (2008) Arbuscular mycorrhiza: the mother of plant root endosymbioses. *Nat Rev Microbiol* 6: 763–775.

Paul S, Kundu A, Pal A (2011) Identification and validation of conserved microRNAs along with their differential expression in roots of *Vigna unguiculata* grown under salt stress. *Plant Cell Tiss Org* 105: 233–242.

Perry J, Brachmann A, Welham T, *et al.* (2009) TILLING in *Lotus japonicus* identified large allelic series for symbiosis genes and revealed a bias in functionally defective ethyl methanesulfonate alleles toward glycine replacements. *Plant Physiol* 151: 1281–1291.

Pfeiffer TW, Pilcher DL (2006) Registration of KY98-2047 and KY98-2932 extra dense pubesence soybean germplasm. *Crop Sci* 46: 480.

Pinheiro C, Kehr J, Ricardo CP (2005) Effect of water stress on lupins temprotein analysed by two-dimensional gel electrophoresis. *Planta* 221: 716–728.

Pnueli L, Hallak-Herr E, Rozenberg M, *et al.* (2002) Molecular and biochemical mechanisms associated with dormancy and drought tolerance in the desert legume *Retama raetam*. *Plant J* 31: 319–330.

Porcel R, Aroca R, Ruiz-Lozano JM (2012) Salinity stress alleviation using arbuscular mycorrhizal fungi. A review. *Agron Sustain Dev* 32: 181–200.

Porch TG, Blair MW, Lariguet P, Galeano C, Pankhurst CE, Broughton WJ (2009) Generation of a mutant population for TILLING common bean genotype BAT 93. *J Am Soc Hort Sci* 134: 348–355.

Poulain D, Almohammad H (1995) Effects of boron deficiency and toxicity on faba bean (*Vicia faba* L.). *Eur J Agron* 4: 127–134.

Rabie GH, Almadini AM (2005) Role of bioinoculants in development of salt-tolerance of *Vicia faba* plants under salinity stress. *Afr J Biotech* 4: 210–222.

Ramos MLG, Gordon AJ, Michin FR, Sprent JI, Parsons R (1999) Effect of water stress on nodule physiology and biochemistry of a drought tolerant cultivar of common bean (*Phaseolus vulgaris* L.). *Annals Bot* 83: 57–63.

Rapparini F, Peñuelas J (2014) Mycorrhizal fungi to alleviate drought stress on plant growth. In: Miransari M (ed.), Use of microbes for the alleviation of soil stresses. Springer Science+ Business Media, New York pp. 139–159.

Rasool S, Ahmad A, Siddiqi TO, Ahmad P (2013) Changes in growth, lipid peroxidation and some key antioxidant enzymes in chickpea genotypes under salt stress. *Acta Physiol Plant* 35: 1039–1050.

Redden B, Leonforte A, Ford R, Croser J, Slattery J (2005) Pea (*Pisum sativum* L.). In: Singh RJ, Jauhar PP (eds), *Genetic Resources, Chromosome Engineering, and Crop Improvement: Grain Legumes*, Vol. 1. CRC Press, Boca Raton, FL, pp. 49–83.

Reddy S, Bhatnagar-Mathur P, Vadez Vincent, Sharma KK (2012) Grain legumes (soybean, chickpea, and peanut): Omics approaches to enhance abiotic stress tolerance. In: Tuteja N, Gill SS, Tiburcio A, Tuteja R (eds), *Improving Crop Resistance to Abiotic Stress*. Wiley-Blackwell-Verlag GmbH & Co., pp. 995–1032.

Reinhart BJ, Bartel DP (2002) Small RNAs correspond to centromere heterochromatic repeats. *Science* 297: 1831.

Repetto O, Bestel-Corre G, Dumas-Gaudot E, Berta G, Gianinazzi-Pearson V, Gianinazzi S (2003) Targeted proteomics to identify cadmium-induced protein modifications in *Glomus mosseae*-inoculated pea roots. *New Phytol* 157: 555–567.

Rodziewicz P, Swarcewicz B, Chmielewska K, Stobiecki M, Wojakowska A (2014) Influence of abiotic stresses on plant proteome and metabolome changes. *Acta Physiol Plant* 36: 1–19.

Rogers C, Wen J, Chen R, Oldroyd G (2009) Deletion-based reverse genetics in *Medicago truncatula*. *Plant Physiol* 151: 1077–1086.

Saha P, Chatterjee P, Biswas AK (2010) NaCl pretreatment alleviates salt stress by enhancement of antioxidant defense system and osmolyte accumulation in mungbean (*Vigna radiata* L. Wilczek). *Indian J Exp Biol* 48: 593–600.

Sassi S, González EM, Aydi S, Arrese-Igor C, Abdelly C (2008) Tolerance of common bean to long-term osmotic stress is related to nodule carbon flux and antioxidant defenses; evidence from two cultivars tolerance. *Plant Soil* 312: 39–48.

Saxena NP, Johansen C, Saxena MC, Silim SN (1993) Selection for drought and salinity tolerance in cool-season food legumes. In: Singh KB, Saxena MC (eds), *Breeding for Tolerance in Cool Season Food Legumes*. John Wiley & Sons, Ltd, Chichester, UK, pp. 245–270.

Schena M, Shalon D, Davis RW, Brown PO (1995) Quantitative monitoring of gene-expression patterns with a complementary DNA microarray. *Science* 270: 467–470.

Schneider KA, Brothers ME, Kelly JD (1997) Marker-assisted selection to improve drought resistance in common bean. *Crop Sci* 37: 51–60.

Serraj R, Krishnamurthy L, Kashiwagi J, Kumar J, Chandra S, Crouch JH (2004) Variation in root traits of chickpea (*Cicer arietinum* L.) grown under terminal drought. *Field Crops Res* 88: 115–127.

Sha A, Chen Y, Ba H, *et al.* (2012) Identification of *Glycine max* microRNAs in response to phosphorus deficiency. *J Plant Biol* 55: 268–280.

Shanker AK, Venkateswarlu B (2011) *Abiotic Stress in Plants – Mechanisms and Adaptations*. InTech Publisher, Janeza Tridne Rijeka, Croatia. ISBN: 9789533073941.

Sharma HC, Srivastava CP, Durairaj C, Gowda CLL (2010) Pest management in grain legumes and climate change. In: Yadav SS, McNeil DL, Redden R, Patil SA (eds), *Climate Change and Management of Cool Season Grain Legume Crops*. Springer, Dordrecht, pp. 115–140.

Sharma KK, Lavanya M (2002) Recent developments in transgenics for abiotic stress in legumes of the semi-arid tropics. JIRCAS Working Report, pp. 61–73.

Sharma KK, Ortiz R (2000) Program for the application of genetic transformation for crop improvement in the semiarid tropics. *In Vitro Cell Dev-Pl* 36: 83–92.

Sharmila P, Phanindra ML, Anwar F, Singh K, Gupta S, Pardha Saradhi P (2009) Targeting prokaryotic choline oxidase into chloroplasts enhances the potential of photosynthetic machinery of plants to withstand oxidative damage. *Plant Physiol Biochem* 47: 391–396.

Siddique KHM (2000) Understanding growth stresses of cool season pulses. In: O'Connell L (ed.), *Shared Solutions Cropping Manual*. Australian Grain, Berekua, Brisbane, pp. 12–16.

Silim, SN, Saxena MC (1993) Adaptation of spring-sown chickpea to the Mediterranean basin. I. Response to moisture supply. *Field Crops Res* 34: 121–136.

Sinclair TR, Serraj R (1995) Legume nitrogen fixation and drought. *Nature* 378: 344.

Singh BB, Matsui T (2002) Cowpea varieties for drought tolerance. In: Fatokun CA, Tarawali SA, Singh BB, Kormawa PM, Tamò M (eds), *Challenges and Opportunities for Enhancing Sustainable Cowpea Production*. IITA, Ibadan, Nigeria, pp. 287–300.

Singh BB, Cambliss OI, Sharma B (1997) Recent advances in cowpea breeding. In: Singh BB, Mohan Raj DR, Dashiell K, Jackai LEN (eds), *Advances in Cowpea Research*. IITA, Ibadan, Nigeria, pp. 30–49.

Singh BB, Mai-Kodomi Y, Terao T (1999) Relative drought tolerance of major rainfed crops of the semi-arid tropics. *Ind J Gene* 59: 1–8.

Singh BB, Solanki RK, Chaubey BK, Verma P (2011) Breeding for improvement of warm season food legumes. In: Pratap A, Kumar J (eds), *Biology and Breeding of Food Legumes*. CAB International, pp: 63–80.

Sinha RP (1988) Early maturity, dwarf mutant of urd bean (*V. mungo* (L.) Hepper). J Nuclear Agri Biol 17: 61–62.

Smethurst CF, Garnet T, Shabala S (2005) Nutrition and chlorophyll fluorescence responses of lucerne (*Medicago sativa*) to waterlogging subsequent recovery. *Plant Soil* 270: 31–45.

Sobhanian H, Razavizadeh R, Nanjo Y, *et al.* (2010) Proteome analysis of soybean leaves, hypocotyls and roots under salt stress. *Proteome Sci* 8: 1–15.

Srinivasarao CH, Ganeshamurthy AN, Ali M (2003) *Nutritional Constraints in Pulse Production*. Indian Institute of Pulses Research, Kanpur, India.

Stoddard FL, Balko C, Erskine W, Khan HR, Link W, Sarker A (2006) Screening techniques and sources of resistance to abiotic stresses in cool-season food legumes. *Euphytica* 147: 167–186.

Streeter JG (2003) Effects of drought on nitrogen fixation in soybean root nodules. *Plant Cell Environ* 26: 1199–1204.

Sunkar R, Jagadeeswaran G (2008) In silico identification of conserved microRNAs in large number of diverse plant species. *BMC Plant Biol* 8: 37.

Svabova L, Lebeda A (2005) *In vitro* selection for improved plant resistance to toxin-producing pathogens. *J Phytopathol* 153: 52–64.

Szittya G, Moxon S, Santos DM, *et al.* (2008) High-throughput sequencing of *Medicago truncatula* short RNAs identifies eight new miRNA families. *BMC Genomics* 9: 593.

Tadege M, Wang TL, Wen J, Ratet P, Mysore KS (2009) Mutagenesis and beyond! Tools for understanding legume biology. *Plant Physiol* 151: 978–984.

Tavakkoli E, Rengasamy P, McDonald GK (2010) High concentrations of Na^+ and Cl^- ions in soil solution have simultaneous detrimental effects on growth of faba bean under salinity stress. *J Exp Bot* 61: 4449–4459.

Tickoo JL, Lal SK, Chandra N, Dikshit HK (2006) Mung bean breeding. In: Ali M, Kumar S (eds), *Advances in Mung Bean and Urd Bean*. Indian Institute of Pulses Research, Kanpur, India, pp. 110–148.

Tiwari KN, Dwivedi BS (1990) Response of eight winter crops to zinc fertilizer on a Typic Ustochrept soil. *J Agri Sci* (Cambridge) 115: 383–387.

Toker C, Cagirgan MI (1998) Assessment of response to drought stress of chickpea (*Cicer arietinum* L.) lines under rainfed conditions. *Turk J Agri Fores* 22: 615–621.

Toker C, Mutlu N (2011) Breeding for abiotic stress. In: Pratap A, Kumar J (eds) *Biology and Breeding of Food Legumes*. CAB International, pp. 241–260.

Toker C, Yadav SS (2010) Legume cultivars for stress environments. In: Yadav SS, McNeil DL, Redden R, Patil SA (eds), *Climate Change and Management of Cool Season Grain Legume Crops*. Springer, Dordrecht, pp. 351–376.

Toker C, Lluch C, Tejera NA, Siddique KHM (2007a) Abiotic stresses. In: Yadav SS, Redden R, Chen W, Sharma B (eds) *Chickpea Breeding and Management*. CAB International, Wallingford, UK, pp. 474–496.

Toker C, Canci, H, Yildirim T (2007b) Evalution of perennial wild *Cicer* species for drought resistance. *Gen Res Crop Evol* 54: 1781–1786.

Toker C, Yildirim T, Canci H, Inci EN, Ceylan FO (2010) Inheritance of resistance to iron deficiency chlorosis in chickpea (*Cicer arietinum* L.). *J Plant Nutr* 33: 1366–1373.

Toker C, Canci H, Siddique KHM (2011) Nutrient disorders. In: Chen W, Sharma HC. Muehlbauer FJ (eds), *Compendium of Chickpea and Lentil Diseases and Pests*. APS Press, Minnesota, pp. 126–131.

Trindade I, Capitao C, Dalmay T, Fevereiro MP, Santos DM (2010) miR398 and miR408 are up-regulated in response to water deficit in *Medicago truncatula*. *Planta* 231: 705–716.

Umezawa T, Mizuno K, Fujimura T (2002) Discrimination of genes expressed in response to the ionic or osmotic effect of salt stress in soybean with cDNA-AFLP. *Plant Cell Environ* 25: 1617–1625.

Vadez V, Krishnamurthy L, Kashiwagi J, *et al.* (2007) Exploiting the functionality of root systems for dry, saline, and nutrient deficient environments in a changing climate. *SAT eJournal* 4(1): 1–61.

Vadez V, Sinclair TR, Serraj R (2000) Asparagine and ureide accumulation in nodules and shoot as feedback inhibitors of N2 fixation in soybean. *Physiol Plant* 110: 215–223.

Valdes-Lopez O, Hernandez G (2008) Transcriptional regulation and signaling in phosphorus starvation: what about legumes? *J Integr Plant Biol* 50: 1213–1222.

Valentine AJ, Benedito VA, Kang Y (2011) Legume nitrogen fixation and soil abiotic stress: From physiology to genomic and beyond. *Annual Plant Rev* 42: 207–248.

Varshney RK, Hiremath P, Lekha P, *et al.* (2009) A comprehensive resource of drought- and salinity-responsive ESTs for gene discovery and marker development in chickpea (*Cicer arietinum* L.). *BMC Genomics* 10: 523.

Varshney RK, Roorkiwal M, Nguyen HN (2013) Legume genomics: From genomic resources to molecular breeding. *The Plant Genome* 6: 1–7.

Velculescu VE, Zhang L, Vogelstein B, Kinzler KW (1995) Serial analysis of gene-expression. *Science* 270: 484–487.

Verdoy D, Coba del Peña T, Redondo FJ, Lucas MM, Pueyo JJ (2006) Transgenic *Medicago truncatula* plants that accumulate proline display nitrogen-fixing activity with enhanced tolerance to osmotic stress. *Plant Cell Environ* 29: 1913–1923.

Wang T, Chen L, Zhao M, Tian Q, Zhang WH (2011) Identification of drought-responsive microRNAs in *Medicago truncatula* by genome-wide high-throughput sequencing. *BMC Genomics* 12: 367.

Wang X, Zhang J, Li F, *et al.* (2005) MicroRNA identification based on sequence and structure alignment. *Bioinformatics* 21: 3610–3614.

Wang Y, Long L (2010) Identification and isolation of the cold-resistance related miRNAs in *Pisum sativum* Linn. *J Liaoning Normal University (Nat Sci Edn)*, 2.

Wightman B, Ha I, Ruvkun G (1993) Posttranscriptional regulation of the heterochronic gene lin-14 by lin-4 mediates temporal pattern formation in C. elegans. *Cell* 75: 855–862.

Xu D, Duan X, Wang B., Hong B, Ho THD, Wu R (1996) Expression of a late embryogenesis abundant protein gene, HVA1, from barley confers tolerance to water deficit and salt stress in transgenic rice. *Plant Physiol* 110: 249–257.

Xu F, Liu Q, Chen L, *et al.* (2013) Genome-wide identification of soybean microRNAs and their targets reveals their organ-specificity and responses to phosphate starvation. *BMC Genomics* 14: 66.

Yamamoto Y, Kobayashi Y, Devi SR, Rikiishi S, Matsumoto H (2002) Aluminum toxicity is associated with mitochondrial dysfunction and the production of reactive oxygen species in plant cells. *Plant Physiol* 128: 63–72.

Yang H, Shankar M, Buirchell BJ, Sweetingham MW, Caminero C, Smith PMC (2002) Development of molecular markers using MFLP linked to a gene conferring resistance to *Diaporthe toxica* in narrow-leafed lupin (*Lupinus angustifolius* L.). *Theor Appl Genet* 105: 265–270.

Yang H, Boersma JG, You MP, Buirchell BJ, Sweetingham MW (2004) Development and implementation of a sequence specific PCR marker linked to a gene conferring resistance to anthracnose disease in narrow-leafed lupin (*Lupinus angustifolius* L.). *Mol Breed* 14: 145–151.

Yeo A (1999) Prediction of the interaction between the effects of salinity and climate change on crop plants. *Sci Hort* 78: 159–174.

Zeng HQ, Zhu YY, Huang SQ, Yang ZM (2010) Analysis of phosphorus-deficient responsive miRNAs and cis-elements from soybean (*Glycine max* L.). *J Plant Physiol* 167: 1289–1297.

Zeng QY, Yang CY, Ma QB, Li XP, Dong WW, Nian H (2012) Identification of wild soybean miRNAs and their target genes responsive to aluminum stress. *BMC Plant Biol* 12: 182.

Zhao B, Ge L, Liang R, *et al.* (2009) Members of miR-169 family are induced by high salinity and transiently inhibit the NF-YA transcription factor. *BMC Mol Biol* 10: 29.

Zhao CZ, Xia H, Frazier TP, *et al.* (2010) Deep sequencing identifies novel and conserved microRNAs in peanuts (Arachis hypogaea L.). *BMC Plant Biol* 10: 3.

Zhou ZS, Huang SQ, Yang ZM (2008) Bioinformatic identification and expression analysis of new microRNAs from *Medicago truncatula*. *Biochem Biophys Res Commun* 374: 538–542.

Zhu BG, Ye CJ, Lu H, *et al.* (2006) Identification and characterization of a novel heat shock transcription factor gene, GMHsfA1, in soybeans (*Glycine max*). *J Plant Res* 119: 247–256.

Zhu H, Choi H, Cook DR, Shoemaker RC (2005) Bridging model and crop legumes through comparative genomics. *Plant Physiol* 137: 1189–1196.

Zhu JK. (2007) *Plant Salt Stress*: John Wiley & Sons, Ltd.

Zhu Y, Zeng H, Dong C, Yin X, Shen Q, Yang Z (2010) MicroRNA expression profiles associated with phosphorus deficiency in white lupin (*Lupinus albus* L.). *Plant Sci* 178: 23–29.

CHAPTER 2

Salt stress and leguminous crops: Present status and prospects

P.S. Sha Valli Khan[1] and P. Osman Basha[2]

[1] Department of Botany, Yogi Vemana University, Vemanapuram, India

[2] Department of Genetics and Genomics, Yogi Vemana University, Vemanapuram, India

2.1 Introduction

Salinity is the build-up of mineral salts in soils and waters. The dissolved mineral salts form electrolytes of cations such as Na^+, Ca^{2+}, Mg^{2+} and K^+, and anions such as Cl^-, SO_4^{2-}, HCO_3^-, CO_3^{2-} and NO_3^-. Other elements contributing to salinity in hypersaline soils and waters comprise Al^{3+}, B, Ba^{2+}, Sr^{2+}, SiO_2 and Mo (Hu & Schmidhalter, 2002). Soil salinity is assessed by the soil's electrical conductivity, and the SI unit of electrical conductivity (ECe) is dS/m (decisiemens per metre). Richards (1954) classified salt-affected areas into non-saline (<2 dS/m), slightly saline (2–4 dS/m), moderately saline (4–8 dS/m), highly saline (8–16 dS/m) and extremely saline (>16 dS/m) based on the ECe.

Ghassemi *et al.* (1995) categorized salinity as 'primary' or 'secondary' salinity. Primary salinity occurs due to the accumulation of salts over long periods of time by two natural processes. The first natural process is the weathering of rocks comprising soluble salts of several types, primarily Cl^- of Na^+, Ca^{2+}, Mg^{2+} and to a lesser extent, SO_4^{2-}, HCO_3^- and CO_3^{2-}. The second process follows due to the deposition of oceanic salt carried inland by wind and rain. The deposited salt mainly contains NaCl, which is a component of oceanic salt (Munns, 2005). Secondary salinity results from anthropogenic activities that alter the hydrological balance in the soil between water applied (irrigation or rainfall) and water consumed by crops (transpiration). The most common causes for secondary salinity are (i) land clearance and the substitution of perennial vegetation with annual crops, and (ii) irrigation schemes using salt-rich irrigation water or causing poor drainage (Ghassemi *et al.*, 1995). Salt-affected areas occur widely in arid and semi-arid areas (Rengasamy *et al.*, 2003).

Based on soil and ground water processes found all over the world, Rengasamy (2006) has further classified salinity into three different types:

1. groundwater-associated salinity (GAS);
2. non-groundwater-associated salinity (NAS);
3. irrigation-associated salinity (IAS).

GAS occurs in discharge areas of the landscape. Water permeates from groundwater to the soil surface carrying the salts dissolved in it. The motivating force for upward movement of water and salts is evaporation from the soil plus plant transpiration. Generally, the water table in the landscape is at or very close to the soil surface. Salt accumulation is high when the water table is less than 1.5 m below the soil surface (Talsma, 1963).

NAS is found in landscapes where the water table is deep and drainage is insufficient. Salts carried by rain, weathering and aeolian deposits are stored within the soil solum.

IAS results from irrigation water accumulating within the root zone because of inadequate leaching. Poor quality irrigation water, low hydraulic conductivity of soil layers as found in heavy clay soils and sodic soils, and high evaporative conditions all promote IAS. Utilization of highly saline effluent water and inadequate drainage and soil management enhance the risk of salinity in irrigated soils. In many irrigation regions, rising saline groundwater interacting with the soils in the root zone can compound the problem (Rengasamy, 2006).

Legumes under Environmental Stress: Yield, Improvement and Adaptations, First Edition. Edited by Mohamed Mahgoub Azooz and Parvaiz Ahmad.

Salinity often leads to the development of other problems in soils such as soil sodicity and alkalinity. Soil sodicity is the result of the binding of Na^+ to the negatively charged clay particles, which leads to clay swelling and dispersal. Hydrolysis of the Na–clay complex results in soil alkalinity. Salt-affected lands occur in practically all climatic regions, from the humid tropics to the polar regions. Saline soils can be found at different altitudes, from below sea level (e.g. around the Dead Sea) to mountains rising above 5000 m, such as the Tibetan Plateau or the Rocky Mountains. However, the salinization process is more widespread in arid and semi-arid regions. Over 800 million hectares of land throughout the world is salt-affected, either by salinity (397 million ha) or the associated condition of sodicity (434 million ha) (FAO, 2005). This is over 6% of the world's total land area. Out of 230 million ha of irrigated land, 45 million ha is affected by salinity or sodicity (20%) (FAO, 2005). Out of 1500 million ha of land farmed by dryland agriculture, 32 million ha (2%) is affected by secondary salinity to varying degrees. Irrigated land has at least twice the productivity of rain-fed land; it produces one-third of the world's food (Munns, 2005). This problem is likely to become more severe in the future as we face climate change at a global level.

Climate change is distressing ecosystems, freshwater supplies and human health (Raman *et al.*, 2012). Emissions of heat-trapping GHGs such as CO_2, methane (CH_4) and nitrous oxides (NO_x) are responsible for large increases in global temperatures, observed in the second half of the 20th century, as reported by the Intergovernmental Panel on Climate Change (IPCC, 2007). The higher the atmospheric concentrations of GHGs, the more the earth will warm in the decades and centuries ahead. Scientists have also observed that during the 20th century, the mean global surface temperature increased by 0.8 °C per year (IPCC, 2007). The effects of global warming will not only be restricted to increasing mean annual temperatures around the globe, but also will have profound consequences on rainfall patterns, natural ecosystems, agriculture, wildlife and life in general (WWF, 2008). Because of climatic changes and global warming, oceanic water is intruding into coastal regions of arid and semi-arid regions and causing salinization (Kundzewicz *et al.*, 2005). Thus, understanding salt stress responses is essential in attempts to breed crops with salt resistance to feed the growing population.

Anticipated dehydration and increases in temperature will cause high evapotranspiration. The drier regions of the world may become even drier. Accordingly, it will become very difficult for water-limited countries to face these challenges. Surface water shortage will increase pressure on use of groundwater, the major part of which is not of a safe and usable quality. Increased soil and water salinity/sodicity negatively affects the chemical and physical characteristics of soil and subsequently reduces the growth and yield of crops. Legumes are the most salt sensitive group in this regard and are anticipated to be significantly affected. The changing circumstances will also necessitate prudent decisions around the selection of crops that can endure salinity stress and addition of legumes in crop rotations. Understanding of genetic variability with respect to salt tolerance will be required. This will involve implementing robust breeding programmes to achieve this objective supported by modern approaches such as biotechnology, mutagenesis and genetic engineering.

This chapter provides an overview of 'salt stress and leguminous crops: present status and prospects' and also briefly discusses the effects of salinity and responses of plants to salinity. Efforts are also made to review the lessons learned from studies of leguminous crops and to highlight recent progress in understanding the physiological, biochemical and molecular mechanisms of salt stress responses and tolerance. Finally we summarize research on 'omics' approaches to a better understanding of salt stress responses and tolerance in leguminous crops.

2.2 Effects of salinity

Natural soil salinity predates human civilization. When early humans, looking for better sources of livelihood, moved into arid lands along the riverbanks, they resorted to irrigated agriculture. With the practice of irrigation came salinity, the first environmental problem caused by humans. The earliest written account of salt lands dates back to 2400 BC and was recorded in the Tigris-Euphrates alluvial plains of Iraq. Salinity destroyed the ancient societies that had thrived for several centuries (Hillel, 2005; Rengasamy, 2006). Sustained salinization of arable land is becoming more widespread and thus decreasing crop yields from formerly productive soils throughout the world.

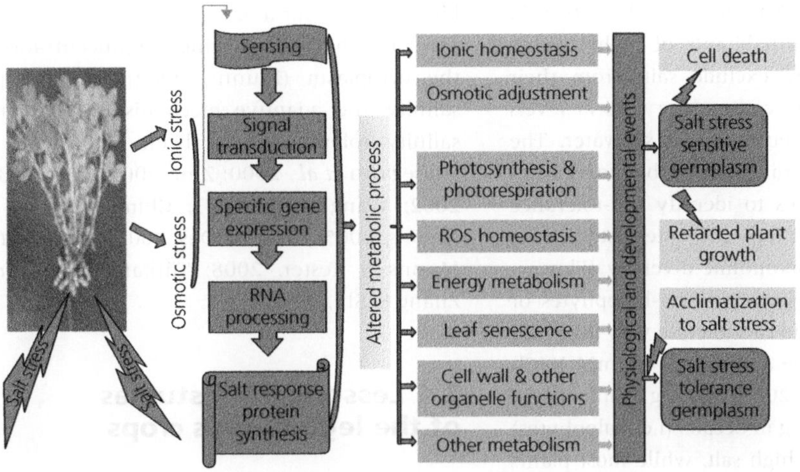

Figure 2.1 Effect of salt stress and responses of plant.

The harmful impacts of salinity include:

- reduced agricultural production;
- lower economic returns due to higher costs of cultivation, reclamation, management, etc.;
- soil erosion due to high dispersibility of soil;
- ecological imbalance due to halophytes and marine life forms invading as fresh waters become brackish;
- adverse impact on human health due to toxic effects of elements such as B, F and Se.

Most crops are sensitive to salt stress at all stages of plant development, from seed germination through vegetative and reproductive growth. Thus, soil salinity is a major factor limiting sustainable agriculture (Hu & Schmidhalter, 2002). Salt stress instigates molecular damage in plant cells that primarily causes hyperosmotic stress, homeostasis disruption and ionic toxicity (Figure 2.1). Osmotic stress inhibits a plant's ability to take up water, and this leads to what is termed 'water deficit' or 'physiological drought' or 'cellular dehydration' (Munns, 2005). Water deficit causes a reduction in the rate of photosynthesis, and exposure of chloroplasts to excess excitation energy limits carbon dioxide fixation, leading to the generation of active/reactive oxygen species (A/ROS). Electrolytes of mineral salts, but especially Na^+ and Cl^-, can enter plant cells through the transpiration stream and accumulate in cellular and extracellular compartments. Although Na^+ represents the major ion causing toxicity related to high salinity, some plant species are also sensitive to chloride, the major anion found in saline soils. Eventually excess build-up of internal concentrations of Na^+ and Cl^- injures cells and affects growth (Chinnusamy *et al.*, 2005).

Salt stress also causes secondary stresses like oxidative stress, which accelerates the production of A/ROS and subsequently alters the balance between the formation and removal of such species (Zhu, 2002). Many plants under salt stress also suffer with other kinds of secondary stresses like mechanical stress and nutritional imbalance. The elevated levels of Na^+, a major ion in saline environments, can induce deficiency of the essential element K^+, impose toxic effects by perturbing K^+-dependent processes, and induce deleterious changes in protein conformation (Mahajan & Tuteja, 2005). During the onset and development of salt stress, all major processes are affected, including photosynthesis, protein synthesis, energy production and lipid metabolism. The effects of mild cases of salt stress are primarily limited to plant growth, development and crop productivity, but in extreme cases, salt stress can lead to plant death (Parida & Das, 2005).

2.3 Responses of plants to salinity

Plants are traditionally classified as halophytes or non-halophytes (glycophytes) referring to their capacity to grow on highly saline environments (Flowers *et al.*, 1977). The exact definition of halophytes is still disputable.

Halophytes complete their life cycle in environments where the concentrations of NaCl lie in the 200–500 mM range. Being the natural inhabitants of highly saline soils, halophytes efficiently exclude salts from their roots and leaves, and some can endure salts at levels more than twice the concentration of seawater. The remarkable levels of salt tolerance shown by the legumes make them good candidates to identify salt-tolerance genes. Halophytes represent only 2% of terrestrial plant species but have a wide taxonomic diversity (Flowers *et al.*, 2010). At the other extreme non-halophytes or glycophytes are sensitive and can tolerate very little salt, suffering irreparable damage at 4 dS/m (≈40 mM NaCl) or more (Chinnusamy *et al.*, 2005). During adaptive evolution in nature, plants living in oceans (i.e. halophytes) retained their tolerance to high salt, while most plants that migrated to land lost their ability and adopted a glycophytic lifestyle. Many agricultural crops, including legumes, are glycophytes and are sensitive to salinity.

Salt tolerance is a complex trait that is controlled by multiple genes and involves various biochemical and physiological mechanisms. The functions of the distinct sets of genes involved in specific biochemical and physiological mechanisms must be combined to achieve substantial increases in salt tolerance. In the last 10 years, significant progress towards understanding the mechanisms of plant salt tolerance has been made in several crop plants including legumes. Plants adapt developmental, morphological, physiological, biochemical and molecular strategies to cope with the detrimental effects of salt stress (Flowers *et al.*, 2010). Moreover, salt-tolerant plants (halophytes) have evolved a number of mechanisms for adaptation to stressful saline environments. Adaptive mechanisms of plants exposed to salinity are based upon utilization of one or more of the following major mechanisms (Parida & Das, 2005):

1. phenological avoidance;
2. salt avoidance through low root permeability for certain salt ions (e.g. Na^+);
3. salt avoidance by secretion using salt bladders or glands;
4. succulence for distribution and dilution of high salt accumulations;
5. compartmentalization of salts in vacuoles;
6. biochemical tolerance (e.g. synthesis of compatible osmolytes, induction of antioxidative enzymes and stimulation of phytohormones);
7. nutritive tolerance.

In summary, salt tolerance mechanisms are of two kinds: those minimizing the entry of salts into the plant and those minimizing the concentration of salt in the cytoplasm (Munns, 2002). Plant responses to salinity and adaptive mechanisms that confer plant salinity tolerance have been extensively reviewed (Hasegawa *et al.*, 2000; Zhu, 2001; Apse & Blumwald, 2002; Wang *et al.*, 2003; Chinnusamy *et al.*, 2005; Munns, 2005; Parida & Das, 2005; Hussain *et al.*, 2008; Munns & Tester, 2008; Türkana & Demiral, 2009; Zhang & Shi, 2013).

2.4 Lessons from studies of the leguminous crops

The Leguminosae (Fabaceae) represent the second largest and most diverse family of dicotyledonous flowering plants, with 20,000 species classified into around 700 genera (Raven & Pothill, 1981; Doyle & Luckow, 2003). Legumes form a symbiotic association in their root system (sometimes on their shoots) with nitrogen-fixing microbes (diazotrophs: in association with legumes collectively called rhizobia) in specialized organs called nodules (Bruning & Rozema 2013).

Legumes as food crops play an important nutritional role in the diet of millions of people living in the developing countries, and legumes are sometimes referred to as the 'poor man's meat'. They provide a significant part of the diet of vegetarians, being vital sources of protein, calcium, iron, phosphorus and other minerals. In addition to those legumes cultivated for human consumption, many yield important fodders, forages and green manures since they assist in nitrogen fixation. Green manures have been used since the beginning of agriculture, but their use has diminished since industrially produced fertilizer became available (Zahran, 1999). Green manure adds nitrogen to the soil and improves soil quality by increasing organic matter content of the soil (Sullivan, 2003). Legumes are known to survive and compete effectively in nitrogen-poor conditions. Root nodules are generally found in the Mimosoideae and the Papilionoideae, but rarely formed in the Caesalpinoideae. Legumes are also used for a variety of other purposes including timber, medicine, tanins and gums. The International Legume Database and Information Service (ILDIS) states that the economic importance of the Leguminosae family is likely to increase as human pressure places greater demand on marginal land. Many Leguminosae species are characteristic of open and disturbed places and are thus well adapted to grow under poor conditions (www.ildis.org).

The yields and productivity of legume crops have been reported to be low in developing countries over the past few decades. Agricultural research and development efforts in many of these countries have focused mainly on improving cereal yields and production to attain food security. However, reorientation of the research spotlight on legumes will have important impacts on nutritional security and soil fertility. Recently observed rising prices reflect the escalating demand for legumes worldwide due to increasing incomes and population. The swelling demand for livestock feed, even in developing countries, is a noteworthy systemic change in the demand structure. In addition, the significant demand from the biodiesel industry for soybean due to the current policy initiatives in Europe and the USA, has also contributed to the upsurge in demand and prices of substitute crops. These factors mean that in the near future there will be substantive shifts in the utilization patterns and price structure for legumes (Gowda *et al.*, 2009).

The Leguminosae family is usually divided into three subfamilies: Papilionoideae, Caesalpinoideae and Mimosoideae (Raven & Pothill, 1981; Doyle & Luckow, 2003). The Papilionoideae is the largest of the three subfamilies, with about two-thirds of all the genera and species of the family. It is also the most widespread, extending further into temperate regions than the other two subfamilies. The two major groups of cultivated species in the Papilionoideae are the tropical or 'phaseoloid' legumes (*Phaseolus, Vigna, Glycine* and *Cajanus)* and the temperate or 'galegoid' legumes (*Melilotus, Trifolium, Medicago, Pisum, Vicia, Lotus, Cicer, Lens* and *Lathyrus*). Groundnut is somewhat distinct from the phaseoloid and galegoid groups of grain legumes (Upadhyaya *et al.*, 2009). The United Nations Food and Agriculture Organization (FAO, 1994) recognizes 11 categories of annual leguminous crops or pulses (http://www.fao.org). These are: dry beans, dry broad beans, dry peas, chickpea, dry cowpea, pigeon peas, lentils, Bambara beans, vetches, lupins and pulses nes. Crops like soybean and groundnut (used mainly for oil extraction) are excluded by the FAO from the list of pulses (Table 2.1).

The Leguminosae (Fabaceae) plant family seems to be particularly sensitive to salinity (Maas & Hoffman, 1977). Salinity tolerance was placed into four categories:

- salt sensitive (80% of biomass production compared to control at ~3 dS/m (equivalent to about 30 mM NaCl);
- moderately sensitive (80% of biomass production at ~6 dS/m (~60 mM NaCl);
- moderately tolerant (80% biomass production at ~11 dS/m (~110 mM NaCl);
- tolerant (80% of biomass production at ~16 dS/m (~160 mM NaCl).

Using these four categories defined by Maas and Hoffman (1977), the majority of legumes are sensitive or moderately sensitive. Legumes not only suffer directly from salt stress like all plants, but also their symbiotic associations show greater sensitivity to salinity than either partner alone. Salinity impacts on various phases of the initiation of symbiosis, nodule formation and functionality (Maas & Hoffman, 1977; Bruning & Rozema, 2013).

A comprehensive list of 1554 halophytes was compiled by James Aronson in 1980. His HALOPH database of salt-tolerant plants of the world was published in 1984 by the Office of Arid Land Studies at the University of Arizona, Tucson, Arizona. This database originally was compiled according to the maximum salinity tolerated by the plant, where this information was available, and expressed as the ECe for a solution in which the plants were growing together with its photosynthetic pathway where known (http://www.sussex.ac.uk/affiliates/halophytes/index.php). Recently Bruning and Rozema (2013) focused on the potential use of legumes in saline agriculture as crops and for fodder, forage or other economically viable uses. There are 46 legumes that are tolerant to salinity levels ranging from 8 dS/m to seawater salinity in the eHALOPH database. This includes annuals and perennials, and legume species from tropical, subtropical and temperate regions. However, most of the species are trees and herbaceous perennials. Many legumes are sensitive to salinity and as a group contain a relatively lower proportion (0.2%) of tolerant species than the background percentage (1%) of all land plants (Rozema & Flowers, 2008). Of four of the most commonly grown legumes in agriculture, namely pea, chickpea, common bean and soybean, soybean is the most salt tolerant (Delgado *et al.*, 1994). Manchanda and Garg (2008) reviewed the responses of legumes to salinity stress with emphasis on

Table 2.1 List of leguminous crops (pulses) as classified by FAO (1994).

S. No	FAOSTAT Code	Commodity	Species name	Common name	Economic use	Remarks
1	0176	Dry beans	*Phaseolus vulgaris* L.	Common bean, kidney bean, haricot bean	Human food (fruits are edible)	–
			P. lunatus L.	Lima, butter bean	Human food	*P. limensis* Macfad.
			P. coccineus L.	Scarlet runner bean	Starchy roots, green pods and dry seeds used as food	---
			P. acutofolius A.Gray	Tepary bean	Major food staple, used as snacks	*P. tenuifolius* A.Gray
			P. angularis (Willd)	Adzuki bean, Azuki bean	Human food	*V. angularis* (Willd)
			P. aureus Roxb	Mung bean, golden, green gram	Human food and forage crop	*V. radiata* (L.) Wikzek
			P. mungo	Black gram, urd	Highly prized pulse	*V. mungo* (L) Hepper
			P. calcaratus Roxb.	Rice bean	Minor food and fodder crop	*V. umbellata* (Thunb.) Ohwi and Ohasji
			P. aconitifolius Jacq.	Moth bean	Food, forage and soil conditioner	*V. aconitifolia* (Jacq.) Marechal
2	0181	Dry broad beans	*Vicia faba* var. equina	Horse-bean	Human food	*Faba sativa* Moench
			V. faba var. major	Broad bean		
			V. faba var. minor	Field bean		
3	0187	Dry peas	*Pisum sativum* L.	Garden pea, field pea	Major human food crop	*P. arvense* L.
4	0191	Chickpeas	*Cicer arietinum* L.	Chickpea, Bengal gram, garbanzos	Major human food crop and animal feed	–
5	0195	Dry cow peas	*Vigna unguiculata* (L). Walp.	Cow pea, black-eyed pea, yardlong bean	Major human food and animal feed	*Dolichus sinensis* *V. sinensis*
6	0197	Pigeon peas	*Cajanus cajan* (L.) Millsp.	Pigeon pea, cajan pea, Congo bean	Major food crop and forage/cover crop	–
7	0201	Lentil	*Lens culinaris* Medicus	Lentil	Human food	*Ervum lens* *Lens esculenta*
8	0203	Bambara beans	*Vigna subterranea* (L.) Verdc.	Bambara ground nut, earth pea, ground bean	Major grain legume in semi-arid Africa	*Voandzeia subterranea* (L.). Thouars *Arachis africana* Burm
9	0205	Vetches	*Vicia sativa L.*	Spring/common vetch	Food, forage and animal feed	–
10	0210	Lupins	*Lupinus* spp.	Lupins	Animal feed and human food in Africa, forage crop	–
11	0211	Pulses nes	*Dalichus* spp.	Lablab, hyacinth bean	Human food, animal feed, forage crops and green manure	–
			Canavalia spp.	Jack or sword bean		
			Posophocarpus tertragonolobus	Winged bean		
			Cymopsis tetragonoloba	Guar bean		
			Stizolobium spp.	Velvet bean		
			Pachyrrhizus erosus	Yam bean		

physiological and biochemical mechanisms of salt tolerance. This part of the chapter covers physiological and molecular mechanisms that have been extensively studied in leguminous crops especially beans, broad beans, peas, chickpea, cowpea, pigeon pea, lentil, lupins, groundnut and soybean.

2.4.1 Beans

The dry bean category in FAO includes all species of *Phaseolus* and *Vigna*. The genus *Phaseolus* comprises about 70 species, all native to the Americas, primarily Mexico. At least four species of this genus have been domesticated since pre-Columbian times for their beans. Most prominent among these is the common bean, *P. vulgaris*, which today is cultivated worldwide in tropical, semitropical and temperate climates. Previous classifications placed a number of other well-known species in this genus, but these have now been removed to the genus *Vigna*, sometimes necessitating a change of species name (Table 2.1).

Salt stress inhibited growth, nodule dry weight and acetylene reduction activity (ARA) in *P. vulgaris* (Ferri *et al.*, 2000). Serraj *et al.* (2001) examined the short-term effects of 0.1 M NaCl on ARA and growth of four genotypes (W, B, R and C) of common bean. After 5 days' exposure to stress, the shoot dry weight of two genotypes, W and B, was 83% of that of control plants not subjected to NaCl; there was no major effect of the salt treatment on the growth of other two genotypes, R and C. For all genotypes, the ARA was reduced dramatically during the first hour following NaCl application, which was followed by a slower decrease during the next 5 h during which genotypical differences in response to salinity were found.

Saadallah *et al.* (2001) studied the genotypic variability for tolerance to salinity of common bean lines. They found variability among common bean lines in the response to salt stress. Lines ABA16 and BRB17 were found to be the most salt-sensitive lines whereas BAT477 and Flamingo were relatively tolerant. Some lines that displayed vigorous growth under the control treatment were seriously affected by salt treatment. Further, the comparative tolerance of some lines to salt appeared to depend on the capacity to maintain a sufficient leaf area and to develop a profuse and well-organized nodular system. Bouhmouch *et al.* (2001) investigated the infectiveness and effectiveness of 19 most osmotolerant (growing at NaCl concentrations >340 mM) rhizobial strains on common bean. The results showed that RP119 was the most effective osmotolerant strain, forming a high number of nodules with greater nodule dry weight. In another study, Bouhmouch *et al.* (2005) also reported that nodule differentiation of *P. vulgaris* plants was affected by salt stress; white nodules emerged that lost their pink colour (leghaemoglobin (Lb) content).

Jebara *et al.* (2001) assessed the effects of NaCl on interactions between native rhizobia from Tunisian soil and common bean lines. This study reported that the nodular dry weight was usually more affected than shoot dry weight by the salt stress. The NAG310 line, in contrast with Flamingo and DOR 364, in symbiosis with 12a3 was further affected by the salt for nodular efficiency, in spite of its high tolerance in regard of nodule growth. In another study, Jebara *et al.* (2005) evaluated antioxidant enzyme expression in response to salt stress (50 mM NaCl) in *P. vulgaris* genotype BAT477 inoculated with strain CIAT899. The results revealed that superoxide dismutase (SOD) and catalase (CAT) nodular isozymes had bacterial and root origins. The same Cu-, Zn-, Fe- and Mn-SOD isoforms were expressed in nodules and roots, whereas in free rhizobia they found only one Fe- and Mn-SOD. Ascorbate peroxidase (APX) and peroxidase (POX) nodule and root profiles had only root origins, as no rhizobial band was identified. Thus, NaCl salinity caused a differential regulation of distinct SOD and POX isozymes, and the enzymes seemed to protect the symbiosis from stress.

The addition of proline to NaCl-treated seedlings of *P. vulgaris* cv. Kizilhac effected enhanced growth in comparison to NaCl-treated seedlings (Demir & Kocakalikan, 2002). Jungklang *et al.* (2003) also observed an increase in the amino acid and proline contents in *P. vulgaris* under salt stress. Ashraf and Bashir (2003) have observed significant reductions in fresh and dry weights of shoots and roots, and shoot length and leaf area of *P. vulgaris* plants under salt stress. Higher amounts of Na^+ and Cl^- accumulated in the nodules, whereas higher amounts of Cl^- accumulated in the leaves. The contents of free amino acids, proline and glycine betaine increased whereas soluble proteins declined in all plant parts of common bean under salt stress. However, significant reductions in nodule fresh mass and nodule number were also observed, but nodule dry weight did not decrease significantly (Ashraf & Bashir, 2003).

Salinity reduced shoot growth more than root growth, increased the root:shoot ratio, and reduced the number of leaves, leaf area and relative growth rate (RGR) in four wild and two cultivated species of *Phaseolus*. The leaf water, osmotic and turgor potentials were also decreased in all the species. Tissue concentrations of Cl^- and Na^+ ions increased significantly, while K^+ ion concentrations were reduced. Further, Ca^{2+} concentrations in roots and stem stayed nearly constant, whereas leaf Ca^{2+} concentrations were raised considerably with salinity. Rates of uptake of K^+, Ca^{2+} and Mg^{2+} ions were significantly disrupted by increasing NaCl (Jimenez *et al.*, 2003). Verdoy *et al.* (2004) reported the greatest expression of proline-rich proteins (PRPs) and late embryogenesis abundant (LEA) proteins in nodule cortex cells and in the cells of vascular bundles in responses to salt stress and water deficit in *P. vulgaris*.

Under salinity conditions plant growth was reduced in salt-tolerant mutants of common bean plants when inoculated with salt-tolerant *Rhizobium tropici* wild-type strain CIAT899. Common bean plants formed a more active symbiosis with salt-tolerant *Rhizobium tropici* wild-type strain CIAT899 than with its decreased salt toleraant (DST) mutant derivatives. The DST mutants formed nodules that exhibited lower antioxidant enzyme activities than wild-type nodules. Levels of nodule CAT appeared to correlate with symbiotic nitrogen-fixing efficiency, and SOD and dehydroascorbate reductase seem to function in the molecular mechanisms underlying the tolerance of the nodules to salinity. Nitrogenase activity declined under salt stress in *P. vulgaris* nodules when inoculated with a salt-tolerant wild-type *Rhizobium tropici* strain rather than with its DST mutant derivative (Tejera *et al.*, 2004). In another study, Tejera *et al.* (2005) investigated the effect of NaCl on growth responses in common bean plants. The adverse effects of salinity were observed on shoot dry weights and shoot:root ratio in all the treatments. An increase in the contents of total soluble sugars and free amino acids was noted. It was suggested that these metabolites might be related to a nodule osmotic adjustment response under saline conditions, or the increase of amino acids could be a consequence of protein degradation. The common bean plants also responded to salinity by decreasing nitrogen fixation parameters (ARA per plant and nodule number) (Tejera *et al.*, 2005).

Shamseldin and Werner (2005) observed that at 0.2% NaCl stress, no significant difference was noticed between the nodule number formed by the salt-sensitive strain EBRI 2 without salt and under salt stress. However, at 0.4% NaCl, stressed *P. vulgaris* plants showed reductions in nodule number, nodule fresh weight and acetylene production of 45%, 51% and 53%, respectively. Jebara *et al.* (2005) subjected *Phaseolus* genotype BAT477 to 50 mM NaCl and reported that ARA was considerably decreased by salt treatments. Cabot *et al.* (2005) compared the salt tolerance of *P. vulgaris* cv. Contender plants, grown in spring and summer seasons at less than 75 mM NaCl and achieved a 2.6-fold greater RGR in summer than in spring under non-saline conditions.

The effect of abscisic acid (ABA) and 100 mM NaCl on growth, nitrogenase activity and nodule metabolism was studied in *P. vulgaris* cv. Coco. The plant dry weight, nodule dry weight, nitrogen fixation (acetylene reduction activity and ureide content), and most enzymes of ammonium and ureide metabolism were affected by both ABA and NaCl. The addition of 1 μM ABA to the nutrient solution before the exposure to salt stress reduced the negative effect of NaCl. It was suggested that ABA application improves the response of common bean plants' symbiosis under saline stress conditions, including the nitrogen fixation process and enzymes of ammonium assimilation and purine catabolism (Khadri *et al.*, 2006). In another study, Khadri *et al.* (2007) compared two cultivars – one salt-sensitive (cv. Coco) and one salt-tolerant (cv. Africa) – of *P. vulgaris* in symbiosis with *Rhizobium tropici* CIAT899 under controlled conditions versus pretreatment individually with ABA or salt or both. The ABA pretreatment in salt-stressed plants improved the growth parameters and normalized nodule weight of Coco and induced lower nitrogenase activity. Plants of tolerant cultivars seem to limit translocation of sodium to shoots to maintain a high K/Na ratio. It was also observed that ABA and proline were involved in the osmotic adjustment to reduce the impact of salt stress.

CAT activity during development of the *Rhizobium*-legume symbiosis as well as its response in salinized plants of *P. vulgaris* was studied. Leaf CAT activity was decreased throughout the ontogeny, but root nodule catalase maintained a high, stable value. In contrast, under salt stress conditions, leaf and nodule catalase activities were decreased (Garcia *et al.*, 2007). The effect of salinity was determined on *Rhizobium tropici* strain CIAT899, native *Rhozobium* and growth of a commercial cultivar (Akman 98) of *P. vulgaris*. The results indicated that populations of *R. tropici* strain CIAT899 and natural *Rhizobium* were decreased with raised salinity levels. The plant root and shoot dry weight, chlorophyll content, plant height, root length, total nitrogen, symbiotic efficiency and efficiency rate were affected by salt stress in tested plants with both inoculations. Total nitrogen content decreased significantly at the highest level of salinity (Uyanöz & Karaca, 2011).

Flavonoids from root exudates of *P. vulgaris* cv. Bush Blue Lake (BBL) obtained under control conditions, salt stress (50 mM) and/or the presence of the plant growth-promoting rhizobacterium (PGPR) *Chryseobacterium balustinum* Aur9 were analysed by high-performance liquid chromatography (HPLC) coupled to electrospray ionization-tandem mass spectrometry (ESI-MS/MS). Six flavonoids – naringenin, isoliquiritigenin, quercetin, umbelliferone, 7′,4-dihydroxyflavone and hesperetin – were identified in the root exudates. The latter three flavonoids have not previously been reported in bean root exudates. The presence of *C. balustinum*, but not salt stress, modified the pattern of flavonoids exuded by the bean roots (Dardanelli *et al.*, 2012).

Faghire *et al.* (2013) studied the effect of salt stress, under glasshouse conditions, on plant biomass, nodulation, and activities of acid phosphatases (APase, EC 3.1.3.2) and trehalose 6-phosphate phosphatase (TPP, EC 3.1.3.12) in the symbiosis between common bean and rhizobia nodules. Salinity stress generally reduced acid phosphatase and trehalose phosphate phosphatase activities in nodules, which were less affected in plants inoculated with RhM11. It was noted that nodule phosphatase activity may be involved in salinity tolerance in common beans, and that levels of salt tolerance depend principally on the specific combination of the rhizobial strain and the host cultivar (Faghire *et al.*, 2013).

The genus *Vigna* includes some well-known cultivated species, including former members of the genus *Phaseolus* such as the adzuki bean (*V. angularis*), the black gram (*V. mungo*), the moth bean (*V. aconitifolia*) and the mung bean (*V. radiata*) (Table 2.1). The other food species include cowpea and Bambara beans. In mung bean salinity induced negative effects on nodulation, Lb content and nitrogenase; these were metabolically regulated and operated through the intervention of some key regulatory substances (Garg *et al.*, 1988).

Nandwal *et al.* (2000) subjected two phenotypically different mung bean genotypes, K-851 (trifoliate) and a mutant (pentafoliate), to increasing levels of salt stress. Salinity led to decreases in the osmotic potential and relative water content (RWC) of nodules. The osmotic potential was reduced more in the mutant than in genotype K-851, while the reverse was true for RWC. A sharp rise in the proline content of leaves, roots and nodules was noticed. Lb levels also fell in both the cultivars after the application of saline irrigation (Nandwal *et al.*, 2000). Panda (2001) reported increases in the activity of antioxidative enzymes like CAT, guaiacol peroxidase and SOD in root and shoot tissues of *V. radiata* under salinity stress.

Kabir *et al.* (2004) studied the effect of potassium on salinity tolerance of mung bean and reported that salinity decreased RWC and water retention capacity, while water saturation deficit and water uptake capacity increased. Reduced dry matter distribution in different plant parts as well as decreased total dry matter in mung bean plants was also observed. Similarly, yield and yield-contributing characters were also affected by salinity, except number of seeds per pod.

A reproducible and efficient transformation system utilizing the nodal regions of the embryonal axis of black gram has been established via *Agrobacterium tumefaciens*. Genetic transformation of *V. mungo* using the glyoxalase I gene driven by a novel constitutive *Cestrum* yellow leaf curling viral promoter has been performed for alleviating salt stress. Exposure to salt stress (100 mM NaCl) of T1 transgenic plants as well as untransformed control plants revealed that the transgenic plants survived under salt stress and set seed whereas the untransformed control plants failed to survive. The higher level of glyoxalase I activity in transgenic lines was directly correlated with their ability to withstand salt stress (Bhomkar *et al.*, 2008).

A comparative study was conducted by Nazar *et al.* (2011) to assess the physiological processes in salt-tolerant (Pusa Vishal) and salt-sensitive (T44) cultivars of mung bean under the influence of salicylic acid and salt stress. Cultivar T44 accumulated higher leaf Na^+ and Cl^- content and exhibited greater oxidative stress than Pusa Vishal. Activity of the antioxidant enzymes ascorbate peroxidase (APX) and glutathione reductase (GR) was greater in Pusa Vishal than T44. In contrast, the activity of SOD was greater in T44. The greater accumulation of leaf nitrogen and sulphur through higher activity of their assimilating enzymes, nitrate reductase (NR) and ATP-sulfurylase (ATPS), increased reduced glutathione

(GSH) content more conspicuously in Pusa Vishal than T44. Application of 0.5 mM salicylic acid (SA) increased nitrogen and sulphur assimilation, GSH content and activity of APX and GR. These results indicate that SA application alleviates the salt-induced decrease in photosynthesis mainly through inducing the activity of NR and ATPS, and increases antioxidant metabolism to a greater extent in Pusa Vishal than T44 (Nazar *et al.*, 2011).

Enhancement of salt (NaCl) tolerance by pretreatment with a sublethal dose (50 mM) of NaCl was investigated in *V. radiata* seedlings. NaCl stress had drastic effects on roots compared to shoots. Accompanying observations were also made on reductions in length and number of root hairs and branches – roots became stout, brittle and brown in colour. Salt stress also caused a gradual reduction in chlorophyll, carotenoid pigment contents and chlorophyll fluorescence intensity. SOD and catechol peroxidase activities increased under stress conditions in both roots and leaves, whereas CAT activity showed an increase in roots and a decrease in leaves. But pretreatment with a sublethal dose of NaCl overcame the adverse effects of stress imposed by NaCl to variable extents, by increasing growth and photosynthetic pigments of the seedlings, modifying the activities of antioxidant enzymes, reducing malondialdehyde and H_2O_2 content, and increasing accumulation of osmolytes like proline (Saha *et al.*, 2010). In another study, Saha and his associates (2012) studied the effect of salt stress on ion contents, Krebs cycle intermediates and Krebs cycle regulatory enzymes in growing mung bean seedlings. Sodium and chloride ion contents increased in both root and shoot whereas potassium ion content decreased in shoots of test seedlings with increasing concentrations of NaCl. Organic acids like pyruvate and citrate levels increased whereas malate levels decreased under stress in both roots and shoots. Salt stress also variedly affected the activities of different enzymes of the respiratory chain (Saha *et al.*, 2012).

The response of eight parents, namely LGG 450, VRMGG Local, LGG 460, EC 30072, VRMGG 1, LGG 410, PMB 27 and VBN 1, and five hybrids, namely LGG450/LGG410, VRMGG Local/PMB27, LGG460/PMB27, EC30072/PMB27 and VRMGG1/VBN1, were studied by Sunil Kumar *et al.* (2012) under different salt stress conditions (control, 4 dS/m and 8 dS/m). A range of characters were reduced with increased salinity levels compared to controls. These included all biometric characters such as number of branches per plant, number of clusters per plant, number of pods per plant, 100 seed weight, seed yield per plant and leaf area index; biochemical characters and biophysical characters, namely total chlorophyll content, nitrate reductase activity, photosynthetic rate, transpiration rate and stomatal conductance. All the hybrids recorded higher levels of salinity tolerance for all the characters compared to their respective parents, except for transpiration rate and stomatal conductance. Interaction effects between the genotypes and salinity levels appeared significant for most of the studied characters. From the study it can be opined that salt stress reduced all the parameters under different salinity levels. Again among the parents and hybrids, VRMGG1/VBN1 was found to be relatively salt tolerant (Sunil Kumar *et al.*, 2012).

2.4.2 Faba bean

The genus *Vicia* comprises about 140 species of flowering plants. Several species of this genus are native to Europe, North America, South America, Asia and Africa. Faba bean, or broad bean (*Vicia faba* L.), is an important pulse crop in the Mediterranean region, west Asia, China and Australia where it is adapted to neutral-alkaline soils. In many cases faba bean is grown as a rain-fed crop on saline-sodic soils where growth and yield are limited by salinity (Tavakkoli *et al.*, 2012). Further, faba bean is considered particularly sensitive to abiotic stress (in particular osmotic stress due to lack of water or enhanced soil salinity) and often suffers from severe yield losses (Hanafy *et al.*, 2013).

Delgado *et al.* (1994) and Cordovilla *et al.* (1996, 1999) compared the effect of salinity on growth and productivity and found pea and faba bean to be significantly affected and soybean and common bean to be moderately affected. Cordovilla *et al.* (1996, 1999), working on the growth and symbiotic performance of faba bean inoculated with *R. leguminosarum* bv. Viciae salt-tolerant (GRA19) or salt-sensitive (GRL19) strains and treated with 100 mM NaCl, showed that salinity significantly decreased the dry matter yield of both shoots and roots. The results also showed higher levels of nodule mass and reduced nitrogen fixation in response to salinity (Cordovilla *et al.*, 1999).

A study was conducted by Amira and Qados (2011) to see the effect of NaCl concentrations (0.0, 60, 120 and 240 mM) on growth, osmotic potential, chlorophyll content and protein content of faba bean seedlings. NaCl caused a rise in plant height at low and medium concentrations and a reduction at the highest concentration, in both measurement periods. No significant effect was noticed in the number of leaves or leaf area with a low concentration, while a reduction was observed for each of the two higher concentrations and in both measurement periods. Salinity enhanced both fresh and dry weights of the shoot in the two measurement periods. Osmotic potential showed a significant decrease with increase in concentrations, and with the duration of the stress periods. Salinity significantly reduced chlorophyll *a* content in both measurement periods. It also significantly diminished chlorophyll *b*, total chlorophyll (chl), and carotenoid contents after 10 days of treatment. An increase was perceived in the protein content over the two measurement periods due to the influence of salinity stress. A directly proportional relationship was found between protein content and the uplift in salt concentrations over the first measurement period, while it was inversely proportional in the second (Amira & Qados, 2011).

Tavakkoli *et al.* (2012) assessed the genotypic variation of faba bean plants for tolerance to salinity stress under controlled conditions and in the field, and the suitability of various physiological traits to screen 11 faba bean genotypes for salt tolerance. The importance of Na^+ and Cl^- exclusion as mechanisms of salt tolerance was diminished at 150 mM NaCl, and salt tolerance was more strongly related to the osmotic potential of the leaves: lower salt tolerance was associated with a lower osmotic potential. The reduction in salt tolerance as the external salt stress increased varied among varieties, and this was associated with the ability to maintain high levels of tissue water content. This study also suggested salt exclusion coupled with a synthesis of organic solutes as important components of salt tolerance in the tolerant genotypes (Tavakkoli *et al.*, 2012).

The PR10a gene isolated from the potato cultivar Desirée was known for its role in enhancing salt and/or drought tolerance in potato. This gene was transferred into faba bean cv. Tattoo by an *Agrobacterium tumefaciens*-mediated transformation system based upon direct shoot regeneration after transformation of meristematic cells derived from embryo axes. Fertile transgenic faba bean plants were recovered. The results suggested that introducing a novel PR10a gene into faba bean could be a promising approach to improve its drought and salt tolerance ability (Hanafy *et al.*, 2013).

2.4.3 Peas

The genus *Pisum* contains major human food crops and is native to southwest Asia and northeast Africa. This genus contains species like *Pisum abyssinicum* (syn. *P. sativum* subsp. *abyssinicum*), *P. fulvum*, *P. sativum* (pea) and *P. arvense* (the field or garden pea), depending on taxonomic interpretation (International Legume Database). Sodium chloride reduced ARA by 77% and Lb by 50% in pea nodules treated with 50 mM NaCl (Delgado *et al.*, 1993, 1994).

Hernandez *et al.* (1993) reported that in the leaf mitochondria from pea plants grown with NaCl, the salt-induced enhanced rate of mitochondrial O_2^- production was concomitant with a strong decrease in the mitochondrial Mn-SOD in NaCl-sensitive plants, whereas in NaCl-tolerant plants this isozyme was induced by salt treatments. The metabolism of chloroplasts and mitochondria under NaCl stress favours the formation of O_2^- and H_2O_2 in two pea cultivars differing in NaCl sensitivity (Hernandez *et al.*, 1995). Hernandez *et al.* (1999) suggested that the salt stress was accompanied by oxidative stress in pea. The tolerant cultivars ensured cell turgor and enhancement of the antioxidant defence system in mitochondria, chloroplast and cytosolic enzymes, with an increase in all components of the ascorbate-glutathione (ASC-GSH) cycle and SOD isozymes (Hernandez *et al.*, 2000, 2001). The growth of pea cv. Lincoln plants (shoot fresh weight and shoot dry weight) was not affected by 50 mM NaCl in the nutrient medium, but was reduced by about 30% and 50% in the presence of 70 and 90 mM NaCl, respectively; the pea cultivar Puget was relatively tolerant to 70 and 90 mM NaCl (Hernandez *et al.*, 2001). In another study, Hernandez and Almansa (2002) also studied the short-term effects of salt stress on antioxidant systems of pea leaves. Salt stress brought about a decrease in osmotic potential, Na^+ concentration and stomatal conductance, but at 48 and 24 h post-stress, respectively, both parameters recovered control values.

Olmos and Hellin (1996) studied the effect of salt stress on two lines of *P. sativum* cv. Challis, one salt-sensitive and the other adapted to 85.5 mM NaCl. The results showed that average concentrations of reducing sugars, total free amino acids and ascorbic acid increased in the line adapted to NaCl and suggested their role in osmotic adjustment. Pea and faba bean plants inoculated with the same salt-tolerant (GRA19) or salt-sensitive (GRL19) *Rhizobium leguminosarum* strain and treated with 100 mMl-1 NaCl showed higher levels of nodule mass and reduced nitrogen fixation in response to salinity (Cordovilla *et al.*, 1999).

El-Hamdaoui *et al.* (2003) reported that salinity produced a reduction of B, Ca, K and Fe contents in the shoots of pea plants. Nodulation, calculated as number of nodules and nitrogen fixation (measured as ARA), was almost completely repressed in pea plants. When examined under the electron microscope the salt-stressed nodules seemed to have a very changed structure, without tissue differentiation and with erratically shaped cells (El-Hamdaoui *et al.*, 2003). Figueira and Caldeira (2005) investigated the influence of different nitrogen forms on salt tolerance of *P. sativum* and reported that it is probable that a functional symbiosis can be established under saline conditions, provided that a salt-tolerant *Rhizobium* isolate with good N_2-fixing ability is utilized. Ahmad and Jhon (2005) reported a reduction in the RWC in *P. sativum* leaves subject to different concentrations of NaCl. Okcu *et al.* (2005) reported that all cultivars of pea were able to germinate at all NaCl levels without significant decrease in germination. Bolaños *et al.* (2006) examined pea root nodules under salinity stress. The nodule number and weight diminished in salt-stressed nodules of pea plants and most of them appeared pale in contrast with the control pink nodules. ARA was not detected in pea nodules developing in plants treated with 75 mM NaCl.

Salt stress distinctly increased the activities of SOD and peroxidase (POD), levels of total phenolics and γ- and Δ-tocopherols, and reduced the total soluble proteins and CAT activity, while the internal levels of H_2O_2 continued unaffected in pea cultivars. Salt-induced oxidative stress occurred in all pea cultivars. However, the response of salt-tolerant and salt-sensitive cultivars with respect to the generation of enzymatic and non-enzymatic metabolites was found to be inconsistent. Of the various antioxidant enzymes and metabolites examined, only CAT activity was found to be a dependable marker of salt tolerance in the set of pea cultivars studied (Noreen & Ashraf, 2009).

The effect of salicylic acid (SA) treatment on the response of pea plants to salinity was studied (Barba-Espin *et al.*, 2011). NaCl-induced damage to leaves was increased by SA, which was correlated with a reduction in plant growth and the contents of ascorbate and glutathione in leaves of salt-treated plants. An increase in hydrogen peroxide also occurred in leaves of salt-exposed plants. Salinity reduced glutathione reductase (GR) activity, but enhanced glutathione S-transferase (GST) and CAT activity. The results showed that SA treatment harmfully affected the response of pea plants to NaCl, and this response was connected with an imbalance in antioxidant metabolism. The data also showed that SA treatment could increase the resistance of salt-stressed plants to probable opportunistic pathogen attack, as offered by increased PR-1b gene expression (Barba-Espin *et al.*, 2011).

A study was conducted to document parental germplasm that could be utilized for development of tolerance to NaCl in field pea. An initial screening experiment of 780 globally distributed *Pisum* L. accessions identified significant variation in response to applied NaCl, based on plant symptoms. Lines with relatively higher tolerance as compared to commercial varieties grown in Australia were most frequently recognized within landraces originating from the central, eastern and southern provinces of China. The most tolerant classified accession was an unadapted landrace 'ATC1836' originating from Greece. Salinity-induced toxicity symptoms were closely connected with decreases of plant growth rate, height, and shoot and root dry matter, and with enhanced concentration of Na^+ at the plant growing tip (Leonforte *et al.*, 2013).

2.4.4 Chickpea

The genus *Cicer* is native to the Middle East and Asia. Its best known and only domesticated member is *Cicer arietinum* L., also commonly known as chickpea, garbanzo bean or Bengal gram, which is commonly used for making dhal. Elsheikh and Wood (1990) reported that salinity significantly reduced the dry weights of both shoots and roots in chickpea. Further, roots appeared to be more sensitive to salinity than shoots even at very low salinity levels of 1.0 dS/m. Sharma *et al.* (1990) observed the negative effect of chloride and sulphate types of salinity on the starch and protein contents of leaves in two chickpea cultivars, whereas accumulation of amino acids, particularly proline, was enhanced with increases in salinity levels (Sharma *et al.*, 1990). Chickpea seedlings when grown in NaCl solution culture suffered a decrease in their carbohydrate concentration at the two highest salinity levels (Ahmed & Bano, 1992; Sheokand *et al.*, 1995).

Mamo *et al.* (1996) discovered chloride concentrations in chickpea plant parts at salt levels (0–8 dS/m) two to five times that of Na^+ with substantial reductions in K^+ concentration. Roots appeared to be more sensitive to salinity than shoots, but only at the higher concentration of salt. On the other hand, chickpea plants on submission to salt stress showed a greater drop in shoot dry weights than the roots (Dua, 1998). Zurayk *et al.* (1998) also observed significant reductions in the dry weight of above-ground biomass in chickpea on treatment with $NaCl+Na_2SO_4$ salts. Nodule dry weight and N-fixation also declined due to salinity. The symbiosis was found to be more salt-sensitive than both the rhizobium and the host plant themselves (Zurayk *et al.*, 1998).

Soussi *et al.* (1998, 1999) found that chickpea plant growth was reduced by salt. However, salt reportedly boosted the accumulation of proline, amino acids and carbohydrates in the leaves of chickpea (Soussi *et al.*, 1998). Soussi *et al.* (1999) compared the two cultivars of chickpea with differential tolerance to salinity with regard to nodulation and nitrogen fixation. The effect of salt on nodulation and nitrogen fixation was more pronounced in Pedrosillano (sensitive) than ILC 1919 (tolerant). Soussi *et al.* (2001) also observed reduced nitrogenase activity in chickpea plants grown under 50 mM NaCl salinity. A rise in fermentative metabolism in nodules suggested that the accumulation of proline, lactate or ethanol may play a vital role in providing energy-yielding substrates for bacteroids in chickpea. Similarly Soussi *et al.* (2003) reported that salt treatment in *C. arietinum* cultivars triggered a build-up of proline and total soluble sugars in the radicles of the salt-tolerant cultivar ILC1919.

Seeds of chickpea when germinated in the presence of NaCl displayed a reduction in K content, while higher levels of Na eventually repressed shoot and seed yield of all the cultivars (Sekeroglu *et al.*, 1999). However, chickpea grown under saline conditions exhibited high yields in spite of low Na and Cl contents in the tissue (Richter *et al.*, 1999). Singh and Singh (1999) and Ozcan *et al.* (2000) reported a build-up of proline in the tolerant genotypes of chickpea. Gholipoor *et al.* (2000) reported a reduction in shoot water content with increasing salinity in chickpea cultivars Jam, Hashem, Kaka and Pirooz.

When two chickpea cultivars, differing in drought tolerance, were exposed to salinity they showed insignificant effects on the leaf water potential and osmotic adjustment, which were higher for the drought-tolerant variety (Katerji *et al.*, 2001). Singh *et al.* (2001) investigated variability in the four genotypes of chickpea under salinity and found higher activities of POX and CAT in the tolerant genotypes. Sadiki and Rabih (2001) screened 200 Moroccan lines of chickpea subjected to salt stress and studied nitrogenase activity, dry weight, nodule mass, plant colour and total nitrogen under NaCl treatment. The study revealed salt tolerance in MCA 103, MCA 131 and MCA 250 lines. Welfare *et al.* (2002) determined the effect of a saline environment on two chickpea cultivars. The study revealed a considerable reduction in plant height, number of leaves and dry weights of leaves, stems and roots. Rao *et al.* (2002) have reported that with rising salinity there was a slight decline in nodule number and nodule biomass in the different genotypes of chickpea above 3 dS/m, but to a lesser extent in the high nodulation selection, which was characteristically superior under both non-saline and stress conditions.

Garg (2004) reported higher amino acids, total soluble sugars and chlorophyll pigments (*a*, *b* and total chlorophyll) in the tolerant cultivars of chickpea as compared to sensitive cultivars. Kukreja *et al.* (2005) reported increased activities of SOD, POX, ascorbate peroxidase (APX), glutathione transferase (GTase), GR and CAT in roots of *C. arietinum* plants exposed to salinity stress. Singla and Garg (2005) studied the influence of salinity on growth and yield attributes in chickpea cultivars and concluded that root growth was more adversely affected than shoot growth, which also had an impact on the root:shoot ratio and plant productivity. Singh *et al.* (2005) analysed 10 genetically diverse chickpea lines for salt tolerance. They reported a decrease in shoot, root and single plant weights with increasing levels of salt.

Nodule conductance to O_2 diffusion has been implicated as a major factor in the suppression of N_2 fixation by soil salinity that severely reduces the production of grain legumes. L'taief *et al.* (2007) studied the effect of salinity on nodule conductance and oxygen uptake by the nodulated roots (inoculated with the strain *Mesorhizobium ciceri* UPMCa7) of Amdoun 1 and INRAT 93-1 varieties of chickpea. Salinity induced a significant reduction in shoot (30% vs 14%), root (43% vs 20%) and nodule biomass (100% vs 43%) for Amdoun 1 relative to INRAT 93-1. Although salinity completely inhibited nodule formation in the sensitive variety Amdoun 1, nodule number and biomass were higher in

Amdoun 1 than in INRAT 93-1 in the absence of salt. This effect was associated with a significantly higher O_2 uptake by nodulated roots (510 vs 255 µmol O_2 per plant/h) and nodule conductance (20 vs 5 µm/s) in Amdoun 1 compared with INRAT 93-1. Salinity did not significantly change the nodule conductance and nodule permeability for INRAT 93-1 (L'taief *et al.*, 2007).

A screening of 263 accessions of chickpea, including 211 accessions from ICRISAT's mini-core collection (10% of the core collection and 1% of the entire collection), showed a six-fold range of variation for seed yield under salinity (1.9 L of 80 mM NaCl per 7.5 kg Vertisol), with several genotypes yielding 20% more than a formerly released salt-tolerant cultivar. The range of difference in yields under salinity was similar in both kabuli and desi chickpeas, showing that breeding for salinity tolerance can be undertaken in both groups. However, no significant correlation was found between the shoot dry weight ratio and the yield ratio, signifying that differences in salinity tolerance among genotypes could not be concluded from measurements in the vegetative stage. The major trait connected to salinity tolerance was the capacity to produce a large number of filled pods, whereas seed size was similar in tolerant and sensitive genotypes. Salinity tolerance was not associated with the shoot Na^+ or K^+ concentrations (Vadez *et al.*, 2007). Similarly, Samineni *et al.* (2011) reported that chickpea is sensitive to salinity in both the vegetative and reproductive phases, with pod formation being particularly sensitive.

The effects of different levels of Na salinity (0, 3, 6 and 9 dS/m on growth, yield and yield component of kabuli (Hashem and Jam) and desi (Kaka and Pirooz) chickpea cultivars were evaluated (Sohrabi *et al.*, 2008). Seeds of four chickpea cultivars were nurtured under 0, 3, 6 and 9 dS/m levels of salinity until maturity. Salinity diminished the plant growth, flower, pod and seed number and seed weight. As salinity increased the negative effect of Na^+ was more pronounced and reached the highest value at 9 dS/m in all cultivars. Four chickpea cultivars have different responses to salinity and the kabuli cultivars appeared to have a greater capacity for salt tolerance compared to desi cultivars. The Hashem cultivar has the highest salinity tolerance among all the cultivars (Sohrabi *et al.*, 2008).

Six hundred accessions of chickpea landraces and wild relatives from 28 different countries, available in the Australian Temperate and Field Crops Collection (ATFCC), were screened for tolerance to salt under greenhouse conditions. The degree of salt tolerance was based on necrosis scores and shoot biomass reduction relative to unstressed controls. There was a wide variation in salinity tolerance evaluated by both measures. Accessions from the Middle East and south Asia (regions with salinity problems, a long history of chickpea cultivation and high diversity) showed greater potential for getting salt-tolerant accessions (Moses *et al.*, 2008).

Neil *et al.* (2013) conducted a study to determine the degree of salt tolerance among chickpea genotypes, and the relationship between salt tolerance and ion accumulation in leaves and reproductive tissues. The enhanced salt tolerance was positively associated with higher pod and seed numbers, and higher shoot biomass, but not with time to 50% flowering nor with the number of filled pods in the non-saline treatment. Pod abortion was higher in the salt-sensitive genotypes, but pollen viability, *in vitro* pollen germination and *in vivo* pollen tube growth were not affected by salinity in either the salt-tolerant or salt-sensitive genotypes. The concentrations of sodium and potassium ions, but not chloride, in the seed were significantly higher in the sensitive than in the tolerant genotypes. Sodium and potassium, but particularly chloride, ions accumulated in leaves and in pod walls, whereas accumulation in the seed was much lower. Considerable genotypic variation for salt tolerance exists in chickpea germplasm. The study suggested that selection of genotypes with high pod and/or seed numbers that accumulate low concentrations of salt in the seed would be beneficial (Neil *et al.*, 2013).

2.4.5 Cowpea

The cowpea, *Vigna unguiculata* (L.) Walp., is one of several species of the widely cultivated genus *Vigna*. Four subspecies are recognized, of which three are cultivated: (i) *Vigna unguiculata* subsp. *dekindtiana* is the wild relative of the cultivated subspecies; (ii) *Vigna unguiculata* subsp. *cylindrica* is a cultivated species, common name catjang; (iii) *Vigna unguiculata* subsp. *sesquipedalis* is a cultivated species, common name yardlong bean; and (iv) *Vigna unguiculata* subsp. *unguiculata* is a cultivated species, common name black-eyed pea. Cowpeas are one of the most important food legume crops in the semi-arid tropics covering parts of Asia, Africa, southern Europe and Central and South America. It is a drought-tolerant and warm-weather crop and well adapted to the drier regions of the tropics, where other food legumes do not perform well. This species also has the useful capacity to fix atmospheric nitrogen through its root nodules, and it grows well in poor and marginal soils containing more than 85% sand, 0.2% organic matter and low levels of phosphorus. In addition this crop is considered an important component of traditional intercropping systems, especially in the complex and elegant subsistence farming systems of the dry savannahs in sub-Saharan Africa (Sanginga *et al.*, 2003; Singh *et al.*, 2003).

Corpas *et al.* (1991), in studies with cowpea plants under salt stress, recognized the subcellular location of three SOD isozymes present in leaves (Mn-SOD, CuZn-SODI and CuZn-SODII). Hernandez *et al.* (1995) reported that in the leaves of salt-stressed cowpea plants, mitochondrial Mn-SOD was considerably reduced, whereas cytosolic and mitochondrial CuZn-SODI was marginally reduced and chloroplastic CuZn-SODII did not display a response. In protoplasts, Mn-SOD showed considerably reduced activity, CuZn-SODII was distinctly inhibited and CuZn-SODI exhibited an intermediate sensitivity.

Sousa *et al.* (2003) studied the protein pattern in cowpea seedlings subjected to salinity (100 mM NaCl) by isolating the albumins from stems and leaves; they reported that in stem, 19 proteins (14.6–76.3 kDa) had their relative concentration augmented by salinity, eight (31.2–65.0 kDa) had their relative concentration decreased by salinity and nine (16.3–39.8 kDa) were apparently synthesized *de novo*, while in the leaves nine (18.2–33.2 kDa) proteins increased in concentration, one (17.1 kDa) decreased in concentration and one (21.2 kDa) was apparently synthesized *de novo*.

Twenty-five genotypes of cowpea were tested for salt tolerance at the vegetative growth stage at salinity levels of 0, 85 and 170 mM NaCl. The genotypes were classified based on plant survival, ion concentration (Na^+ and Cl^-) in root and shoot, and biomass accumulation. Biomass accumulation was affected by both 85 and 170 mM NaCl in all groups of genotypes; however, salt-tolerant and moderately salt-tolerant genotypes displayed more biomass than genotypes classified as moderately salt-sensitive and salt-sensitive. In all genotypes Cl^- concentration was greater in shoots than roots, and became enhanced as salinity increased. Similarly Na^+ concentration amplified with increasing salinity. However, in salt-tolerant and moderately salt-tolerant genotypes, Na^+ concentration was more in roots than shoots, while in moderately salt-sensitive and salt-sensitive genotypes, Na^+ was higher in shoots than roots (Murillo-Amador *et al.*, 2006).

A study was conducted by Wilson *et al.* (2006) to see the effect of salinity on leaf gas exchange by determining net photosynthetic rate per unit leaf mass (Pnm) and per unit leaf area (Pna), and stomatal conductance (gs) of the four cowpea cultivars. The study revealed a highly significant reduction of Pnm, Pna and gs due to salinity. Pnm was more sensitive to salinity than Pna. It was found that increasing stomatal closure with increasing salinity might limit Pnm or Pna. There was no significant difference of Pnm and Pna among the four cultivars, while a significant difference was found only for gs. No significant salt × cultivar interaction effect was found with Pnm, Pna and gs, demonstrating that the four cowpea cultivars have the same response pattern of leaf gas exchange to salinity (Wilson *et al.*, 2006).

Cavalcanti *et al.* (2007) reported that cowpea roots and leaves present different mechanisms of response to lipid peroxidation and CAT and SOD activities during salt stress and recovery. However, these responses were not connected with the growth reduction in cowpea. The alternative oxidase (Aox) was studied at different levels (transcript, protein and capacity) in response to an osmotic shock applied to roots of two cowpea cultivars, Vita 3 and Vita 5, tolerant and sensitive to drought/saline stress respectively. The VuAox1 and VuAox2a mRNA were not detected in either cultivar under all tested conditions while the VuAox2b gene was differentially expressed. Upon salt stress, the VuAox2b gene was overexpressed. At 100 mM NaCl, this VuAox2b gene overexpression led to a higher amount and capacity of Aox. This effect was reduced at 200 mM NaCl (Costa *et al.*, 2007).

2.4.6 Pigeon pea

Pigeon pea, *Cajanus cajan* (L.) Millsp., is the sixth most important grain legume crop of rain-fed agriculture. Pigeon peas are cultivated in more than 25 tropical and subtropical countries either as sole crops or intermixed with cereals. Pigeon peas are both a food crop (for dried peas, flour or green vegetable peas) and a forage/cover crop. In combination with cereals, pigeon peas make a well-balanced human food. The plant plays an important role in subsistence agriculture as fodder, fuel and symbiotic nitrogen-fixing crop. Pigeon peas are very drought resistant and the crop is cultivated on marginal land by resource-poor farmers.

Shiva Prakash and Sarin (1993) have described a positive correlation between the Na^+ level and proline concentration in the salt-tolerant cell lines of *C. cajan*. Increased Na^+ levels in shoots and roots of *C. cajan* plants exposed to salinity have been reported by Gill and Sharma (1993). Accelerated nodule greening supplemented by a concomitant reduction in Lb content of the nodules and ARA were observed when 45-day-old chickpea plants were subjected to 50 and 100 mM NaCl treatments for 7 and 14 days, respectively (Sheokand *et al.*, 1995). Similar observations have been reported in pigeon pea (Garg & Dua, 1996; Swaraj *et al.*, 2000). Bishnoi *et al.* (2006) observed changes in the protein profile of pigeon pea genotypes Manak (salt-tolerant) and ICPL 88039 (salt-sensitive) to stress of 3 mM boron + 60 mM NaCl. The appearance was reported of 95.6 kDa proteins in plumule and 67.5 kDa proteins in the radicle of Manak under NaCl treatment with no specific protein in ICPL 88039.

Arbuscular mycorrhizal (AM) fungi occur extensively in natural ecosystems as well as in salt-affected soils and are proven suitable candidates for bioamelioration of saline soils. Plants respond to salinity by building up sugars and other low-molecular-weight compatible solutes. An investigation was carried out to evaluate the effect of interactions between an AM fungus and salinity stress on growth, nitrogen fixation and trehalose metabolism in two genotypes of pigeon pea, namely Sel 85N (salt-tolerant) and ICP 13997 (salt-sensitive) (Garg & Chandel, 2011). Salinity reduced plant biomass (shoot and root) in both genotypes and also caused a decline in shoot-to-root ratio (SRR). However, a slight decline was observed in Sel 85N compared to ICP 13997. Colonization by AM fungus was reduced with increasing salinity levels but mycorrhizal responsiveness (MR) was enhanced. Genotypic variability in nitrogen fixation and trehalose metabolism in response to salinity and mycorrhization was also noticed. An increment in nodule number was associated with a reduction in dry mass. Subsequently, nodular activity (leghaemoglobin, ARA, nitrogen content) was reduced under soil salinity, the reduction being more profound in ICP 13997 than in Sel 85N. A symbiotic relationship between plant roots and *Glomus mosseae* caused salinity tolerance in a genotype-dependent manner in pigeon pea (Garg & Chandel, 2011).

Proline functions as an osmoprotectant and plays an important role in osmotic balancing, protecting subcellular structures and enzymes, and in improving cellular osmolarity to provide the turgor necessary for cell expansion under stress conditions. The enzyme delta-1-pyrroline-5-carboxylate synthetase (P5CS), a rate-limiting enzyme in proline biosynthesis that is known for convening improved salt and drought stress, is subject to feedback inhibition by proline. Transgenic pigeon pea plants were developed using efficient *in vitro* transformation of embryonic structures and *Agrobacterium tumefaciens* strain and the *Vigna aconitifolia P5CSF129A* genes under a constitutive 35S promoter. The resultant primary transgenic plants exhibited more build-up of proline than their non-transformed plants. Levels of proline were also raised in T1 transgenic plants when grown in the presence of 200 mM NaCl. In addition to their improved growth performance, elevated chlorophyll and relative water content under high salinity, these plants also had reduced levels of lipid peroxidation. This study revealed that overproduction of proline might play an important role in protecting against salt shock and cellular integrity in pigeon pea (Surekha *et al.*, 2014).

2.4.7 Lentils

The genus *Lens* contains four species of small, erect or climbing herbs with pinnate leaves and small inconspicuous white flowers and small flattened pods. The lentil (*Lens culinaris*) is a bushy annual plant grown for its lens-shaped seeds. Ashraf and Waheed (1993) reported that leaf soluble proteins decreased due to salt stress in all lines of lentil, irrespective of their salt tolerance. Studies done by Rai and Singh (1999) indicated that the interactions between salt-tolerant lentil genotypes and *Rhizobium* strains were significant and resulted in greater nodulation and N_2 fixation (nitrogenase activity) in sodic soils compared to uninoculated controls.

Bandeoglu *et al.* (2004) evaluated the growth performance of lentil under NaCl-salinity stress and reported considerable reduction in shoot/root length and fresh/dry weights of the plants. Levels of H_2O_2, lipid peroxidation and electrolyte leakage were enhanced under NaCl salinity in leaves of lentil, whereas roots were less affected. Leaves had CuZn-SOD, Fe-SOD and Mn-SOD activity, whereas Fe-SOD activity was missing in root extracts. Salt stress did not cause a significant increase in total SOD activity of leaf tissues but a significant enhancement (88%) was observed in roots due to increased CuZn-SOD isoforms. Roots had relatively higher APX and GR activities than leaves; upon salt stress no significant changes in APX, CAT and GR activities were observed in roots, but APX activity was higher than in leaves, suggesting that root tissues of lentil were better protected from NaCl stress-induced oxidative damage due to enhanced total SOD activity together with higher levels of APX activity under salinity stress (Bandeoglu *et al.*, 2004).

2.4.8 Lupins

The genus *Lupinus* (commonly known as lupin or lupine) includes over 200 species, with centres of diversity in North and South America. Yu and Rengel (1999) reported that salinity stress enhanced the activity of CuZn-SOD by 14.5% without affecting the activities of the other SOD forms in narrow-leafed *Lupinus angustifolius* L.

2.4.9 Groundnut

The genus *Arachis* comprises about 70 species of annual and perennial flowering plants. At least one species, the peanut (*Arachis hypogaea* L.), is a major food crop species of global importance; some of the other species are cultivated for food to a small extent in South America. Other species such as *A. pintoi* are cultivated worldwide as forage and soil conditioning plants, with the leaves providing a high-protein feed for grazing livestock and a nitrogen source in agroforestry and permaculture systems. Jain *et al.* (2001) observed increased levels of proline in the NaCl-selected lines of groundnut, which were probably connected with osmotic adjustments and the protection of membrane integrity. It was further observed that addition of proline to the culture medium improved fresh weight accumulation and reduced peroxidative damage to the lipid membranes. Changes in the ionic relations of salt-sensitive and salt-tolerant cell lines of groundnut showed a slight increase in Na^+ concentration with an increase in NaCl levels (1.3–2.3-fold); however, the concentration of K^+ remained essentially constant (Jain *et al.*, 2001).

2.4.10 Soybean

The best known species of the genus *Glycine* is the soybean (*Glycine max* L.). The majority of *Glycine* species are found only in Australia, and the soybean's native range is in East Asia. A few species extend from Australia to East Asia (e.g., *G. tomentella* and *G. tabacina*). With cultivation largely concentrated in Latin America and North America, soybean is an important source of protein for human and animal nutrition, as well as a major source of vegetable oil. High levels of salt have negative impacts on growth, nodulation, agronomy traits, and seed quality and quantity and thus reduce crop yields. To cope with salt stress, soybean has evolved several tolerance mechanisms, such as (i) maintenance of ion homeostasis; (ii) osmotic adjustment; (iii) restoration of osmotic balance; and (iv) other metabolic and structural adaptations. A review by Phang *et al.* (2008) integrated the relevant studies on soybean and proposed a working model to describe its salt stress responses at the molecular level. Serraj *et al.* (1994) exposed soybean plants to short-duration salt stress and reported that the ARA of salt-stressed nodulated roots fell to 80% within 60 min after application of 0.1 M NaCl relative to controls. Salinity-induced dry matter decline is more severe in the shoot than the root (Shalhevet *et al.*, 1995).

Comba and Benavides (1997) studied the effect of salt stress on 377 salt-tolerant and 411 relatively sensitive cultivars of soybean. It was observed that 150 mM NaCl caused a decrease of 25% in GSH, 50% in GR, 30% in ascorbate peroxidase (APX), 20% in CAT and 25% in SOD activities in the sensitive cultivars while there were increases of 15, 20, 30 and 18% in the tolerant cultivars, respectively, thereby indicating a correlation between nitrogen fixation and oxidative stress induced by salt treatment. Tramontano and Jouve (1997) reported a five-fold proline increase and a two-fold trigonelline increase after salt stress in soybean. NaCl significantly inhibited nitrogenase activity, nodule number and dry matter accumulation per plant in four soybean cultivars (Abd-Alla *et al.*, 1998). Comba *et al.* (1998) reported that under mild salt stress, the elevated levels of antioxidant enzymes and decreased glutathione protect nodules against the activated oxygen species (AOS) thus circumventing lipid and protein peroxidation and Lb breakdown, while severe saline treatment produced a permanent decay in Lb and nitrogenase activity despite diminished glutathione levels and GR activity.

Liu and van Staden (2000) selected NaCl-tolerant lines from callus cultures initiated from soybean cotyledons and reported that salt tolerance in this culture was characterized by accumulation of Na^+, Cl^-, proline and sugars when the plants were grown under salt stress, as well as on normal media. Garg (2002) reported that soybean displayed higher tolerance than chickpea towards saline conditions and the index for tolerance appeared to be related to the increased levels of sugars, amino acids, proteins and nucleic acids as compared to chickpea (sensitive). Malencic *et al.* (2003) investigated salt stress tolerance parameters in various genotypes of soybean and reported high SOD activity and low lipid peroxidation in the tolerant genotypes. Luo *et al.* (2005) studied differential sensitivity to chloride and sodium ions in seedlings of *G. max* and *G. soja* under NaCl stress and reported that the control plants had higher osmotic potential than plants treated with Cl, Na and NaCl. The studies by Luo *et al.* (2005) revealed that Cl^- was more toxic than Na^+ to seedlings of *G. max* than to *G. soja*; the injury was positively correlated with the content of Cl^- in the leaves, and negatively with that in the roots. A study was conducted by Lee *et al.* (2009) to determine inheritance of salt tolerance in wild soybean (*G. soja* Sieb. and Zucc.). The results showed that *G. soja* line PI483463 had a single dominant gene for salt tolerance that was different than the gene in *G. max* line S-100. The symbol *Ncl2* was designated for this new salt tolerance allele (Lee *et al.*, 2009).

The effects of salinity become exacerbated when the soil is deficient in oxygen. Therefore oxygenation has been used to reduce the impact of salinity on plants under oxygen-limiting soil environments. Oxygenation increased above-ground biomass yield and water use efficiency (WUE) by 13% and 22%, respectively, compared with controls in soybean. Higher yield with oxygenation was accompanied by greater plant height and stem diameter and reduced specific leaf area and leaf Na^+ and Cl^- concentrations. Oxygenation also led to a greater rate of photosynthesis, higher relative water content in the leaf, reduced crop water stress index and lower leaf water potential. Oxygenation enhanced yield and may make an important impact on irrigated agriculture where saline soils pose restraints to crop production (Bhattarai & Midmore, 2009).

A study conducted by Hakeem *et al.* (2012b) evaluated growth, lipid peroxidation and antioxidant enzyme activities in response to salt stress in 10 genotypes of soybean. The results suggested that Pusa-24 and Pusa-37 are salt-sensitive and salt-tolerant genotypes of soybean, respectively, and that the antioxidant defence system is associated with conferring the sensitiveness and tolerance in these genotypes (Hakeem *et al.*, 2012b). The influence of genetic background on salt tolerance was investigated using 10 soybean genotypes, namely Pusa-20, Pusa-40, Pusa-37, Pusa-16, Pusa-24, Pusa-22, BRAGG, PK-416, PK-1042 and DS-9712. Plant growth in all genotypes was decreased by salt stress as compared to controls. The differences in osmotic adjustment between all the genotypes were associated with the concentrations of ions such as Na^+ and the leaf proline concentration (Khan *et al.*, 2013).

Nitric oxide (NO) is a gaseous signalling molecule involved with the regulation of diverse processes in plants. Certain studies have demonstrated a role for exogenous application of NO in mediating responses to abiotic stress. The role of exogenously applied 2,2′-(hydroxynitrosohydrazono) bis-ethanimine (DETA/NO) in upgrading long-term salinity stress on soybean has been studied by Egbichi *et al.* (2014). Long-term salinity stress drastically affected the plants as indicated by decreased biomass of shoots, roots and nodules of soybean plants. In contrast, supplementation with 10 μM DETA/NO improved growth of soybean plants under NaCl as evidenced by increased shoot, root and nodule weights and nodule number. Further analysis showed that long-term salinity stress led to increased cellular hydrogen peroxide (H_2O_2) content and high levels of cell death in the soybean. Treatments with NO, either as DETA/NO alone or in combination with NaCl, resulted in reversal of H_2O_2 to basal levels. It was also observed that application of DETA/NO resulted in increased enzymatic activity of ascorbate peroxidase (APX). The role of NO in increasing tolerance to salinity stress in soybean may result from either its antioxidant capacity by direct scavenging of H_2O_2 or its role in activating APX activity, which is crucial in scavenging H_2O_2 (Egbichi *et al.*, 2014).

2.5 Omics technologies for understanding salt stress responses in legumes

Plant cells may alter their gene expression in order to tolerate salt stress, resulting in an upregulation, downregulation,, induction or total suppression of some stress-responsive genes and proteins. Previously, a one-gene approach was applied to explain abiotic stresses, including salt stress responses, in a 'cause and effect' manner. But as salt stress involves polygenic and complex mechanisms, a range of molecular tools are required to understand and dissect the cascades of events that are involved in salt stress responses and mechanisms. In addition, many studies have shown that environmental stresses that cause cellular dehydration, like salt, water and cold stress, often lead to similar changes in plant gene expression and metabolism, and strong cross-talk can be seen in their signalling pathways. Therefore, more comprehensive approaches, including quantitative and qualitative analysis of gene expression products, are necessary for better analysis at the transcriptome, proteome and metabolome levels (Rossignol *et al.*, 2006).

The omic technologies such as transcriptomics, proteomics and metabolomics have generated major interest among researchers and also opened up new perspectives in studying stress physiology of many plants (Langridge & Fleury, 2011). The 'omic' technologies also generate huge amounts of information, which have enriched the field of bioinformatics, with thousands of new algorithms and new software published every year. The development of these tools has allowed the inference of the causal relationship among the analysed elements. Another significant technological development has been the improvement in data storage and computational capabilities associated with the acquisition and processing of large datasets. In addition, numerous web and software platforms intended to share, assimilate and visualize in a biological context the vast amount of data have been developed.

Numerous large germplasm collections have been founded, which contain large amounts of genetic diversity, including landraces and varieties as well as wild relatives and modern varieties no longer in use. Legumes have large variations in the size of their nuclear genomes, ranging from 370 million base pair (Mbp) in *Lablab niger* to 13,000 Mbp in *Vicia faba* (Aramuganathan & Earle, 1991a,b). Black gram, mung bean, common bean, lima bean, tepary bean and cowpea have the smallest genomes (574 to 647 Mbp); pigeon pea (784, 882 Mbp) and chickpea (738 Mbp) have slightly larger genomes; soybean has a relatively large genome (1115 Mbp); while pea and lentil (4063 to 4397 Mbp) and broad bean (12,603 Mbp) have massive genomes. Genomic resources include: markers; expressed sequence tag (EST) and bacterial artificial chromosome (BAC) libraries; genetic, cytogenetic and physical maps; and identification of quantitative trait loci (QTL) associated with beneficial traits; all of these resources are used in applied breeding. In recent years, tremendous progress has been made towards developing genetic markers, especially short sequence repeats (SSRs) and single nucleotide polymorphisms (SNPs), and/or construction of high-density genetic linkage maps in chickpea, groundnut and pigeon pea (Varshney *et al.*, 2009). Molecular markers have enabled researchers to more rapidly and precisely characterize genetic diversity, identify trait-based genetically diverse germplasm, target genes underlying key agronomic traits, and develop molecular assays that are both relevant and of appropriate scale for breeding applications. More importantly, high-throughput and cost-effective genotyping platforms, combined with automation in phenotyping methodologies, will encourage the application of genomic tools into breeding programmes, and thus usher in an era of genomics-enabled molecular breeding in legumes (Varshney *et al.*, 2009).

Next generation sequencing (NGS) permits the mass sequencing of genomes, which is producing a vast array of genomic information. NGS data can be further analysed by bioinformatics tools enabling the discovery of new genes and regulatory sequences and their positions, and making available large collections of molecular markers. Genome-wide expression studies provide breeders with a better understanding of the molecular basis of complex traits as they usually reveal their multigene nature and the importance of environmental influences. Genome approaches also include targeting induced local lesions in genomes (TILLING) and EcoTILLING, which have potential to screen mutants and germplasm collections for allelic variants of target genes aiding in their functional analysis.

Medicago truncatula and *Lotus japonicus* have emerged as models for plant genomic research in legumes. Scientists now study these species to investigate a range of questions from disease resistance to environmental stress tolerance and from bacterial and fungal symbiosis to complex secondary metabolism. However, both are temperate legumes. For tropical legumes, soybean has emerged as a model genome. Genomes of several legumes have been sequenced and some species are currently the subject of independent large-scale sequencing projects (Table 2.2). In addition, large-scale transcriptomics, proteomics, metabolomics, phenomics and bioinformatics resources and reverse genetic tools have been developed. The characterization of these three legume genomes (*M. truncatula*, *L. japonicus* and soybean) will undoubtedly enhance ongoing comparative genomic

Table 2.2 Summary of leguminous crops genome sequence information.

S. No.	Scientific name	Estimated genome size	Sequencing strategy	Reference
1	*Cajanus cajan* (L.) Millsp.	833.07 Mb	Scaffold/bacterial artificial chromosome (BAC)	Varshney *et al.*, 2012
2	*Cicer arietinum* L.	740 Mb	Draft assembly, whole genome shotgun sequence	Varshney *et al.*, 2013
3	*Glycine max* (L.) Merr.	1115 Mb	Whole-genome shotgun approach integrated with physical and high-density genetic maps	Schmutz *et al.*, 2010
4	*G. soja* Sieb. & Zucc.	~1100 Mbp	Whole-genome shotgun sequencing	Kim *et al.*, 2010
5	*Lotus japonicus* L.	472 Mb	Clone-by-clone sequencing and shotgun sequencing of selected regions of the genome	Sato *et al.*, 2008
6	*Lupinus angustifolius* L.	1.153 Gb	Draft assembly from a whole-genome shotgun sequencing	Yang *et al.*, 2013
7	*Medicago truncatula* Godr. & Gnem.	~454 to 526 Mbp	Euchromatin based on a recently completed BAC-by-BAC sequencing approach	Young *et al.*, 2011
8	*Phaseolus vulgaris* (L.)	~650 Mb	V1.0, the first chromosome-scale version includes BAC and fosmid end sequence whole-genome shotgun (WGS) method	http://www.phytozome.net/commonbean.php
9	*Vigna unguiculata* L. (Walp.)	620 Mb	160 Mb completed and project is underway	http://cowpeagenomics.med.virginia.edu/
10	*Arachis hypogaea* L.	2890 Mb	In progress	http://www.peanutbioscience.com

analyses. Interestingly, in leguminous crops genome-wide studies are important to identify and establish network(s) involved in stress response pathways, which could eventually be manipulated to minimize crop losses due to abiotic stresses (Urano *et al.*, 2010).

The analysis of gene transcripts is probably the most developed field. Indeed, there are several platforms available for gene transcript analysis. The quantitative real-time polymerase chain reaction (qRT-PCR) technique is considered highly accurate but permits analysis of only a limited number of genes. In contrast, gene microarray technology allows the analysis of thousands of genes at a time. RNA-Seq is oriented to the performance of unbiased analysis of RNA transcripts, which generates gigabyte-sized readouts with all the RNA transcripts of a given cell, organism or tissue (Brady *et al.*, 2007; Steibel *et al.*, 2009). This increasing amount of data generated has led researchers to create databases of experiments in environmental stresses (Dinneny *et al.*, 2008).

Mantri *et al.* (2007) studied transcriptional profiling in the leaf, root and/or flower tissues in tolerant and susceptible genotypes in chickpea under salinity, drought and cold, and reported over two-fold differential expression in hundreds of transcripts. The differentially expressed genes coded for several functional and regulatory proteins, which indicates that multiple genes regulate the abiotic stress response mechanisms. A very limited number of molecular markers and candidate genes are available for undertaking molecular breeding in chickpea to tackle salinity stresses. The study conducted by Varshney *et al.* (2009) reports on the generation and analysis of a comprehensive resource of drought- and salinity-responsive expressed sequence tags (ESTs) and gene-based markers. A total of 20,162 (18,435 high quality) drought- and salinity-responsive ESTs were generated from 10 different root tissue cDNA libraries of chickpea. Hierarchical clustering of 105 selected contigs provided clues about stress-responsive candidate genes, and their expression profiles showed predominance in specific stress-challenged libraries. Such a set of chickpea ESTs serves as a resource of high-quality transcripts for gene discovery and development of functional markers associated with abiotic stress tolerance that will facilitate chickpea breeding. Mapping of gene-based markers in chickpea will also add more reference points to align genomes of chickpea and other legume species (Varshney *et al.*, 2009).

A study by Molina *et al.* (2011) reported characterization of salt-stress responses of chickpeas at the transcriptome level using DeepSuperSAGE. From a total of 144,200 analysed 26 bp tags in roots and nodules together, 21,401 unique transcripts were identified. Of these, only 363 and 106 specific transcripts, respectively, were commonly upregulated or downregulated (>3.0-fold) under salt stress in both organs, evidencing a differential organ-specific response to stress. It was found that transcripts were associated with the generation and scavenging of reactive oxygen species (ROS), as well as being involved in Na^+ homeostasis. Although both processes are already very well characterized in other plants, the information generated in this study is very valuable. Additionally, a set of more than 15 candidate transcripts were proposed to be potential components of the salt overly sensitive (SOS) pathway in chickpea. Newly identified transcript isoforms are potential targets for breeding novel cultivars with high salinity tolerance. This study proves that these targets can be combined into breeding schemes by using microarrays and RT-PCR assays downstream of the generation of 26 bp tags by SuperSAGE (Molina *et al.*, 2011).

A study was conducted to isolate and characterize a novel JAZ family gene, *GsJAZ2*, from *G. soja* following exposure to salt, alkali, cold and drought. Subcellular localization studies using a green fluorescent protein (GFP) fusion protein showed that GsJAZ2 was localized to the nucleus. These results suggest that the newly isolated wild soybean GsJAZ2 is a positive regulator of plant salt stress tolerance (Zhu *et al.*, 2012). A study by Song *et al.* (2012) described the identification by microarray analysis of 49 soybean transcription factors thought to be inducible by salinity stress. A semi-quantitative RT-PCR-based expression assay confirmed the salinity stress inducibility of 45 of these 49 transcription factors. Salinity stress was revealed to affect the methylation status of four of these ten transcription factors (one MYB, one b-ZIP and two AP2/DREB family members). Chromatin immunoprecipitation (ChIP) analysis indicated that the activation of three of the four DNA methylated transcription factors was correlated with an increased level of histone H3K4 trimethylation and H3K9 acetylation, and/or a reduced level of H3K9 demethylation in various parts of the promoter or coding regions. This study indicated a vital role for gene activation/repression by DNA methylation and/or histone modification of some transcription factor genes in soybean tolerance to salinity stress (Song *et al.*, 2012).

Another study was conducted to characterize the biological functions of a receptor-like cytoplasmic serine/threonine protein kinase gene, *GsRLCK*, which was previously identified as a putative salt-alkali stress-related gene from the transcriptome profiles of *G. soja*. Bioinformatic analysis indicated that GsRLCK protein comprises a conserved kinase catalytic domain and two transmembrane domains at the N-terminus, but no typical extracellular domain. Consistently, GsRLCK-eGFP fusion protein was observed on the plasma membrane, but eGFP alone was distributing throughout the cytoplasm in onion epidermal cells. Quantitative real-time PCR analysis revealed the induced expression of GsRLCK by ABA, salt, alkali and drought stresses. The study indicated that GsRLCK was a novel receptor-like cytoplasmic protein kinase and played a crucial role in plant responses to salt stress (Sun *et al.*, 2013).

The transcriptional profiles of genes in leaves and roots of seedlings (two-leaf stage) of the soybean inbred line HJ-1 were studied after 48 h under various stress conditions including salt stress (120 mM NaCl) to identify responsible genes. Under salt stress, 874 genes were upregulated in leaves, and 1822 genes were upregulated in roots, compared with expression in the corresponding organ in control plants. Furthermore, comparison of gene expression among salt-, salt-alkali- and drought-treated plants revealed that genes associated with calcium signalling and nucleic acid pathways were upregulated in the responses to all three stresses, indicating a degree of cross-talk among these pathways. This study provided new insights into the stress tolerance mechanisms of soybean (Fan *et al.*, 2013).

The estimated size of the *Glycine max* genome is 1115 Mb in 20 chromosomes. A total of 950 Mb of the genome sequence has been completed (Schmutz *et al.*, 2010) giving a predicted 46,430 protein-coding genes with high confidence, which is 70% more than *Arabidopsis thaliana*. A total of 78% of the predicted genes are present at the ends of chromosomes, and 75% of the genes are present in multiple copies. A total of 12.2% of the predicted genes were identified as encoding transcription factors.

Following transcript analyses, proteins are the second most important aspect in defining an organism's phenotype, representing the products of gene expression. To encompass all the translational events from mRNA formation to functional proteins, analytical platforms need to be able to evaluate not only the presence/absence of a given protein but also the potential post-translational modifications (e.g., phosphorylations, glycosylations or prenylations) and also have the potential to assess protein–protein interactions. Available platforms include the traditional 2D gel electrophoresis – combined or not with fluorescent dyes, as in the difference in gel electrophoresis (DIGE) technique – which is useful for protein fingerprinting when coupled to mass spectrometry (MS) for protein identification. The shotgun proteomics approach is based on nano-liquid chromatography (nanoLC) separation of matrices and MS detection; it offers a deeper and less biased coverage of the proteome including low abundant proteins (Zhao *et al.*, 2013). Future challenges in proteomics are the development of new analytical techniques and workflows to overcome the lack of reproducibility, and the implementation of new features for data export and comparison in databases (Kosva *et al.*, 2013). Several researchers have reported that proteins respond to salt stress differently in different plants including leguminous crops (Hakeem *et al.*, 2012a; Komatsu & Hossain 2013; Kosva *et al.*, 2013; Zhao *et al.*, 2013) (Table 2.3).

Kav *et al.* (2004) studied the proteome-level changes in the roots of *Pisum sativum* in response to salinity. This study identified 35 protein spots that exhibited significant changes in abundance due to NaCl treatment. These proteins were identified as pathogenesis-related (PR) 10 proteins, antioxidant enzymes such as superoxide dismutase (SOD) as well as nucleoside diphosphate kinase (NDPK). The study suggested the possible existence of a novel signal transduction pathway with a potentially important role in abiotic stress response.

Jain *et al.* (2006) studied the proteome of a salinity-tolerant *Arachis hypogea* L. callus cell line in comparison with sensitive counterparts. Several low-molecular-weight proteins were identified or significantly upregulated in the tolerant line, and tandem mass spectrometry analysis revealed the presence of PR10 proteins. The study recognized the role of the differentially

Table 2.3 Summary of proteomic publications in leguminous crops.

S. No.	Species	Variety	Tissue	Salt treatment	Proteomic approach	IDs	Unique proteins	References
1	*Arachis hypogea* L.	JL-24	Callus cell	200 mM, 12 d	2DE/ESI-LC-MS-MS	25	6	Jain *et al.*, 2006
2	*Glycine max* L.	Enrei	Hypocotyls, root	100 mM, 16 d	2DE/ ESI-QTOF-MS-MS	7	7	Aghaei *et al.*, 2009
3		Enrei	Hypocotyls, root, leaves	40 mM, 16 d	2DE/MALDI-TOF-MS	38	31	Sobhanian *et al.*, 2010b
4		LEE 68 (ST) N2899 (ST)	Germinating seeds	100 mM until radicle protrusion from the seed coat	2DE/MALDI-TOF-MS	18	18	Xu *et al.*, 2011
5		Pusa-24 (SS) Pusa-37 (ST)	Seedling	25, 50, 75, 100, 125 and 150 mM, 10 d	2DE/ MALDI-TOF-MS	173	40	Hakeem et al 2012
6		Jackson (SS) Lee 68 (ST)	Leaf	150 mM, 1, 12, 72, 144 h	2DE/ MALDI-TOF-TOF-MS	91	78	Ma *et al.*, 2012
7	*Lathyrus sativus* L.	LP 24	Leaf	500 mM, 12 h, 24h, 3 d, 6 d, 12 d	2DE/ESI-LC-MS/MS	44	30	Chattopadhyay *et al.*, 2011
8	*Lupine luteus* L.	Mister	Mitochondria	250, 500 mM, 12 h	2DE/ESI-LC-MS/MS	21	21	Wojtyla *et al.*, 2013
9	*Pisum sativum* L.	Cutlans	Root	75, 150 mM, 7 d; 75, 150 mM, 42 d	2DE/ ESI-QTOF-MS-MS	35	24	Nat *et al.*, 2004

expressed phosphorylated PR10 proteins in mediating salinity stress responses.

Chattopadhyay *et al.* (2011) have developed differential proteomes to understand the molecular basis of salt stress in *Lathyrus sativus* by subjecting it to high salinity and other abiotic stresses. About 400 protein spots were detected in each of the stress proteomes with one-fourth showing more than a two-fold difference in expression values. Eighty such proteins were subjected to liquid chromatography (LC)-tandem MS/MS analyses, which resulted in the identification of 48 stress-responsive proteins (SRPs) presumably involved in a variety of functions including metabolism, signal transduction, protein biogenesis and degradation and cell defence in a stress-specific manner. The study provided insights into understanding the range of stress-regulated proteins and the biological processes they control as well as having consequences for strategies to advance stress adaptation in plants.

Protein expression levels were evaluated using a proteomics approach in hypocotyls and roots of soybean plants subjected to salt stress. In response to salt stress 321 proteins were detected. Seven were reproducibly found to be upregulated or downregulated. Late embryogenesis abundant protein, beta conglycinin, elicitor peptide three precursor and basic/helix-loop-helix protein were upregulated, while protease inhibitor, lectin and stem 31-kDa glycoprotein precursor were downregulated. The study revealed that salt stress can alter the expression level of certain proteins in the hypocotyl and root of soybean, and that these may play a role in the adaptation to salt stress (Aghaei *et al.*, 2009). Sobhanian *et al.* (2010b) analysed the proteome of soybean leaves, hypocotyls and roots under salt stress. Nineteen, 22 and 14 proteins out of 340, 330 and 235 proteins in the leaves, hypocotyls and roots, respectively, were upregulated or downregulated by NaCl treatment. Glyceraldehyde-3-phosphate dehydrogenase was downregulated in the leaf/hypocotyls and fructokinase 2 was downregulated in the hypocotyls/root with NaCl treatment.

Xu *et al.* (2011) studied the responses of soybean seeds of two cultivars, Lee68 and N2800, to salt stress during germination at both physiological and proteomic levels. Proteomic approaches identified 350 protein spots, of which 18 spots showed changes in abundance as a result of salt stress in both cultivars. Ferritin and 20S proteasome subunit β-6 were upregulated. Glyceraldehyde-3-phosphate dehydrogenase, glutathione S-transferase (GST) 9, GST 10 and seed maturation protein PM 36 were all downregulated by salt stress in Lee68. These proteins were upregulated in N2800 under salt stress. The study indicates the role played by the proteins in defence against salt stress during soybean seed germination. Ma *et al.* (2012) investigated the comparative proteomic analysis of soybean seedling leaves of the salt-sensitive genotype Jackson and the salt-tolerant genotype Lee68. About 800 proteins were detected and 91 were found to be differentially expressed. The identified proteins were involved in 14 metabolic pathways and cellular processes such as ROS production and scavenging, accelerated proteolysis and reduced biosynthesis of proteins, impaired photosynthesis, abundant energy supply and enhanced biosynthesis of ethylene.

The effect of salt stress on the mitochondrial proteome of lupin (*L. luteus* Mister) embryo axes was investigated by Wojtyla *et al.* (2013). Twenty-one spots showed significant alterations in protein expression profiles. The identified proteins were associated with enzymes of the Krebs cycle, mitochondrial electron transport chain, and enzymes and proteins involved in mitochondrial biogenesis and stress responses.

The term metabolomics has been described as the identification and quantitation of all low-molecular-weight metabolites in a given organism, at a given developmental stage and in a given organ, tissue or cell type. This is an interesting task due to the broad array of molecules with different structures and chemical properties. Unlike transcriptomics, there is no single approach to identify all compounds, and the strategic combination of extraction and identification techniques is key to enhancing the coverage of the technique. The most prevalent metabolomics techniques place emphasis on metabolites with comparable and specific chemical properties; globally known as metabolite profiling, these cover only a fraction of the metabolome. To achieve complete coverage of the wide range of metabolites present in the plant kingdom numerous analytical techniques consisting of a separation technique coupled to a detection device (usually MS) are shared. However, there are substitutes that dispense with a separation technique; such is the case of flow injection analysis coupled to MS (FIA/MS) or the application of different analyzers such as nuclear magnetic resonance (NMR) or Fourier transform infrared spectroscopy (FTIR), which are utilized only for fingerprinting purposes. The separation part delivers the selectivity required for certain groups of metabolites. Among all analysers that can be used with the separation techniques mentioned, the most popular in metabolomics are MS analysers, particularly those providing accurate mass measures such as hybrid quadrupole/time-of-flight (TOF) or orbitraps. However, more targeted techniques are still extensively used for the quantitation of several plant metabolites and hormones due to their enhanced sensitivity and specificity (Arbona *et al.*, 2013).

Comparative targeted metabolic profiling shows marked differences in the metabolite composition between salt-sensitive and salt-tolerant soybean varieties. Principal components analysis clearly confirmed that it is possible to use secondary metabolites, for example, isoflavones and saponins, to discriminate between closely related soybean genotypes. Genistin and group B saponins were recognized as the key secondary metabolites connected with salt tolerance. These individual metabolites may provide additional insight into salt tolerance and adaptation of plants (Wu *et al.*, 2008). The future challenge is the integration of transcriptomics, proteomics and metabolomics within a single framework that will allow a better understanding of how plants respond to a changing environment.

2.6 Conclusions and future prospects

Leguminous crops are often subject to various environmental stresses that can severely affect plant development and eventually productivity. Production of stress-tolerant cultivars of leguminous crops is a pressing need today because of the growing threat to crop productivity of global warming and changing climatic patterns, in addition to the anticipated human population explosion in the near future. Plant salinity tolerance is a multifaceted physiological trait involving adaptations to signalling and metabolic networks. Studies on salt stress responses and tolerance in many crop plants have progressed rapidly in recent years. Morphological, physiological, biochemical, genetic and genomic analyses have made important discoveries of salt-responsive genes, proteins and metabolites involved in different cellular pathways. In this review, we have attempted to summarize significant contributions related to salt stress responses and tolerance in leguminous crops, particularly grain legumes.

Many agriculturally important legumes are salt sensitive. However, there is variability in the salt tolerance of the legumes within and between species, and of the rhizobia and their functional symbiosis with the plant. A number of curious physiological and molecular components in the salt response pathways of leguminous crops have been exposed. They provide crucial information for testing new hypotheses in ROS scavenging, ion/osmotic homeostasis, protein synthesis/turnover, cell structure modulation and energy metabolism. Despite the significant progress, large gaps remain in our knowledge of transmemberane ion transport, cellular compartmentalization, sensors/receptors in signal transduction and metabolites in energy supply. Therefore, further physiological and molecular studies are needed to focus research in this area. Based on the results presented here, it also seems that the responses to salinity of nitrogen fixation closely follow biomass production in the most tolerant legumes. Symbionts like rhizobia and arbuscular mycorrhizal fungi are generally found to be more salt tolerant than their host plants. Functional symbiosis sometimes makes an important difference in the salt tolerance and productivity of plants. Therefore this line of research should be continued to develop agriculturally important legumes with improved salt tolerance.

More systematic studies using both forward and reverse genetics tools in leguminous crops are required. Reverse genetics study of the homologous genes in leguminous crops would provide information about the similarity and differences of the functions of the transporters important for salt tolerance. Forward genetics screening for mutants that are sensitive or tolerant to salt stress would classify novel genes and/or salt-tolerance mechanisms unique to leguminous plants. It can be envisaged that more and more genome sequences of leguminous crops will be available in the near future along with rapid developments in sequencing technologies and falls in the associated costs. Genome sequence information will also be very beneficial in identifying the natural variant genes bestowing salt tolerance in salt-tolerant lines within a species. Such information on salt tolerance will prove important for molecular breeding of salt tolerance in leguminous crops. The link with functional genomics may also enable identification of candidate genes as tools to further interpret the mechanisms involved in the efficiency of rhizobial symbiosis and its adaptation to salt stress.

Proteomics has clearly brought important discoveries for our innate understanding of the molecular mechanisms and complex signalling and metabolic networks in leguminous crops under salt stress. By integrated analysis of current proteomics results available for five plant species (*A. hypogea, G. max, L. sativus, L. luteus* and *P. sativum*) we found 452 protein identities representing 255 unique proteins (UPs). Most of the proteomic studies have reported the abundant salt-responsive proteins rather than the low-abundance proteins, including salt-responsive transcription factors, protein kinases and membrane/vesicle transport proteins. Clearly proteomics still has much to fulfil of its presumed potential in this area. Integration of 'omics' tools such as transcriptomics, proteomics and metabolomics will give us a more holistic view of the molecular mechanisms underlying salt stress responses and tolerance. This can then be used to predict and validate how diverse components generate responses and control different pathways toward the ultimate goal of improving salt tolerance in leguminous crops for enhanced productivity.

References

Abd-Alla MH, Vuong TD, Harper JE (1998) Genotypic differences in dinitrogen fixation response to NaCl stress in intact and grafted soybean. *Crop Sci* 38: 72–77.

Aghaei K, Ehsanpour AA, Shah AH, Komatsu S (2009) Proteome analysis of soybean hypocotyl and root under salt stress. *Amino Acids* 36: 91–98.

Ahmad P, Jhon R (2005) Effect of salt stress on growth and biochemical parameters of *Pisum sativum* L. *Arch Agron Soil Sci* 51: 665–672.

Ahmed J, Bano M (1992) The effect of sodium chloride on the physiology of cotyledons and mobilization of reserved food in *Cicer arietinum*. *Pak J Bot* 24: 40–48.

Amira MS, Qados A (2011) Effect of salt stress on plant growth and metabolism of bean plant *Vicia faba* (L.). *J Saudi Soc Agric Sci* 10: 7–15.

Apse MP, Blumwald E (2002) Engineering salt tolerance in plants. *Curr Opin Biotechnol* 13: 146–150.

Arbona V, Manzi M, de Carlos O, Aurelio GC. (2013) Metabolomics as a tool to investigate abiotic stress tolerance in plants. *Int J Mol Sci* 14: 4885–4911.

Arumuganathan K, Earle ED (1991a) Nuclear DNA content of some important plant species. *Plant Mol Biol Rep* 9: 208–218.

Arumuganathan K, Earle ED (1991b) Estimation of nuclear DNA content of plants by flow cytometry. *Plant Mol Biol Rep* 9: 229–241.

Ashraf M, Bashir A (2003) Salt stress induced changes in some organic metabolites and ionic relations in nodules and other plant parts of two crop legumes differing in salt tolerance. *Flora* 198: 486–498.

Ashraf M, Waheed A (1993) Responses of some local/exotic accessions of lentil (*Lens culinaris* Medic.) to salt stress. *J Agron Soil Sci* 170: 103–112.

Bandeoglu E, Eyidogan F, Yucel M, Oktem HA (2004) Antioxidant responses of shoots and roots of lentil to NaCl–salinity stress. *Plant Growth Regul* 42: 69–77.

Barba-Espín G, Clemente-Moreno MJ, Alvarez S, García-Legaz MF, Hernández JA, Díaz-Vivancos P (2011) Salicylic acid negatively affects the response to salt stress in pea plants. *Plant Biol (Stuttg)* 13: 909–917.

Bhattarai SP, Midmore DJ (2009) Oxygation enhances growth, gas exchange and salt tolerance of vegetable soybean and cotton in a saline vertisol. *J Integr Plant Biol* 51: 675–688.

Bhomkar P, Upadhyay CP, Saxena N, *et al.* (2008) Salt stress alleviation in transgenic *Vigna mungo* L. Hepper (blackgram) by overexpression of the *glyoxalase I* gene using a novel *Cestrum* yellow leaf curling virus (CmYLCV) promoter. *Mol Breeding* 22: 169–181.

Bishnoi SK, Kumar B, Rani C, *et al.* (2006) Changes in protein profile of pigeon pea genotypes in response to NaCl and boron stress. *Biol Plant* 50: 135–137.

Bolaños L, Martín M, El-Hamdaoui A, Rivilla R, Bonilla I (2006) Nitrogenase inhibition in nodules from pea plants grown under salt stress occurs at the physiological level and can be alleviated by B and Ca. *Plant Soil* 280: 135–142.

Bouhmouch I, Brhada F, Filali-Maltouf A, Aurag J (2001) Selection of osmotolerant and effective strains of Rhizobiaceae for inoculation of common bean (*Phaseolus vulgaris*) in Moroccan saline soils. *Agronomie* 21: 591–599.

Bouhmouch I, Souad-Mouhsine B, Brhada F, Aurag J (2005) Influence of host cultivars and Rhizobium species on the growth and symbiotic performance of *Phaseolus vulgaris* under salt stress. *J Plant Physiol* 162: 1103–1113.

Brady SM, Orlando DA, Lee JY, *et al*, (2007) A high-resolution root spatiotemporal map reveals dominant expression patterns. *Science* 318: 801–806.

Bruning B, Rozema J (2013) Symbiotic nitrogen fixation in legumes: Perspectives for saline agriculture. *Environ Exp Bot* 92: 134–143.

Cabot C, Garcia MC, Sibole JV (2005) Relationship between xylem ion concentration and bean growth responses to short-term salinisation in spring and summer. *J Plant Physiol* 162: 327–334.

Cavalcanti FR, Lima JPMS, Silva SLF, Viégas R, Silveira JAG (2007) Roots and leaves display contrasting oxidative response during salt stress and recovery in cowpea. *J Plant Physiol* 164: 591–600.

Chattopadhyay A, Subba P, Pandey A, *et al.* (2011) Analysis of the grass pea proteome and identification of stress-responsive proteins upon exposure to high salinity, low temperature, and abscisic acid treatment. *Phytochemistry* 72: 1293–1307.

Chinnusamy V, Jagendorf A, Zhu JK (2005) Understanding and improving salt tolerance in plants. *Crop Sci* 45: 437–448.

Comba ME, Benavides MP (1997) Relationship between nitrogen fixation and oxidative stress induction in nodules of salt-treated soybean plants. *Phyton* 60: 115–126.

Comba ME, Benavides MP, Tomaro ML (1998) Effect of salt stress on antioxidant defence system in soybean root nodules. *Aust J Plant Physiol* 25: 665–671.

Cordovilla M, Ligero P, Ligero F, Lluch C (1996) Growth and assimilation in nodules in response to nitrate levels in *Vicia faba* under salt stress. *J Exp Bot* 47: 203–210.

Cordovilla M, Ligero P, Lluch C (1999) Effects of NaCl on growth and nitrogen fixation and assimilation of inoculated and KNO_3 fertilized *Vicia faba* L. and *Pisum sativum* L. plants. *Plant Sci* 140: 127–136.

Corpas FJ, Sandalio LM, Palma JM, *et al.* (1991) Subcellular distribution of superoxide dismutase in leaves of ureide-producing leguminous plants. *Physiol Plant* 81: 285–291.

Costa JH, Jolivet Y, Hasenfratz-Sauder MP, *et al.* (2007) Alternative oxidase regulation in roots of *Vigna unguiculata* cultivars differing in drought/salt tolerance. *J Plant Physiol* 164l: 718–727.

Dardanelli MS, Córdoba FJFD, Estévez J, *et al.* (2012) Changes in flavonoids secreted by *Phaseolus vulgaris* roots in the presence of salt and the plant growth-promoting rhizobacterium *Chryseobacterium balustinum*. *Appl Soil Ecol* 57: 31–38.

Delgado MJ, Garrido JM, Ligero F, Lluch C (1993) Nitrogen fixation and carbon metabolism by nodules and bacteroids of pea plants under sodium chloride stress. *Physiol Plant* 89: 824–829.

Delgado MJ, Ligero F, Lluch C (1994) Effects of salt stress on growth and nitrogen fixation by pea, faba-bean, common-bean, and soybean plants. *Soil Biol Biochem* 26: 371–376.

Demir Y, Kocakalikan I (2002) Effect of NaCl and proline on bean seedlings cultured in vitro. *Biol Plant* 45: 597–599.

Dinneny JR, Long TA, Wang JY, *et al.* (2008) Cell identity mediates the response of *Arabidopsis* roots to abiotic stress. *Science* 320: 942–945.

Doyle JJ, Luckow MA (2003) The rest of the iceberg. Legume diversity and evolution in a phylogenetic context. *Plant Physiol* 1331: 900–910.

Dua RP (1998) Salinity tolerance in chickpea (*Cicer arietinum* L.). *Indian J Plant Physiol* 3: 102–106.

Egbichi M, Keyster M, Ludidi KN (2014) Effect of exogenous application of nitric oxide on salt stress responses of soybean. *S Afr J Bot* 90: 131–136.

El-Hamdaoui A, Redondo-Nieto M, Torralba B, Rivilla R, Bonilla I, Bolanos L (2003) Influence of boron and calcium on the tolerance to salinity of nitrogen-fixing pea plants. *Plant Soil* 251: 93–103.

Elsheikh EAE, Wood M (1990) Effect of salinity on growth, nodulation and nitrogen yield of chickpea (*Cicer arietinum* L.). *J Exp Bot* 41: 1263–1269.

Faghire M, Mohamed F, Taoufiq K, *et al.* (2013) Genotypic variation of nodules' enzymatic activities in symbiotic nitrogen fixation among common bean (*Phaseolus vulgaris* L.) genotypes grown under salinity constraint. *Symbiosis* 60: 115–122.

Fan XD, Wang JQ, Yang N, *et al.* (2013) Gene expression profiling of soybean leaves and roots under salt, saline-alkali and drought stress by high-throughput Illumina sequencing. *Gene* 512: 392–402.

FAO (1994) Definition and classification of commodities, 4. Pulses and derived products. United Nations Food and Agriculture Organization.

FAO (2005) Salt-affected soils from sea water intrusion: Strategies for rehabilitation and management. Food and Agriculture Organization of the UN.

Ferri A, Lluch C, Ocana A (2000) Effect of salt stress on carbon metabolism and bacteroid respiration in root nodules of common bean (*Phaseolus vulgaris*). *Plant Biol* 2: 396–402.

Figueira EMAP, Caldeira GCN (2005) Effect of nitrogen nutrition on salt tolerance of *Pisum sativum* during vegetative growth. *J Plant Nutr Soil Sci* 168: 359–363.

Flowers TJ, Troke PF, Yeo AR (1977) The mechanism of salt tolerance in halophytes. *Ann Rev Plant Physiol* 28: 89–121.

Flowers TJ, Galal HK, Bromham L (2010) Evolution of halophytes: multiple origins of salt tolerance in land plants. *Funct Plant Biol* 37: 604–612.

García NAT, Iribarne C, Palma F, Lluch C (2007) Inhibition of the catalase activity from *Phaseolus vulgaris* and *Medicago sativa* by sodium chloride. *Plant Physiol Biochem* 45: 535–541.

Garg N (2002) Salinity stress-induced changes in key metabolites in the nodules of *Glycine max* L. (soybean) and *Cicer arietinum* L. (chickpea) and the manoeuvrability of their response through plant growth regulators. *J Plant Biol* 29: 137–142.

Garg NJ (2004) Variability in response of chickpea (*Cicer arietinum* L.) cultivars to salt stress in germination and early growth of the seedlings. *Ind J Plant Physiol* 9: 21–28.

Garg N, Chandel S (2011) The effects of salinity on nitrogen fixation and trehalose metabolism in mycorrhizal *Cajanus cajan* (L.) Millsp. plants. *J Plant Growth Regul* 30: 490–503.

Garg N, Dua IS (1996) Effect of saline conditions on nitrogen fixing potential of pigeonpea (*Cajanus cajan* (L.) Millsp.) and its dependence on the availability of cytokinins. *Curr Res Plant Sci* 2: 131–136.

Garg N, Dua IS, Sharma SK, Garg OP (1988) Nodulation and nitrogen fixing capacity in relation to variable salinity levels in *Vigna radiata* L. (Mung.). *PU Res Bull (Sci)* 39 (III–IV): 187–196.

Ghassemi F, Jakeman AJ, Nix HA (1995) *Salinization of Land and Water Resources: Human Causes, Extent, Management and Case Studies*. University of New South Wales Press Ltd, Sydney.

Gholipoor M, Ghasemi-Golezani K, Khooie FR, Moghaddam M (2000) Effects of salinity on initial seedling growth of chickpea (*Cicer arietinum* L.). *Acta Agron Hung* 48: 337–343.

Gill KS, Sharma PC (1993) Mechanism of salt injury at seedling and vegetative growth stages in *Cajanus cajan* (L.) Millsp. *Plant Physiol Biochem* 20: 49–52.

Gowda C, Rao P, Tripathi P, Gaur S, Deshmukh R (2009) Regional shift in chickpea production in India. In: Ali M, Kumar S (eds), *Milestones in Food Legumes Research*. IIPR, Kanpur, pp. 21–35.

Hakeem KR, Chandaana R, Ahemad RP, Iqbal M, Ozaturk M (2012a) Relevance of proteomic investigations in plant abiotic stress physiology. *OMICS* 16: 621–635.

Hakeem KR, Khan F, Chandana R, Siddiqui TO, Iqbal M (2012b) Genotypic variability among soybean genotypes under NaCl stress and proteome analysis of salt-tolerant genotype. *Appl Biochem Biotech* 168: 2309–2329.

Hanafy MS, El-Banna A, Schumacher HM, Jacobsen HJ, Hassan FS (2013) Enhanced tolerance to drought and salt stresses in transgenic faba bean (*Vicia faba* L.) plants by heterologous expression of the PR10a gene from potato. *Plant Cell Rep* 32: 663–674.

Hasegawa PM, Bressan RA, Zhu JK, Bohnert HJ (2000) Plant cellular and molecular responses to high salinity. *Ann Rev Plant Biol* 51: 463–499.

Hernandez JA, Almansa MS (2002) Short-term effects of salt stress on antioxidant systems and leaf water relations of pea leaves. *Physiol Plant* 115: 251–257.

Hernandez JA, Almansa MS, Del Rio LA, Sevilla F (1993) Effect of salinity on metalloenzymes of oxygen metabolism in two leguminous plants. *J Plant Nutr* 16: 2539–2554.

Hernandez JA, del Rio LA, Sevilla F (1994) Salt stress-induced changes in superoxide dismutase isozymes in leaves and mesophyll protoplasts from *Vigna unguiculata* L. Walp. *New Phytol* 126: 37–44.

Hernandez JA, Olmos E, Corpas FJ, Sevilla F, Del Rio L (1995) Salt induced oxidative stress in chloroplasts of pea plants. *Plant Sci* 105: 151–167.

Hernandez JA, Campillo A, Jimenez A, Alarcon JJ, Sevilla F (1999) Response of antioxidant systems and leaf water relations to NaCl stress in pea plants. *New Phytol* 141: 241–251.

Hernandez JA, Jimenez A, Mullineaux P, Sevilla F (2000) Tolerance of pea (*Pisum sativum* L.) to a long term salt stress is associated with induction of antioxidant defences. *Plant Cell Environ* 23: 853–62.

Hernandez JA, Ferrer MA, Jimenez A, Barcelo AR, Sevilla F (2001) Antioxidant systems and O_2^-/H_2O_2 production in the apoplast of pea leaves. Its relation with salt-induced necrotic lesions in minor veins. *Plant Physiol* 127: 817–831.

Hillel D (2005) Salinity: management. In: Hillel D, Hatfield JH, Powlson DS, Rosenzweige C, Scow KM, Singer MJ, Sparks DL (eds), *Encyclopaedia of Soils in the Environment*. Elsevier/Academic Press, London, pp. 435–442.

Hu Y, Schmidhalter U (2002) Limitation of salt stress to plant growth. In: Hock B, Elstner CF (eds), *Plant Toxicology*. Marcel Dekker Inc., New York, pp. 91–224.

Hussain TM, Chandrasekhar T, Hazara M, Sultan Z, Saleh BK, Gopal GR (2008) Recent advances in salt stress biology – a review. *Biotech Mol Biol Rev* 3: 8–13.

IPCC (2007) Core Writing Team; Pachauri R.K, Reisinger A (eds), *Climate Change 2007: Synthesis Report: Contribution of Working Groups I, II and III to the Fourth Assessment Report of the Intergovernmental Panel on Climate Change*. Available at: http://www.ipcc.ch

Jain M, Mathur G, Koul S, Sarin NB (2001) Ameliorative effects of proline on salt stress-induced lipid peroxidation in cell lines of groundnut (*Arachis hypogaea* L.). *Plant Cell Rep* 20: 463–468.

Jain S, Srivastava S, Sarin NB, Kav NN (2006) Proteomics reveals elevated levels of PR 10 proteins in saline-tolerant peanut (*Arachis hypogaea*) calli. *Plant Physiol Biochem* 44: 253–259.

James MF, Sprent EK, Franco JIAA (1995) Stem and root-nodules on the tropical wet land legume *Aeschynomene fluminensis*. *New Phytol* 130: 531–544.

Jebara M, Drevon JJU, Aouani ME (2001) Effects of hydroponic culture system and NaCl on interactions between common bean lines and native rhizobia from Tunisian soils. *Agronomie* 21: 601–605.

Jebara S, Jebara M, Limam F, Aouani ME (2005) Changes in ascorbate peroxidase, catalase, guaiacol peroxidase and superoxide dismutase activities in common bean (*Phaseolus vulgaris*) nodules under salt stress. *J Plant Physiol* 162: 929–936.

Jimenez JS, Debouck DG, Lynch JP (2003) Growth, gas exchange, water relations, and ion composition of *Phaseolus* species grown under saline conditions. *Field Crop Res* 80: 207–222.

Jungklang J, Usui K, Matsumoto H (2003) Differences in physiological responses to NaCl between salt-tolerant *Sesbania rostrata* Brem & Oberm and non-tolerant *Phaseolus vulgaris* L. *Weed Biol Manag* 3: 21–27.

Kabir ME, Karim MA, Azad MAK (2004) Effect of potassium on salinity tolerance of mungbean (*Vigna radiata* L. Wilczek). *J Biol Sci* 4: 103–110.

Katerji N, Hoorn JW, Hamdy A, Mastrorilli M, Oweis T, Malhotra RS (2001) Response to soil salinity of two chickpea varieties differing in drought tolerance. *Agric Water Manag* 50: 83–96.

Kav NNV, Srivastava S, Goonewardenel L, Blade SF (2004) Proteome-level changes in the roots of *Pisum sativum* in response to salinity. *Ann Appl Biol* 145: 217–230.

Khadri M, Tejera NA, Lluch C (2006) Alleviation of salt stress in common bean (*Phaseolus vulgaris*) by exogenous abscisic acid supply. *J Plant Growth Regul* 25: 110–119.

Khadri M, Tejera NA, Lluch C (2007) Sodium chloride–ABA interaction in two common bean (*Phaseolus vulgaris*) cultivars differing in salinity tolerance. *Environ Exp Bot* 60: 211–218.

Khan F, Hakeem KR, Siddiqi TO, Ahmad A (2013) RAPD marker associated with salt tolerance in soybean genotypes under salt stress. *Appl Biochem Biotechnol* 170: 257–272.

Kim MY, Lee S, Van K, *et al.* (2010) Whole-genome sequencing and intensive analysis of the undomesticated soybean (*Glycine soja* Sieb. and Zucc.) genome. *Proc Natl Acad Sci USA* 107: 22032–22037.

Komatsu S, Hossain Z (2013) Organ-specific proteome analysis for identification of abiotic stress response mechanism in crop. *Front Plant Sci* 4: 1–9.

Kosva K, Prasil IT, Vitamvas P (2013) Protein contribution to plant salinity response and tolerance acquisition. *Int J Mol Sciences* 14: 6757–6789.

Kukreja S, Nandwal AS, Kumar N, Sharma SK, Unvi V, Sharma PK (2005) Plant water status, H_2O_2 scavenging enzymes, ethylene evolution and membrane integrity of *Cicer arietinum* roots as affected by salinity. *Biol Plant* 49: 305–308.

Kundzewicz Z, Ulbrich U, Brucher T, *et al.* (2005) Summer floods in central Europe – Climate change track? *Nat Hazar* 36: 165–189.

Langridge P, Fleury D (2011) Making the most of 'omics' for crop breeding. *Trends Biotechnol* 29: 33–40.

L'taief B, Sifi B, Zaman-Allah M, Drevon JJ, Lachaal M (2007) Effect of salinity on root-nodule conductance to the oxygen diffusion in the *Cicer arietinum-Mesorhizobium ciceri* symbiosis. *J Plant Physiol* 164: 1028–1036.

Lee JD, Shannon JG, Vuong TD, Nguyen HT (2009) Inheritance of salt tolerance in wild soybean (*Glycine soja* Sieb. and Zucc.). *J Hered* 100: 798–801.

Leonforte A, Forster JW, Redden RJ, Nicolas ME, Salisbury PA (2013) Sources of high tolerance to salinity in pea (*Pisum sativum* L.). *Euphytica* 189: 203–216.

Liu T, van Staden J (2000) Selection and characterization of sodium chloride-tolerant callus of *Glycine max* (L.) Merr cv. Acme. *Plant Growth Regul* 31: 195–207.

Luo Q, Yu B, Liu Y (2005) Differential sensitivity to chloride and sodium ions in seedlings of *Glycine max* and *G. soja* under NaCl stress. *J Plant Physiol* 162: 1003–1012.

Ma H, Song L, Shu Y, *et al.* (2012) Comparative proteomic analysis of seedling leaves of different salt tolerant soybean genotypes. *J Proteomics* 75: 1529–1546.

Maas EV, Hoffman GJ (1977) Crop salt tolerance – current assessment. *J Irr Drain Div* 103: 115–134.

Mahajan S, Tuteja N (2005) Cold, salinity and drought stresses: an overview. *Arch Biochem Biophys* 444: 139–158.

Malencic DJ, Popovic M, Miladinovic J (2003) Stress tolerance parameters in different genotypes of soybean. *Biol Plant* 46: 67–72.

Mamo T, Richter C, Heiligtag B (1996) Salinity effects on the growth and ion contents of some chickpea (*Cicer arietinum* L.) and lentil (*Lens culinaris* Medic) varieties. *J Agron Crop Sci* 176: 235–247.

Manchanda G, Garg N (2008) Salinity and its effects on the functional biology of legumes. *Acta Physiol Plant* 30: 595–618.

Mantri NL, Ford F, Coram TE, Pang ECK (2007) Transcriptional profiling of chickpea genes differentially regulated in response to high-salinity, cold and drought. *BMC Genomics* 8: 303.

Molina C, Zaman-Allah M, Khan F, *et al.* (2011) The salt-responsive transcriptome of chickpea roots and nodules via deep Super SAGE. *BMC Plant Biol* 14:, 11– 31.

Moses FA, McNeil MD, James BR, Kollmorgen F, Pittock C (2008) Sampling strategies and screening of chickpea (*Cicer arietinum* L.) germplasm for salt tolerance. *Genet Resour Crop Evol* 55: 53–63.

Munns R (2002) Comparative physiology of salt and water stress. *Plant Cell Environ* 25: 239–250.

Munns RA (2005) Gene and salt tolerance: bringing them together. *New Phytol* 167: 645–663.

Munns R, Tester M. (2008) Mechanisms of salinity tolerance. *Ann Rev Plant Biol* 59: 651–681.

Murillo-Amador B, Troyo-Diéguez E, García-Hernández JL, *et al.* (2006) Effect of NaCl salinity in the genotypic variation of cowpea (*Vigna unguiculata*) during early vegetative growth. *Sci Hortic-Amsterdam* 108: 423–431.

Nandwal AS, Godara M, Kamboj DV, Kundu BS, Mann A, Kumar B (2000) Nodule functioning in trifoliate and penta foliate mung bean genotypes as influenced by salinity. *Biol Plant* 43: 459–462.

Nat KNV, Srivastava S, Goonewardene L, Blade SF (2004) Proteome-level changes in the roots of *Pisum sativum* in response to salinity. *Ann Appl Biol* 145: 217–230.

Nazar R, Iqbal N, Syeed S, Khan NA (2011) Salicylic acid alleviates decreases in photosynthesis under salt stress by enhancing nitrogen and sulfur assimilation and antioxidant metabolism differentially in two mungbean cultivars. *J Plant Physiol* 168: 807–815.

Neil CT, Turner TD, Colmer JQ, *et al.* (2013) Salinity tolerance and ion accumulation in chickpea (*Cicer arietinum* L.) subjected to salt stress. *Plant Soil* 365: 347–361.

Noreen Z, Ashraf M (2009) Assessment of variation in antioxidative defense system in salt-treated pea (*Pisum sativum*) cultivars and its putative use as salinity tolerance markers. *J Plant Physiol* 166: 1764–1774.

Okcu G, Kaya MD, Atak M (2005) Effects of salt and drought stresses on germination and seedling growth of pea (*Pisum sativum* L). *Turk J Agric* 29: 237–242.

Olmos E, Hellin E (1996) Mechanisms of salt tolerance in a cell line of *Pisum sativum*: biochemical and physiological aspects. *Plant Sci* 120: 37–45.

Ozcan H, Turan MA, Koc O, Cibili Y, Taban S (2000) Growth and variations in proline, sodium, chloride, phosphorus and potassium concentrations of chickpea (*Cicer arietinum* L.) cultivars under salt stress. *Turk J Agric For* 24: 649–654.

Panda SK (2001) Oxidative response of green gram under salinity stress. *Indian J Plant Physiol* 6: 438–440.

Parida AK, Das AB (2005) Salt tolerance and salinity effects on plants: a review. *Ecotox Environ* Safe 60: 324–349.

Phang TH, Shao G, Lam HM (2008) Salt tolerance in soybean. *J Integr Plant Biol* 50: 1196–1212.

Prakash SN, Sarin NB (1993) Isolation of salt tolerant cell lines of *Cajanus cajan* and comparative study of their physiological responses with intact plants under salt stress. *Plant Physiol Biochem* 20: 11–15.

Rai R, Singh RP (1999) Effect of salt stress on interaction between lentil (*Lens culinaris*) genotypes and *Rhizobium* spp. strains: symbiotic N_2 fixation in normal and sodic soils. *Biol Fertil Soils* 29: 187–195.

Raman SV, Iniyan S, Goic R (2012) A review of climate change, mitigation and adaptation. *Renew Sust Energ Rev* 16: 878–897.

Rao DLN, Giller KE, Yeo AR, Flowers TJ (2002) The effects of salinity and sodicity upon nodulation and nitrogen fixation in chickpea (*Cicer arietinum*). *Ann Bot* 89: 563–570.

Raven PH, Pothill RM (1981) Biogeography of the Leguminose. In: Pothill RM, Raven PH (eds), *Advances in Legume Systematics*. Royal Botanical Gardens, Kew, pp. 27–34.

Rengasamy P (2006) World salinization with emphasis on Australia. *J Exp Bot* 57: 1017–1023.

Rengasamy P, Chittleborough D, Helyar K (2003) Root zone constraints and plant-based solutions for dry land salinity. *Plant Soil* 257: 249–260.

Richards LA (1954) Diagnosis and improvement of saline and alkali soils. USDA Agric. Handbook 60. U.S. Government Printing Office, Washington, DC.

Richter C, Heiligtag B, Nagieb M, Mukhallablati U (1999) The effects of salinity on growth and ion content of chickpea (*Cicer arietinum* L.), egg fruit (*Solanum melongena* L.) and red pepper (*Capsicum annuum* L.) varieties. *Tropenlandwirt* 100: 3–16.

Rossignol M, Peltier JB, Mock HP, Matros A, Maldonado AM, Jorrin JV (2006) Plant proteome analysis: a 2004–2006 update. *Proteomics* 6: 5529–5548.

Rozema J, Flowers T (2008) Crops for a salinized world. *Science* 322: 1478–1480.

Saadallah K, Drevon J, Mokhtar Abdelly C (2001) Genotypic variability for tolerance to salinity of N_2-fixing common bean (*Phaseolus vulgaris*). *Agronomie* 21: 675–682.

Sadiki M, Rabih K (2001) Selection of chickpea (*Cicer arietinum*) for yield and symbiotic nitrogen fixation ability under salt stress. *Agronomie* 21: 659–666.

Saha P, Chatterjee P, Biswas AK (2010) NaCl pretreatment alleviates salt stress by enhancement of antioxidant defense system and osmolyte accumulation in mungbean (*Vigna radiata* L. Wilczek). *Indian J Exp Biol* 48: 593–600.

Saha P, Kunda P, Biswas AK (2012) Influence of sodium chloride on the regulation of Krebs cycle intermediates and enzymes of respiratory chain in mungbean (*Vigna radiata* L. Wilczek) seedlings. *Plant Physiol Biochem* 60: 214–222.

Samineni S, Siddique KHM, Gaur PM, Colmer TD (2011) Salt sensitivity of the vegetative and reproductive stages in chickpea (*Cicer arietinum* L.): Podding is a particularly sensitive stage. *Environ Exp Bot* 71: 260–268.

Sanginga N, Dashiell K, Diels J, *et al.* (2003) Sustainable resource management coupled to resilient germplasm to provide new intensive cereal-grain legume-livestock systems in the dry savanna. *Agric Ecosyst Environ* 100: 305–314.

Sato S, Nakamura Y, Kaneko T, *et al.* (2008) Genome structure of the legume *Lotus japonicus*. *DNA Res* 15: 227–239.

Schmutz J, Cannon SB, Schleuter J, *et al.* (2010) Genome sequence of the palaeopolyploid soybean. *Nature* 463: 178–183.

Sekeroglu N, Kara MS, Dede O, Askin T (1999) Effect of salinity on germination, early seedling growth, Na and K constituents of chickpea. *Turk J Field Crops* 4: 79–84.

Serraj R, Roy G, Drevon JJ (1994) Salt stress induces a decrease in the oxygen uptake of soybean nodules and in their permeability to oxygen diffusion. *Physiol Plant* 91: 161–168.

Serraj R, VasquezDiaz H, Hernandez G, Drevon JJ (2001) Genotypic difference in the short-term response of nitrogenase activity (C_2H_2 reduction) to salinity and oxygen in the common bean. *Agronomie* 21: 645–651.

Shalhevet J, Huck MG, Schroeder BP (1995) Root and shoot growth responses to salinity in maize and soybean. *Agron J* 87: 512–516.

Shamseldin A, Werner D (2005) High salt and high pH tolerance of new isolated *Rhizobium* etli strains from Egyptian soils. *Curr Microbiol* 50: 11–16.

Sharma KD, Datta KS, Varma SK (1990) Effect of chloride and sulphate salinity on nodulation and nitrogen fixation in chickpea. *Ann Biol* 6: 13–20.

Sheokand S, Dhandi S, Swaraj K (1995) Studies on nodule functioning and hydrogen peroxide scavenging enzymes under salt stress in chickpea nodules. *Plant Physiol Biochem* 33: 561–566.

Shiva Prakash N, Sarin NB (1993) Isolation of salt tolerant cell lines of *Cajanus cajan* and comparative study of their physiological responses with intact plants under salt stress. *Plant Physiol Biochem* 20: 11–15.

Singh A, Kumari R, Garg AP, Varma A (2005) Endomycorrhizal diversity. In: Podila GK, Varma A (eds), *Basic Research and Applications of Mycorrhizae*. IK International, New Delhi, pp. 1–23.

Singh AK, Singh RA (1999) Effect of salinity on sodium, potassium and proline content of chickpea genotypes. *Indian J Plant Physiol* 4: 111–113.

Singh AK, Singh RA, Sharma SG (2001) Salt stress induced changes in certain organic metabolites during seedling growth of chickpea. *Legume Res* 24: 11–15.

Singh BB, Ajeigbe HA, Tarawali SA, Fernandez-Rivera S, Abubakar M (2003) Improving the production and utilization of cowpea as food and as fodder. *Field Crops Res* 84: 169–177.

Singla R, Garg N (2005) Influence of salinity on growth and yield attributes in chickpea cultivars. *Turk J Agric For* 29: 231–235.

Sobhanian H, Motamed N, Jazii FR, Nakamura T, Komatsu S (2010a) Salt stress induced differential proteome and metabolome response in the shoots of *Aeluropus lagopoides* (Poaceae), a halophyte C_4 plant. *J Proteome Res* 9: 2882–2897.

Sobhanian H, Razavizadeh R, Nanjo Y, *et al.* (2010b) Proteome analysis of soybean leaves, hypocotyls and roots under salt stress. *Proteome Sci* 8: 19.

Sohrabi Y, Heidari G, Esmailpoor B (2008) Effect of salinity on growth and yield of Desi and Kabuli chickpea cultivars. *Pak J Biol Sci* 11: 664–667.

Song Y, Ji D, Li S, Wang P, Li Q, Xiang F (2012) The dynamic changes of DNA methylation and histone modifications of salt responsive transcription factor genes in soybean. *PLoS One* 7(7): e41274.

Sousa MF, Campos FAP, Prisco JT, Eneas-Filho J, Gomes-Filho E (2003) Growth and protein pattern in cowpea seedlings subjected to salinity. *Biol Plant* 47: 341–346.

Soussi M, Ocana A, Lluch C (1998) Effects of salt stress on growth, photosynthesis and nitrogen fixation in chick-pea (*Cicer arietinum* L.). *J Exp Bot* 49: 1329–1337.

Soussi M, Lluch C, Ocana A (1999) Comparative study of nitrogen fixation and carbon metabolism in two chickpea (*Cicer arietinum* L) cultivars under salt stress. *J Exp Bot* 50: 1701–1708.

Soussi M, Khadri M, Lluch C, Ocana A (2001) Carbon metabolism and bacteroid respiration in nodules of chick-pea (*Cicer arietinum* L.) plants grown under saline conditions. *Plant Biosyst* 135: 157–164.

Soussi M, Khadri M, Pliego L, Lluch C, Ocana A (2003) Germination of different cultivars of *Cicer arietinum* in saline conditions: profile of proteins and accumulation of compatible solutes. In: *Fixation symbiotique del'azote et development durable dans le Bassin méditerranéen*. INRA, Paris, pp. 321–330.

Steibel JP, Poletto R, Coussens PM, Rosa GJ (2009) A powerful and flexible linear mixed model framework for the analysis of relative quantification RT-PCR data. *Genomics* 94: 146–152.

Sullivan P (2003) Overview of cover crops and green manures. ATTRA. Available from: https://attra.ncat.org/attra-pub/summaries/summary.php?pub=288

Sun X, Sun M, Luo X, *et al.* (2013) A *Glycine soja* ABA-responsive receptor-like cytoplasmic kinase, GsRLCK, positively controls plant tolerance to salt and drought stresses. *Planta* 237: 1527–1545.

Sunil Kumar B, Prakash M, Narayanan S, Gokulakrishnan J (2012) Breeding for salinity tolerance in mungbean. *APCBEE Procedia* 4: 30–35.

Surekha Ch, Kumari KN, Aruna LV, Suneetha G, Arundhati A, Kavi Kishor PB (2014) Expression of the *Vigna aconitifolia P5CSF129A* gene in transgenic pigeon pea enhances proline accumulation and salt tolerance. *Plant Cell Tiss Org* 116: 27–36.

Swaraj K, Dhandi S, Sheokhand S, Sheoran IS (2000) Nodule functioning and recovery under salt stress in Pigeonpea (*Cajanus cajan* L. Millsp). *J Plant Biol* 27: 19–26.

Talsma T (1963) The control of saline groundwater. Doctorate thesis, University of Wageningen. *Reprint of Bulletin of University of Wageningen*, no. 63, pp. 1–68.

Tavakkoli E, Paull J, Rengasamy P, McDonald GK (2012) Comparing genotypic variation in faba bean (*Vicia faba* L.) in response to salinity in hydroponic and field experiments. *Field Crops Res* 127: 99–108.

Tejera NA, Campos R, Sanjuan J, Lluch C (2004) Nitrogenase and antioxidant enzyme activities in *Phaseolus vulgaris* nodules formed by *Rhizobium tropici* isogenic strains with varying tolerance to salt stress. *J Plant Physiol* 161: 329–338.

Tejera NA, Campos R, Sanjuan J, Lluch C (2005) Effect of sodium chloride on growth, nutrient accumulation, and nitrogen fixation of common bean plants in symbiosis with isogenic strains. *J Plant Nutr* 28: 1907–1921.

Tramontano WA, Jouve D (1997) Trigonelline accumulation in salt stressed legumes and the role of other osmoregulators as cell cycle control agents. *Phytochemistry* 44: 1037–1040.

Türkana I, Demiral T (2009) Recent developments in understanding salinity tolerance. *Environ Exp Bot* 67: 2–9.

Turner NC, Colmer TD, Quealy J, *et al.* (2013) Salinity tolerance and ion accumulation in chickpea (*Cicer arietinum* L.) subjected to salt stress. *Plant Soil* 365: 347–361.

Upadhyaya HD, Reddy LJ, Gowda CLL, Singh S (2009) Phenotypic diversity in cold-tolerant peanut (*Arachis hypogaea* L.) germplasm. *Euphytica* 165: 279–291.

Urano K, Kurihara Y, Seki M, Shinozaki K (2010) 'Omics' analyses of regulatory networks in plant abiotic stress responses. *Curr Opin Plant Biol* 13: 132–138.

Uyanöz R, Karaca U (2011) Effects of different salt concentrations and *Rhizobium* inoculation (native and *Rhizobium tropici* CIAT899) on growth of dry bean (*Phaseolus vulgaris* L.). *Eur J Soil Biol* 47: 387–391.

Vadez V, Krishnamurthy L, Serraj R, *et al.* (2007) Large variation in salinity tolerance in chickpea is explained by differences in sensitivity at the reproductive stage. *Field Crops Res* 104: 123–129.

Varshney RK, Hiremath PJ, Lekha P, *et al.* (2009) A comprehensive resource of drought and salinity responsive ESTs for gene discovery and marker development in chickpea (*Cicer arietinum* L.). *BMC Genomics* 10: 523–541.

Varshney RK, Chen W, Li Y, *et al.* (2012) Draft genome sequence of pigeon pea (*Cajanus cajan*), an orphan legume crop of resource-poor farmers. *Nat Biotechnol* 30: 83–89.

Varshney RK, Song C, Saxena RK, *et al.* (2013) Draft genome sequence of chickpea (*Cicer arietinum*) provides a resource for trait improvement. *Nat Biotechnol* 31: 240–246.

Verdoy D, Lucas MM, Manrique E, Covarrubias AA, De Felipe MR, Pueyo JJ (2004) Differential organ-specific response to salt stress and water deficit in nodulated bean (*Phaseolus vulgaris*). *Plant Cell Environ* 27: 757–767.

Wang XW, Vinocur B, Altman A (2003) Plant responses to drought, salinity and extreme temperatures: towards genetic engineering for stress tolerance. *Planta* 218: 1–14.

Welfare K, Yeo AR, Flowers TJ (2002) Effect of salinity and ozone, individually and in combination on the growth and ion contents of two chickpea (*Cicer arietinum* L.) varieties. *Environ Pollut* 120: 397–403.

Wilson C, Liu X, Lesch SM, Suarez DL (2006) Growth response of major USA cowpea cultivars: II. Effect of salinity on leaf gas exchange. *Plant Sci* 170: 1095–1101.

Wojtyla L, Kosmala A, Garnczarska M (2013) Lupine embryo axes under salinity stress. II. Mitochondrial proteome response. *Acta* Physiol Plant 35: 2383–2392.

Wu W, Zhang Q, Zhu Y, Lam HM, Cai Z, Guo D (2008) Comparative metabolic profiling reveals secondary metabolites correlated with soybean salt tolerance. *J Agric Food Chem* 56: 11132–11138.

WWF (2008) *The 2008 Living Planet Report.* World Wild Fund for Nature, Gland, Switzerland.

Xu XY, Fan R, Zheng R, Li CH, Yu DY (2011) Proteomic analysis of seed germination under salt-stress in soybeans. *J Zhejiang Univ Sci B* 12: 507–517.

Yang H, Tao Y, Zheng Z, et al. (2013) Draft genome sequence, and a sequence-defined genetic linkage map of the legume crop species *Lupinus angustifolius* L. *PLoS ONE* 8(5): e64799.

Young ND, Debellée F, Oldroyd GED, *et al.* (2011) The Medicago genome provides insight into the evolution of rhizobial symbioses. *Nature* 480: 520–524.

Yu Q, Rengel Z (1999) Drought and salinity differentially influence activities of superoxide dismutases in narrow-leafed lupins. *Plant Sci* 142: 1–11.

Zahran HH (1999) Rhizobium-legume symbiosis and nitrogen fixation under severe conditions and in an arid climate. *Microbiol Mol Biol Rev* 63: 968–989.

Zhang JL, Shi H (2013) Physiological and molecular mechanisms of plant salt tolerance. *Photosynth Res* 115: 1–22.

Zhao Q, Zhang H, Wang T, Chen S, Dai S (2013) Proteomics based investigation of salt responsive mechanisms in plant roots. *J Proteomics* 82: 230–253.

Zhu D, Cai H, Luo X, *et al.* (2012) Over-expression of a novel JAZ family gene from *Glycine soja*, increases salt and alkali stress tolerance. *Biochem Biophys Res Commun* 426: 273–279.

Zhu JK (2001) Plant salt tolerance. *Trends Plant Sci* 6: 66–71.

Zhu JK (2002) Salt and drought stress signal transduction in plants. *Annu Rev Plant Biol* 53: 247–73.

Zurayk R, Adlan M, Baalbaki R, Saxena MC (1998) Interactive effects of salinity and biological nitrogen fixation on chickpea (*Cicer arietinum* L.) growth. *J Agron Crop Sci* 180: 249–258.

CHAPTER 3

Nutrient deficiencies under stress in legumes: An overview

Fakiha Afzal[1], Tahira Khan[1], Ammarah Khan[1], Saifullah Khan[1], Hammad Raza[1], Ayesha Ihsan[1], Mohammad Abass Ahanger[2] and Alvina Gul Kazi[1]

[1] *Atta-ur-Rahman School of Applied Biosciences, National University of Sciences and Technology (NUST), Islamabad, Pakistan*

[2] *School of Studies in Botany, Jiwaji University, Gwalior, India*

3.1 Introduction

Legumes are a diverse and biologically significant group of plants belonging to the family Fabaceae encompassing about 18,000 to 19,000 species in about 650–750 genera. They are second only to the Gramineae in their value to humans (Bouchenak & Lamri-Senhadji, 2013). Various legumes such as alfalfa, peanuts, peas, green beans, soybeans, dry beans, dry peas, chickpeas, broad beans, lentils and many others represent an important component of human diets in several areas of the world and are a principal source of protein for many. In addition, because of their nitrogen-fixing capability in association with soil bacteria they add to soil fertility and hence help crop growth as well as yield. Many pharmacological products are derived from legumes. The presence of many bioactive compounds including saponins, oligosaccharides, lectins, phytoestrogens and phenolic compounds which have significant physiological and metabolic roles in humans, is well documented (Vavilov, 1995; Postgate, 1998).

Proper mineral supplementation is one of the essential and basic strategies for ensuring proper growth, development and yield in crop plants. Nutritional deficiency in plants leads to a number of physiological, biochemical and molecular changes such as stunted growth, necrosis, synthesis and accumulation of protective compounds, and changes in expression of genes responsible for crop performance (Ahanger *et al.*, 2014).

In the continuously varying environment, plants are exposed to numerous environmental fluctuations that alter every physiological process and biochemical pathway. 'Stress' refers to any environmental condition that affects normal plant growth, metabolism and development. Stress factors may be from natural sources such as temperature, high irradiance (photo-inhibition, photo-oxidation), osmotic imbalance, flooding, mineral deficiency or pest attack, or be anthropogenic (i.e., due to human activities). Plants need to cope with these stresses in order to ensure their survival. Global environmental changes are causing a rise in temperature that will lead to floods, melting ice caps, and variability and unpredictability of rainfall and drought. Surface water scarcity will lead to the utility of groundwater that is not safe. This will contribute towards the increase in salinity of soil and will have a negative impact on crops. In this regard, legumes will be affected the most because of their sensitivity.

The minerals needed by plants are categorized into two groups on the basis of their quantitative requirements. Macronutrients include six elements (N, P, K, S, Mg and Ca) and are found as major components of the plant tissues (George *et al.*, 2008). Micronutrients usually work as regulators of different metabolic processes and the enzymes involved in efficient running of these processes. Many different factors like temperature, water availability, pH and mineral deficiency can affect the nitrogen-fixing ability of the legumes.

Legumes under Environmental Stress: Yield, Improvement and Adaptations, First Edition. Edited by Mohamed Mahgoub Azooz and Parvaiz Ahmad.

3.2 Environmental stresses and crop growth

As already pointed out, both natural and anthropogenically derived stress factors occur in nature. Plants have to cope with a plethora of potentially unfavourable conditions. Nevertheless there are various indigenous mechanisms triggered in plants when exposed to environmental extremes. Nutrient deficiency in legumes under drought and saline conditions affects growth of the plants, altering their physiology including nitrogen fixation, and extreme deficiencies can cause the death of plants. Generalized mechanism of stress is elaborated in figure 3.1.

The concept of stress was developed by Hans Selye in 1936; it refers to the unfavourable environmental constraints experienced by plants. Plants are bound to their environment and unlike animals they cannot run away from environmental and anthropogenic stressors. Stress avoidance and stress tolerance are the two broad categories of avoiding/mitigating mechanisms. Plants lack very specialized responses to environmental stressors but there are a few general responses, which include a high or low rate of growth or adaptation. The rapid responses of plants and the changes in their metabolic pathways should not be regarded as responses to environmental stressors (Figure 3.2). Both biotic and abiotic stresses do affect growth and development of plants. Frequently occurring stresses in nature are noted below.

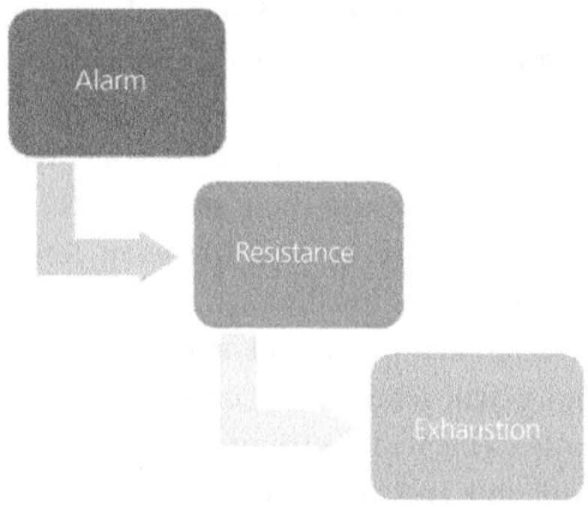

Figure 3.1 Mechanism of stress.

3.2.1 Drought and salt stresses

Water scarcity is a severe environmental constraint to crop production (Masoumi *et al.*, 2010). Drought stress induces several changes, such as reductions in leaf area, stem extension, root proliferation, and water use efficiency. Worldwide, 20% of irrigated land is salt affected. High salinity has a negative impact on the growth of the plant, and in sensitive crops generally can lead to plant death. It disrupts the ionic and osmotic balance of the cell.

3.2.2 Heavy metal stress

Heavy metal toxicity is one of the major stresses experienced by plants. The high reactivity of metals can influence plant growth, senescence and energy production. If large amounts of metals are accumulated in plants, they adversely affect the absorption and

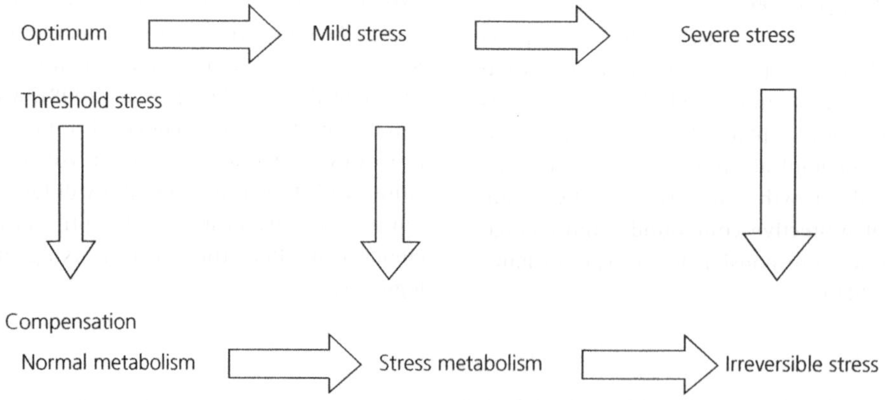

Figure 3.2 Stressor dose and stress effect relationship in plants.

transport of essential elements, disturb metabolism, and have an impact on growth and reproduction (Xu & Shi, 2000). The heavy metals accumulated in plants include lead (Pb), cadmium (Cd), aluminium (Al), selenium (Se), arsenic (As), mercury (Hg), zinc (Zn) and chromium (Cr).

3.2.3 Heat and cold stresses

Temperature is one of the major determinants of plant growth and development. Low temperature can be regarded as one of the most harmful stresses affecting temperate plants. Heat stress can affect the expression of certain genes by causing denaturation of certain vital plant proteins. A prominent effect of high temperature-induced protein denaturation is loss of membrane integrity, which affects cellular structure and transport as well.

3.2.4 Oxidative stress

Environmental stresses lead to the production of highly reactive substances called reactive oxygen species (ROS), which interact with a number of cell molecules and metabolites causing cell damage. Plants possess both enzymatic as well as non-enzymatic antioxidant defense systems, which counteract the effect of ROS by scavenging them and protecting the photosynthetic and cellular functions (Nobuhiro & Mittler, 2006; Jaleel *et al.*, 2008a, 2008b; Ahmad *et al.*, 2010). Overproduction of ROS results in oxidative stress (Ahmad *et al.*, 2010).

Whatsoever the type of stress is, plant follow similar phases of stress and their consequences on the plant (figure 3.3). Plants require optimal conditions for optimal growth and development. Light absorbed by plant pigments is used in photosynthesis or dissipated as heat. White light or laser-induced chlorophyll fluorescence of pre-darkened leaves is a very suitable tool for early stress detection in plants. The fluorescence of plants consists almost exclusively of the fluorescence of leaves, which account for the largest above-ground surface area of plants. A small part of the light energy absorbed in photosynthesis is lost during the migration from the antenna pigment to the reaction centres and is dissipated by a number of non-photochemical processes including heat and re-emission of small but easily detectable amounts (2–5% *in vivo*) of the absorbed radiation. This re-emission occurs at longer wavelengths in the red and far-red spectral regions, and is termed chlorophyll fluorescence (ChlF) (Bradbury & Baker, 1981). The emitted light when observed through the spectrophotometer gives the emission spectrum (Figure 3.4).

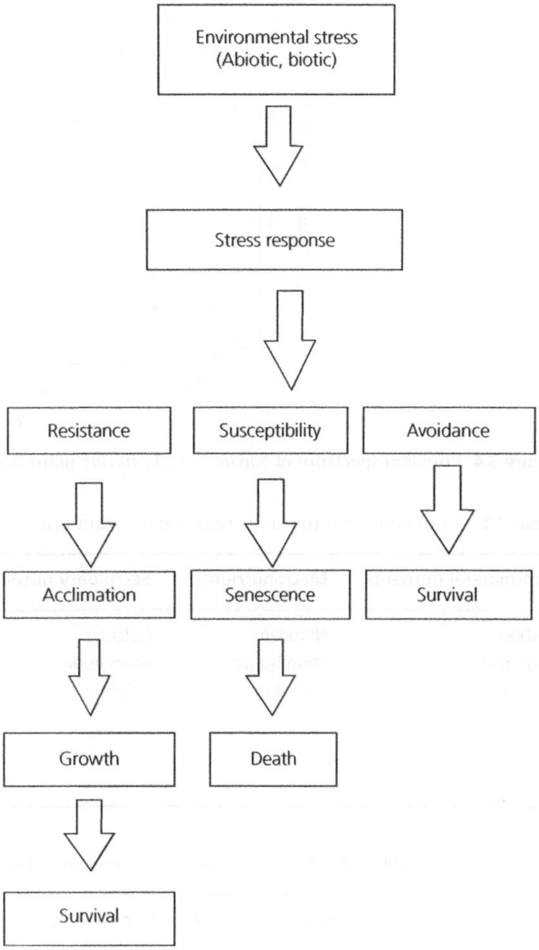

Figure 3.3 Phases of stress and their consequences on the plant.

The spectral reflectance of plants may indicate the degree of stress imposed by various stressors. The leaf spectral responses to environmental conditions that inhibit plant growth generally involve an increased amount of reflectance in the visible or the infrared spectra. As a result of decreased absorption of light by the pigments, reflectance at the visible wavelength is seen to increase in stressed plants compared to unstressed plants (Carter, 1993).

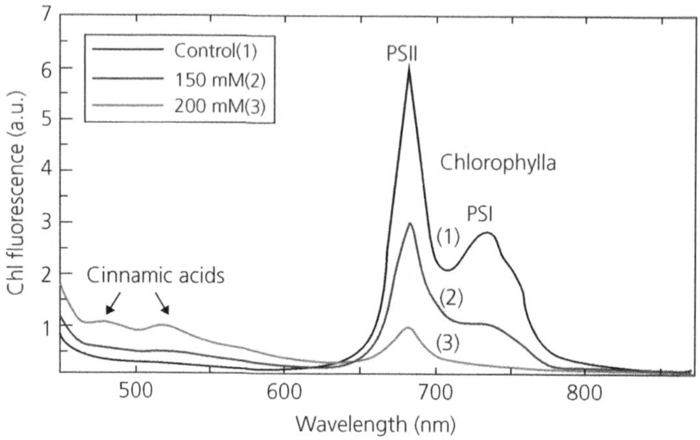

Figure 3.4 Emission spectrum of *Saccharum officinarum* plant under intense salinity.

Table 3.1 List of important nutrients required for plant growth.

Non-mineral nutrients	Macronutrients	Secondary nutrients	Micronutrients	Essential micronutrients for some plants
Carbon	Nitrogen	Calcium	Boron	Sodium
Hydrogen	Phosphorus	Magnesium	Chloride	Nickel
Oxygen	Potassium	Sulphur	Copper	Cobalt
			Iron	Vanadium
			Manganese	Silicon
			Molybdenum	

Table 3.2 Effects of nutrient deficiency on plant development.

Deficient nutrient	Effect on development
Nitrogen	Reduces growth and branching in legumes
Phosphorus	Stunted growth of plant, abnormal dark-green leaves. Also affects the metabolism, germination of seeds and ripening of fruits
Potassium	Legume stands degenerate to weeds and grass
Calcium	Reduces root and plant development
Magnesium	Reduces plant growth
Sulphur	Reduces protein production and nodule formation
Boron	Reduces nitrogen fixation, growth, nodule formation

3.3 Effects of nutrient deficiency

Important nutrients have been classified on the basis of plants' requirements (Table 3.1). Nutrient deficiencies in plants can lead to marked effects on growth and development, and can lead to death (Table 3.2). Macronutrient deficiencies can be observed easily but the micronutrient deficiency symptoms are more challenging to observe and identify.

3.4 Methods to control nutrient deficiency

The principal way to control nutrient deficiency is to first diagnose the plant with the particular deficiency. For this, soil analysis has to be carried out. Using the data from this analysis, the appropriate remedy can be applied to provide the plant with the particular nutrient appropriate to the plant's needs in a timely fashion.

3.5 Micronutrient deficiency in plants

3.5.1 Boron

Boron is one of the essential micronutrients required for proper working of the symbiotic association in legumes (Yamagishi & Yamamoto 1994). Boron plays an important role in maintaining the structural integrity of cell walls and cell membranes (Marschner, 1995; Bolanos *et al.*, 1996). Its deficiency also affects the activity of indole acetic acid oxidase, sugar translocation and nucleic acids resulting in stunted growth.

Boron plays a role in the synthesis of the polysaccharide capsule required for the recognition of bacteria by the legume. When boron is absent, a mutant polysaccharide capsule is formed, which is not properly recognized by the root and hence the plant defense mechanism becomes activated (Abreu *et al.*, 2012). Inside the infection thread, boron controls the interaction between a matrix glycoprotein called arabinogalactan protein-like extension (AGPE) and RG-2 (rhamnogalacturonan 2). This binding of AGPE and RG-2 keeps the bacteria entrapped within the thread and also promotes the apical growth of the infection thread (Reguera *et al.*, 2010). During nodule organogenesis, boron plays an important role in signalling (Reguera *et al.*, 2009). No enzyme discovered so far has been found to use boron as a cofactor, but still the deficiency of boron has a negative effect on the formation and growth of the nodule.

3.5.2 Molybdenum

In legumes, molybdenum is required for the synthesis of proteins associated with nitrogen metabolism (Wiedenhoeft, 2006). It acts as a cofactor for the nitrogenase enzymes (Hansch & Mendel, 2009). Usually soils having very low pH are deficient in molybdenum and therefore have reduced nitrogen-fixing activity. Mo occupies the central position in the cofactor FeMoCo, which is involved in a number of oxidation-reduction processes during nitrogen fixation. Any fluctuation in Mo concentration can affect the activity of the enzyme complex and hence nitrogen fixation (Shah *et al.*, 1999), and upon addition of Mo, nitrogen fixation is significantly enhanced (Delgado *et al.*, 2006).

3.5.3 Cobalt

Cobalt is the cofactor of vitamin B_6 (cobalamine), which acts as a coenzyme in nitrogen fixation and nodule growth (Jordan & Richard, 1998). Proper functioning of the bacterial enzyme complexes also requires cobalt. Deficiency badly affects nodule formation and hence nitrogen fixation (Dilworth *et al.*, 1979). Cobalt is also required for the synthesis of vitamin B_{12}, which is needed for fatty acid metabolism, for the production of haem group cytochromes and for the leghaemoglobin that acts with the help of methylmalonyl coenzyme A mutase. Rhizobia use cobalt for functioning of two enzymes, homocysteine methyltransferase, required for the formation of methionine, and ribonucleotide reductase, which is involved in the catalysis of deoxyribonucleotide formation (Licht *et al.*, 1996). Experiments show that elevated concentrations of cobalt (usually >30 µg/mL) play a very important role in protecting legumes under drought stress by inhibiting the production of ethylene, which is a stress hormone. Cobalt has the ability to reduce polyamine content as well as antioxidamt enzyme activity. This, in turn, increases the ROS level and also increases cell membrane damage by inhibiting ethylene. Higher concentrations of cobalt (>50 mg/kg of soil) caused decreases in antioxidant enzyme activity, reducing sugars, starch, free amino acids and protein content in *Vigna radiata* (Jayakumar *et al.*, 2008). However, Gad (2012) reported that a lower concentration (8 ppm) caused a significant increase in height, number and area of leaves, nitrogenase activity, number of nodules, and nitrogen, phosphorus and potassium content in *Arachis hypogaea*.

3.5.4 Copper

Wilting and the appearance of grey patches in leaves are the major signs of copper deficiency in plants. Copper is the major component of the respiratory proteins required for nitrogen fixation by rhizobia (Delgado *et al.*, 1998). Copper is also required for a particular protein that works in co-operation with nitrogen-fixing genes of rhizobia (Snowball *et al.*, 1980). Because of its role as a cofactor for cytochrome *c* oxidase, it is important in the energetic processes occurring in the cells (Delgado *et al.*, 1998). Copper is also important for the metabolism of carbon and nitrogen, formation of cell walls and the detoxification of ROS. At times of stress, the level of ROS is quite high. The copper chaperone, a metalloprotein, delivers copper ions to the Cu-superoxide dismutase (CuSOD) and hence mediates detoxification of ROS. Copper also activates the enzyme polyphenol oxidase (PPO), hence inhibiting membrane lipid peroxidation. Copper along with molybdenum and a petrin molecule plays an important role in the synthesis of intermediate complexes that help in transporting materials across membranes in legumes (Hansch & Mendel, 2009).

3.5.5 Zinc

Plants require zinc for various physiological processes and its deficiency can lead to severe problems (Weisany *et al.*, 2011). It plays an important role in transcriptional regulation of the Ros-type regulator MucR, which is necessary for establishing and maintaining cell envelope integrity, polysaccharide biosynthesis, iron homeostasis, genome plasticity and transcriptional regulation in many legumes (Caswell *et al.*, 2013). Zinc also acts as a cofactor for different enzymes, including metalloproteases, aminopeptidases, nucleases and CuZn-SOD (Chauhan & O'Brian, 1995; Hansch & Mendel, 2009). Zinc is believed to bring changes in both host and rhizobia at genetic levels and has no direct role in nodule development. In protein–protein interactions it is a constituent of zinc finger motifs.

3.5.6 Iron

Iron is required for nitrogenase and hydrogenase enzyme complexes, and is one of the key components of leghaemoglobin, the most abundant protein in the nodule. Leghaemoglobin maintains the anaerobic environment inside the root nodule so that the activity of nitrogenase is maintained. Iron deficiency hinders nitrogenase activity and hence nitrogen fixation. Besides this, iron is also an important component of the electron carrier ferredoxin, and hence important in energy transfer pathways. It has been shown that nitrogenase and leghaemoglobin constitute about 30% of the total protein content of root nodules, and both these enzymes contain iron as an essential component (O'Hara *et al.*, 1988; Bonilla *et al.*, 2010). As mentioned above, iron is a cofactor of many enzymes of both bacteria and plants that play a vital role in nitrogen fixation (Tang *et al.*, 1990). Nitrogenase, nitrate reductases and the enzymes involved in biosynthesis of nucleic acids and detoxification of ROS also require iron (O'Hara *et al.*, 1988).

Some iron (Fe^{2+}) must be present in the vicinity of root nodules so that the enzyme iron-chelate reductase can perform chelation of iron from soil to either host plant cell or to bacteria. This process is a reduction reaction in which Fe^{2+} is converted to Fe^{3+} with the help of NAD^{+} or $NADP^{+}$ in order to maintain the optimum pH necessary for chelation (Slatni *et al.*, 2012).

3.5.7 Manganese

Manganese has an important role in the detoxification of active oxygen species. It also is important in the formation of polyamines necessary for plant growth and development (Evans & Malmberg, 1989). This element is best known for being a cofactor of superoxide dismutase, whose function is to scavenge superoxide radicals. This scavenging action is necessary to guarantee the stability of the symbiotic relation between the plant and bacteria during stress. Maganese is also responsible for activation of enzymes of the shikimic acid cycle and of the Krebs cycle (Hansch & Mendel, 2009). Signalling events between the bacteria and host plant also involve manganese through its role in maintaining levels of carboxylic acids and flavonoids by controlling Krebs cycle enzymes. Manganese is also a cofactor in the water-splitting reactions of photosystem II of plant cells.

3.5.8 Nickel

Nickel is required for various metabolic processes like enzyme activities, maintaining a proper redox state of cells and various other biochemical, physiological and growth responses, intensive industrial and agricultural activities have often raised its levels in soils, resulting in toxicity (Patrick *et al.*, 1987; Yusuf *et al.*, 2011; Hussain *et al.*, 2013) and hence provoking oxidative stress (Rao & Sresty, 2000; Gonnelli *et al.*, 2001; Boominathan & Doran, 2002; Gajewska *et al.*, 2006). Increased nickel concentrations cause chlorosis, inhibition of root and shoot growth, photosynthesis, protein synthesis and enzyme activity, and alteration of membrane integrity in plants (Hussain *et al.*, 2013). Plants grown in soil containing an optimum amount of nickel show increased weight of the root nodules and enhanced yield. At optimum concentrations, nickel regulates hydrogenase activity, and maximal efficiency of this enzyme results in enhanced nitrogen-fixing activity of rhizobia. Nickel has been found to have a link with plant urease and Ni-metallochaperones. Plants use ureides as nitrogen-transporting molecules, which play a significant role in nodule formation (Dalton *et al.*, 1985). In some rhizobia, the enzyme NiFe-dehydrogenase oxidizes hydrogen gas to obtain energy. Nickel constitutes an important component of NiFe-nitrogenase, thereby having an active role in nitrogen fixation. Hence for the proper functioning of plant systems an optimum concentration of nickel is important.

3.6 Roles of macronutrients in growth of legumes

Every nutrient is of biochemical and physiological importance and has a minimum as well a maximum concentration for plant homeostatic function. The macronutrients are also called 'primary nutrients', except for sulphur, which is called a 'secondary nutrient' because it is needed in lower concentrations than the others. It has been observed by Sanchez (2002) that annual nutrient depletion rates across 37 African countries are 22 kg N/ha, 2.5 kg P/ha and 15 kg K/ha. These nutrients are essential for plants and their deficiency, or complete absence may result in retarded growth and hence reduced yield.

The presence of appropriate concentrations of macronutrients is important in various cycles, such as the nitrogen, phosphorus and potassium cycles. Nitrogen, phosphorus and potassium (NPK) are the principal macronutrients needed for plant growth; nevertheless others also make significant contributions to growth (Ahanger *et al.*, 2014). Mineral nutrients may also influence N_2 fixation in legumes and non-legumes at various stages of the symbiotic process, including infection, nodule development, nodule function and host plant growth (Weria *et al.*, 2013). These macronutrients also regulate the amount of storage proteins in legumes.

3.6.1 Role of carbon

Carbon is an essential element in plant growth. Dry matter of a typical green plant contains a large (35–50%) and relatively constant proportion of carbon. Approximately 40 million tons of nitrogen is injected symbiotically into the soil each year by rhizobia in legumes mainly due to exchange of nitrogen from bacteria for reduced carbon from the plants. Nitrogen fixation is a finely regulated process that is crucial in nutrient cycling and involves significant carbon allocations (Olivera *et al.*, 2004; Valentine *et al.*, 2011). Forests with nitrogen-fixing trees accumulate more carbon compared to non-nitrogen fixing trees (Sigrid *et al.*, 2002). Furthermore, the ability of legumes to exchange carbon for nitrogen with their symbiotic partner gives them a competitive edge over non-legumes. Therefore, raised concentrations of CO_2 will provide more photosynthetic carbon fixation, which can then be exchanged with nitrogen resulting in an increase in the rate of nitrogen fixation. The cost of symbiotic nitrogen fixation varies from 2 to 3 mg of carbon per mg of fixed N_2 (Valentine *et al.*, 2011).

3.6.2 Role of nitrogen

Nitrogen is one of the prime macronutrients, impacting nutrient uptake and N_2 fixation. It is involved in amino acid production, is a component of vitamins and affects energy reactions in plants. Nitrogen is present in very large quantities in the atmosphere as dinitrogen gas (N_2) However, this atmospheric nitrogen cannot be used directly by plants, so they have evolved several mechanisms to overcome this. The most important of these is symbiotic nitrogen fixation. Nodule formation and nitrogenase activity are both inhibited by the presence of mineral nitrogen (Vieira *et al.*, 1998), but the inhibition of symbiotic nitrogen fixation by mineral nitrogen in the soil depends upon the time at which the nodulation starts along with the concentration of nitrogen (Anne-Sophie *et al.*, 2002). Sarah *et al.* (2013) also supported the hypothesis that N_2 fixation is controlled by the availability of nitrogen. N_2 fixation therefore has a direct relationship with the presence or absence of mineral nitrogen in the soil and with its concentration.

3.6.3 Role of phosphorus

Phosphorus is also an important nutrient and nodules are great sinks of it (Hart, 1989). It is important in various biochemical and molecular processes (Epstein & Bloom, 2005). Many developmental processes depend upon phosphorus, including root growth, as well as ATP production. It is actively involved in photosynthesis and respiration, and is vital for seed formation. Phosphorus availability has an influence on N_2 fixation and nodule formation (Saxena & Rewari, 1991). A 35% reduction in the number of legumes in soybean plants was seen by Tsvetkova and Georgiev (2003) by over supplying phosphorus as compared to controls, while its deficiency caused a reduction in the size of nodules rather than number. Hence, phosphorus concentration is critical with respect to nitrogen fixation in legumes, and both higher and lower concentrations can reduce both nitrogen fixation (Weria *et al.*, 2013; Sarah *et al.*, 2013) and the biomass of the symbiotic partners (Sarah *et al.*, 2013). When P is deficient, N_2 fixation decreases as a result of adaptation to low N demand regulated by feedback mechanisms (Almeida *et al.*, 2000). So an optimum concentration of phosphorus is important, even crucial, for better utilization of other nutrients like nitrogen. Islam *et al.* (2013) further confirmed a 38% increase in nitrogen fixation by phosphorus application; however, this effect of phosphorus is indirect and is probably due to enhanced growth and production of dry matter. Both oversupply and undersupply of phosphorus can have a negative impact on plant growth, and the phosphorus concentration in soil must be maintained at a certain level, depending upon the plant species, to ensure proper functioning of the plant.

3.6.4 Role of sulphur

Sulphur, besides having important utility in many biological processes, is also a constituent of amino acids like cysteine and methionine, and of vitamins such as biotin or thiamine. Sulphur deficiency affects plant growth severely but when applied properly, it improves plant growth and seed yield (Hafeez ur Rehman *et al.*, 2013). Sulphur is important for the production of amino acids, seeds, enzymes and vitamins, and also promotes nodule development in legumes. Deficiency of sulphur can lead to reduced production of sulphur-containing amino acids (methionine and cysteine) and hence reduced nutritional value (Sexton *et al.*, 1998). In addition it forms a variety of organic compounds, modulation of which can provide tolerance against salinity. Sulphur deficiency reduces N_2 fixation in legumes by affecting nodule development and function (Scherer, 2008). Varin *et al.* (2009) demonstrated the effect of sulphur on growth of legumes in white clover (*Trifolium repens* L.) and reported that application of sulphur increased the whole-plant dry mass, nodule biomass and volume, nodule protein content and root amino acid concentrations. Plants supplied with little or no sulphur showed poor nodule growth and low nitrogenase activity and leghaemoglobin content, which ultimately reduced nitrogen content, confirming the interactive role of these mineral elements. Proper supplementation of sulphur enhances nitrogen fixation and hence nitrogen content (Fazli *et al.*, 2008). Islam *et al.* (2013) also showed that sulphur has a direct influence on nitrogen fixation, and that sulphur application caused a 33% increase in nitrogen fixation over controls. Therefore, sulphur has an important role to play in nitrogen fixation in root nodules of legumes, affecting the process of nitrogen fixation more or less directly.

3.6.5 Role of potassium

Potassium is the third major macroelement essential for plant growth and is involved in photosynthesis, protein synthesis and enzyme activation, increasing water uptake and improving the quality of seeds and fruits (Maathuis, 2009). Inadequate amounts of potassium result in poor root development, small seeds and slow growth. In alfalfa, Grewala and Williamsa (2002) reported that potassium application has a direct and vital impact on nodulation, yield and leaf-to-stem ratio. Yamada and colleagues (2002) showed the effect of potassium deficiency on two types of crops. In type 1 crops (wheat and soybean) potassium concentration decreases when nitrogen concentration decreases, whereas in type 2 crops (sunflower) potassium concentration remains constant or is independent of decreases in nitrogen concentration. They found that in Type 2 crops potassium deficiency decreased the photosynthetic rate and increased the photorespiratory rate and dark respiratory rate. On the other hand, in Type 1 crops the potassium deficiency had a less marked effect on CO_2 and consequently a less marked effect on the above-mentioned processes. Potassium therefore has an important role to play in regulating the process of nitrogen fixation and also in several other processes that are vital to the proper functioning and growth of plants.

3.7 Storage proteins in legumes and effect of nutritional deficiency

Legume seeds accumulate a large amount of protein within membrane-bound organelles, protein bodies or storage vacuoles. These are non-structural proteins and mostly have no catalytic activity. The most abundant class of proteins in legumes is globulins, which are classified as 7S and 11S according to their sedimentation rates. These are salt-soluble proteins. Legume seeds also contain some anti-nutritional compounds, which are an evolutionary adaptation to help plants survive and complete their life cycle under natural conditions (Duranti, 2006). Storage proteins have many beneficial effects on health, such as plasma cholesterol reduction, anti-cancer and anti-inflammatory properties, weight control, etc. But nutritional deficiency does have an effect on these storage proteins. Tabe *et al.* (2002) showed that the synthesis of proteins with moderate or high contents of sulphur-containing amino acids (legumin and PA1 respectively) was reduced in seeds from plants grown under conditions of sulphur deficiency. On the other hand the production of viclin (a non-sulphur-containing protein) was increased in seeds produced after a long period of sulphur deficiency. Furthermore, these authors also showed that relative abundance of sulphur and nitrogen affects the production of some proteins like sulphate transporter, APR, sulphite reductase (SiR), oligoadenylate synthetase thiollyase (OASTL), serine acetyltransferase (SAT) and S-poor storage protein β-subunit of β-conglycinin. They proposed that OAS is a key indicator of the relative abundance of sulphur and nitrogen and their application results in the increase in mRNAs encoding for the above-mentioned proteins. In another experiment by Steven *et al.* (2002), it was shown that there is a significant increase in the concentration of arginine in plants grown under sulphur deficiency, which again supports the idea that sulphur deficiency increases the amount of non-sulphur proteins. These researches clearly demonstrate that nutritional deficiencies have a direct impact on storage proteins of legumes.

3.8 Protective mechanisms triggered in legumes under stress

3.8.1 Drought and salinity stress

Drought and salinity are serious environmental problems restricting legume plant productivity (Gaur *et al.*, 2008) by affecting photosynthesis and respiration (Kaiser, 1987; Chaves, 1991). Overall changes induced in physiology and biochemistry of plants under water and salinity stress are usually attributed to:

1 Reduction in leaf size, stem extension and root proliferation.
2 Reduced water use efficiency (WUE) and hence altered plant-water relations.
3 Decreased nodule activity and hence nitrogen metabolism.
4 In addition, salinity causes specific ion toxicity due to accumulation of sodium, chloride, and/or boron in the tissue to damaging levels.

Increased production of ROS results in oxidative stress (Foyer & Noctor, 2003; Ahmad *et al.*, 2010). Plants are well equipped with antioxidant defence systems, which include both enzymatic and low-molecular-weight non-enzymatic antioxidant systems (Nobuhiro & Mittler, 2006; Ahmad *et al.*, 2010). Under water and salinity stress increased activities of antioxidant enzymes are well documented (Sairam & Srivastava 2002; Ahmad *et al.*, 2008, 2012; Iqbal & Bano, 2009; Khan & Naqvi, 2010; Li *et al.*, 2011; Chang *et al.*, 2012; Weisany *et al.*, 2012). Rhizobia in the nodules of legumes are slightly more resistant to drought/salt stress than the plants themselves, thereby enhancing resistance to water and salinity stresses (Serraja *et al.*, 2008). However, under severe stress nitrogen metabolism is reduced to a great extent.

Plants, including legumes, accumulate several compatible solutes to maintain water content of the cell. An experiment was carried out on chickpea in which it was subjected to drought for 3, 5 and 7 days. There was a decrease in leaf production, shoot elongation and biomass, along with an increase in proline accumulation. Proline is an alpha-amino acid that serves as a catalyst to many organic reactions (Verbruggen & Hermans, 2008; Macar *et al.*, 2008; Szabados & Savoure, 2010).

Photo-inhibition is the light-induced reduction in the photosynthetic activity of a plant; the extent of light energy absorbed and used in photochemical processes has a direct link with photo-inhibition, which increases under water stress and in turn limits CO_2 assimilation. Stress-induced stomatal closure limits CO_2 availability for carboxylation resulting in excessive excitation of photosystem II (PSII) and hence reducing its efficiency. The protective mechanisms that legumes have evolved in order to minimize these effects are:

1 non-photochemical energy dissipation;
2 movement of chloroplasts;
3 change in chlorophyll concentration;
4 increase in ROS capacity;
5 paraheliotropism (leaf movement).

The extent of paraheliotropism increases with a decrease in water potential, decrease in light interception and maintenance of a high percentage of PSII reaction centres (Pastenes *et al.*, 2005).

3.8.2 Cold stress

Legumes adapt to stresses by changing their metabolic rates, protein synthesis, osmolytes and gene expression. Cold stress is a climatic condition in which the temperature is so low that it harms the normal metabolism of the legume. Low temperatures disturb the photosynthetic ability of legumes, and the effect is more pronounced compared to high temperatures (Tsonev *et al.*, 1991). Freeze tolerance is associated with cellular osmotic stabilization, photosynthesis modification, antioxidant production, and modification in hormone metabolism and cell wall composition and dynamics (Lucau-Danila *et al.*, 2012). MMK4 kinase is a post-translational protective enzyme (Bogre *et al.*, 1997) that is activated in legumes in response to exposure to cold stress (Jonak *et al.*, 1996). An experiment was carried out to investigate the nature of the reduction in stem elongation in response to low temperatures. Gibberellic acid (GA) controls stem elongation, and light deactivates GA, which assists in short-term low temperature adaptation in legumes and the low temperature light stress response. According to this experiment, the stem elongation depends on the timing of the temperature drop. A temperature drop at night with a time span of 3–4 hours showed a decrease in stem elongation by 27%. A temperature drop in the middle of the light period with a time span of 2 hours reduced the rate of cell elongation by 55% (Stavang *et al.*, 2007). Legumes also require stable and competent mitochondria to recover from cold stress (Prasad *et al.*, 1994). The inner membrane of mitochondria in cold-tolerant legumes is also known to have a higher percentage of 18-carbon unsaturated fatty acids, a higher membrane fluidity as well as increased activity of cytochrome *c* oxidase (De Santis *et al.*, 1999). NADH is known for its protection against cold stress by sustaining oxidative phosphorylation at low temperatures.

3.9 Conclusion

Every single organism found on the planet needs a set of particular conditions to survive. If such conditions are not available, the growth and development of that organism is badly affected. Nutritional deficiency in legumes leads to a number of physiological and biochemical changes in the plants including reduced nitrogen fixation and hence yield. Macronutrients perform a variety of functions in legumes. They help in photosynthesis, enzyme activation and vitamin production, improvement of food quality, production of amino acids and N_2 fixation. Changes in their concentrations can have positive or negative effect on the rate of N_2 fixation, which therefore affects the overall growth of the plant. As discussed, macronutrients (nitrogen, phosphorus, potassium, sulphur and carbon) play a vital role in normal growth and development of plants; their deficiency results in overproduction of ROS, and reduced photosynthesis, respiration and water use efficiency. These nutrients also have a role in regulating the amount of storage proteins in legumes and hence the quality of yield.

References

Abreu I, Reguera M, Bonilla A, Bolonas L, Bonilla I (2013) Mineral nutrition in legume-rhizobia nitrogen fixing symbiosis. In: González MBR, Gonzalez-López J (eds), *Beneficial Plant-Microbial Interactions: Ecology and Applications*. CRC Press, pp. 122–140.

Ahanger MA, Tyagi SR, Wani MR, Ahmad P (2014) Drought tolerance: role of organic osmolytes, growth regulators, and mineral nutrients. In: Ahmad P, Wani MR (eds), *Physiological Mechanisms and Adaptation Strategies in Plants Under Changing Environment*, Vol. 1. Springer Science+Business Media, New York, pp. 25–55.

Ahmad P, John R, Sarwat M, Umar S (2008) Responses of proline, lipid peroxidation and antioxidative enzymes in two varieties of *Pisum sativum* L. under salt stress. *Int J Plant Prod* 2: 353–368.

Ahmad P, Umar S, Sharma S (2010) Mechanism of free radical scavenging and role of phytohormones in plants under abiotic stresses. In: Ashraf M, Ozturk M, Ahmad MSA (eds), *Plant Adaptation and Phytoremediation*, pp. 99–118.

Ahmad P, Hakeem K, Kumar A, Ashraf M, Akram NA (2012) Salt induced changes in photosynthetic activity and oxidative defense system of three cultivars of mustard (*Brassica juncea* L). *African J Biotech* 11: 2694–2703.

Almeida JP, Hartwig UA, Frehner M, Nösberger J, Luscher A (2000) Evidence that P deficiency induces N feedback regulation of symbiotic N_2 fixation in white clover (*Trifolium repens* L.). *J Exp Bot* 51: 1289–1297.

Anne-Sophie V, Christophe S, Nathalie G, Munier-Jolain, Ney B (2002) Effect of mineral nitrogen on nitrogen nutrition and biomass partitioning between the shoot and roots of pea (*Pisum sativum* L.). *J Plant and Soil* 242: 251–262.

Bogre L, Ligterink W, Meskiene I, Barker PJ, Heberle-Bors E, Huskisson NS (1997) Wounding induces the rapid and transient activation of a specific MAP kinase pathway. *J Plant Cell* 9: 75–83.

Bolanos L, Brewin NJ, Bonilla I (1996) Effects of boron on rhizobium-legume cell-surface interaction on nodule development. *J Plant Physiol* 110: 1249–1256

Bonilla I, Bolanos L, Bastias E, Carlos AL, Carmen MB, Micaela C (2010) Interactions between salinity and boron toxicity in tomato plants involve apoplastic calcium. *J Plant Physiol* 167: 54–60.

Boominathan R, Doran PM (2002) Ni-induced oxidative stress in roots of the Ni hyperaccumulator, *Alyssum bertolonii*. *New Phytol* 156: 205–215.

Bouchenak M, Lamri-Senhadji M (2013) Nutritional quality of legumes and their role in cardiometabolic risk prevention: a review. *J Med Food* 16: 185–198.

Bradbury M, Baker NR (1981) Analysis of the slow phases of the *in vivo* chlorophyll fluorescence induction curve. Changes in the redox state of Photosystem II electron acceptors and fluorescence emission from Photosystems I and II. *Biochim Biophys Acta –Bioenergetics* 635: 542–551.

Carter GA (1993) Responses of leaf spectral reflectance to plant stress. *Am J Bot* 80: 239–243.

Chang ML, Liao LJ, Lin JI, Liu ZH, Hsu YT, Lee TM (2012) Modulation of antioxidant defense system and NADPH oxidase in *Pluchea indica* leaves by water deficit stress. *Bot Stud* 53: 45–54.

Chauhan S, O'Brian MR (1995) A mutant *Bradyrhizobium japonicum*-aminolevulinic acid dehydratase with an altered metal requirement functions *in situ* for tetrapyrrole synthesis in soybean root nodules. *J Biol Chem* 270: 19823–19827.

Chaves MM (1991) Effects of water deficits on carbon assimilation. *J Exp Bot* 42: 1–16.

Dalton DA, Evans HJ, Hanus FJ (1985) Stimulation by nickel of soil microbial urease activity and urease and dehydrogenase activity in soyabeans grown in a low nickel soil. *Plant Soil* 88: 245–258.

Delgado MJ, Eulogio JB, Downie JA (1998) Genes involved in the formation and assembly of rhizobial cytochromes and their role in symbiotic nitrogen fixation. *Adv Microb Physiol* 40: 191–231.

Delgado MJ, Alvaro TA, Chouhara T, Eulogio JB (2006) Functional characterization of *Bradyrhizobium japonicum* modA and modB genes involved in molybdenum transport, *Microbiology* 152: 199–207.

De Santis A, Landi P, Genchi G (1999) Changes of mitochondrial properties in maize seedlings associated with selection for germination at low temperature. Fatty acid composition, cytochrome c oxidase, and adenine nucleotide translocase activities. *J Plant Physiol* 119: 743–754.

Dilworth MJ, Robson AD, Chattel DL (1979) Cobalt and nitrogen fixation in *Lupinus angustifolius* L. II. Nodule formation and function. *New Phytol* 83: 63–79.

Duranti M (2006) Grain legume proteins and nutraceutical properties. *Fitoterapia* 77: 67–82.

Epstein E, Bloom AJ (2005) *Mineral Nutrition of Plants: Principles and Perspectives*, 2nd edn. Sinauer Associates, Sunderland, MA.

Evans PT, Malmberg RL (1989) Do polyamines have roles in plant development? *Annu Rev Plant Physiol Plant Mol Biol* 40: 235–269.

Fazli IS, Jamal A, Ahmad S, Masoodi M, Khan JS, Abdin MZ (2008) Interactive effect of sulphur and nitrogen on nitrogen accumulation and harvest in oilseed crops differing in nitrogen assimilation potential. *J Plant Nutr* 31: 1203–1220.

Foyer CH, Noctor G (2003) Redox sensing and signaling associated with reactive oxygen in chloroplasts, peroxisomes and mitochondria. *Physiol Plantarum* 119: 355–364.

Gad N (2012) Response of groundnut (*Arachis hypogaea*) plants to cobalt and molybdenum mixture. *Middle East J Agri Res* 1: 19–25.

Gajewska E, Sklodowska M, Slaba M, Mazur J (2006) Effect of nickel on antioxidative enzyme activities, proline and chlorophyll content in wheat shoots. *Biol Plant* 50: 653–659.

Gaur PM, Krishnamurthy L, Kashiwagi J (2008) Improving drought-avoidance root traits in chickpea (*Cicer arietinum* L.) – Current status of research at ICRISAT. *Plant Prod Sci* 11: 3–11.

George EF, Hall MA, De Klerk GJ (2008) The components of plant tissue culture media i: macro- and micro-nutrients. In: George EF, Hall MA, De Klerk GJ (eds), *Plant Propagation by Tissue Culture*. Exegetics, Basingstoke, UK, pp. 65–113.

Gonnelli C, Galardi F, Gabbrielli R (2001). Nickel and copper tolerance and toxicity in three Tuscan populations of *Silene paradoxa*. *Physiol Plant* 113: 507–514.

Grewala HS, Williamsa R (2002) Influence of potassium fertilization on leaf to stem ratio, nodulation, herbage yield, leaf drop and common leaf spot disease of alfalfa. *J Plant Nutr* 781–795.

Hafeez ur Rehman, Qumer I, Farooq M, *et al.* (2013) Sulphur application improves the growth, seed yield and oil quality of canola. *Acta Physiol Plant* doi: I:10.1007/s11738-013-1331-9.

Hansch R, Mendel RR (2009) Physiological functions of mineral micronutrients (Cu, Zn, Mn, Fe, Ni, Mo, B, Cl). *Curr Opin Plant Biol* 12: 259–266.

Hart AL (1989) Distribution of phosphorus in nodulated white clover plants. *J Plant Nutr* 12: 159–171.

Hussain MB, Ali S, Azam A, *et al.* (2013) Morphological, physiological and biochemical responses of plants to nickel stress: A review. *Afr J Agric Res* 8: 1596–1602.

Iqbal I, Bano A (2009) Water stress induced changes in antioxidant enzymes, membrane stability and seed protein profile of different wheat accessions. *Afr J Biotechnol* 8: 6576–6587.

Islam M, Mohsan S, Ali S (2013) Interactive effect of sulfur and phosphorus on nodulation, nitrogen fixation and nutrient uptake by chickpea grown on calcareous soils in Pakistan. *J Plant Nutr* doi: 10.1080/01904167.2013.805220.

Jaleel CA, Gopi R, Manivannan P, *et al.* (2008a) Endogenous hormonal and enzymatic responses of *Catharanthus roseus* with triadimefon application under water deficits. *Comptes Rendus Biologi* 331: 844–852.

Jaleel CA, Gopi R, Gomathinayagam M, Panneerselvam R (2008b) Effects of Calcium chloride on metabolism of salt stressed *Dioscorea rotundata*. *Acta Biologi Cracoviensia Series Botan* 50: 63–67.

Jayakumar K, Vijayarengan P, Chang-Xing Z, Jaleel CA (2008) Soil applied cobalt alters the nodulation, leg-haemoglobin content and antioxidant status of *Glycine max* (L.) Merr. *Colloids Surfaces B: Biointerfaces* 67: 272–275.

Jonak C, Geiger S, Ligterink W, Barker PS, Huskisson NS, Hirt H (1996) Stress signaling in plants: a mitogen activated protein kinase pathway is activated by cold and drought. *Proc Natl Acad Sci USA* 93: 11274–11279.

Jordan A, Richard P (1998) Ribonucleotide reductases. *Annu Rev Biochem* 67: 71–98.

Kaiser WM (1987) Effects of water deficit on photosynthetic capacity. *Physiol Plant* 71: 142–149.

Khan N, Naqvi FN (2010) Effect of water stress on lipid peroxidation and antioxidant enzymes in local bread wheat hexaploids. *J Food Agric Environ* 8: 521–526.

Li X, Zhang L, Li Y (2011) Preconditioning alters antioxidative enzyme responses in rice seedlings to water stress. *Procedia Environmental Sciences* 11: 1346–1351.

Licht S, Gary JG, Joanne S (1996) Thiol radicals in ribonucleotide reductase. *Science* 271: 477–481.

Lucau-Danila A, Toitot C, Goulas E, *et al.* (2012) Transcriptome analysis in pea allows to distinguish chilling and acclimation mechanisms. *Plant Physiol Biochem* 58: 236–244.

Maathuis FJM (2009) Physiological functions of mineral macronutrients. *Plant Biol* 12: 250–258.

Macar TK, Ekmekci YZ, Naturforsch C (2008) PSII photochemistry and antioxidant responses of a chickpea variety exposed to drought. *Z Naturforsch C* 63: 583–594.

Marschner H (1995) *Mineral Nutrition of Higher Plants*, 2nd edn. Academic Press, London.

Masoumi A, Kafi M, Hamidreza Khazaei H, Davari K (2010) Effect of drought stress on water status, electrolyte leakage and enzymatic antioxidants of kochia (*Kochia scoparia*) under saline condition. *Pak J Bot* 42: 3517–3524.

Nobuhiro S, Mittler R (2006) Reactive oxygen species and temperature stresses: A delicate balance between signaling and destruction. *Physio Plant* 126: 45–51.

O'Hara GW, Dilworth MJ, Boonkerd N, Parkpian P (1988) Iron deficiency specifically limits nodule development in peanut inoculated with *Bradyrhizobium* sp. *New Phytol* 108: 51–57.

Olivera M, Tejera N, Iribarne C, Ocana A, Lluch C (2004) Growth, nitrogen fixation and ammonium assimilation in common bean (*Phaseolus vulgaris*); effects of phosphorus. *J Physiol Plant* 121: 498–505.

Pastenes C, Pimentel P, Lillo JJ (2005) Leaf movements and photoinhibition in relation to water stress in field-grown beans. *J Exp Bot* 56: 425–433.

Patrick BH, Welch RM, Cary EE, Checkai RT (1987) Micronutrients: Beneficial effects of nickel on plant growth. *J Plant Nutr* 10: 2125–2135.

Postgate J (1998) *Nitrogen Fixation*, 3rd edn. Cambridge University Press, Cambridge, UK.

Prasad TK, Anderson MD, Stewart CR (1994) Acclimation, hydrogen peroxide, and abscisic acid protect mitochondria against irreversible chilling injury in maize seedlings. *Plant Physiol* 105: 619–627.

Rao KVM, Sresty TV (2000) Antioxidative parameters in the seedlings of pigeon pea (*Cajanus cajan* (L.) Millspaugh) in response to Zn and Ni stresses. *Plant Sci* 157: 113–128.

Reguera M, Villarejo A, Abreu A, Bonilla I, Bolanos L (2009) Protein N-glycosylation patterns are altered during development of legume symbiotic nodules under boron deficiency. In: Becana M (ed.), *Biological Nitrogen Fixation and Plant-Associated Microorganisms*. Gráficas Alós, Huesca, Spain, pp. 169–172.

Reguera M, Wimmer M, Bbustos P, Goldbach HE, Bolanos L, Bonilla I (2010) Ligands of boron in Pisum sativum nodules are involved in regulation of oxygen concentration and rhizobial infection. *Plant Cell Environ* 33: 1039–1048.

Sairam RK, Srivastava GC (2002) Changes in antioxidant activity in subcellular fractions of tolerant and susceptible wheat genotypes in response to long term salt stress. *Plant Sci* 162: 897–904.

Sanchez PA (2002) Soil fertility and hunger in Africa. *Science* 295: 2019–2020.

Sarah A. Batterman, Wurzburger N, Hedin LO (2013) Nitrogen and phosphorus interact to control tropical symbiotic N_2 fixation: A test in *Inga punctata*. *J Ecol* doi:10.1111/1365-2745.1213.

Saxena AK, Rewari RB (1991) Influence of phosphate and zinc on growth, nodulation and mineral composition of chickpea (*Cicer arietinum* L.) under salt stress. *World J Microbiol Biotech* 7: 202–205.

Scherer HW (2008) Impact of sulfur on N_2 fixation of legumes: sulfur assimilation and abiotic stress in plants. In. Khan NA, Singh S, Umar S (eds), *Sulphur Assimilation and Abiotic Stress in Plants*. Springer-Verlag, Berlin, Heidelberg, pp. 43–54.

Serraja R, Vasquez Diaz BH, Drevonc JJ (2008) Effects of salt stress on nitrogen fixation, oxygen diffusion, and ion distribution in soybean, common bean and alfalfa. *J Plant Nutr* 21: 475–488.

Sexton PJ, Paek NC, Richard MS (1998) Effects of nitrogen source and timing of sulfur deficiency on seed yield and expression of 11S and 7S seed storage proteins of soybean. *Field Crops Res* 59: 1–8.

Shah VK, Rangarg P, Chatterjee R, *et al.* (1999) Requirement of NifX and other nif proteins for *in vitro* biosynthesis of the iron-molybdenum cofactor of nitrogenase. *J Bacteriol* 181: 2797–2801.

Sigrid CR, Binkley D, Parrotta JA (2002) Greater soil carbon sequestration under nitrogen-fixing trees compared with *Eucalyptus* species. *Ecosystems* 5: 217–231.

Slatni T, Marta DO, Imem BS, *et al.* (2012) Immunolocalization of H^+ATPase and IRT1 enzymes in N_2-fixing common bean nodules subjected to iron deficiency. *J Plant Physiol* 169: 242–248.

Snowball K, Robson AD, Loneragan JF (1980) The effect of copper on nitrogen fixation in subterranean clover (*Trifolium subterraneum*). *New Phytol* 1: 63–72.

Stavang JA, Junttila O, Moe R, Olsen JE (2007) Differential temperature regulation of GA metabolism in light and darkness in pea. *J Exp Bot* 58: 3061–3069.

Steven GT, Bilsborrow PE, Hocking TJ, Bennett J (2002) Effect of sulphur deficiency on the growth and metabolism of sugar beet (*Beta vulgaris* cv Druid). *J Sci Food Agric* 80: 2057–2062.

Szabados L, Savoure A (2010) Proline: a multifunctional amino acid. *Trends Plant Sci* 15: 89–97.

Tabe L, Hagan N, Higgins TJV (2002) Plasticity of seed protein composition in response to nitrogen and sulfur availability. *Curr Opin Plant Biol* 5: 212–217.

Tang C, Alan AD, Michael JD (1990) The role of iron in nodulation and nitrogen fixation in *Lupinus angustifolius*. *New Phytol* 114: 173–182.

Tsonev T, Velikova V, Georgieva K, Hyde PF, Jones HG (1991) Low temperature enhances photosynthetic down-regulation in French bean (*Phaseolus vulgaris* L.) plants. *Ann Bot* 91: 343–352.

Tsvetkova GE, Georgiev GI (2003) Effect of phosphorus nutrition on the nodulation, nitrogen fixation and nutrient-use efficiency of *Bradyrhizobium japonicum*-soybean. *Bulg J Plant* (special issue), pp. 331–335.

Valentine AJ, Benedito VA, Kang Y (2011) Legume nitrogen fixation and soil abiotic stress: from physiology to genomics and beyond. *Annu Plant Rev* 42: 207–248.

Varin S, Cliquet JB, Personeni E, Avice JC, Lemauviel-Lavenant S (2009) How does sulphur availability modify N acquisition of white clover (*Trifolium repens* L.)? *J Exp Bot* 61: 225–234.

Verbruggen N, Hermans C (2008) Proline accumulation in plants: a review. *Amino Acids* 35: 753–759.

Vieira RF, Cardoso EJBN, Vieira C, Cassini STA (1998) Foliar application of molybdenum in common beans: nitrogenase and reductase activities in a soil of high fertility. *J Plant Nutr* 21: 169–180.

Weisany W, Yousef S, Gholamreza H, Adel H, Kazem G (2011) Changes in antioxidant enzymes activity and plant performance by salinity stress and zinc application in soybean (*Glycine max* L.). *Plant Omics J* 5: 60–67.

Weria W, Yaghoub R, Kaveh HA (2013) Role of some mineral nutrients in biological nitrogen fixation. *Bull Env Pharma Life Sci* 2: 77–84.

Wiedenhoeft AC (2006) Introduction to plants and plant nutrition. In: Wiedenhoeft AC, Hopkins WG (eds), *Plant Nutrition*. Infobase Publishing, New York, pp. 2–13.

Xu Q, Shi G (2000) The toxic effects of single Cd and interaction of Cd with Zn on some physiological index of [*Oenanthe javanica* (Blume) DC]. *J Nanjing Normal University (Natural Science)* 23: 97–100.

Yamagishi M, Yamamoto Y (1994) Effects of boron on nodule development and symbiotic nitrogen fixation in soybean plants. *Soil Sci Plant Nutr* 40: 265–74.

Yusuf M, Fariduddin Q, Hayat S, Ahmad A (2011) Nickel: an overview of uptake, essentiality and toxicity in plants. *Bull Environ Contam Toxicol* 86: 1–17.

CHAPTER 4

Chickpea: Role and responses under abiotic and biotic stress

Saiema Rasool[1], Arafat Abdel Hamed Abdel Latef[2] and Parvaiz Ahmad[3]

[1]*Forest Biotech Laboratory, Department of Forest Management, Faculty of Forestry, Universiti Putra Malaysia, Serdang, Selangor, Malaysia*

[2]*Department of Botany, Faculty of Science, South Valley University, Qena, Egypt*

[3]*Department of Botany, S.P. College, Srinagar, Jammu and Kashmir, India*

4.1 Introduction

Chickpea is a cool season legume crop grown worldwide as a food crop. The seed is the main edible part of the plant. It is also called garbanzo gram or Bengal gram. It ranks third (FAO, 2008) among the food legumes after beans and pea. More than 50 countries are reported to grow chickpea; 22 cultivate more than 20,000 ha, and 19 cultivate 10,000 to 20,000 ha. Major chickpea-producing countries are: India (65% of annual production), Pakistan (10%), Turkey (7%), Iran (3%), Myanmar (2%), Mexico (1.5%) and Australia (1.5%) (FAO, 2008).

Chickpea is a cheap and important source of protein for those people who cannot afford animal protein or who are largely vegetarian. Furthermore, chickpea is also a good source of minerals (calcium, phosphorus, magnesium, zinc and iron), unsaturated fatty acids, fibre and β-carotene). Chickpea also plays an important role in maintaining soil fertility by fixing nitrogen at rates of up to 140 kg/ha/year (Flowers *et al.*, 2010). Therefore, this crop requires relatively low inputs of nitrogen as it derives 70% of its N through symbiotic N_2 fixation and benefits other cereal crops as well (Siddique *et al.*, 2005). Chickpea contributes a significant amount of residual nitrogen to the soil and adds organic matter thereby improving soil health and fertility.

On the basis of seed colour and geographical distribution, chickpea is grouped into two types: desi (Indian origin) and kabuli (Mediterranean and Middle Eastern origin) (Bampidisa & Christodoulou, 2011). Desi seeds are brown, black or green in colour and are cultivated as a legume crop. Kabuli seeds are white to cream coloured and are used for cooking purposes for humans (Iliadis, 2001). The seed weight generally ranges from 0.1 to 0.3 g for desi and 0.2 to 0.6 g for kabuli (Frimpong *et al.*, 2009). The desi type accounts for about 80–85% of total chickpea production, and is grown mostly in Asia and Africa (Jukanti *et al.*, 2012), while the kabuli types are largely grown in West Asia, North Africa, North America and Europe.

Plants, including chickpea, often experience external environmental stress (abiotic and biotic stress) that hampers their growth and development (Ahmad *et al.*, 2010, 2011, 2012a,b, 2014a,b). Crop yield is also decreased enormously due to these stresses. However, plants are equipped with some mechanisms that can enhance their tolerance to these stresses (Ahmad *et al.*, 2010, 2011, 2012a,b, 2014a,b). This chapter will throw light on the responses of chickpea to biotic and abiotic stress. Biotechnological approaches to enhance tolerance to these stresses in chickpea are also discussed.

4.2 Origin and occurrence

Chickpea is one of the oldest legume crops and is consumed widely across the world. Vavilov (1926) was the first to recognize the Near Eastern, Central Asian, Indian and Mediterranean regions as the probable centres of origin for chickpea. The crop most probably originated

Legumes under Environmental Stress: Yield, Improvement and Adaptations, First Edition. Edited by Mohamed Mahgoub Azooz and Parvaiz Ahmad.

in south-eastern Turkey and neighbouring parts of Syria and may have been first grown in Turkey 7500 years ago (Oplinger *et al.*, 1990; Singh & Ocampo, 1997).

Chickpea (*Cicer arietinum* L.) is the only cultivated species in the genus *Cicer* and is a self-pollinated diploid ($2n=2x=16$) crop. It has been domesticated from *C. reticulatum* Ladizinsky, a closely related wild species. After its domestication in the Middle East this crop spread throughout the Mediterranean region, India and Ethiopia (van der Maesan, 1987). It has now been introduced to Mexico, Argentina, Peru, Chile, Australia and the USA (Duke, 1981). And other upcoming chickpea producers include Ethiopia, Iraq, Israel, Jordan, Syria, Canada, Morocco, Malawi and Tanzania. The genus includes 34 perennial herbs and nine annuals (Muehlbauer, 1993).

4.3 General botany

Chickpea seedlings are hypogeal and growth of the plumule produces an erect shoot. The primary root is long and produces lateral roots. Chickpea stems are branched, erect or dispersed, at times shrubby and much branched, 0.2-metre tall, glandular pubescent, dark green or bluish green in colour. Leaves are compound, arranged in an alternate phyllotaxy, and generally are imparipinnate with 11 to 13 leaflets arranged on a rachis with a small petiole. The foliage is covered with glandular hairs, which secrete highly acidic exudates, and this is considered important in conferring tolerance to insect pests, such as the pod borer. The plant has a deep root system and is considered a hardy crop. In deep Vertisols, roots have penetrated deeper than 120 cm (Sheldrake & Saxena, 1979). These workers have also reported the development of lobed nodules. Plants attain a height of 20 to 100 cm, although tall cultivars under favourable conditions can grow up to 130 cm (Reddy *et al.*, 1985).

The two distinct types of chickpea, 'kabuli' (known as *macrosperma*) and 'desi' (known as *microsperma*), differ in their geographical distribution and plant type. Desi types are found in Central Asia and in the Indian subcontinent while the kabuli types are found in the Mediterranean region. Kabuli types are usually taller than the desi types.

Flowers are axillary, solitary or in an inflorescence of two or three. They are white, pink, purplish or blue in colour. The corolla is generally purple in the desi type and white in the kabuli type, and is rarely blue. Plants in the genus *Cicer* have only one carpel per flower. The small papilionaceous flowers of chickpea have diadelphous stamens – nine stamens have fused filaments and the tenth is free. During the growth of the flower bud, the filaments extend and anthers deposit pollen on the pistil. Both the pistil and the anther remain inside the keel. Self-pollination is the rule, although natural cross-pollination and cleistogamous pollination have also been found, which take place before opening of the bud (Singh & Ibrahim, 1990).

The pod is a rhomboid ellipsoid, with normally one or two seeds, three at maximum, and is inflated and glandular-pubescent. Pod size varies greatly, but the trait is least influenced by environment. Pod-filling is highly dependent on weather and varies from 8.97 to 56.53% (Pundir *et al.*, 1992). The seed colour is cream, yellow, brown, black, or green, and seeds are rounded to angular. The seed coat is smooth, wrinkled or tuberculate, and the seed is laterally compressed with a median groove around two-thirds of the seed, which is beaked anteriorly. Germination is cryptocotylar (Duke, 1981; van der Maesen, 1987).

Because of the deep taproot system, chickpea can grow to full maturity in conditions that are unsuitable for most other crops. The long taproot enhances the plant's capacity to withstand drought conditions, and makes chickpea well suited for cultivation in cooler areas with low rainfall.

Chickpea is very sensitive to excess water availability, and therefore is grown in sandy loam soils with an appropriate drainage system. The productivity of chickpea is greatly reduced by very cold conditions. Chickpea is a rabi crop grown in the months of September–November and harvested in the months of February–April. After sowing, the maturity period ranges from 95 to 110 days.

4.4 Nutritional uses

Chickpea is consumed mostly as seed in different forms and preparations are determined by ethnic and regional factors (Ibrikci *et al.*, 2003; Jukanti *et al.*, 2012). On the Indian subcontinent it is ground to make flour (besan) used to prepare different snacks (Hulse, 1991). In Asia and Africa, chickpea is used in stews, soups and salads and consumed in boiled, salted, roasted and fermented forms (Gecit, 1991). These different forms of consumption provide consumers with valuable nutrition and potential health benefits.

Although it is one of the founder crops (Zohary & Hopf, 2000) with potential nutritional or medicinal qualities, chickpea has not received the amount of research devoted to other founder crops such as wheat and barley. It has been consumed by humans since ancient times due to its good nutritional properties. In addition, chickpea is of interest as a functional food with potential beneficial effects on human health. Chickpea is a source of zinc, folate and protein, is rich in dietary fibre, and a good source of carbohydrates for persons with insulin sensitivity. The total content of carbohydrate fat and sugar in chickpea is higher than in other pulses (Jukanti *et al.*, 2012).

The International Crops Research Institute for the Semi-Arid Tropics (ICRISAT) has reported that chickpea seeds contain on average 23% protein, 64% total carbohydrates (47% starch, 6% soluble sugars), 5% fat, 6% crude fibre and 3% ash. Studies have also shown that chickpea is helpful for lowering blood cholesterol (Pittaway *et al.*, 2008). Unlike other oilseed pulses, the oil content in chickpea is relatively low (Gül *et al.*, 2008). But chickpea oil contains many medicinal and nutritionally important tocopherols, sterols and tocotrienols (Zia-Ul-Haq *et al.*, 2007). The human need for minerals and vitamins is fulfilled through a well-balanced daily diet of cereals, pulses, vegetables, fruits, meat and dairy products. The chickpea like many other pulses not only brings variety to the cereal-based daily diet of millions of people in Asia and Africa, but also provides essential vitamins and minerals (Duhan *et al.*, 1999; Cabrera *et al.*, 2003). Chickpea is relatively economical and a good source of folic acid and tocopherols (Ciftci *et al.*, 2010). Chickpea is an excellent source of folic acid and contains other water-soluble vitamins like riboflavin, pantothenic acid and pyridoxine. These levels are similar or higher than those observed in other pulses (Lebiedzińska & Szefer, 2006). The most important and widely distributed carotenoid that is converted to vitamin A is β-carotene (Abbo *et al.*, 2005). On a dry weight basis, the chickpea seed has a greater concentration of β-carotene than golden rice endosperm (Ye *et al.*, 2000; Abbo *et al.*, 2005) or red coloured wheat (USDA, 2010).

Physiological benefits from the consumption of chickpea have also been reported, which reduce the risk of chronic diseases and optimize health. Therefore, we can say that chickpea is a functional food in addition to its accepted role as a source of protein and fibre.

The present-day yield potential of chickpea exceeds 5.0 t/ha, but the average yield is stagnating at approximately 0.8 t/ha. This wide gap between average and potential yield is due to abiotic and biotic stresses. Therefore in this chapter, we focus on chickpea under abiotic and biotic stresses.

4.5 Abiotic stress

Abiotic stresses are the major adverse environmental factors that limit plant productivity (Doupis *et al.*, 2011; Ahmad *et al.*, 2010, 2011, 2012a,b, 2014a,b). Plants are subjected to various abiotic stresses, such as drought, salinity, extreme temperatures and flooding, that adversely affect their growth and development and ultimately lead to physio-morphological, biochemical and molecular changes in the plant (Potters *et al.*, 2007). It is clear that chickpea is treasured for its nutritive value, in particular its high protein content, and hence study of problems like drought, cold and salinity stresses are very important in relation to chickpea (Singh *et al.*, 1994). The most crippling economic losses in agriculture occur due to these environmental stressors, which affect plants in many ways. All these abiotic stresses affect the water relations of the plant, both at the cellular level as well as the whole-plant level, causing both specific and non-specific reactions, damage and adaptation reactions. These stresses share common features.

Drought is a common abiotic stress that reduces the yield of chickpea (Leport *et al.*, 2006). The second most important abiotic stress in chickpea is cold stress. Susceptibility is greater at the late vegetative stage than at the seedling stage (Singh *et al.*, 1984). Drought stress in chickpea causes a 40–50% reduction in yield globally (Ahmad *et al.*, 2005). Among pulse crops chickpea is the only legume that is sown in spring in the Mediterranean Basin without irrigation (Toker & Çağirgan, 1998). Late sowing increases the risk that this crop is affected by drought stress. Towards maturity, terminal drought stress is normally accompanied by increasing ambient temperatures of more than 30 °C, which may affect pod filling (Johansen *et al.*, 1994; Wery *et al.*, 1994). In Mediterranean environments, the change in sowing period from spring to winter has increased the yield by efficient utilization of rainwater, but improved tolerance for low temperatures is needed to further improve yields (Millan *et al.*, 2006).

Chickpea is considered drought tolerant as it is grown in semi-arid regions, and many cultivars with different levels of tolerance to drought have been bred. Chickpea was found to be superior to pea for dryland crop production in semi-arid climates due to an adaptive root distribution (Benjamin & Nielsen, 2006). Under drought stress, dry matter yield in faba bean, pea and chickpea was reduced by 36.4, 23.9 and 14.5%, respectively (Amede *et al.*, 2003).

Indian varieties especially are adapted to drought and high temperature stress, making them ideal for the study of the molecular mechanisms of drought tolerance (Boominathan *et al.*, 2004). ICRISAT has released a kabuli chickpea genotype ICCV 2, and this variety has proved successful due to its drought resistance mechanism. In chickpea, ICC 8261 and FLIP 87-59C are the most popular drought-tolerant germplasm lines (Saxena *et al.*, 2002; Singh *et al.*, 1997). Another germplasm accession, ICC 4958 (desi), has also been identified as one of the best genotypes with a highly desirable root system (Kashiwagi *et al.*, 2005). This genotype is used to develop drought-tolerant varieties by incorporating the trait of an extensive and deep root system into a well-adapted genetic background (Kashiwagi *et al.*, 2007). KAK 2 and JGK 1 are already released kabuli cultivars that are drought tolerant (Kashiwagi *et al.*, 2007).

Tolerance to cold stress is acquired only after a period of cold acclimation (Mahajan & Tuteja, 2005; Apostolova *et al.*, 2008). This process involves the removal of the limitations that normally occur when plants grown at higher temperatures are suddenly exposed to cold stress. Although adapted to dry areas of the Middle East, chickpea has been grown worldwide.

In chickpea, mean daily temperatures below 15 °C lead to flower and pod abortion in some parts of India and Australia (Savithri *et al.*, 1980; Srinivasan *et al.*, 1999; Clarke *et al.*, 2004). When it is sown in autumn or early spring chickpea encounters chilling stress at the reproductive stage (Clarke *et al.*, 2008), while it faces freezing stress at the vegetative stage (Materne *et al.*, 2007; Toker *et al.*, 2007; Saeed *et al.*, 2010; Toker & Yadav, 2010). Winter hardiness of chickpea is less than that of lentil and faba bean (Murray *et al.*, 1988).

Malhotra and Singh (1991) reported that cold tolerance at the vegetative stage is controlled by at least five genes in chickpea with both additive and non-additive gene effects, and was dominant over susceptibility. They suggested that selection for cold tolerance would be more effective if dominance and epistatic effects were reduced after selfing generations (Toker & Mutlu, 2011). Clarke *et al.* (2008) underlined that there were no published data on the genetics of tolerance to chilling at the reproductive stage in chickpea, despite the fact that molecular markers were linked to chilling tolerance and susceptibility in some varieties. Nevertheless, these markers are absent from marker-assisted selection (MAS) in other chilling-tolerant chickpea (Clarke *et al.*, 2008; Toker & Mutlu 2011).

The advantage of winter sowing of chickpea over traditional spring sowing is the increase in yield, greater water use efficiency and better moisture conditions (Millan *et al.*, 2006; Heidarvand *et al.*, 2011). And the disadvantage of winter sowing is the risk of winter killing due to cold stress. To expand the range of potential winter sowing of chickpea we have to use both conventional and molecular breeding techniques to obtain improved cold tolerance. The best sources for cold tolerance in chickpea are ILC 8262 (Singh *et al.*, 1992) and ILC 8617 (Singh *et al.*, 1997) with rosette-type and dark-green leaves in the seedling stage, plus late flowering. Clarke *et al.* (2004) developed two chilling-tolerant genotypes, 'Rupali' and 'Sonali'. The cold-tolerant chickpea varieties developed by ICRISAT are ICCV 88502, ICCV 88503, ICCV 88506, ICCV 88510, ICCV 88516 and ICC 8923. These varieties can set pods at low temperatures and restrict plant growth, thus giving higher yields than conventional cultivars, which fail to set pods at similar temperatures.

Singh *et al.* (1989) proposed a screening and selection technique for cold tolerance in chickpea, which in turn has been combined with screening for resistance to ascochyta blight. The technique involves: (i) early sowing (in October) of test materials; (ii) using at least one known cold-susceptible (ILC 533) and cold-tolerant accession (ILC 8617 is cold tolerant and ascochyta blight resistant); (iii) using at least one known ascochyta blight-susceptible but cold-tolerant accession (ILC 8262); (iv) inoculation with ascochyta-infected crop debris prior to flowering and ensuring proper moisture provision; and (v) evaluating the test materials for resistance to ascochyta blight and tolerance to cold using a visual scale scored from 1 to 9 (Toker & Canci, 2003; Toker *et al.*, 2007; Toker & Mutlu, 2011). This technique can be useful for a large number of test materials and could easily be adopted for cold and chilling tolerance in other food legumes (Toker & Mutlu, 2011).

Soil salinity is another of the abiotic stresses that adversely affect plant growth and development. Soluble salts in surplus quantity in the soil lead to the osmotic stress that results in specific ion toxicity and ionic imbalance (Munns, 2003) and the consequences of these can lead to the death of the plant (Rout & Shaw, 2001). Therefore, to overcome the salinity threat, we have to find and select relatively salt-tolerant genotypes within a species, through conventional selection or breeding techniques.

Chickpea has been reported as salt sensitive due to pre-dawn leaf water potential expressed as a water stress day index during the growing period and according to soil salinity (Katerji *et al.* 2003). B. Singh *et al.* (2005) reported that salt stress affects nodulation, carbon fixation, plant water status, and H_2O_2 scavenging enzymes in chickpea cultivars. Recently, Rasool *et al.* (2013) reported that the tolerance of the chickpea genotypes SKUA-06 and SKUA-07 to salinity seems to be related to the efficiency of the antioxidant enzymes superoxide dismutase (SOD), catalase (CAT), ascorbate peroxidase (APX) and glutathione reductase (GR) in preventing accumulation of reactive oxygen species (ROS), thereby maintaining redox homeostasis and the integrity of cellular components.

A number of salt-tolerant chickpea genotypes were identified that differed in the uptake and distribution of Na^+ and Cl^- ions (CSG 88101, CSG 8927, CSG 8977, CSG 8962 and CSG 8943) (Dua & Sharma, 1995). Another tolerant variety that can be grown in saline soils with an electrical conductivity (EC) of 4 to 6 dS/m is the desi variety Karnal Chana-1 (CSG 8963), which was was released in India (Millan *et al.*, 2006). On the whole, the kabuli type was more tolerant than the desi type after screening 252 genotypes and breeding lines (Serraj *et al.*, 2004). Maliro *et al.* (2007) screened about 200 genotypes and wild *Cicer* genotypes for salt tolerance in Australia. The most tolerant were CPI 060546, ILC 01302 (from Turkey), ICC 06772 and ICC 6474 (from Iran), ICC 8294, ICC 438, CPI 53008 (from India) and UC 5 (from USA); none of the wild relatives screened were tolerant. This indicates that selection and breeding of some cultivated genotypes on saline soils has occurred, but screening for salt tolerance is limited by its huge potential for interaction with other environmental stresses, which makes it difficult to separate genetic and environmental variations (Flowers, 2004; Toker *et al.*, 2007).

Deficiencies of some elements in agricultural soils decrease yields of chickpea. For example, nitrogen (N) and phosphorus (P) deficiencies in chickpea have been reported to cause worldwide yield losses of 709,000 and 653,000 t/year, respectively. Similarly, yield losses caused by micronutrient deficiencies have been estimated at about 360,000 t/year (Ryan, 1997; Toker & Mutlu, 2011). Iron-induced deficiency has been reported in chickpea (Wallace, 1960; Toker *et al.*, 2010; Toker & Mutlu, 2011). Chickpea was found to be more sensitive to Zn deficiency than oilseeds and cereals (Tiwari & Dwivedi, 1990; Toker & Mutlu, 2011).

Anthropogenic activities are generally responsible for heavy metal toxicity on agricultural land. About 235 million hectares of land has been polluted by toxic heavy metals (Giordani *et al.*, 2005). Arsenic is one of the primary toxic metals, mostly found in the oxidized state as arsenite (As(III)) and arsenate (As(V)). Arsenic is a non-essential element for both plants and animals. Soils enriched with arsenic are considered to be the main sources of contamination in the food chain and water supplies, and are of great environmental concern because arsenic is known to be a carcinogen and mutagen (Fayiga & Ma, 2006). Arsenic is mainly taken up as arsenate by the plant, arsenate exposure causes stress in plants including inhibition of growth, physiological disorders and ultimately death of the plant (Stoeva & Bineva, 2003; Stoeva *et al.*, 2005). Cytoplasmic arsenate is reduced to arsenite and interferes with metabolic processes involving phosphate, giving it the potential to be toxic to the plant (Meharg & Hartley-Whitaker, 2002; Stoeva & Bineva, 2003). Though the arsenic is not a redox metal, there is still the chance that the exposure of plants to inorganic arsenic results in the generation of ROS, which readily cause conversion of arsenate to arsenite in plants (Meharg & Hartley-Whitaker, 2002). This problem is severe if arsenic can contaminate the food chain; the solution is to localize the metalloid in non-edible parts of the crops and seedlings (Sahu *et al.*, 2007). So, thorough screening is needed to select varieties that can live on contaminated sites with good growth and yield and have the potential to localize metalloids in their roots.

A few varieties of chickpea, like cv.CSG-8962, are tolerant to heavy metal stress (Gupta *et al.*, 2002a). It has been found that chickpea is very capable of producing phytochelatins (PCs) and homo-phytochelatins (hPCs) in response to low concentrations of cadmium in hydroponics culture and in soils contaminated by fly-ash

(Gupta *et al.*, 2002a,b). These PCs are part of the detoxifying mechanism for heavy metals in higher plants (Mehra & Tripathi, 2000; Cobbett & Goldsbrough, 2002). PCs are low-molecular-weight peptides chemically related to glutathione (GSH) that make complexes with metals and metalloids and sequester them into vacuoles and thus detoxify the cell (Grill *et al.*, 2006); hPCs contain homoglutathione instead of GSH, having the N-terminal β-alanine (Grill *et al.*, 2006). Manifestation of such metal-binding peptides in plants could be an important biochemical indicator of heavy metal contamination under various environments (Gupta *et al.*, 2002a,b), even if their changes in response to various metal ions have not been fully examined. Several reports have revealed that arsenic induces the synthesis of PCs as well as hPCs (Gupta *et al.*, 2004; Tripathi *et al.*, 2007), and complexes of arsenic with PCs and GSH have also been demonstrated in various plants (Raab *et al.*, 2004, 2005).

4.6 Chickpea and abiotic stress: The 'omics' approach

The 'omics' include proteomics, transcriptomics, genomics and transgenomics. Recent advances in these areas have contributed greatly to a better understanding of the molecular and genetic bases of stress responses, which has been an important bottleneck for molecular and transgenic breeding (Reddy *et al.*, 2012). Here we summarize significant progress in chickpea research regarding abiotic stress tolerance.

4.6.1 Proteomics

Proteomics is the systematic analysis of the proteins expressed by the genome. It not only describes entire proteomes at the cell, organ or tissue level (Porubleva *et al.*, 2001) but also compares proteomes under different stressful environmental factors (Ahsan *et al.*, 2007).

Comparative proteomics analysis in JG-62, a drought-tolerant variety of chickpea, resulted in the identification of 134 differentially expressed proteins that include predicted and novel dehydration-responsive proteins (Bhushan *et al.*, 2007). Another study provided insights into the complex metabolic network operating in the nucleus during dehydration in the chickpea (Pandey *et al.*, 2008). Approximately 205 protein spots were found to be differentially regulated under dehydration. Mass spectrometry analysis allowed the identification of 147 differentially expressed proteins, presumably involved in a variety of functions including gene transcription and replication, molecular chaperones, cell signalling and chromatin remodelling. The dehydration-responsive nuclear proteome of chickpea revealed a coordinated response, which involves both regulatory and functional proteins (Reddy *et al.*, 2012).

4.6.2 Transcriptomics

Transcriptomes are defined as the set of all the messenger RNAs (mRNAs) in a cell/tissue/organism, and investigation of populations of mRNAs is thus called 'transcriptomics'. Genome-wide expression profiling is a useful tool for studying genes involved in various biological phenomena, identifying the candidate genes, and revealing the molecular cross-talk of gene regulatory networks among abiotic stress responses (Reddy *et al.*, 2012).

Transcriptional profiling in chickpea under drought, cold and high salinity was undertaken using cDNA microarray analysis to look at the gene expression in the leaf, root and/or flower tissues in tolerant and susceptible genotypes (Mantri *et al.*, 2007). The transcripts differentially expressed in response to each particular stress were analysed and a transcriptional change of over two-fold was pronounced for 109, 210 and 386 genes after drought, cold and high-salinity treatments, respectively. Among these, 2, 15 and 30 genes were differentially expressed between tolerant and susceptible genotypes studied for drought, cold and high salinity, respectively. The differentially expressed genes coded for various functional and regulatory proteins, highlighting the multiple gene control and complexity of abiotic stress response mechanisms in chickpea (Reddy *et al.*, 2012).

4.6.3 Genomics

The use of genetic and genomic analysis to assist in identifying DNA regions tightly linked to agronomic traits in crops, the so-called 'molecular markers, can assist breeding strategies for crop enhancement. Genomics includes the development of molecular markers for genetic diversity analysis and it provides novel opportunities to manipulate quantitative trait loci (QTL) via marker-assisted selection (MAS) to develop improved cultivars (Reddy *et al.*, 2012).

MAS is being deployed in chickpea at ICRISAT (India) to introgress QTL alleles associated with a large root size into elite germplasm (Saxena *et al.*, 2002). Terminal drought can cut chickpea yield by amounts ranging from 20% to more than 50%. Hence, a deep root system capable of extracting additional soil moisture should positively impact yield in drought-prone areas (Crouch & Serraj, 2002).

4.6.4 Transgenomics

The use of transgenic technology, or 'transgenomics', potentially offers a more targeted gene-based approach for understanding the mechanisms governing stress tolerance, providing a complementary means for genetic enhancement of field crops, thereby alleviating some of the major constraints to crop productivity in developing countries (Sharma & Ortiz, 2000).

Various transgenic technologies for improving stress tolerance have been developed involving the expression of functional genes including those encoding for enzymes required for the biosynthesis of osmoprotectants (Tarczynski *et al.*, 1993; Kavi Kishore *et al.*, 1995; Hayashi *et al.*, 1997) or modifying membrane lipids (Kodama *et al.*, 1994; Ishizaki-Nishizawa *et al.*, 1996), late embryogenesis proteins (Xu et al 1996) and detoxification enzymes (McKersie *et al.*, 1996).

Osmoregulation is one of the best mechanisms for abiotic stress tolerance, especially if osmoregulatory genes could be triggered in response to drought, salinity and high temperatures. A prokaryotic osmoregulatory choline oxidase gene (*codA*) has been targeted at the chloroplasts to enhance the potential of the photosynthetic machinery of chickpea to withstand oxidative damage (Reddy *et al.*, 2012). Chloroplasts from plants of transgenic lines were evaluated for their efficacy to withstand photo-inhibitory damage. The the loss in photosystem II (PSII) activity in chloroplasts of wild-type plants exposed to high light intensity was significantly greater than that in chloroplasts of transgenic chickpeas. The results indicated that H_2O_2 produced by choline oxidase as a by-product during synthesis of glycine-betaine is responsible for building a stronger antioxidant system in chloroplasts of transgenic chickpea plants (Sharmila *et al.*, 2009). Similarly at ICRISAT, the *P5CSF129A* gene encoding the mutagenized D1-pyrroline-5-carboxylate synthetase (P5CS) for the overproduction of proline was introduced into chickpea (Reddy *et al.*, 2012). The accumulation of proline in several of these transgenic lines was more pronounced and increased significantly in the leaves when exposed to water stress along with a decrease in free radicals as measured by a decrease in levels of malonaldehyde (MDA), a lipid peroxidation product (Bhatnagar-Mathur *et al.* 2009). However, the overexpression of proline appeared to have no beneficial effect on biomass accumulation since only a few transgenic lines showed a significant increase in biomass production towards the end of the progressive drying period. In any case, the overexpression of *P5CSF129A* gene resulted in only a modest increase in the transpiration efficiency (TE), thereby indicating that the enhanced proline had little bearing on the components of yield architecture that are significant in overcoming the negative effects of drought stress in chickpea (Reddy *et al.*, 2012). These results agree with earlier reports in other crops (Turner *et al.*, 1980; Morgan, 1984; Serraj & Sinclair, 2002).

4.7 Biotic stress

Diseases, insects, pests and plant-parasitic nematodes are the major biotic stressors that can drastically affect chickpea yield (Basandrai *et al.*, 2011).

4.7.1 Chickpea and diseases

The major diseases are ascochyta blight (*Ascochyta rabiei*), fusarium wilt (*Fusarium oxysporum* f. sp. *Ciceri*), phytophthora root rot (*Phytophthora medicaginis*) and botrytis grey mould (*Botrytis cinerea*) (Ahmad *et al.*, 2005; Knights *et al.*, 2008; Singh *et al.*, 2008).

4.7.2 Chickpea and insect pests

The major pests include helicoverpa pod borer (*Helicoverpa armigera* and *Helicoverpa punctigera*) and leaf miner (*Liriomyza cicerina*) (Ahmad *et al.*, 2005; Materne *et al.*, 2011).

4.7.3 Chickpea and plant-parasitic nematodes

Plant-parasitic nematodes reported in the major chickpea-growing areas include reniform nematode (*Rotylenchulus reniformis*), root-knot nematodes (*Meloidogyne* spp.), root-lesion nematodes (*Pratylenchus* spp.) and cyst-forming nematodes (*Heterodera* spp.); they are estimated to cause annual yield losses of 14% (Castillo *et al.*, 2008; Materne *et al.*, 2011).

4.8 Breeding of chickpea for biotic stress

Annual *Cicer* species have been assessed for their reaction to ascochyta blight, fusarium wilt, cyst nematode, leaf miner and seed beetle at ICARDA (International Centre for Agricultural Research in the Dry Areas), and a high level of resistance to each stress has been identified (reviewed by Kumar *et al.*, 2011).

4.8.1 Breeding chickpea for resistance to diseases

Varieties with enhanced ascochyta blight resistance have been released in India, Pakistan, Syria, the USA, Canada and Australia (Ahmad *et al.*, 2005; Materne *et al.*, 2011).

Basandrai *et al.* (2011) reported that sources resistant to ascochyta blight have been identified and used in breeding programmes (Malhotra *et al.*, 2003; Pande *et al.*, 2005, 2007b; Basandrai *et al.*, 2008). Lines with moderate resistance to this disease have continually been delivered to national programmes from ICARDA and ICRISAT (Malhotra *et al.*, 2003; Pande *et al.*, 2005, 2006; Basandrai *et al.*, 2008; Kaur *et al.*, 2008; Sarker *et al.*, 2008). Pande *et al.* (2006) reported three accessions as being moderately resistant to ascochyta blight, but to date resistance sources have not been identified against pathotypes III and IV, as identified in Syria (Sarker *et al.*, 2008). Gene pyramiding in lines has resulted in higher resistance (Kaur *et al.*, 2008; Gaur *et al.*, 2010).

Materne *et al.* (2011) reported that varieties with stable resistance to fusarium wilt have been released in India and also in other countries, and recent advances in understanding of the genetic control of resistance are likely to result in successful pyramiding of resistant genes (Singh *et al.*, 2008). Pande *et al.* (2006) reported high levels of resistance to this disease, with 21 accessions free from the disease and 25 resistant. Chaudhary (2009) showed that the varieties JG 315, Avrodhi, DCP 92-3, JG 74, BG 372 and KWR 108 were reported as being resistant against fusarium wilt (Basandrai *et al.*, 2011).

A number of varieties moderately resistant to dry root rot have been identified (Pande *et al.*, 2006; Kaur *et al.*, 2008). Pande *et al.* (2006) reported six accessions with moderate resistance to dry root rot among 211 accessions in the desi chickpea mini-core collection. Under natural epiphytotic conditions, lines GL 84102, GL 88223, GLK 88114 and GF 89-75 showed moderate resistance to stem rot. The wild *Cicer* species *C. judaicum*, *C. reticulatum*, *C. pinnatifidum* and *C. yamashitae* are reported as being tolerant to stem rot (Kaur *et al.*, 2008).

Forty genotypes and some accessions of wild *Cicer* species (*C. judaicum*, *C. reticulatum*, *C. echinospermum* and *C. pinnatifidum*) have shown resistance to botrytis grey mould (BGM; Basandrai *et al.*, 2006, 2008; Pande *et al.*, 2006, 2007a). Some lines derived through interspecific hybridization have shown a high level of resistance to BGM (Kaur *et al.*, 2008). *C. judaicum* is reported to have resistant genes for ascochyta blight, fusarium wilt and botrytis grey mould (van der Maesen & Pundir, 1984).

Genotypes ICC 11284 and ICC 13441 showed combined resistance against ascochyta blight and BGM, and dry root rot and fusarium wilt, respectively (Anonymous, 2010).

4.8.2 Breeding chickpea for resistance to insects and pests

Sharma (2004) evaluated 150 accessions of wild chickpea for resistance to helicoverpa pod borer under field and greenhouse conditions. Basandrai *et al.* (2011) reported that a large amount of germplasm, including cultigens and wilds, was evaluated for resistance to pod borer, and low to moderate levels of resistance were reported, with line ILL 506 possessing a good level of resistance (Pratap *et al.*, 2002; Sharma *et al.*, 2003). Few resistance sources for leaf miner have been identified at ICARDA (Malhotra *et al.*, 1996).

4.8.3 Breeding chickpea for resistance to plant-parasitic nematodes

Greco and Di Vito (1993) reported valuable sources of resistance to cyst nematode in *C. bijugum*, *C. pinnatifidum* and *C. reticulatum*.

Some wild accessions have shown resistance to more than one stress (Singh *et al.*, 1994; Ahmad *et al.*, 2005). For example, ILWC 7-1 of *C. bijugum* showed resistance to ascochyta blight, fusarium wilt, leaf miner, cyst nematode and cold, and ILWC 33/S-4 of *C. pinnatifidum* to ascochyta blight, fusarium wilt, seed beetle and cyst nematode (Kumar *et al.*, 2011).

Various authors (Jaiswal & Singh, 1989; Singh & Ocampo, 1997; S. Singh *et al.*, 2005; Kumar *et al.*, 2011) have shown that accessions of *C. reticulatum* potentially can supply genes for high yield.

4.9 Conclusion and future prospects

Chickpea is acknowledged as the poor man's meat because of its high nutritional value, especially its protein content. Chickpea is affected by various abiotic stresses but drought is considered the most important environmental constraint to crop productivity. There is a lack of simple and accurate screening procedures to screen parental genotypes and breeding populations for various abiotic stresses, and this is the major bottleneck in the development of stress-tolerant cool season pulse crops. Genotypes resistant to various abiotic stresses are not widely available to breeders. According to some researchers wild species of cool season pulse crops possess desired traits for a number of abiotic stresses, and conventional or biotechnological approaches are needed to develop cultivars resistant to abiotic stress. The tolerant varieties mentioned in this chapter shall help the ongoing and future investigations of abiotic stress in chickpea with the aim of developing durable stress tolerance.

References

Abbo S, Molina C, Jungmann R (2005) Quantitative trait loci governing carotenoid concentration and weight in seeds of chickpea (*Cicer arietinum* L.). *Theor Appl Genet* 111: 185–195.

Ahmad F, Gaur P, Croser J (2005) Chickpea (*Cicer arietinum* L.). In: Singh R, Jauhar PP (eds), *Genetic Resources, Chromosome Engineering and Crop Improvement – Grain Legumes*, Vol. 1. CRC Press, pp. 187–217.

Ahmad P, Sarwat M, Sharma S (2008) Reactive oxygen species, antioxidants and signaling in plants. *J Plant Biol* 51: 167–173.

Ahmad P, Jaleel CA, Sharma S (2010) Antioxidative defense system, lipid peroxidation, proline metabolizing enzymes and biochemical activity in two genotypes of *Morus alba* L. subjected to NaCl stress. *Russ J Plant Physiol* 57: 509–517.

Ahmad P, Nabi G, Ashraf M (2011) Cadmium-induced oxidative damage in mustard [*Brassica juncea* (L.) Czern. & Coss.] plants can be alleviated by salicylic acid. *South Afr J Bot* 77: 36–44.

Ahmad P, Hakeem KR, Kumar A, Ashraf M, Akram NA (2012a) Salt-induced changes in photosynthetic activity and oxidative defense system of three cultivars of mustard (*Brassica juncea* L.). *Afr J Biotechnol* 11: 2694–2703.

Ahmad P, Ozturk M, Gucel S (2012b) Oxidative damage and antioxidants induced by heavy metal stress in two cultivars of mustard (L) plants. *Fresen Environ Bull* 21: 2953–2961.

Ahmad P, Ashraf M, Azooz MM, Rasool S, Akram NA (2014a) Potassium starvation-induced oxidative stress and antioxidant defense responses in *Brassica juncea*. *J Plant Interact* 9: 1–9.

Ahmad P, Ozturk O, Sharma S, Gucel S (2014b) Effect of sodium carbonate-induced salinity-alkalinity on some key osmoprotectants, protein profile, antioxidant enzymes, and lipid peroxidation in two mulberry (*Morus alba* L.) cultivars. *J Plant Interact* 9: 460–467.

Ahsan N, Lee DG, Lee SH, Lee KW, Bahk JD, Lee BH (2007) A proteomic screen and identification of waterlogging-regulated proteins in tomato roots. *Plant Soil* 295: 37–51.

Amede, T, Schubert S, Starh K (2003) Mechanisms of drought resistance in grain legumes. I. Osmotic adjustment. *Ethiopian J Sci* 26: 37–46.

Anonymous (2010) Project Coordinator's Report 2009–10 of All-India Coordinated Research Project on Chickpea. Indian Institute of Pulses Research, Kanpur, India.

Apostolova P, Yordanova R, Popova L (2008) Response of antioxidative defense system to low temperature stress in two wheat cultivars. *Gen Appl Plant Physiol* 34: 281–294.

Bampidisa VA, Christodoulou V (2011) Chickpeas (*Cicer arietinum* L.) in animal nutrition: A review. *Animal Feed Sci Technol* 168: 1–20.

Basandrai AK, MacLeod WJ, Siddique K, Pande S, Payne P (2006) Evaluation of wild chickpea germplasm against Australian isolates of *Botrytis cinerea* the causal organism of *Botrytis* grey mould of chickpea. *BGM Newsletter* November.

Basandrai AK, Pande S, Basandrai D (2008) Recent advances in host-parasite interaction, chickpea-*ascochyta rabiei* and *Botrytis cinerea* systems. In: Setia RC, Nayyar H, Setia N (eds), *Crop Improvement, Strategies and Applications*. IK International, New Delhi, pp. 265–301.

Basandrai AS, Basandrai D, Duraimurugan P, Srinivasan T (2011) Breeding for biotic stress. In: Pratap A, Kumar J (eds), Biology and Breeding of Food Legumes. CAB International, pp. 220–240.

Benjamin JG, Nielsen DC (2006) Water deficit effects on root distribution of soybean, field pea and chickpea. *Field Crops Res* 97: 248–253.

Bhatnagar-Mathur P, Vadez V, Jyostna Devi M, Lavanya M, Vani G, Sharma KK (2009) Genetic engineering of chickpea (*Cicer arietinum* L.) with the P5CSF129A gene for osmoregulation with implications on drought tolerance. *Mol Breed* 23: 591–606.

Bhushan D, Pandey A, Choudhary M, Datta A, Chakraborty S, Chakraborty N (2007) Comparative proteomics analysis of differentially expressed proteins in chickpea extracellular matrix during dehydration stress. *Mol Cell Proteomics* 6: 1868–1884.

Boominathan P, Shukla R, Kumar A, *et al.* (2004) Long term transcript accumulation during the development of dehydration adaptation in *Cicer arietinum*. *Plant Physiol* 135: 1608–1620.

Cabrera C, Lloris F, Giménez R (2003) Mineral content in legumes and nuts: contribution to the Spanish dietary intake. *Sci Total Environ* 308: 1–14.

Castillo P, Navas-Cortes JA, Landa BB, Jimenez-Diaz RM, Vovlas N (2008) Plant-parasitic nematodes attacking chickpea and their *in planta* interactions with rhizobia and phytopathogenic fungi. *Plant Dis* 92: 840–853.

Chaudhary RG (2009) Chickpea diseases. In: *25 Years of Research at IIPR*. Indian Institute of Pulses Research, Kanpur, India, pp. 85–115.

Ciftci H, Ozkaya A, Cevrimli BS (2010) Levels of fat-soluble vitamins in some foods. *Asian J Chem* 22: 1251–1256.

Clarke HJ, Khan TN, Siddique KHM (2004) Pollen selection for chilling tolerance at hybridization leads to improved chickpea cultivars. *Euphytica* 139: 65–74.

Clarke HJ, Khan T, Croser J, *et al.* (2008) Temperature tolerance in food legumes. In: Kharkwal MC (ed.), *Food Legumes for Nutritional Security and Sustainable Agriculture*. Indian Journal of Genetics and Plant Breeding, IARI, New Delhi, India, pp. 678–710.

Cobbett C, Goldsbrough P (2002) Phytochelatins and metallothioneins: roles in heavy metal detoxification and homeostasis. *Annu Rev Plant Biol* 53: 159–182.

Crouch JH, Serraj R (2002) DNA marker technology as a tool for genetic enhancement of drought tolerance at ICRISAT. In: *Field Screening for Drought Tolerance in Crop Plants with Emphasis on Rice: International Workshop on Field Screening for Drought Tolerance in Rice*. ICRISAT, Patancheru, India, pp. 155–170.

Doupis G, Chartzoulakis K, Beis A, Patakas A (2011) Allometric and biochemical responses of grapevines subjected to drought and enhanced ultraviolet-B radiation. *Aust J Grape Wine Res* 17: 36–42.

Dua RP, Sharma PC (1995) Salinity tolerance of Kabuli and Desi chickpea genotypes. *International Chickpea and Pigeonpea Newsletter* 2: 19–22.

Duhan A, Khetarpaul N, Bishnoi S (1999) In starch digestibility (in vitro) of various pigeonpea cultivars through processing and cooking. *Ecol Food Nutr* 37: 557–568.

Duke JA (1981) *Handbook of Legumes of World Economic Importance*. Plenum Press, New York.

FAO (2008) Land and plant nutrition management service. Available at: www.fao.org/ag/agl/agll/spush

Fayiga AO, Ma LQ (2006) Using phosphate rock to immobilize metals in soil and increase arsenic uptake by hyperaccumulator Pteris vittata. *Sci Total Environ* 359: 17–25.

Flowers TJ (2004) Improving crop salt tolerance. *J Exp Bot* 55: 307.

Flowers TJ, Gaur PM, Laxmipathigowda CL, *et al.* (2010) Salt sensitivity in chickpea. *Plant Cell Environ* 33: 490–509.

Frimpong A, Sinha A, Tafan B (2009) Genotype and growing environment influence chickpea (*Cicer arientinum* L.) seed composition. *J Sci Food Agric* 89: 2052–2063.

Gaur P, Pande S, Khan T, *et al.* (2010) Multiple disease resistance in chickpea. In: *4th Workshop on Medicago truncatula and 5th International Congress on Legume Genetics and Genomics Pathogens, Pests & Symbionts*, 2–8 July 2010, Washington State University, Washington.

Gecit HH (1991) Chickpea utilization in Turkey. In: *Proceedings of a Consultants Meeting*. ICRISAT, Andhra Pradesh, India, pp. 69–74.

Giordani C, Cecchi S, Zanchi C (2005) Phytoremediation of soil polluted by nickel using agricultural crops. *Environ Manag* 36: 675–681.

Greco N, Di Vito M (1993) Selection for nematode resistance in cool season food legumes. In: Singh KB, Saxena MC (eds), *Breeding for Stress Tolerance in Cool Season Food Legumes*. John Wiley & Sons/ICARDA, Chichester, UK, pp. 157–166.

Grill E, Mishra S, Srivastava S, Tripathi RD (2006) Role of phytochelatins in phytoremediation of heavy metals. In: Singh SN, Tripathi RD (eds), *Environmental Bioremediation Technologies*. Springer, Heidelberg, pp. 101–145.

Gül MK, Ömer EC, Turhan H (2008) The effect of planting time in fatty acids and tocopherols in chickpea. *Eur Food Res Technol* 226: 517–522.

Gupta DK, Tohoyama H, Joho M, Inouhe M (2002a) Possible roles of phytochelatins and glutathione metabolism in cadmium tolerance in chickpea roots. *J Plant Res* 115: 429–437.

Gupta DK, Rai UN, Tripathi RD, Inouhe M (2002b) Impacts of fly-ash on soil and plant responses. *J Plant Res* 115: 401–409.

Gupta DK, Tohoyama H, Joho M, Inouhe M (2004) Changes in the levels of phytochelatins and related metal-binding peptides in chickpea seedlings exposed to arsenic and different heavy metal ions. *J Plant Res* 117: 253–256.

Hayashi H, Mustardy L, Deshnium P, Ida M, Murata N (1997) Transformation of Arabidopsis thaliana with the codA gene for choline oxidase: accumulation of glycine betaine and enhanced tolerance to salt and cold stress. *Plant J* 12: 133–142.

Heidarvand L, Maali-Amiri R, Naghavi MR, Farayedi Y, Sadeghzadeh B, Alizadeh K (2011) Physiological and morphological characteristics of chickpea accessions under low temperature stress. *Russian J Plant Physiol* 58: 157–163.

Hulse JH (1991) Nature, composition and utilization of pulses. In: *Uses of Tropical Grain Legumes, Proceedings of a Consultants Meeting*. ICRISAT, Andhra Pradesh, India, pp. 11–27.

Ibrikci H, Knewtson SJB, Grusak MA (2003) Chickpea leaves as a vegetable green for humans: evaluation of mineral composition. *J Sci Food Agric* 83: 945–950.

Iliadis C (2001) Evaluation of six chickpea varieties for seed yield under autumn and spring sowing. *J Agric Sci Camb* 137: 439–444.

Ishizaki-Nishizawa O, Fujii T, Azuma M, *et al.* (1996) Low-temperature resistance of higher plants is significantly enhanced by a nonspecific cyanobacterial desaturase. *Nat Biotechnol* 14: 1003–1006.

Jaiswal HK, Singh BD (1989) Analysis of gene effects for yield traits in crosses between *C. arietinum* and *C. reticulatum*. *Indian J Genet Pl Br* 49: 9–17.

Johansen B, Baldev B, Brouwer JB, *et al.* (1994) Biotic and abiotic stresses constraining productivity of cool season food legumes in Asia, Africa and Oceania. In: Muehlbauer FJ, Kaiser WJ (eds), *Expanding the Production and Use of Cool Season Food Legumes*. Kluwer Academic, pp. 175–194.

Jukanti AK, Gaur PM, Gowda CLL, Chibbar RN (2012) Nutritional quality and health benefits of chickpea (*Cicer arietinum* L.): a review. *Brit J Nutr* 108: 11–26.

Kashiwagi J, Krishnamurthy L, Upadhyaya HD, Krishna H, Chandra S, Vadez V, Serraj R (2005) Genetic variability of drought-avoidance root traits in the mini-core germplasm collection of chickpea (*Cicer arietinum* L.). *Euphytica* 146: 213–222.

Kashiwagi J, Upadhyaya HD, Krishnamurthy L, Singh S (2007) Identification of high-yielding large-seeded Kabuli chickpeas with drought avoidance root traits. *J SAT Agric Res* 3(1): 1–2; available at: http://oar.icrisat.org/id/eprint/2952

Katerji N, van Hoorn JW, Hamdy A, Mastrorilli M (2003) Salinity effect on crop development and yield analysis of salt tolerance according to several classification methods. *Agric Water Manag* 62: 37–66.

Kaur L, Tripathi HS, Vishwa D, Reddy MV, Singh G, Kharkwal MC (2008) Management of diseases in food legumes. In: Kharkwal MC (ed.), *Proceedings of the 4th International Food Legumes Research Conference*, 18–22 October 2005. ICRISAT, New Delhi, India, Vol. 1, pp. 608–637.

Kavi Kishore PB, Hong Z, Miao GH, Hu CAA, Verma DPS (1995) Overexpression of Δ^1-pyrroline-5-carbolyate synthetase increases proline production and confers osmotolerance in transgenic plants. *Plant Physiol* 108: 1387–1394.

Knights EJ, Southwell RJ, Schwinghamer MW, Harden S (2008) Resistance to *Phytophthora medicaginis* Hansen and Maxwell in wild *Cicer* species and its use in breeding root rot resistance chickpea (*Cicer arietinum L.*). *Aust J Agr Res* 59: 383–387.

Kodama H, Hamada T, Horiguchi G, Nishimura M, Iba K (1994) Genetic enhancement of cold tolerance by expression of a gene for chloroplast ω-3 fatty acid desaturase in transgenic tobacco. *Plant Physiol* 105: 601–605.

Kumar S, Imtiz M, Gupta S, Pratap A (2011) Distant hybridization and alien gene introgression. In: Pratap A, Kumar J (eds), *Biology and Breeding of Food Legumes*. CAB International, pp. 81–110.

Lebiedzińska A, Szefer P (2006) Vitamins B in grain and cereal-grain food, soy-products and seeds. *Food Chem* 95: 116–122.

Leport L, Turner NC, Davies SL, Siddique KHM (2006) Variation in pod production and abortion among chickpea cultivars tinder terminal drought. *Europ J Agron* 24: 236–246.

Mahajan S, Tuteja N (2005) Cold, salinity and drought stresses: an overview. *Arch Biochem Biophys* 444: 139–158.

Malhotra RS, Singh KB (1991) Gene action for cold tolerance in chickpea. *Theor Appl Genet* 82: 598–601.

Malhotra RS, Singh KB, Van Rheenen HA, Pala M (1996) Genetic improvement and agronomic management of chickpea with emphasis on the Mediterranean region. In: Saxena NP, Saxena MC, Johnsen C, Virmani SM, Harris H (eds), *Adaptation of Chickpea in the West Asia and North Africa Region*. ICRISAT, Patancheru, India, and ICARDA, Aleppo, Syria, pp. 217–232.

Malhotra RS, Baum M, Udupa SM, Bayaa B, Kabbabe S, Khalaf G (2003) *Ascochyta* blight resistance in chickpea with emphasis on the Mediterranean region. In: Sharma RN, Srivastava GK, Rathore AL, Sharma ML, Khan MA (eds), *Proceedings of the International Chickpea Conference on Chickpea Research for the Millennium*, 20–22 January, Indira Gandhi Agricultural University, Raipur, India, pp. 108–117.

Maliro MFA, McNeil D, Redden B, Kollmorgen JF, Pittock C (2007) Sampling strategies and screening of chickpea (*Cicer arietinum* L.) germplasm for salt tolerance. *Genet Resour Crop Ev* doi: 10.1007/s10722-007-9214-9.

Mantri NL, Ford R, Coram TE, Pang ECK (2007) Transcriptional profiling of chickpea genes differentially regulated in response to high-salinity, cold and drought. *BMC Genomics* 8: 303.

Materne M, McNeil D, Hobson K, Ford R (2007) Abiotic stresses. In: Yadav SS, McNeil D, Stevenson PC (eds), *Lentil: An Ancient Crop for Modern Times*. Springer, Dordrecht, The Netherlands, pp. 315–329.

Materne M, Leonforte A, Hobson K, Paull J, Gnanasambandam A (2011) Breeding for improvement of cool season food legumes. In: Pratap A, Kumar J (eds), *Biology and Breeding of Food Legumes*. CAB International, pp. 49–62.

McKersie BD, Bowley SR, Harjanto E, Leprince O (1996) Water-deficit tolerance and field performance of transgenic alfalfa over-expressing superoxide dismutase. *Plant Physiol* 111: 1177–1181.

Meharg AA, Hartley-Whitaker J (2002) Arsenic uptake and metabolism in arsenic resistant and non resistant plant species. *New Phytol* 154: 29–43.

Mehra RK, Tripathi RD (2000) Phytochelatins and metal tolerance. In: Agrawal M, Agrawal SB (eds), *Environmental Pollution and Plant Responses*. CRC Press, Boca Raton, pp. 367–382.

Millan T, Clarke HJ, Siddique KHM, *et al.* (2006) Chickpea molecular breeding: new tools and concepts. *Euphytica* 147: 81–103.

Morgan JM (1984) Osmoregulation and water stress in higher plants. *Annu Rev Plant Physiol* 35: 299–348.

Muehlbauer FJ (1993) Food and grain legumes. In: Janick J, Simon JE (eds), *New Crops*. John Wiley & Sons, Inc., New York, pp. 256–265.

Munns R (2003) Comparative physiology of salt and water stress. *Plant Cell Environ* 25: 239–250.

Murray GA, Eser D, Gusta LV, Eteve G (1988) Winter hardiness in pea, lentil, faba bean and chickpea. In: Summerfield RJ (ed.), *World Crops: Cool Season Food Legumes*. Kluwer Academic Publishers, Dordrecht, The Netherlands, pp. 831–843.

Oplinger ES, Hardman LLS, Oelke EA, Kaminski AR, Schulte EE, Doll JD (1990) Chickpeas (garbanzo bean). Alternative Field Crops Manual. University of Wiconsin Coop. Ext., University of Minnesota Center for Alternative Plant and Animal Products and the Minnesota Ext. Service.

Pande S, Siddique KHM, Kishore GK, *et al.* (2005) *Ascochyta* blight of chickpea: Biology, pathogenicity, and disease management. *Aust J Agr Res* 56: 317–322.

Pande S, Krishna Kishore G, Upadhyay HD, Narayana Rao J (2006) Identification of sources of multiple disease resistance in mini-core collection of chickpea. *Plant Dis* 90: 1214–1218.

Pande S, Galloway J, Gaur PM, *et al.* (2007a) Botrytis grey mould of chickpea, a review of biology, epidemiology, and disease management. *Aust J Agr Res* 57: 1137–1150.

Pande S, Gaur PM, Sharma M, Rao JN, Rao BV, Krishna Kishore G (2007b) Identification of single and multiple disease resistance in desi chickpea genotypes to *Ascochyta* blight, Botrytis gray mold and *Fusarium* wilt. *J SAT Agr Res* 3(1): 1–3; available at: http://oar.icrisat.org/id/eprint/2617

Pandey A, Chakraborty S, Datta A, Chakraborty N (2008) Proteomics approach to identify dehydration responsive nuclear proteins from chickpea (*Cicer arietinum* L). *Mol Cell Proteomics* 7: 88–107.

Pittaway JK, Robertson IK, Ball MJ (2008) Chickpeas may influence fatty acid and fiber intake in an ad libitum diet, leading to small improvements in serum lipid profile and glycemic control. *J Am Diet Assoc* 108: 1009–1013.

Porubleva L, Vander VK, Kothari S, Oliver DJ, Chitnis PR (2001) The proteome of maize leaves: use of gene sequences and expressed sequence tag data for identification of proteins with peptide mass fingerprints. *Electrophoresis* 22: 1724–1738.

Potters G, Pasternak TP, Guisez Y, Palme KJ, Jansen MAK (2007) Stress-induced morphogenic responses: growing out of trouble. *Trends Plant Sci* 12: 98–105.

Pratap A, Basandrai D, Mehta PK, Basandrai AK (2002) Evaluation of chickpea (*Cicer arietinum* L.) genotypes against pod borer. *Indian J Genet Pl Br* 62: 131–134.

Pundir RPS, Reddy KN, Mengesha MH (1992) Pod volume and pod filling as useful traits of chickpea. *International Chickpea Newsletter* 17: 18–20.

Raab A, Feldmann J, Meharg AA (2004) The nature of arsenic-phytochelatin complexes in *Holcus lanatus* and *Pteris cretica*. *Plant Physiol* 134: 1113–1122.

Raab A, Schat H, Meharg AA, Feldmann J (2005) Uptake, translocation and transformation of arsenate and arsenite in sunflower (*Helianthus annuus*): Formation of arsenic-phytochelatin complexes during exposure to high arsenic concentrations. *New Phytol* 168: 551–558.

Rasool S, Ahmad A, Siddiqi TO, Ahmad P (2013) Changes in growth, lipid peroxidation and some key antioxidant enzymes in chickpea genotypes under salt stress. *Acta Physiol Plant* 35: 1039–1050.

Reddy DS, Bhatnagar-Mathur P, Vadez V, Sharma KK (2012) Grain legumes (soybean, chickpea, and peanut): omics approaches to enhance abiotic stress tolerance. In: Tuteja N, Gill SS, Tiburcio AF, Tuteja R (eds), *Improving Crop Resistance to Abiotic Stress*. Wiley-VCH Verlag GmbH & Co., pp. 995–1032.

Reddy MV, Singh KB, Saxena MC, Malhotra RS, Hanounik S (1985) Tallest chickpea. *International Chickpea Newsletter* 12: 7–8.

Rout NP, Shaw BP (2001) Salt tolerance in aquatic macrophytes: Ionic relation and interaction. *Biol Plant* 55: 91–95.

Ryan JG (1997) A global perspective on pigeonpea and chickpea sustainable production systems: Present status and future potential. In: Asthana AN, Ali M (eds), *Recent Advances in Pulses Research*. Indian Society of Pulses Research and Development, Kanpur, India, pp. 1–31.

Saeed A, Darvishzadeh R, Hovsepyan H, Asatryan A (2010) Tolerance to freezing stress in *Cicer* accessions under controlled and field conditions *Afr J Biotech* 9: 2618–2626.

Sahu RK, Katiyar S, Tiwari J, Kisku GC (2007) Assessment of drain water receiving effluent from tanneries and its impact on soil and plants with particular emphasis on bio-accumulation of heavy metals. *J Environ Biol* 28: 685–690.

Sarker A, Singh BB, Malhotra RS, *et al.* (2008) Insect pest management in food legumes, the future strategies. In: Kharkwal MC (ed.), *Proceedings of the 4th International Food Legumes Research Conference*, 18–22 October 2005. ICRISAT, New Delhi, India, vol. 1, pp. 588–607.

Savithri KS, Ganapathy PS, Sinha SK (1980) Sensitivity to low temperature in pollen germination and fruit-set in *Cicer arietinum* L. *J Exp Bot* 31: 475–481.

Saxena S (1993) Micropropagation of adult plants of two bamboo species: *Dendrocalamus longispathus* Kurz and *Bambusa vulgaris* Schrader. PhD thesis, University of Delhi.

Saxena NP, Krishanamurthy L, Johansen C (2002) Genetic improvement of drought in chickpea at ICRISAT. In: Saxena NP, O'Toole JC (eds), *Field Screening for Drought Tolerance in Crop Plants with Emphasis on Rice: International Workshop on Field Screening for Drought Tolerance in Rice, ICRISAT*. ICRISAT, Patancheru, India, pp. 128–137.

Serraj R, Sinclair TR (2002) Osmolyte accumulation: can it really help increase crop yield under drought conditions? *Plant Cell Environ* 25: 333–341.

Serraj R, Krishnamurthy L, Upadhyaya HD (2004) Screening chickpea mini-core germplasm for tolerance to soil salinity. *International Chickpea and Pigeonpea Newsletter*, 11: 29–32.

Sharma HC (2004) A little help from wild: exploiting wild relatives of chickpea for resistance to *Helicoverpa armigera*. ICRISAT, Patancheru, India.

Sharma KK, Ortiz R (2000) Program for the application of the genetic engineering for crop improvement in the semi-arid tropics. *In Vitro Cell Dev Biol Plant* 36: 83–92.

Sharma RN, Srivastava G, Rathore AL, Sharma ML, Khan MA (2003) Chickpea research for the millennium. In: *Proceedings of the International Chickpea Conference*, 20–22 January 2003. Indira Gandhi Agricultural University, Raipur, India.

Sharmila P, Phanindra MLV, Anwar F, Singh K, Gupta S, Pardha Saradhi P (2009) Targeting prokaryotic choline oxidase into chloroplasts enhances the potential of photosynthetic machinery of plants to withstand oxidative damage. *Plant Physiol Biochem* 47: 391–396.

Sheldrake AR, Saxena NP (1979) The growth and development of chickpea under progressive moisture stress. In: Mussel H, Staples RC (eds), *Stress Physiology in Crop Plants*. John Wiley & Sons, Ltd., Chichester, pp. 465–485.

Siddique KH, Johansen C, Kumar Rao JVDK, Ali M (2005) Role of legumes in sustainable cropping systems. In: *Abstracts, Fourth International Food Legumes Research Conference – Food legumes for Nutritional Security and Sustainable Agriculture*, 18-22 Oct, New Delhi, India, p 31.

Singh B, Singh BK, Kumar J, Yadav SS, Usha K (2005) Effects of salt stress on growth, nodulation, and nitrogen and carbon fixation of ten genetically diverse lines of chickpea (*Cicer arietinum* L.). *Aust J Agric Res* 56: 491–495.

Singh G, Kapoor S, Singh K (1982) Screening of chickpea for grey mold resistance. *International Chickpea Newsletter* 7: 13–14.

Singh KB, Ibrahim H (1990) *Hybridization Techniques in Chickpea*. ICARDA, Aleppo, Syria.

Singh KB, Ocampo B (1997) Exploitation of wild *Cicer* species for yield improvement in chickpea. *Theor Appl Genet* 95: 418–423.

Singh KB, Reddy MV (1993) Sources of resistance to *Ascochyta* blight in wild *Cicer* species. *Netherlands J Plant Pathol* 99: 163–167.

Singh KB, Reddy MV, Nene YL (1984) International testing of chickpeas for resistance to Ascochyta blight. *Plant Dis* 68: 782–784.

Singh KB, Malhotra RS, Saxena MC (1989) Chickpea evaluation for cold tolerance under field conditions. *Crop Sci* 29: 282–285.

Singh KB, Malhotra RS, Saxena MC (1992) Registration of ILC 8262, a cold-tolerant germplasm line of chickpea. *Crop Sci* 32: 508.

Singh KB, Malhotra RS, Halila H, Knights EJ, Verma, MM (1994) Current status and future strategy in breeding chickpea for resistance to biotic and abiotic stresses. *Euphytica* 73: 137–149.

Singh KB, Omar M, Saxena MC, Johansen C (1997) Screening for drought resistance in spring chickpea in the Mediterranean region. *J Agron Crop Sci* 178: 227–235.

Singh R, Sharma P, Varshney RK, Sharma SK, Singh NK (2008) Chickpea improvement: Role of wild species and genetic markers. *Biotechnol Genet Eng Rev* 25: 267–314.

Singh S, Gumber RK, Joshi N, Singh K (2005) Introgression from wild *Cicer reticulatum* to cultivated chickpea for productivity and disease resistance. *Plant Breeding* 124: 477–480.

Srinivasan A, Saxena NP, Johansen C (1999) Cold tolerance during early reproductive growth of chickpea (*Cicer arietinum* L.): Genetic variation in gamete development and function. *Field Crops Res* 60: 209–222.

Stoeva N, Berova M, Zlatev Z (2005) Effect of arsenic on some physiological parameters in bean plants. *Biol Plant* 49: 293–296.

Stoeva N, Bineva T (2003) Oxidative changes and photosynthesis in oat plants grown in As-contaminated soil. *Bulg J Plant Physiol* 29: 87–95.

Tarczynski MC, Jensen RG, Bohnert HJ (1993) Stress protection of transgenic tobacco by production of the osmolyte mannitol. *Science* 259: 508–510.

Tiwari KN, Dwivedi BS (1990) Response of eight winter crops to zinc fertilizer on a Typic Ustochrept soil. *J Agric Sci (Camb)* 115: 383–387.

Toker C, Ça irgan MI (1998) Assessment of response to drought stress of chickpea (*Cicer arietinum* L.) lines under rainfed conditions. *Turk J Agric For* 22: 615–621.

Toker C, Canci H (2003) Selection of chickpea (*Cicer arietinum* L.) genotypes for resistance to ascochyta blight [*Ascochyta rabiei* (Pass.) Labr.], yield and yield criteria. *Turk J Agric For* 27: 277–283.

Toker C, Mutlu N (2011) Breeding for abiotic stress. In: Pratap A, Kumar J (eds), *Biology and Breeding of Food Legumes*. CAB International, pp. 241–261.

Toker C, Yadav SS (2010) Legume cultivars for stress environments. In: Yadav SS, McNeil DL, Redden R, Patil SA (eds), *Climate Change and Management of Cool Season Grain Legume Crops*. Springer, Dordrecht, The Netherlands, pp. 351–376.

Toker C, Lluch C, Tejera NA, Siddique KHM (2007) Abiotic stresses. In: Yadav SS, Redden R, Chen W, Sharma B (eds), *Chickpea Breeding and Management*. CAB International, Wallingford, UK, pp. 474–496.

Toker C, Yildirim T, Canci H, Inci EN, Ceylan FO (2010) Inheritance of resistance to iron deficiency chlorosis in chickpea (Cicer arietinum L.). *J Plant Nutr* 33: 1366–1373.

Tripathi RD, Srivastava S, Mishra S , Singh N, Tuli R, Gupta DK, Maathuis FJM (2007) Arsenic hazards: Strategies for tolerance and remediation by plants. *Trends Biotechnol* 25: 158–165.

Turner NC, Jones MM (1980) Turgor maintenance by osmotic adjustment: a review and evaluation. In: Turner NC, Kramer PJ (eds), *Adaptation of Plants to Water and High Temperature Stress*. Wiley Interscience, New York, pp. 38–42.

USDA (2010) National Nutrient Database for Standard Reference, Release 22 (2009). United States Department of Agriculture. Available at: http://www.nal.usda.gov/fnic/foodcomp/search/ [accessed 20 June 2014].

van der Maesen LJG (1987) Origin, history and taxonomy of chickpea. In: Saxena MC, Singh KB (eds), *The Chickpea*. CAB International, Wallingford, UK, pp. 11–34.

van der Maesen, LLG, Pundir RPS (1984) Availability and use of wild *Cicer* germplasm. *Plant Genetic Resources Newsletter* 57: 19–24.

Vavilov NI (1926) *Studies on the Origin of Cultivated Plants*. Leningrad, pp. 129–238.

Wallace A (1960) Iron chlorosis in horticultural plants: A review. *Proc Am Soc Horticul Sci* 75: 819–841.

Wery J, Slim SN, Knights EJ, Malhotra RS, Cousin R (1994) Screening techniques and sources of tolerance to extremes of moisture and air temperature in cool season food legumes. In: Muehlbauer FJ, Kaiser WJ (eds), *Expanding the Production and Use of Cool Season Food Legumes*. Kluwer Academic Publishers, pp. 439–456.

Xu D, Duan X, Wang B, Hong B, Ho THD, Wu R (1996) Expression of a late embryogenesis abundant protein gene, HVA1, from barley confers tolerance to water deficit and salt stress in transgenic rice. *Plant Physiol* 110: 249–257.

Ye X, Babili A, Kioti A (2000) Engineering the pro-vitamin A biosynthetic pathway into (carotenoid free) rice endosperm. *Science* 287: 303–305.

Zia-Ul-Haq M, Ahmad M, Iqbal S (2007) Characterization and compositional study of oil from seeds of desi chickpea (*Cicer arietinum* L.) cultivars grown in Pakistan. *J Am Oil Chem Soc* 84: 1143–1148.

Zohary D, Hopf M (2000) *Domestication of Plants in the Old World*, 3rd edn. Clarendon Press, Oxford.

CHAPTER 5

Chickpea and temperature stress: An overview

Viola Devasirvatham[1], Daniel K.Y. Tan[1], Pooran M. Gaur[2] and Richard M. Trethowan[1]
[1] *Faculty of Agriculture and Environment, Plant Breeding Institute, The University of Sydney, Cobbitty, NSW, Australia*
[2] *International Crops Research Institute for the Semi-Arid Tropics, Patancheru, Telangana, India*

5.1 Introduction

Chickpea is an important food grain legume and an essential component of crop rotations throughout the world. However, the adaptation and productivity of chickpea is often limited by low and high temperatures. Cold stress generally occurs in the late vegetative and reproductive stages across the geographical areas of chickpea production. Cold and freezing temperatures (−1.5°C to 15°C) are considered a major problem during the seedling stage of winter-sown chickpea in Mediterranean areas and autumn-sown crops in temperate regions (Singh, 1993). South Australia and parts of north India are most affected by chilling temperatures at flowering (Berger *et al.*, 2011). On the other hand, high day and night temperatures (>30/16°C) may cause damage during the reproductive stage on winter-sown chickpea in Mediterranean in-season rainfall areas, south Asia and spring-sown regions (Berger *et al.*, 2011). In chickpea, temperature is a major environmental factor regulating the timing of flowering thus influencing grain yield (Summerfield *et al.*, 1990; Berger *et al.*, 2004). Both low and high temperatures can limit the growth and grain yield of chickpea at all phenological stages.

The FAO climate change technical paper and the Intergovernmental Panel on Climate Change (IPCC) have provided evidence of climate change linked to human activity. Global temperature has been increasing at the rate of 0.74°C per 100 years (IPCC, 2007a). Over the past 50 years, the linear warming trend has been nearly twice the rate of the previous 100 years (FAO, 2009). Projections to the end of the 21st century estimate a rise in global mean temperature of between 1.8 and 4°C, depending on greenhouse emissions and changes in rainfall patterns (IPCC, 2007a,b). Such changes in climate will impact crop production and some estimates suggest a grain yield decrease of between 8 and 30% (ICRISAT, 2009).

Changes in seasonal temperature and rainfall patterns and their subsequent impact on yield may change the geographic distribution of chickpea production. In Australia, chickpea could expand in new production areas where the frequency of low temperatures (<15°C) is higher during early crop growth (Maqbool *et al.*, 2010). However, temperatures lower than 10°C at flowering can reduce grain yield by 15–20% (Chaturvedi *et al.*, 2009). In contrast, the frequency of high temperatures (>30°C) during the reproductive stage is often higher in the Australian chickpea production areas of northern New South Wales (NSW) (Devasirvatham *et al.*, 2012a) and any increase in the frequency and duration of these temperatures will limit productivity. A decrease in chickpea yields of 53 kg/ha was observed in north India per 1°C increase in seasonal temperature (Kalra *et al.*, 2008). In south India, the yield loss was estimated to be 10–15% for every 1°C increase beyond the optimum temperature (Upadhyaya *et al.*, 2011). The effect of high and low temperatures on grain quality (grain size and seed coat colour) is also a recognized problem (Wery *et al.*, 1994).

Considerable progress in the improvement of chickpea adaption to stressful environments has been made. Screening the germplasm in the field and controlled

Legumes under Environmental Stress: Yield, Improvement and Adaptations, First Edition. Edited by Mohamed Mahgoub Azooz and Parvaiz Ahmad.

environments for stress response has increased our knowledge of plant responses to stress in chickpea, and this information has been used in crop improvement. Physiological response (e.g. canopy temperature) and male (pollen) and female (ovary) reproductive function under stress have been investigated to determine their suitability as stress screening techniques (Clarke & Siddique, 2004; Ibrahim, 2011). Sources of tolerance to temperature stress identified using these methods can be used to develop genetic populations to increase our understanding of inheritance. These populations can be used to map quantitative trait loci (QTL) and the resulting linked markers used for marker-assisted selection (MAS). This chapter explores plant responses to high and low temperatures and the implications for stress tolerance breeding in chickpea.

5.2 Impacts on productivity

5.2.1 Temperature stresses during the vegetative period

Cold temperature (<15°C) at emergence reduces crop establishment and results in plants with low vigour. In some sensitive genotypes, cold temperature causes whole plant necrosis and plant death. During chickpea germination, cold temperature also increases susceptibility to soil-borne pathogens thus retarding plant growth and reducing dry matter production (Wery *et al.*, 1994). A plant survival score can be used as an index to describe genotype tolerance to low temperature under field conditions. Chickpea genotypes (Sel 95Th1716 and Sel 96Th11439) were identified as cold tolerant based on plant survival scores in northwestern Iran (Heidarvand *et al.*, 2011). Genotypic differences in chickpea establishment were identified in Australia, and the cold-tolerant cultivars CPI 562896, Semsen and Sombrero showed improved establishment under low temperatures (Wery *et al.*, 1994).

Cold temperatures decrease membrane stability, modify proteins and lipids, and cause changes in respiration and photosynthesis (Croser *et al.*, 2003). Abscisic acid (ABA) content was observed to increase in seedlings at temperatures of 1–7°C compared with the control (23°C) (Nayyar *et al.*, 2005a). The sugar and proline contents also increased under cold stress in the same study. These observations suggest that manipulation of ABA could improve cold tolerance in chickpea.

Field screening for cold tolerance during the late vegetative stage where plants were exposed to −7.4°C for 3 weeks was effective in identifying cold-tolerant genotypes (Malhotra & Singh, 1991). These authors scored materials on a scale of 1 to 9 and concluded that the method was effective in identifying tolerant, moderately tolerant and sensitive genotypes. This method was used to screen cultivated and wild chickpea in the Mediterranean region (Toker, 2005).

Cold temperature generally encourages prolonged vegetative growth in chickpea. Temperature is the main determinant for flower initiation in most environments, although some authors have linked flower initiation in chickpea to a photothermal response (Roberts *et al.*, 1985). In northern NSW, Australia, flower initiation can commence at ≤15°C, although the occurrence of flower abortion will likely be high (Jenkins & Brill, 2011).

The minimum temperature for germination is 10–15°C (Ellis *et al.*, 1986). At high temperatures, greater than 42.5°C, germination decreases significantly (Ibrahim, 2011) and above 45°C no germination is observed due to lack of embryo growth (Singh & Dhaliwal, 1972). Similarly, high temperature affects photosynthesis, transpiration rate and plant growth (Singh & Dhaliwal, 1972) and the length of vegetative period is generally reduced. In other words, phenology can be modified under high temperature (Summerfield *et al.*, 1984). At high temperatures (>35°C), the vegetative period was reduced by 10 to 15 days compared with optimum temperature (28°C) at Kanpur, India (ICRISAT, 2011). High temperature therefore accelerates flowering and reduces the overall crop growth period.

High temperatures can cause cellular abnormalities such as oxidative stress, and denaturation of proteins and enzymes. Oxidative injury occurs as lipid peroxidation, and hydrogen peroxide content tends to increase in heat-sensitive genotypes at day and night temperatures of over 40/30°C compared with heat-tolerant genotypes (Kumar *et al.*, 2012a). ABA remains high at 40/35°C but was observed to decline at 45/40°C (Kumar *et al.*, 2012b). A membrane injury test based on electrolyte leakage from leaves was shown by Ibrahim (2011) to be an effective measure of high temperature sensitivity in chickpea, with sensitive types displaying high degrees of membrane injury. Therefore, heat stress injury can be measured using a combination of oxidative stress assessments, ABA level and membrane injury in chickpea.

Both high and low temperature stresses can affect seed germination, seedling survival, photosynthesis, membrane function, and protein and hormone function. Cold stress encourages a prolonged vegetative period but high temperatures reduce the vegetative period.

5.2.2 Temperature stresses during the reproductive period

Temperature stress at or around flowering is considered a major challenge to yield in many chickpea production areas. Mean daily temperatures at or exceeding 15°C can cause flower abortion (Clarke & Siddique, 1998). Temperatures of less than 10°C during flowering induce flower shedding, low pod set and ultimately poor seed set. Poor pollen viability and germination are the main reasons for low pod set (Savithri *et al.*, 1980). The field and controlled environment screening at ICRISAT identified chickpea genotypic variation for temperature stress during the flowering stage. Plants exposed to mean daily temperatures of 20°C produced more pods than at 15°C. These experiments identified cold-tolerant genotypes such as ICCV 88502 and ICCV 88503 (Srinivasan *et al.*, 1998). Cold temperature also reduces partitioning of assimilates to the vegetative parts, resulting in reduced harvest index (HI). This reduction in harvest index is more common in south Asia and Australia than other production areas (Siddique & Sedgley, 1986; Saxena, 1990).

Poor pod set in chickpea can occur due to the failure of male or female floral parts, or both. Low temperatures at flowering can affect anther dehiscence. Mean daily temperatures of 15°C can also reduce anther dehiscence and pollen load on stigma (Srinivasan *et al.*, 1999). However, at a similar temperature, pollen viability and pollen germination on the stigma were higher in the tolerant lines ICCV 88501, ICCV 88502 and ICCV 88503 than in the sensitive cultivars Chafa and Annigeri (Srinivasan *et al.*, 1999). Pollen function was clearly more sensitive to temperature change than pistil function (esterase activity). Clarke and Siddique (2004) showed that pollen viability and pollen germination on the stigma were the primary reasons why pod set in chickpea was reduced during low-temperature stress. Pollen sensitivity to low temperatures was identified at 5 and 9 days before anthesis (Clarke, 2001). Clarke and Siddique (2004) and Srinivasan *et al.* (1999) also observed that low temperature did not affect the pistil function, i.e. esterase activity. However, pollen tube growth on the styles of sensitive genotypes was retarded due to cold temperatures. Short pollen tube length at low temperature (15°C) in *in vitro* germination tests was observed at 20°C and 25°C by Savithri *et al.* (1980). Ultimately pod set is reduced by low temperature as observed by Srinivasan *et al.* (1999). They found that pod set was reduced at a low temperature regime of 15/5°C compared with the control (25/15°C).

High temperature during the reproductive stage is a major cause of yield loss due to partial or complete pollen sterility. In chickpea, temperatures at or exceeding 35°C affected male reproductive tissue (anther and pollen), function (pollen germination and tube growth) and pod set. Both anther and pollen showed more structural abnormalities under stress including changes in anther locule number, anther epidermis wall thickening, and pollen sterility rather than functional abnormalities (e.g. *in vivo* pollen germination) (Figure 5.1) (Devasirvatham *et al.*, 2013). Pollen abnormalities can also be found at high temperature, including leakage of pollen protoplast, zigzag pollen tube growth, pseudo-germination and bulbous tip formation in the pollen tube (Devasirvatham *et al.*, 2013). Heat-tolerant chickpea genotypes had clear pollen tube growth on the style following pollen germination and this was confirmed by pod set (Figure 5.2). In heat-sensitive genotypes no pollen germination on the stigma was observed due to complete pollen sterility at temperatures at or exceeding 35°C (Devasirvatham *et al.*, 2012b, 2013) (Figure 5.2). Pollen sterility in the heat-sensitive genotypes is a function of lower sucrose levels, resulting in poor pollen function and pod set (Kaushal *et al.*, 2013).

At very high temperature (45/35°C) both pollen fertility and stigma function can be affected. Observations by Kumar *et al.* (2012b) indicate that oxidative stress in the leaves results in poor fertilization. Devasirvatham *et al.* (2012b, 2013) concluded that the critical temperature affecting pod set was ≥37°C for heat-tolerant genotypes (such as ICCV 92944, ICC 1205 and ICC 15614) and >33°C for heat-sensitive genotypes (ICC 5912, ICC 4567 and ICC 10685).

5.2.3 Temperature stresses during post-anthesis period

Post-anthesis temperature stress, particularly after commencement of pod set, can cause significant pod abortion and decreased grain filling. In chickpea, cold stress decreased the rate and duration of grain filling and

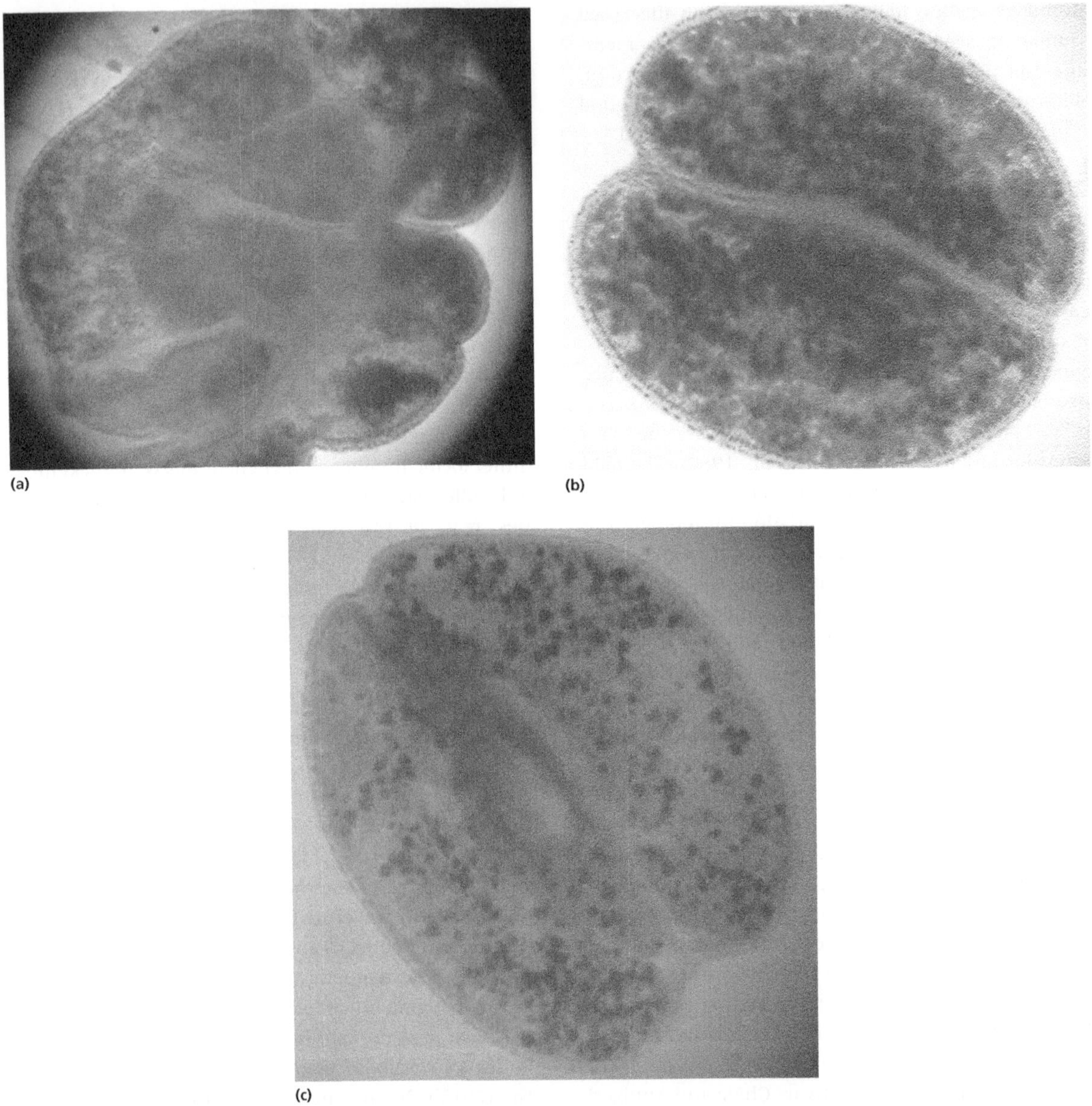

Figure 5.1 Heat-sensitive genotype anther structural abnormalities: anther stained with Alexander's stain. (a) Locule number changed (ICC 4567). (b) Anther epidermis wall is thickened (ICC 4567). (c) Anther shows fertile and sterile pollen grain (ICC 5912). Fertile – red in colour; sterile – green in colour. Scale: 10 μm.

produced smaller seeds (Nayyar *et al.*, 2007; Kaur *et al.*, 2008). At 13/5°C, chickpea average seed weight and size decreased by 41% and 24%, respectively, compared with 28/17°C, largely because seed filling duration reduced from 20 days (non-stressed) to 14 days (cold stressed) (Nayyar *et al.*, 2005b). Similarly, low post-anthesis temperature reduced yield by 1.3 t/ha in northern NSW, Australia, during 2009 (Moore *et al.*, 2010).

In a controlled environment study, Wang *et al.* (2006) reported a grain yield reduction of 33–39% for post-anthesis heat stress compared with pre-anthesis heat stress. This was possibly due to poor remobilization of

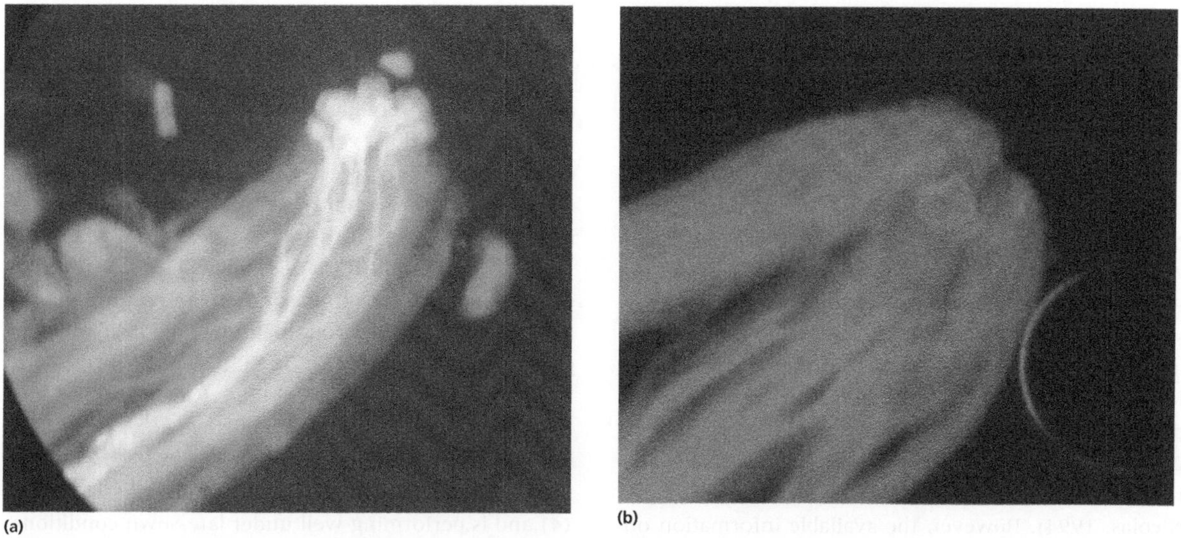

Figure 5.2 Effect of high temperatures on the pollen germination on the stigma. (a) Heat-tolerant: ICC 15614 – pollen germination on the stigma. (b) Heat-sensitive: ICC 10685 – no pollen germination on the stigma. Reproduced from Devasirvatham *et al.* (2013).

Figure 5.3 Comparison of seed size under heat stress. Larger seeds (left side) from non-stressed and smaller seeds (right side) from heat-stressed conditions. Photo courtesy of V. Devasirvatham.

photosynthates to the grain, thus lowering seed weight and seed number per plant (Wang *et al.*, 2006). Both temperature stress extremes influenced seed shape and seed coat colour (Figure 5.3). Generally, temperature stress reduces cotyledon cell number, cell expansion, grain filling rate and ultimately seed weight (Munier-Jolain & Ney, 1998).

Temperature stress can influence grain filling by altering the concentration of hormones, particularly abscisic acid (ABA) and enzymes, in plant tissue. As discussed earlier, ABA plays a significant role in both cold and heat stress tolerance in chickpea (Nayyar *et al.*, 2005a; Kumar *et al.*, 2012a) and is generally downregulated under stress. Exogenous application of ABA increased tolerance to cold stress by improving survival rate through the reproductive stage (Nayyar *et al.*, 2005a). The exogenous application of ABA decreased electrolyte leakage and increased pollen viability, germination, flower retention, pod set, seed size and grain yield (Kumar *et al.*, 2008). Similarly, Kumar *et al.* (2012b) showed that ABA treatment reduced oxidative injury in chickpea under high temperature. Clearly, exogenous application of ABA will improve grain filling under temperature stress and hence grain yield.

5.3 Impacts on nutritional and processing quality

Environmental stresses during seed development have a negative effect on the quality of chickpea seeds (Behboudian *et al.*, 2001). However, comparatively few studies have dealt with the effect of temperature on seed development and quality in chickpea. Nayyar *et al.* (2005b) reported that under cold stress grain sugar concentration increased in chickpea but the accumulation of storage proteins, starch and several amino acids decreased. However, the effect was influenced by the stage of seed development. There was a greater reduction of starch, proteins, soluble sugars, fat, crude fibre and storage protein fractions when cold stress occurred in late pod-filling compared to early pod-filling stages (Kaur *et al.*, 2008). However, seed germination was inhibited when plants were stressed at early pod-filling.

The effects of high temperature stress were generally similar to cold stress. High sucrose synthase and low invertase activity were observed in the seeds of heat-tolerant genotypes compared with heat-sensitive types during early pod filling (Chickpea Technical Report, 2011). Generally, high temperature during grain filling reduces dough and baking quality in grain crops (Stone & Nicolas, 1994). However, the available information on grain quality under temperature stresses in chickpea is limited. There is clearly a need to extend our knowledge of grain quality including baking quality under both high and low temperature stresses.

5.4 Breeding for tolerance to temperature stresses

Chickpea improvement has focused on yield potential and regional adaptation through resistance and tolerance to abiotic and biotic stresses, plant type and grain characteristics. At present, the selected bulk method is the most common selection technique used in chickpea breeding (Gaur *et al.*, 2007). The selected bulk method is relatively inexpensive to employ and the response to selection is generally not inferior to more labour-intensive methods such as pedigree selection (Salimath *et al.*, 2007). This section describes some of the breeding strategies used to improve temperature tolerance in chickpea and explores options for future breeding.

5.4.1 High temperature tolerance

A simple but effective field screening technique for heat tolerance at the reproductive stage in chickpea has been developed at ICRISAT (Gaur *et al.*, 2013, 2014). It involves advancing sowing date to synchronize the reproductive phase of the crop with the occurrence of higher temperatures (≥35°C). This method was effective in identifying heat-tolerant germplasm at ICRISAT and several other locations in India (Gaur *et al.*, 2013, 2014).

A few heat-tolerant chickpea cultivars (ICCV 88512 and ICCV 513) were identified more than a decade ago (Dua, 2001). However, heat tolerance research in chickpea has only received significant attention in recent years. More recently, Krishnamurthy *et al.* (2011) identified 18 stable heat-tolerant genotypes (e.g. ICC 1205, ICC 637 and ICC 15618) by field screening a reference set of chickpea from southern and central Indian field trials. Short-duration, high-yielding, heat-tolerant genotypes (ICC 5597, ICC 5829, ICC 6121, ICC 7410, ICC 11916, ICC 13124, ICC 14284, ICC 14368 and ICC 14653) were identified by Upadhyaya *et al.* (2011). A heat-tolerant breeding line, ICCV 92944, has been released in Myanmar (as Yezin 6) and in India (as JG 14) and is performing well under late-sown conditions (Gaur *et al.*, 2013). Several breeding lines with higher yields under heat stress than the standard cultivar ICCV 929944 have been identified (Gaur *et al.*, 2013, 2014). Outside India, Kaloki (2010) identified ICCV 92318 as a source of heat tolerance in the semi-arid environments of Kenya through the African Climate Change Breeding Program.

Devasirvatham *et al.* (2012b) confirmed the heat tolerance of ICCV 92944 using a pollen selection method. Devasirvatham *et al.* (2013) also confirmed the heat tolerance of germplasm identified earlier by Krishnamurthy *et al.* (2011) (ICC 1205, ICC 15614) using pollen viability in the field and controlled environment studies, and suggested using this technique to develop heat-tolerant cultivars. These materials have been incorporated into chickpea improvement at ICRISAT and new heat-tolerant progeny are under development as genetic mapping populations (Gaur *et al.*, 2013). Diversity arrays technology (DArT) (Mace *et al.*, 2008) markers with good genome coverage were associated with traits targeted for high temperature tolerance in chickpea, and many genomic regions linked with phenology and grain yield have been identified (Devasirvatham, 2012), thus demonstrating the feasibility of applying genetic association analysis to explore complex traits in future. While there is clearly significant variation for high temperature tolerance in adapted chickpea, there is a compelling need to extend the search for new genetic diversity to provide additional allelic variation for temperature tolerance. The wild annual *Cicer* sp. is a possible source of variation and could be exploited. This new allelic variation would allow plant breeders to lift the current reproductive temperature limits on chickpea.

5.4.2 Low temperature tolerance

Low temperature stress breeding generally aims to develop materials adapted to the temperature range −1.5 to 15°C at the reproductive stage and less than −1.5°C at the vegetative growth (Croser *et al.*, 2003). Different sources of resistance to cold tolerance are reported by Chaturvedi *et al.* (2009), and several cold-tolerant breeding lines such as ICCVs 88502, 88503, 88506, 88510 and 88516 have been developed that set pods at less than 15°C in India (ICRISAT, 1994). The Indian Agricultural Research Institute (IARI) has also developed a few cold-tolerant genotypes (BGD 112 green, BG 1100, BG 1101, PUSA 1103, BGD 1005, PUSA 1108, DG 5025, DG 5027, DG 5028, DG 5036 and DG 5042) (Gaur *et al.*, 2007). Using pollen as a selection method, Clarke *et al.* (2004) confirmed the cold tolerance of ICCV 88516 and 88510 and the sensitivity of Amethyst, Dooen, Tyson and FLIP84-15C in Western Australia. Accessions of cultivated and wild *Cicer* sp. were screened for cold tolerance at ICARDA (Singh *et al.*, 1995). These authors reported cold tolerance in the lines ILC 8262, ILC 8617 (a mutant) and a FLIP 97-82C from cultivated *Cicer* along with wild annual chickpea such as *C. bijugum* and *C. reticulatum*.

Later, Toker (2005) identified chilling tolerance (<−1.5°C) in annual wild *Cicer* sp. of *yamashitae*. Heidarvand *et al.* (2011) identified the genotypes Sel 95Th1716 and Sel 96Th11439 as chilling tolerant based on field screening at the vegetative stage where plants were exposed to −11°C to −25°C at the Dryland Agriculture Research Institute (DARI) of Iran.

Both additive and non-additive gene effects govern cold tolerance in chickpea. Cold tolerance was observed to be dominant over susceptibility for at least five sets of genes (Malhotra & Singh, 1990). Breeding at ICARDA has resulted in the expansion of genetic variability for flowering at low temperatures using cultivated × wild *Cicer* crosses. The genes responsible for flowering at low temperature have been transferred from wild to cultivated lines (Chaturvedi *et al.*, 2009). These reports suggested that wild relatives of chickpea can be used as a source of tolerance to low temperatures in applied breeding.

5.5 Conclusions

Both high and low temperature stresses cause grain yield loss. Cold stress encourages a prolonged vegetative period while high temperatures reduce the duration of the vegetative period. Reduced pollen viability and pollen germination on the stigma are the primary causes of poor pod set in chickpea following low temperature stress. Similarly, high temperature stress disrupts pollen viability and anther dehiscence. However, stigma receptivity is not affected by either stress. The rate and duration of seed filling are both decreased by cold and high temperature stresses.

Recent chickpea breeding programmes targeting both high and low temperature stresses have been initiated by many countries including India, Australia and Canada, with global centres such as ICARDA and ICRISAT supporting the wider effort through the characterization and exploitation of genetic resources. Screening for tolerance to temperature stresses has identified many promising sources of tolerance to both high and low temperature in chickpea. However, field-based screening is generally based on delayed sowing, and biomass development and the length of the vegetative phase are reduced in such treatments, thus reducing the fitness of plants to survive temperature extremes at flowering. Field-based methods that impose a temperature stress on a normally grown plant should be developed to confirm and validate the response of chickpea lines already identified. The identification of QTLs for temperature stress tolerance and linked molecular markers will undoubtedly improve rates of genetic advance and marker-assisted selection can easily be incorporated into most breeding methods.

Rapid progress has been made in the development of genomic resources for chickpea, and breeders have already started integrating molecular breeding strategies such as marker-assisted backcrossing (MABC) and marker-assisted recurrent selection (MARS) to improving drought tolerance in chickpea (Gaur *et al.*, 2012). Advances in marker systems and genotyping technologies such as DArT and single nucleotide polymorphisms (SNPs) and genotyping by sequencing (GBS) have made genotyping large numbers of materials cost efficient. The integration of genomic technologies in chickpea breeding will greatly improve efficiency of developing chickpea cultivars that are more resilient to changes in temperature. For example, MARS recombines significant gene effects found among the progeny of a single population in chickpea (Cobos *et al.*, 2007) and will be useful in enhancing tolerance to temperature extremes. Another potential approach is genomic selection (Nayak *et al.*, 2010), where training populations that are representative of the wider gene pool are assembled, genotyped and phenotyped and all the estimated gene effects used to assemble new chickpea cultivars. However, these breeding-by-design approaches are completely dependent upon the accuracy of the phenotyping data used in estimating the gene effects. The screening methods outlined in this chapter offer scope for rapid and accurate phenotyping for chickpea temperature stress tolerance and when integrated in a molecular breeding scheme, should provide the temperature-tolerant chickpea cultivars required for an increasingly hostile production environment.

References

Behboudian MH, Ma Q, Turner NC, Palta JA (2001) Reactions of chickpea to water stress: yield and seed composition. *J Sci Food Agric* 81: 1288–1291.

Berger JD, Turner NC, Siddique KHM, *et al.* (2004) Genotype by environment studies across Australia reveal the importance of phenology for chickpea (*Cicer arietinum* L.) improvement. *Crop Pasture Sci* 55: 1071–1084.

Berger JD, Milroy SP, Turner NC, Siddique KHM, Imtiaz M, Malhotra R (2011) Chickpea evolution has selected for contrasting phenological mechanisms among different habitats. *Euphytica* 180: 1–15.

Chaturvedi SK, Mishra DK, Vyas P, Mishra N (2009) Breeding for cold tolerance in chickpea. *Trends Biosci* 2: 1–6.

Chickpea Technical Report (2011) Improving heat tolerance in chickpea for enhancing its productivity in warm growing conditions and mitigating impacts of climate change. ICRISAT, Patancheru, Andhra Pradesh, India, pp. 1–46

Clarke HJ (2001) Improving tolerance to low temperature in chickpea. In: *4th European Conference on Grain Legumes. Towards the Sustainable Production of Healthy Food, Feed and Novel Products.* Cracow, Poland, July 8–12, 2001, pp. 34–35.

Clarke HJ, Siddique KHM (1998) Growth and development. In: Loss S, Brandon N, Siddique KHM (eds), The chickpea. Western Australia Agriculture Bulletin 1326. Department of Western Australia Agriculture.

Clarke HJ, Siddique KHM (2004) Response of chickpea genotypes to low temperature stress during reproductive development. *Field Crops Res* 90: 320–334.

Clarke HJ, Khan TN, Siddique KHM (2004) Pollen selection for chilling tolerance at hybridization leads to improved chickpea cultivars. *Euphytica* 139: 65–74.

Cobos MJ, Rubio J, Fernandez-Romero MD, *et al.* (2007) Genetic analysis of seed size, yield and days to flowering in a chickpea recombinant inbred line population derived from a kabuli × desi cross. *Ann App Biol* 151: 33–42.

Croser JS, Clarke HJ, Siddique KHM, Khan TN (2003) Low-temperature stress: Implications for chickpea (*Cicer arietinum* L.) improvement. *Crit Rev Plant Sci* 22: 185–219.

Devasirvatham V (2012) The basis of chickpea heat tolerance under semi-arid environments. PhD thesis, The University of Sydney, NSW, Australia.

Devasirvatham V, Tan DKY, Gaur PM, Raju TN, Trethowan RM (2012a) High temperature tolerance in chickpea and its implications for plant improvement. *Crop Pasture Sci* 63: 419–428.

Devasirvatham V, Gaur PM, Mallikarjuna N, Raju TN, Trethowan RM, Tan DKY (2012b) Effect of high temperature on the reproductive development of chickpea genotypes under controlled environments. *Funct Plant Bio* 39: 1009–1018.

Devasirvatham V, Gaur PM, Mallikarjuna N, Raju TN, Trethowan RM, Tan DKY (2013) Reproductive biology of chickpea response to heat stress in the field is associated with the performance in controlled environments. *Field Crops Res* 142: 9–19.

Dua RP (2001) Genotypic variations for low and high temperature tolerance in gram (*Cicer arietinum*). *Indian J Agri Sci* 71: 561–566.

Ellis RH, Covell S, Roberts EH, Summerfield RJ (1986) The influence of temperature on seed germination rate in grain legumes. XI. Intraspecific variation in chickpea (*Cicer arietinum* L.) at constant temperatures. *J Exp Bot* 37: 1503–1515.

FAO (2009) *The State of World Fisheries and Aquaculture 2008*. Also available at: ftp://ftp.fao.org/docrep/fao/011/i0250e/i0250e01.pdf. UN Food and Agriculture Organization, Rome.

Gaur PM, Gowda CLL, Knights EJ, *et al.* (2007) Breeding achievements. In: Yadav SS, Redden RJ, Chen W, Sharma B (eds), *Chickpea Breeding and Management*. pp. 391–416. CAB International, Wallingford, Oxon, UK.

Gaur PM, Jukanthi AK, Varshney RK (2012) Impact of genomic technologies on chickpea breeding strategies. *Agron* 2: 199–221.

Gaur PM, Chaturvedi SK, Kumar S, *et al.* (2013) High temperature tolerance in food legumes to mitigate impact of climate change. In: *Proceedings of the National Conference on Crop Improvement and Adaptive Strategies to Meet Challenges of Climate Change*, 22–24 February 2013, University of Agricultural Sciences, Bangalore, India, pp. 29–32.

Gaur PM, Jukanti AK, Srinivasan S, *et al.* (2014) Climate change and heat stress tolerance in chickpea. In: Tuteja N, Gill SS (eds), *Climate Change and Plant Abiotic Stress Tolerance*. Wiley-VCH, Weinheim, Germany, pp. 839–855.

Heidarvand L, Amri RM, Naghavi MR, Farayedi Y, Sadeghzadeh B, Alizadeh KH (2011) Physiological and morphological characteristics of chickpea accessions under low temperature. *Russian J Plant Physi* 58: 157–163.

Ibrahim HM (2011) Heat stress in food legumes: evaluation of membrane thermostability methodology and use of infra-red thermometry. *Euphytica* 180: 99–105.

ICRISAT (1994) Cold tolerant chickpea varieties. Plant material description no. 53. International Crops Research Institute for the Semi-arid Tropics, Patancheru, Hyderabad, Andhra Pradesh, India. pp. 1–4.

ICRISAT (2009) Climate change in the semi-arid tropics. Available at: www.icrisat.org./rds/Climate_Change_SAT_flyer.pdf.

ICRISAT (2011) Improving heat tolerance in chickpea for enhancing its productivity in warm growing conditions and mitigating impacts of climate change. Technical report. International Crops Research Institute for the Semi-arid Tropics, Patancheru, Hyderabad, Andhra Pradesh, India.

IPCC (2007a) *Climate Change 2007: the Physical Science Basis. Summary for Policymakers*. WMO/UNEP, Paris.

IPCC (2007b) Summary for policymakers. In: Parry ML, Canziani M, Palutikof O, van der Linden J, Hanson C (eds), *Climate Change 2007: Impacts, Adaptation and Vulnerability. Contribution of Working Group II to the Fourth Assessment Report of the Intergovernmental Panel on Climate Change*. Cambridge University Press, Cambridge, UK, pp. 7–22.

Jenkins L, Brill R (2011) Chickpea time of sowing trial. DAN00156: Northern Pulse Agronomy Project. NSW DPI, Australia.

Kaloki P (2010) Sustainable climate change adaptation options in agriculture: The case of chickpea in the semi-arid tropics of Kenya. Report of the African Climate Change Fellowship Program. International Crops Research Institute for the Semi-Arid Tropics, Nairobi, Kenya, pp. 1–54.

Kalra N, Chakraborty D, Sharma A, *et al.* (2008) Effect of temperature on yield of some winter crops in northwest India. *Curr Sci* 94: 82–88.

Kaur G, Kumar S, Nayyar H, Upadhyaya HD (2008) Cold stress injury during the pod-filling phase in chickpea (*Cicer arietinum* L.): Effects on quantitative and qualitative components of seeds. *J Agron Crop Sci* 194: 457–464.

Kaushal N, Awasthi R, Gupta K, Gaur PM, Siddique KHM, Nayyar H (2013) Heat stress induced reproductive failures in chickpea (*Cicer arietinum* L.) are associated with impaired sucrose metabolism in leaves and anthers. *Funct Plant Bio* 40: 1334–1349.

Krishnamurthy L, Gaur PM, Basu PS, *et al.* (2011) Large genetic variation for heat tolerance in the reference collection of chickpea (*Cicer arietinum* L.) germplasm. *Plant Gene Res* 9: 59–61.

Kumar S, Kaur G, Nayyar H (2008) Exogenous application of abscisic acid improves cold tolerance in chickpea (*Cicer arietinum* L.). *J Agron Crop Sci* 194: 449–456.

Kumar S, Thakur P, Kaushal N, Malik JA, Gaur P, Nayyar H (2012a) Effect of varying high temperatures during reproductive growth on reproductive function, oxidative stress and seed yield in chickpea genotypes differing in heat sensitivity. *Arch Agron Soil Sci* 59: 823–843.

Kumar S, Kaushal N, Nayyar H, Gaur PM (2012b) Abscisic acid induces heat tolerance in chickpea (*Cicer arietinum* L.) seedlings by facilitated accumulation of osmoprotectants. *Acta Phys Plan* 34: 1651–1658.

Mace ES, Xia L, Jordan DR, *et al.* (2008) DarT markers: diversity analyses and mapping in *Sorghum bicolor*. *BMC Genomics* 9: 1–11.

Malhotra RS, Singh KB (1990) The inheritance of cold tolerance in chickpea. *J Gene Plant Bre* 44: 227–230.

Malhotra RS, Singh KB (1991) Gene action for cold tolerance in chickpea. *Theor Appl Genet* 82: 598–601.

Maqbool A, Shafiq S, Lake L (2010) Radiant frost tolerance in pulse crops – a review. *Euphytica* 172: 1–12.

Moore K, Jenkins L, Hertel K, Callaghan G (2010) Chickpea in 2010: Low temperature effects in 2009. GRDC update paper. Available at: www.grdc.com.au

Munier-Jolain NG, Ney B (1998) Seed growth rate in grain legumes II. Seed growth rate depends on cotyledon cell number. *J Exp Bot* 49: 1971–1976.

Nayak SN, Zhu H, Varghese N, *et al.* (2010) Integration of novel SSR and gene-based SNP marker loci in the chickpea genetic map and establishment of new anchor points with *Medicago truncatula* genome. *Theor Appl Genet* 20: 1415–1441.

Nayyar H, Bains T, Kumar S (2005a) Low temperature induced flower abortion in chickpea: relationship to abscisic acid and cryoprotectents in reproductive organs. *Environ Exp Bot* 53: 39–47.

Nayyar H, Bains TS, Kumar S (2005b) Chilling effects during seed filling on accumulation of seed reserves and grain yield. *J Sci Food Agric* 85: 1925–1930.

Nayyar H, Kaur G, Kumar S, Upadhyaya HD (2007) Low temperature effects during seed filling on chickpea genotypes (*Cicer arietinum* L.): Probing mechanisms affecting seeds reserves and yield. *J Agron Crop Sci* 193: 336–344.

Roberts EH, Hadley P, Summerfield RJ (1985) Effects of temperature and photoperiod on flowering in chickpeas (*Cicer arietinum* L.). *Ann Bot* 55: 881–892.

Salimath PM, Toker C, Sandhu JS, *et al.* (2007) Conventional breeding methods. In: Yadav SS (ed.), *Chickpea Breeding and Management*. CAB International, Wallingford, UK, pp. 369–390.

Savithri KS, Ganapathy PS, Sinha SK (1980) Sensitivity to low temperature in pollen germination and fruit-set in *Cicer arietinum* L. *J Exp Bot* 31: 475–481.

Saxena MC (1990) Problems and potentials in chickpea production in the nineties. In: Van Rheenen HA, Saxena MC (eds), *Chickpea in the Nineties. Proceedings of the 2nd International Workshop on Chickpea Improvement*, ICRISAT, Patancheru, India, pp. 13–25.

Siddique KHM, Sedgley RH (1986) Chickpea (*Cicer arietinum* L.), a potential grain legume for southwestern Australia: seasonal growth and yield. *Crop Pasture Sci* 37: 245–261.

Singh KB (1993) Problems and prospects of stress resistance breeding in chickpea. In: Singh KB, Saxena MC (eds), *Breeding for Stress Tolerance in Cool-Seasons Food Legumes*. John Wiley & Sons, Ltd., Chichester, UK, pp. 17–37.

Singh KB, Malhotra RS, Saxena MC (1995) Additional sources of tolerance to cold in cultivated and wild *Cicer* species. *Crop Sci* 35: 1491–1497.

Singh NH, Dhaliwal GS (1972) Effect of soil temperature on seedling emergence in different crops. *Plant Soil* 37: 441–444.

Srinivasan A, Johansen C, Saxena NP (1998) Cold tolerance during early reproductive growth of chickpea (*Cicer arietinum* L.): Characterization of stress and genetic variation in pod set. *Field Crops Res* 57: 181–193.

Srinivasan A, Saxena NP, Johansen C (1999) Cold tolerance during early reproductive growth of chickpea (*Cicer arietinum* L.): genetic variation in gamete development and function. *Field Crops Res* 60: 209–222.

Stone PJ, NicoLas ME (1994) Wheat cultivars vary widely in their responses of grain yield and quality to short periods of post-anthesis heat stress. *Aust J Plant Physiol* 21: 887–900.

Summerfield RJ, Hadley P, Roberts EH, Minchin FR, Rawsthrone S (1984) Sensitivity of chickpea (*Cicer arietinum* L.) to hot temperatures during the reproductive period. *Exp Agric* 20: 77–93.

Summerfield RJ, Virmani SM, Roberts EH, Ellis RH (1990) Adaption of chickpea to agroclimatic constraints. In: van Rheenen HA, Saxena MC (eds), *Chickpea in the Nineties. Proceedings of the Second International Workshop on Chickpea Improvement*, 4–8 December 1989. ICRISAT Center, Patancheru, Andhra Pradesh, India, pp. 50–61.

Toker C (2005) Preliminary screening and selection for cold tolerance in annual wild *Cicer* species. *Genet Resour Crop Evol* 52: 1–5.

Upadhyaya HD, Dronavalli N, Gowda CLL, Singh S (2011) Identification and evaluation of chickpea germplasm for tolerance to heat stress. *Crop Sci* 51: 2079–2094.

Wang J, Gan YT, Clarke F, McDonald CL (2006) Response of chickpea yield to high temperature stress during reproductive development. *Crop Sci* 46: 2171–2178.

Wery J, Silim SN, Knights EJ, Malhotra RS, Cousin R (1994) Screening techniques and sources of tolerance to extremes of moisture and air temperature in cool season food legumes. *Euphytica* 73: 73–83.

CHAPTER 6

Effect of pesticides on leguminous plants: An overview

Harpreet Kaur, Renu Bhardwaj, Vinod Kumar, Anket Sharma, Ravinder Singh and Ashwani Kumar Thukral
Department of Botanical and Environmental Sciences, Guru Nanak Dev University, Amritsar, India

6.1 Introduction

In terms of their importance to humans, the legumes are second only to the Gramineae. They include important grain species like pigeon pea (*Cajanus cajan*), pea (*Pisum sativum*), broad bean (*Vicia faba*), chick pea (*Cicer arietinum*), cowpea (*Vigna unguiculata*) and soybean (*Glycine max*); pasture species like alfalfa (*Medicago sativa*), clovers (*Trifolium* spp.), sweet clovers (*Meliotus* spp.), trefoil (*Lotus corniculatus*) and vetches (*Vicia* spp.); and agroforestry species like members of the genera *Dalbergia, Pterocarpus, Gliricidia, Anadenathera, Erythrina, Calliandra, Parkea* and *Samanea* (Graham & Vance, 2003). Legumes contribute about 33% of the dietary protein nitrogen requirement of humans (Vance *et al.*, 2000). The quality and yield of legumes are drastically reduced by pest infestations. So, to combat pest infestations and to maintain the quantity and quality of legumes, a broad spectrum of pesticides are applied (DDT, chlordane, aldrin, chlorpyrifos, etc.). Some of these pesticides (e.g. DDT) are highly persistent and non-biodegradable. There are three phases of pesticide metabolism in plants (Karthikeyan *et al.*, 2003). In phase I reactions, pesticides are transformed by addition of functional groups. In phase II reactions, phase I products undergo conjugation with sugars, glutathione, phosphate, amino acids and sulphate. Phase III reactions include the formation of bound residues by co-polymerization with molecules such as lignins (Singh, 2012).

Leguminous plants form symbiotic associations with *Rhizobium* bacteria, which live inside root nodules of the plant and reduce atmospheric nitrogen. This reduced form of nitrogen is used by the plant to form organic compounds such as proteins, nucleic acids, etc. This nitrogen fixation plays an important role in the growth and development of leguminous plants. The application of pesticides negatively influences the growth of the bacteria, the interaction between the symbiotic bacteria and the legumes, the number of nodules and the activity of nitrogenase enzyme, thereby reducing the rate of nitrogen fixation. This reduction in nitrogen fixation reduces the growth and yield of leguminous plants (Fox *et al.*, 2007). However, pesticide-tolerant strains of bacteria can enhance the growth of legumes under pesticide stress.

6.2 Uptake, metabolism and persistence of pesticides

Legumes such as cowpea, chickpea, field pea, pigeon pea, green gram, faba bean, black gram, beans, lentil and grass pea are important food sources. These crops are susceptible to attack by various types of insect pest, such as *Maruca vitrata* (spotted pod borer), *Helicoverpa armigera* (pod borer), *Melanagromyza obtusa* (pod fly), *Etiella zinckenella* (spiny pod borer), *Sitona* sp. (pea and bean weevil), *Ophiomyia phaseoli* (stem fly), *Empoasca* spp. (leafhoppers), *Aphis falae* and *Aphis craccivora* (aphids), *Spodoptera exigua, Spodoptera litura* and *Amsacta* spp. (defoliators), *Bruchus pisorum* and *Callosobruchus chinensis* (bruchids), *Caliothrips indicus* and *Megaleurothrips dorsalis* (thrips) and *Mylabris* spp. (blister beetles) (Sharma, 2005).

Legumes under Environmental Stress: Yield, Improvement and Adaptations, First Edition. Edited by Mohamed Mahgoub Azooz and Parvaiz Ahmad.

To control the pests of leguminous plants various types of pesticides are sprayed on them such as carbofuran, chlorfenvinphos, chlordane, benomyl, carbaryl, dieldrin, aldrin, carbendazim and chlorpyrifos (Lama, 2008). Uptake of pesticides in plants can occur through roots to any of the aerial parts such as leaves and fruits. Uptake of pesticides depends on various factors including the dissociation constant and lipophilicity, the physicochemical properties of the pesticides and soil, and environmental conditions such as temperature, precipitation, sunlight and relative humidity (Finlayson & MacCarthy, 1973; Schroder & Collins, 2002). Once taken up by plants, the pesticides translocate to the shoots either through xylem or phloem, or in roots via the phloem alone. Pesticides' mobility depends on their water solubility; they can be stored adjacent to the absorption site, in storage cells adjoining the translocation pathway or in the areas that are very actively metabolizing. As the conditions in plants change, pesticides may be released and transported to other plant parts.

Metabolism can alter pesticide structure and the pesticide may be detoxified or activated. As previously mentioned, three phases of plant metabolism can be recognized (Karthikeyan *et al.*, 2003). Phase I reactions are the initial reactions of metabolism and are catalysed by cytochrome P450 enzymes (Mougin *et al.*, 2001). These reactions transform pesticides by addition of functional groups. Various types of transformation reactions of pesticides occur in plants, including hydrolysis, oxidation, epoxidation, O-dealkylation, N-dealkylation and desulphuration (Casida & Lykken, 1969). The functional groups generated in phase I reactions such as hydroxyl groups act as 'handles' for further conjugation and storage reactions catalysed by phase II and phase III enzymes. In pesticide hydrolysis reactions, plant hydrolytic enzymes such as esterases, proteases and lipases detoxify and degrade the pesticide (Singh, 2012).

Oxidation reactions are the reactions in which an electron is lost by an element or there is an increase in the proportion of oxygen in a molecule. These are the most important reactions of phase I metabolism. Oxidation of pesticides can be divided into two groups: microsomal oxidation and extramicrosomal oxidation. In microsomal oxidation reactions, an atom of an oxygen molecule is added to the substrate while the second atom of the oxygen molecule is reduced to water. NADPH acts as the source of the electron, which is used to reduce FAD or cytochrome P450. Epoxidation is an important microsomal reaction catalysed by the enzyme epoxidase. It leads to the formation of epoxides, which are stable and persistent in the environment. It also leads to the production of arene oxides, which are epoxide intermediates in aromatic compound hydroxylation. In N-dealkylation reactions alkyl groups in carbamates, amides or amines or the alkyl groups attached to the electronegative nitrogen atoms are oxidatively removed by converting them to the corresponding aldehydes. The reaction leads to the detoxification of the compound. There is spontaneous rearrangement of this α-hydroxy intermediate to form the aldehyde. Ether and ester structures of organophosphorus insecticides undergo O-dealkylation and are detoxified. Desulphuration reactions are also called phosphorothioate oxidations. These reactions increase the insecticidal activity and mammalian toxicity of organophosphorus compounds such as phosphorodithioate and phosphorothionates. Microsomal mono-oxygenase oxidatively desulphurates the P=S groups of these compounds to their respective P=O analogues. The P=O analogues have more affinity to acetylcholinesterase and thus inhibit acetylchlolinesterase more potently (Singh, 2012). In phase II reactions soluble conjugates are formed (mainly with amino acids and sugars). Glutathione, amino acids, glucose and malonic acid are the plant products involved in conjugation. Soluble conjugates formed during phase II reactions are stored in vacuoles (Sandermann, 1992; Korte *et al.*, 2000). In phase III reactions bound or non-extractable residues are formed by conjugation to insoluble structures in the plant. One such reaction is co-polymerization with lignin. By turning the toxic pesticides into bound residues, their water solubility is reduced and so their reactivity and toxicity are also reduced.

Many of the pesticides have a long half-life and persist for a long time in plants. This persistence is harmful to the plants as well as to the animals and humans that ingest them as food. Neetu (2013) detected pesticide residues of DDT in four types of pulses, namely red lentils, pigeon pea, green gram and black gram. The pesticide pirimiphos-methyl was detected in toasted chickpea flour samples collected from the markets of Tenerife in the Canary Islands (Gonzalez-Curbelo *et al.*, 2012). Organophosphorus pesticides such as omethoate, isocarbophos and triazophos were detected in cowpea and kidney pea samples collected from farmers' markets (Ping-Sheng & Qi, 2010). Various techniques can be used to determine the types and quantity of pesticides in plant samples, such as GC-MS (gas chromatography-mass spectrometry), GC-NPD (gas chromatography with nitrogen phosphorus detection), GLC (gas liquid chromatography) and UHPLC-MS (ultra-high performance liquid chromatography-mass spectrometry) (Kaphalia *et al.*, 1990; Jian *et al.*, 2013).

6.3 Effects of pesticides on leguminous plants

Pesticides have an inhibitory effect on the growth of legumes due to the reduced activity of the microorganisms that promote rhizospheric growth (De Bertoldi *et al.*, 1978; Siti *et al.*, 1982; Koch *et al.*,1997). Studies have found that there is a progressive decline in plant growth linked with the application of pesticides (Ahemad, 2014). Lewis *et al.* (1996) found that pesticides have a fatal effect on the soil microbe population that is useful for plant growth. However, certain microorganisms, such as *Rhizobium* sp. strain MRL3, have shown a remarkably high resistance to pesticides of many chemical families. Physiological changes induce the formation of novel metabolic pathways to bypass a biochemical reaction obstructed by a specific pesticide. This produces temporary resistance or tolerance against the pesticide (Bellinaso *et al.*, 2003). The effects of various pesticides on growth, yield and photosynthesis are summarized below:

6.3.1 Effect on growth

Accumulation of insecticides was found to cause a reduction in the population of nitrifiers as well as enzymatic activity of soil (Das *et al.*, 2003). High concentrations of insecticides led to a loss in agricultural output and fertility of the soil (Fox *et al.*, 2007). Fipronil reduced shoot and root growth in legumes. This reduction occurred as result of its effect on plant organs, mainly the nodules, which led to disruption of the symbiotic legume-*Rhizobium* association and hence of nitrogen fixation resulting in a growth decline (Evans *et al.*, 1991). The inhibitory effect of insecticides might be due to inhibition of the enzymes involved in growth and development or to the disruption of signalling between phytochemicals, such as apigenin, luteolin and NodD *Rhizobium* receptor, that are important for nodule initiation and fixation of nitrogen (Fox *et al.*, 2007). Reduction in the activity of nitrogenase, nodulation and total biomass content of alfalfa plants was reported when they were treated separately with DDT, pentachlorophenol, methylparathion and bisphenol A (Fox *et al.*, 2007). The effect of Brominal, Gramoxone, Afugan, Sumi oil and Selecron was determined on the nodulation and growth of *Glycine max* (L.) (Abd-Alla *et al.*, 2000). Application of Selecron, Brominal, Afugan and Gramoxone inhibited the growth of *Phaseolus vulgaris* L., *Vigna sinensis* and *Lupinus albus* L. The chlorophyll content in leaves and shoots, number of root nodules, dry weight of whole plant and yield of grains declined due to imazaquin phototoxicity affecting the growth of the soybean plants (Alonge, 2000). The legumes tested with fipronil insecticide showed varied responses; and this was mainly because the toxic effect of this insecticide depends on both the physiology and genetics of the plants, which change from species to species (Ahemad & Khan, 2011b). A decreased concentration of chlorpyrifos had a significant effect, which was non-phototoxic and effective, resulting in increased growth, whereas high concentration of chlorpyrifos had an inhibitory effect on growth parameters such as height of the plant, number of leaves, total area of leaves and biomass of plant. Similarly, Bashir *et al.* (2007) noted the effect of mancozeb on different morphological parameters in *Lens culinaris* L. at various developmental stages. They concluded that at low concentrations of mancozeb, the morphological parameters increased whereas at high concentrations a decline was observed. A study of *Vigna radiata* L. under chlorpyrifos showed a reduction in growth parameters with increased concentration, exhibited as a decline in the leaf number and leaf area, which could be due to obstruction in translocation of photosynthates and decline in chlorophyll content (Parween *et al.*, 2012). Similar results were observed in bean leaves (Ruano *et al.*, 1988). Under the application of dimethoate insecticide the seedlings of cowpea showed drastic variations in the photosynthetic machinery and reduced rate of photorespiration hence increasing the chances of damage, which resulted in decline of growth and biomass production. Pesticides showed an adverse effect on synthesis of proteins and the enzymes involved in metabolism (Kapoor *et al.*,1996; Boldt & Jacobsen 1998). It was further emphasized by Pham *et al.* (2004) that pesticides were responsible for genotoxicity. High concentrations of pesticides lead to reduced functioning and organism survival (Kumar *et al.*, 2010).

6.3.2 Effect on yield

Pesticides have inhibitory effects on crop yield, and on the soil flora and fauna (Fox *et al.*, 2007; Ahemad & Khan, 2012). Fungicides and insecticides possess a quick and effective action and are reliable, and many chemicals have been reported to be effective against insect pests and various crop diseases (Allen, 1982; Cook, 1986; Adejumo, 2005). Experiments have shown significant differences between damaged (pod borer) parts in plants treated with pesticides, and untreated plants. Damaged parts may fail to produce seeds or may produce low quality or non-viable seeds. Thus pesticides protect the pods from pod borer and enhance seed yield (Mugo, 1989). They also bring about effective control of foliar fungal diseases and insect pests resulting in higher biomass yields, which are desirable for efficient use of nutrients and water. As a result, there is accumulation of photoassimilates, which are used for kernel development (Okeyo-Owuor & Khamala, 1980; Olupot *et al.*, 2004). Abulo *et al.* (2005) observed that high biomass yield contributes to weed suppression and light utilization. All these factors collectively increase the grain yield. Okeyo-Owuor (1978) reported that pesticides like dimethoate reduced yield losses. Experiments on cowpea (a grain legume) have shown that pesticides effectively impact the legume-insect pest complex and help in increasing the yield (Kyamanywa 1996). The pesticide contaminates the soil and influences soil microbe population, which further affects the fertility of soil. However, Dubey *et al.* (2012) studied the effect of different pesticides such as carbosulfan, phorate, thiomethoxam and carbofuran on the soil microflora of leguminous crops and found that no significant change occurred in the viability of bacteria. They further concluded that bacteria dwelling in the soil possess the ability to degrade the pesticides.

Pesticides disrupt the phytoestrogen signalling system between rhizobia and legume that regulates their symbiotic relationship. They also disrupt the initial binding of rhizobia to lectins present on the root hairs (Musarrat & Haseeb, 2000; Fox *et al.*, 2004). As a result, pesticides affect the symbiotic relationship between leguminous plants and rhizobia due to competition of pesticides for binding sites of rhizobia on the root hairs of legumes. Insecticides inhibit nodulation in the legumes by reducing the number of available sites on the host plant for the binding of rhizobia by causing reduction in the supply of carbohydrates to existing nodules. Thus, insecticides decrease the growth and survival of rhizobia and also disrupt the signalling required to initiate the development of nodules in the plants. Legume plants grown under pyriproxyfen-amended soils showed a decrease in the chlorophyll content. The most drastic effect of this insecticide was observed in chickpea plants, where chlorophyll content decreased by 14% compared to control plants. In lentil the reduction in chlorophyll was found to be 9% compared to control plants. Chlorophyll content showed a marginal decrease in pea and green gram. Under the application of pyriproxyfen, the seed protein of green gram and chickpea was significantly decreased, by 4% and 5% respectively, as compared to control plants, while a marginal reduction was observed in the seed protein of lentil and pea, by 3% and 1%, as compared to the control (Anderson *et al.*, 2004). The maximum reduction in seed yield was observed in lentil while pea showed the least reduction in seed yield under pyriproxyfen treatment. Various agrochemicals including insecticides induce the symbiotic phenotype by disrupting the attachment of rhizobia to the roots of host plants, reducing the production of root nodules and nitrogenase activity, which in turn decreases the nitrogen content and reduces the plant yield (Fox *et al.*, 2007). The decrease in phosphorus content and seed yield under pesticide application is mainly due to the reduction of the proteins and enzymes in metabolic pathways that are involved in synthesis of proteins and uptake of phosphorus (Boldt & Jacobsen, 1998; Nare *et al.*, 2010; Ahemad & Khan, 2011a).

6.3.3 Effect on photosynthetic activity

Any reduction in leaf area lessens photosynthetic activity, which results in decline of yield. Metabolic enzymes are adversely affected by pesticides; therefore it appears that the functioning of various enzymes and cofactors involved in the photosynthetic carbon reduction cycle, such as NADP, Rubisco, 3-phosphoglyceric acid (3-PGA), aldolase and NAD-glyceraldehyde-3-P-dehydrogenase, is inhibited by insecticides (Mapose & Cossa, 2005). In leguminous plants, the phosphorus and nitrogen contents are among the most important factors for their growth. A reduction in the nitrogen content of legumes follows from any adverse impact of the plants' symbiotic relationship with rhizobia bacteria (Boldt & Jacobsen, 1998). The chlorophyll content of mung bean decreased with increased concentrations of herbicides applied. Moreover, it further resulted in stimulation of the oxidation process (Kaushik & Inderjit, 2006). Increased concentration of insecticides on spinach resulted in chlorophyll reduction, which consequently damaged the photosynthetic apparatus (Sresen *et al.*, 2000). However, in some cases positive effects of pesticides were also observed, such as increments in the plant height, area of leaves and the uptake of nutrients (Lagana *et al.*, 2000; Nakamura *et al.*, 2000).

6.3.4 Effect on nitrogen fixation

6.3.4.1 Nodule formation

Pesticide application decreases the process of nodulation in legume plants by reducing the carbohydrate supply to already formed nodules. This results in a decrease in sustenance of rhizobia and lessening of biochemical signalling, which leads to inhibition of nodule development due to decreased cell division (Anderson *et al.*, 2004). Decrease in nodule formation was observed in *Cicer arietinum, Lens esculentum, Pisum sativum* and *Vigna radiata* after pesticide application (Boldt & Jacobsen, 1998). A reduction in nitrogen fixation was observed in legume plant roots due to the decrease in root nodule formation and late rhizobial association (Fox *et al.*, 2007). Reduced nodule formation was observed in *Glycine max* after the application of Afugan, Brominal, Gramoxone and Selecron (Abd-Alla *et al.*, 2000). Carbamate application in *Pisum sativum* and *Vigna sinensis* resulted in suppression of nodule formation (Aggarwal *et al.*, 1986). In *Glycine max* decline of nodule formation was noticed after the application of imazaquin (Alonge, 2000). Reduced nodulation was observed in *Pisum sativum* and *Lens esculentum* after treatment with the pesticide pyriproxyfen. The decreasing trend of nodule number was *Pisum sativum* (44), *Vigna radiata* (14), *Cicer arietinum* (5) and *Lens esculentum* (5) (Ahmed, 2014). The insecticide pentachlorophenol is reported to prevent nodulation between the legume *Medicago sativa* and the bacterium *Sinorhizobium meliloti* (Fox et *al.*, 2007). In the legumes *Trifolium pratense, Medicago sativa* and *Ornithopus compressus,* a significant decrease in nodule formation was observed with the bacteria *Rhizobium* and *Bradyrhizobium* after the application of the pesticides carbendazim, thiram and imazethapyr. The nodulation process was greatly affected by the use of these pesticides and is summarized in Table 6.1 (Niewiadomska & Klama, 2005).

The pesticides Funaben T and Pivot 100SL were reported to have negative impacts on nodule formation in *Medicago sativa* and *Trifolium pratense* (Niewiadomska and Sawicka, 2002; Niewiadomska, 2004). Nodule formation in *Pisum sativum, Phaseolus vulgaris* and *Lupinus* decreased after the application of the pesticides Afugan, Brominal, Gramoxone and Selecron (Abd-Alla *et al.*, 2000). Application of three pesticides, 2,4-D amine salt, glyphosate (Roundup) and atrazine, to the leguminous plants *Crotalaria hebecarpa, Desmodium cephalotes* and *Tephrosia strigosa* resulted in a significant decline in the number of root nodules (Shankar *et al.*, 2012). Nodule formation in *Pisum sativum* is negatively affected by the pesticides terbuthylazine, simazine, prometryn and bentazone (Singh & Wright, 1999, 2002). In *Cicer arietinum* L., the application of Pyrifos resulted in the

Table 6.1 Effect of pesticides on nodulation by symbiotic bacteria in legumes. Reproduced with permission of the *Polish Journal of Microbiology*.

Legume	Bacterium + pesticide treatment	Average nodule number	Colour
Trifolium pratense	*Rhizobium leguminosarum* bv. *trifolii* KGL	8	Pink
Trifolium pratense	*Rhizobium leguminosarum* bv. *trifolii* KGL + herbicide	3	Pink
Trifolium pratense	*Rhizobium leguminosarum* bv. *trifolii* KGL + fungicide	5	Pink
Trifolium pratense	*Rhizobium leguminosarum* bv. *trifolii* KGL + herbicide + fungicide	3	Pale pink
Medicago sativa	*Rhizobium melilotii* Bp	15	Pink
Medicago sativa	*Rhizobium melilotii* Bp + herbicide	7	Pink
Medicago sativa	*Rhizobium melilotii* Bp + fungicide	8	Pink
Medicago sativa	*Rhizobium melilotii* Bp + herbicide + fungicide	6	Pink
Ornithopus compressus	*Bradyrhizobium* sp. *Ornithopus* B	11	Pink
Ornithopus compressus	*Bradyrhizobium* sp. *Ornithopus* B + herbicide	7	Pink
Ornithopus compressus	*Bradyrhizobium* sp. *Ornithopus* B + fungicide	6	Pink
Ornithopus compressus	*Bradyrhizobium* sp. *Ornithopus* B + herbicide + fungicide	5	Pink

inhibition of nodule formation (rhizobial bacterium) but other pesticides, namely Lorsban, Decis, Karate and Ripcord were considered safe for the rhizobial population. For *Azotobacter,* Lorsban was seen to have no side effect on the nodulation process whereas other four pesticides (Pyrifos, Decis, Karate and Ripcord) were observed to have a negative impact on nodulation (Khan *et al.*, 2009). In *Vigna sinensis* L., *Phaseolus vulgaris* L. and *Lupinus albus* L., a decline in the nodulation process was observed after treating the plants with Afugan, Brominal, Gramoxone, Selecron and sumi oil (Abd-Alla *et al.,* 2000). Atrazine, isoproturon and metribuzin application to *Vigna radiata* resulted in a reduction of nodule number (Khan *et al.*, 2006). Application of the pesticide fipronil to *Pisum sativum, Cicer arietinum, Lens esculentum* and *Vigna radiata* has caused a reduction in nodule formation. This pesticide also has a negative effect on the leghaemoglobin content of nodules (Ahemad & Khan, 2011c).

6.3.4.2 Effect on nitrogenase activity

A decline in nitrogenase activity was observed in the legume *Medicago sativa* L. under pesticide stress (methyl parathion, DDT, bisphenol A and pentachlorophenol), which resulted in a reduction of nitrogen content and yield of the plant (Fox *et al.*, 2007). Reduced nitrogenase activity was noted in *Cicer arietinum, Lens esculentum, Pisum sativum* and *Vigna radiata* after pyriproxyfen application. This was due to the decline in leghaemoglobin content inside root nodules (Ahemed, 2014). In *Medicago sativa,* a significant reduction in nitrogenase activity was observed after treating the plant with DDT, methyl parathion, pentachlorophenol and chrysin (Fox *et al.*, 2007). Decline in nitrogenase activity of the bacteria *Rhizobium* and *Bradyrhizobium* was seen in the legumes *Trifolium pratense, Medicago sativa* and *Ornithopus compressus* after the use of the pesticides Funaben T and Pivot 100SL. In *Arachis hypogaea,* application of endosulfan (conc. 250 ppm) resulted in decreased nitrogenase activity (Darure *et al.*, 2012).

Effective symbiosis by rhizobia enhanced nodulation and nitrogen fixation, and the availability of more nitrogen to the plants increased their yield (Graham & Vance, 2000). Differences in nodule number induced differences in the fixed nitrogen and affected the yield of the host plant (Amijee & Giller, 1998). It has been demonstrated that pesticides result in the reduction of attachment of rhizobia to the roots of leguminous plants, which in turn diminishes the fixed nitrogen content and plant yield (Fox *et al.*, 2007). The pesticide tebuconazole negatively affected the symbiotic association between pea and *Rhizobium,* consequently decreasing the growth, nutrient uptake and seed yield of the plants (Ahemad & Khan, 2011d). The survival and viability of *Bradyrhizobium* on peanut seeds was inhibited by the pesticides Vitavax and benomyl. Due to this, the symbiosis of *Bradyrhizobium* with peanut was inhibited, resulting in the reduction in nitrogen fixation and seed yield of the plants (Hashem *et al.,* 1997). All these studies demonstrate that pesticide stress weakens the association between legumes and nitrogen-fixing bacteria, which in turn decreases nodulation and nitrogenase activity. Under these conditions, the amount of fixed nitrogen declines, which reduces the yield of the host plants (Figure 6.1).

6.4 Pesticide tolerance in legumes

Rhizobacteria enhance the plant growth and productivity by the production of siderophores, phytohormones, antibiotics, etc. They also decrease ethylene concentration and induce systemic resistance (Lugtenberg & Kamilova, 2009; Zaidi *et al.*, 2009; Tank & Saraf, 2010). Rhizobacteria may degrade or hydrolyse pesticides enzymatically (Herman *et al.*, 2005; Yang & Lee, 2008). The symbiotic association between rhizobia and bradyrhizobia enhances nitrogen fixation throughout the root length, which reduces the requirement for inorganic nitrogen and improves the soil nitrogen content (Sanginga *et al.*, 1996; Bellone *et al.*, 1997). *Rhizobium* increases the nitrogen-fixing capacity of plants through symbiotic association with them (Fox *et al.*, 2007). Legume plants were inoculated with competent rhizobial strains to enhance the nodulation and nitrogen fixation (Ahemad & Khan, 2009). Phosphorus is one of the most important elements for plant growth and productivity. Some rhizobial strains of *Bradyrhizobium* and *Rhizobium leguminosarum* bv. phaseoli can solubilize insoluble phosphate and enhance plant growth (Halder & Chakrabartty, 1993). When *Rhizobium* and phosphate-solubilizing bacteria were used together, a synergistic effect on grain yield and symbiotic effectiveness of mungbean was observed. Phosphate-solubilizing bacteria improved the symbiotic effectiveness and competitive ability of inoculated *Rhizobium* sp. under field conditions in lentil (Kumar & Chandra, 2008). Symbiotic bradyrhizobia

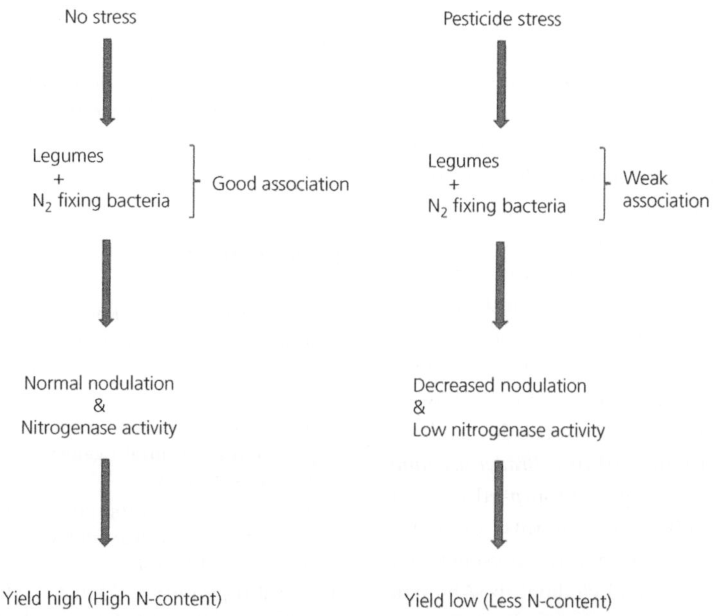

Figure 6.1 Effect of pesticide stress on nodulation and nitrogenase activity of nitrogen-fixing bacteria.

promote plant growth through an increase in the production of phytohormones like indole acetic acid (IAA) (Tien *et al.*, 1979; Hussain *et al.*, 1987). Most of the nitrogen-fixing rhizobia produce IAA (Prevost *et al.*, 1987; Arora *et al.*, 2001). Many reports showed that IAA production enhanced plant growth as well as nutrient uptake of plants by directly increasing the production of plant growth-promoting rhizobacteria (PGPR) (Lippman *et al.*, 1995). Plant growth-promoting rhizobacteria also released 2,3-butanediol and acetoin, which promoted the growth of *Arabidopsis* (Ryu *et al.*, 2003). Rhizobacterial genera like *Mesorhizobium*, *Bradyrhizobium* and *Rhizobium* isolated from the pea, produced substantial amount of IAA in Luria broth medium. Exopolysaccharides produced by rhizobacteria helped in providing protection to the cell against phagocytosis, phage attack and desiccation. They also helped in N_2 fixation by reducing oxygen tension (Tank & Saraf, 2003). HCN released by *Rhizobium* (Beauchamp *et al.*, 1991) was toxic to many pathogens and helped in controlling the growth of various types of pathogens (Bagnasco *et al.*, 1998). Rhizobia isolated from tomato rhizosphere, *Rhizoctonia batalicota*, *Pythium* sp. and *Bradyrhizobium* strains, all produced HCN (Alvarez *et al.*, 1995). HCN enhanced growth by inhibiting plant pathogens as well as fungal growth (Chakraborty & Purkayastha, 1984). Rhizobitoxine[1] promoted nodulation in legume species and also helped in the symbiosis between rhizobia and legumes (Duodu *et al.*, 1999) as well as inhibiting the synthesis of ethylene (Xie *et al.*, 1996). Siderophores are other microbial metabolites that supply iron to the plant under stress conditions (Indiragandhi *et al.*, 2008). Siderophores enhanced antimicrobial biosynthesis by microorganisms, local and systematic host resistance (Joseph *et al.*, 2007; Sinha & Mukherjee, 2008). The root nodule bacteria, *Rhizobium* sp. and *Bradyrhizobium* enhanced the production of siderophores (Carson *et al.*, 2000; Arora *et al.*, 2001; Deshwal *et al.*, 2003). Some siderophores have structural characteristics like carboxylate (rhizobatin from *Sinorhizobium meliloti*) (Smith *et al.*, 1985).

The nodule microorganisms synthesize plant growth substances and siderophores, which help the growth of legumes in pesticide-contaminated soil (Wani *et al.*, 2008). By the production of all these substances, the

[1]Rhizobitoxine is a phytotoxin synthesized by *Bradyrhizobium elkanii* (which forms mutualistic symbioses with legumes). It causes chlorosis on a number of legume hosts. Nodule development is inhibited or downregulated by ethylene, but rhizobitoxine inhibits ethylene biosynthesis and plays a positive role in symbiosis between the rhizobia and legumes.

bacteria enhance plant growth and tolerance to pesticide stress. The effects of different concentrations of fipronil and pyriproxyfen on lentil plants were studied by Ahemad and Khan (2011e). *Rhizobium* sp. strain MRL3, which was tolerant to pyriproxyfen and fipronil, could produce biomolecules responsible for plant growth promotion in sufficient amounts, both in the presence and absence of insecticide. Insecticides at the recommended and higher doses reduced the symbiotic properties, plant dry biomass, seed yield and nutrient uptake of lentil plants. The application of *Rhizobium* sp. strain MRL3 enhanced the measured parameters significantly in comparison to the plants grown in soils having the same concentrations of both insecticides and no inoculant (Ahemad & Khan, 2011e). *Bradyrhizobium* strain MRM6, tolerant to quizalofop-*p*-ethyl and clodinafop, produced sufficient amounts of plant growth-promoting substances both in the presence and absence of quizalofop-*p*-ethyl and clodinafop (Ahemad & Khan, 2011f). Similar plant growth-promoting substances are produced by *Mesorhizobium ciceri* (Wani *et al.*, 2008). The rhizobacteria isolated from the chickpea, pea, green gram and lentil nodule and mustard rhizosphere showed tolerance to herbicides (quizalofop-*p*-ethyl, clodinafop, metribuzin and glyphosate), insecticides (fipronil, pyriproxyfen, imidacloprid and thiamethoxam) and fungicides (tebuconazole, hexaconazole, metalaxyl and kitazin) ranging from 400 to 3200 µg/mL (Ahemad & Khan, 2011g). The genetic make-up and physiology in microbes are implicated in making plant-microbe systems resistant to pesticides by forming new bypass metabolic mechanisms that are not inhibited by pesticides (Bellinaso *et al.*, 2003).

6.5 Conclusion

Pesticides taken up by plants from the environment move to the various plant parts either through xylem or phloem. By metabolic conversions pesticides are ultimately changed into soluble conjugates by phase II reactions or to bound residues by phase III reactions. Pesticides can persist in plants for a long time due to their long half-lives. Their application adversely affects the activity, efficiency and survival of symbiotic bacteria of legumes. Nitrogen fixed by these bacteria and the other compounds produced by them, like HCN, siderophores and phytohormones, are very important for the growth and yield of these plants, which are negatively affected due to the adverse effects of pesticides on the bacteria. Bacterial tolerance to pesticides increases the growth and performance of legume plants under pesticide stress.

References

Abd-Alla MH, Omar SA, Karanxha S (2000) The impact of pesticides on arbuscular mycorrhizal and nitrogen-fixing symbioses in legumes. *Appl Soil Ecol* 14: 191–200.

Abulo P, Olupot G, Ochwoh VA, Adipala E (2005) Effect of phosphate fertilizer and pesticide application on the yield of groundnuts in central Uganda. *African Crop Science Conference Proceedings* 7: 37–40.

Adejumo TO (2005) Crop protection strategies for major diseases of cocoa, coffee and cashew nut in negeria. *Afr J Biotechnol* 4: 143–150.

Aggarwal TC, Narula N, Gupta KG (1986) Effect of some carbamate pesticides on nodulation, plant yield and nitrogen fixation by *Pisum sativum* and *Vigna sinensis* in the presence of their respective rhizobia. *Plant Soil* 94: 125–132.

Ahemad M (2014) Growth suppression of legumes in pyriproxyfen stressed soils: A comparative study. *Emir J Food Agric* 26: 66–72.

Ahemad M, Khan MS (2009) Effect of insecticide tolerant and plant growth promoting *Mesorhizobium* on the performance of chickpea grown in insecticide stressed alluvial soils. *J Crop Sci Biotechnol* 12: 217–226.

Ahemad M, Khan MS (2011a) Ecotoxicological assessment of pesticides towards the plant growth promoting activities of lentil (*Lens esculentus*) specific *Rhizobium* sp. strain MRL3. *Ecotoxicology* 20: 661–669.

Ahemad M, Khan MS (2011b) Pesticide interactions with soil microflora: importance in bioremediation. In: Ahmad I, Ahmad F, Pichtel J (eds), *Microbes and Microbial Technology: Agricultural and Environmental Applications*. Springer, New York, pp. 393–413.

Ahemad M, Khan MS (2011c) Comparative study of the growth parameters of legumes grown in fipronil-stressed soils. *Eurasian J Biosci* 5: 29–36.

Ahemad M, Khan MS (2011d) Effect of tebuconazole-tolerant and plant growth promoting *Rhizobium* isolate MRP1 on pea-*Rhizobium* symbiosis. *Sci Hortic* 129: 266–272.

Ahemad M, Khan MS (2011e) Insecticide-tolerant and plant growth promoting *Rhizobium* improves the growth of lentil (*Lens esculentus*) in insecticide stressed soils. *Pest Manag Sci* 67: 423–429.

Ahemad M, Khan MS (2011f) Response of greengram [*Vigna radiata* (L.) Wilczek] grown in herbicide-amended soil to inoculation with *Bradyrhizobium* sp. (Vigna) MRM6. *J Agric Sci Technol* 13: 1209–1222.

Ahemad M, Khan MS (2011g) Effects of insecticides on plant-growth-promoting activities of phosphate solubilizing *Rhizobacterium klebsiella* sp. strain PS19. *Pestic Biochem Phys* 100: 51–56.

Ahemad M, Khan MS (2012) Productivity of greengram in tebuconazole-stressed soil, by using a tolerant and plant growth promoting *Bradyrhizobium* sp. MRM6 strain. *Acta* Physiol Plant 34: 245–254.

Allen DJ (1982) *The Pathology of Tropical Food Legumes. Disease Resistance in Crop Improvement*. John Wiley & Sons, Ltd., University of Cambridge.

Alonge SO (2000) Effect of imazaquin applications on the growth, leaf chlorophyll and yield of soybean in the Guinea Savanna of Nigeria. *J Environ Sci Health [B]* 35: 321–336.

Alvarez MAB, Gagne S, Antoun H (1995) Effect of compost on rhizosphere microflora of the potato and on the incidence of plant growth promoting *Rhizobacteria*. *Appl Environ Microbiol* 61: 194–199.

Amijee F, Giller KE (1998) Environmental constraints to nodulation and nitrogen fixation of *Phaseolus vulgaris* L. in Tanzania, I. A survey of soil fertility, root nodulation and multilocational responses to rhizobium inoculation. *African Crop Sci* 6: 159–169.

Anderson A, Baldock JA, Rogers SL, Bellotti W, Gill G (2004) Influence of chlorsulfuron on rhizobial growth, nodule formation, and nitrogen fixation with chickpea. *Aust J Agric Res* 55: 1059–1070

Arora NK, Kang SC, Maheshwari DK (2001) Isolation of siderophore producing strains of *Rhizobium meliloti* and their biocontrol potential against *Macrophomina phaseolina* that causes charcoal rot of groundnut. *Curr Sci* 81: 673–677.

Bagnasco P, De La Fuente L, Gualtieri G, Noya F, Arias A (1998) Fluorescent *Pseudomonas* sp. as biocontrol agents against forage legume root pathogenic fungi. *Soil Biol Biochem* 30: 1317–1322.

Bashir F, Siddiqi TO, Mahmooduzzafar IM (2007) Effect of different concentrations of mancozeb on the morphology and anatomy of *Lens culinaris* L. *Int J Environ Sci* 11: 71–74.

Beauchamp CJ, Dion P, Kloepper JW, Antoun H (1991) Phytological characteristic of opine utilizing rhizobacteria for traits related to plant growth activity. *Plant Soil* 132: 273–279.

Bellinaso ML, Greer CW, Peralba MC, Henriques, JA, Gaylarde CC (2003) Biodegradation of the herbicide trifluralin by bacteria isolated from soil. *FEMS Microbial Ecol* 43: 191–194.

Bellone CH, De Bellone SDVC, Pedraza RO, Monzon MA (1997) Cell colonization and infection thread formation in sugar cane roots by *Azotobacter diazotrophicus*. *Soil Biol Biochem* 29: 965–967.

Boldt TS, Jacobsen CS (1998) Different toxic effects of the sulphonylurea herbicides metsulfuron methyl, chlorsulfuron and thifensulfuron methyl on fluorescent Pseudomonads isolated from an agricultural soil. *FEMS Microbiol Lett* 161: 29–35.

Carson KC, Meyer JM, Dilworth MJ (2000) Hydroxamate siderophore of root nodule bacteria. *Soil Biol Biochem* 32: 11–21.

Casida JE, Lykken L (1969) Metabolism of organic pesticide chemicals in higher plants. *Annu Rev Plant Phys* 20: 607–636.

Chakraborty U, Purkayastha RP (1984) Role of rhizobi-toxine in protecting soybean roots from *Macrophomina phaseolina*. *Can J Microbiol* 30: 285–289.

Cook RJ (1986) Inter-relationship of plant health and the sustainability of agriculture with special reference to plant diseases. *Am J Alternative Agr* 1: 19–24.

Darure MV, Chonde SG, Bhosale PR, Raut PD (2012) Effect of residual concentration of endosulfan on the nitrogenase enzyme activity in *Arachis hypogea*. *Ann Biol Res* 3: 192–195.

Das AC, Chakravarty A, Sukul P, Mukherjee D (2003) Influence and persistence of phorate and carbofuran insecticides on microorganisms in rice field. *Chemosphere* 53: 1033–1037.

De Bertoldi M, Rambelli A, Griselli M (1978) Effects of benomyl and captan on rhizosphere fungi and the growth of *Allium cepa*. *Soil Biol Biochem* 10: 265–268.

Deshwal VK, Dubey RC, Maheshwari DK (2003) Isolation of plant growth promoting strains of *Bradyrhizobium, Arachis* sp. with biocontrol potential against *Macrophomina phaseolina* causing charcoal rot of peanut. *Curr Sci* 84: 443–448.

Dubey V, Singh D, Shukla A, Shukla S, Singh N (2012) Effect of application of different pesticides to leguminous crops on soil microflora of Sidhi district (M.P.). *Int J Eng Res Dev* 3: 1–3.

Duodu S, Bhuvaneshwari TV, Stokkermans TJ, Peters NK (1999) A positive role for rhizobitoxine in *Rhizobium* legume symbiosis. *Mol Plant Microbe In* 12: 1082–1089.

Evans J, Seidel J, Connor GE, Watt J, Sutherland M (1991) Using omethoate insecticide and legume inoculant on seed. *Aust J Exp Agric* 31: 71–76.

Finlayson DG, MacCarthy HR (1973) Pesticides residues in plants. In: Edwards CA (ed.), *Environmental Pollution by Pesticides*. Plenum Press, London and New York, pp. 57–86.

Fox JE, Gulledge J, Engelhaupt E, Burow ME, McLachlan JA (2007) Pesticides reduce symbiotic efficiency of nitrogen-fixing rhizobia and host plants. *Proc Natl Acad Sci USA* 104: 10282–10287.

Fox JE, Starcevic M, Jones PE, Burow ME, McLachlan JA (2004) Phytoestrogen signaling and symbiotic gene activation are disrupted by endocrine-disrupting chemicals. *Environ Health Persp* 112: 672–677.

Gonzalez-Curbelo MA, Dionis-Delgado S, Asensio-Ramos M, Hernandez-Borges J (2012) Pesticide analysis in toasted barley and chickpea flours. *J Sep Sci* 35: 299–307.

Graham PH, Vance CP (2000) Nitrogen fixation in perspective: An overview of research and extension needs. *Field Crop Res* 65: 93–106.

Graham PH, Vance CP (2003) Legumes: importance and constraints to greater use. *Plant Physiol* 131: 872–877.

Halder A, Chakrabartty PK (1993) Solubilization of inorganic phosphate by *Rhizobium*. *Folia Microbiol* 38: 325–330.

Hashem FM, Saleh SA, van Berkum P, Voll M (1997) Survival of *Bradyrhizobium* sp. (Arachis) on fungicide-treated peanut seed in relationship to plant growth and yield. *World J Microbiol Biotechnol* 13: 335–340.

Herman PL, Behrens M, Chakraborty S, Crastil BM, Barycki J, Weeks DP (2005) A three component dicamba O-demethylase from *Pseudomonas maltiphilia* strain DI-6: Gene isolation, characterization and heterologous expression. *J Biol Chem* 280: 24759–24767.

Hussain AM, Arshad AM, Hussain A, Hussain F (1987) Response of maize (*Zea mays*) to *Azotobacter* inoculation under fertilized and unfertilized conditions. *Biol Fert Soils* 4: 73–77.

Indiragandhi P, Anandham R, Madhaiyan M, Sa TM (2008) Characterization of plant growth promoting traits of bacteria isolated from larval guts of Diamondback moth *Plutella xylostella* (Lepidoptera: Plutellidae). *Curr Microbiol* 56: 327–333.

Jian W, Wendy C, Willis C (2013) Chromatography/electrospray ionization-tandem mass spectrometry determination of 151 pesticides in soybeans and pulses. *J AOAC Int* 96: 1114–1133.

Joseph B, Patra RR, Lawrence R (2007) Characterization of plant growth promoting rhizobacteria associated with chickpea (*Cicer arietinum* L.). *Int J Plant Prod* 2: 141–152.

Kaphalia BS, Takroo R, Mehrotra S, Nigam U, Seth TD (1990) Organochlorine pesticide residues in different Indian cereals, pulses, spices, vegetables, fruits, milk, butter, desi ghee, and edible oils. *J Assoc Off Anal Chem* 73: 509–512.

Kapoor K, Leenta, A (1996) Observations on growth responses of cyanobacteria under the influence of herbicides. *Pollut Res* 15: 343–351.

Karthikeyan R, Davis LC, Erickson LE, *et al.* (2003) Studies on responses of non-target plants to pesticides: A review. Hazardous Substance Research Center, Kansas State University. Manhattan, KS.

Kaushik SI (2006) Phytotoxicity of selected herbicides to mung bean (*Phaseolus aureus* Roxb.). *Environ Exp Bot* 22: 41–48.

Khan H, Zeb A, Ali Z, Shah SM (2009) Impact of five insecticides on chickpea (*Cicer arietinum* L.) nodulation, yield and nitrogen fixing rhizospheric bacteria. *Soil Environ* 28: 56–59.

Khan MS, Chaudhry P, Wani PA, Zaidi A (2006) Biotoxic effects of the herbicides on growth, seed yield and grain protein of green gram. *J Appl Sci Environ Manag* 3: 141–146.

Koch M, Tanami Z, Bodani H, Winnger S (1997) Field application of vesicular-arbuscular mycorrhizal fungi improved garlic yield in disinfected soil. *Mycorrhiza* 7: 47–50.

Korte F, Kvesitadze G, Ugrekhelidze D, *et al.* (2000) Organic toxicants and plants. *Ecotox Environ Safe* 47: 1–26.

Kumar N, Anubhuti JI, Amb MK (2010) Chronic toxicity of the triazole fungicide tebuconazole on a heterocystous, nitrogen fixing rice paddy field cyanobacterium, Westiellopsis prolific Janet. *J Microb Biotech* 20: 1134–1139.

Kumar R, Chandra R (2008) Influence of PGPR and PSB on *Rhizobium leguminosarum* Bv. *viciae* strain competition and symbiotic performance in Lentil. *J Agr Sci* 4: 297–301.

Kyamanywa S (1996) Influence of time of insecticide application on control of insect pests of cowpea and grain yield of cowpea at Mtwapa coastal province of Kenya. *Afr Crop Sci* 4: 373–382.

Lagana A, Fasciani GFL, Marino A, Mosse M (2000) Determination of diphenyl ether herbicides and metabolites in natural waters using HPLC with diode array tandem mass spectrophotometric detection. *Anal Chim Acta* 414: 74–94.

Lama JP (2008) Standard setting on pesticide residues to ensure food safety. *Agr Environ* 9: 46–53.

Lewis JA, Lumsden RD, Locke JC (1996) Biological control of damping-off diseases caused by *Rhizoctonia solani* and *Pythium ultimum* with various food bases. *J Biol Sci* 6: 163–173.

Lippman B, Leinhos V, Bergmann H (1995) Influence of auxin producing rhizobacteria on root morphology and accumulation in maize *Zea mays* L. caused by inoculation with Indole-3-acetic acid (IAA) producing *Pseudomonas* and *Actinobacter* strains or IAA applied exogenously. *J Appl Bot* 68: 31–36.

Lugtenberg B, Kamilova F (2009) Plant growth promoting rhizobacteria. *Annu Rev Microbiol* 63: 541–556.

Mapose IC, Cossa LS (2005) Yield and quality responses of cowpea varieties to partial defoliations as vegetable. *Afr Crop Sci Conf Proc* 7: 41–43.

Mougin CP, Corio-Costet MF, Werck-Reichhart D (2001) Plant and fungal cytochrome P-450s: their role in pesticide transformation. In: Hall JC, Hoagland RE, Zoblotowicz RM (eds), *Pesticide Biotransformation in Plants and Microorganisms: Similarities and Divergences*. ACS Symposium Series 777, Washington DC, pp. 166–181.

Mugo HM (1989) Studies of insect pests of pigeon pea (*Cajanus cajan* Mill sp.) during the flowering and post flowering stage and their impact on seed yield in Kenya. MSc thesis, University of Nairobi.

Musarrat J, Haseeb A (2000) Agrichemicals as antagonists of lectin-mediated Rhizobium legume symbiosis: Paradigms and prospects. *Curr Sci* 78: 793–797.

Nakamura A, Yohorti Y, Watanabe K, Sato Y, Boeger P, Wakabayashi K (2000) Peroxidation formation of lipid hydroperoxidase in etiolated leaves. *Pestic Biochem Phys* 66: 206–212.

Nare RWA, Savadogo PW, Gnankambary Z, Sedogo MP (2010) Effect of endosulfan, deltamethrin and profenophos on soil microbial respiration characteristics in two land use systems in Burkina Faso. *J Environ Sci* 4: 261–270.

Neetu T (2013) Determination of chlorinated pesticide in vegetables, cereals and pulses by gas chromatography in East National Capital Region, Delhi, India. *J Agricult Forestry Sci* 1: 27–28.

Niewiadomska A (2004) Effect of carbendazim, imazetapir and thiram on nitrogenase activity, number of microorganisms in soil and yield of Red clover (*Trifolium pratense* L.). *Pol J Environ Stud* 13: 403–410.

Niewiadomska A, Klama J (2005) Pesticide side effect on the symbiotic efficiency and nitrogenase activity of *Rhizobiaceae* bacteria family. *Pol J Microbiol* 54: 43–48.

Niewiadomska A, Sawicka A (2002) Effect of carbendazim, imazetapir and thiram on nitrogenase activity, number of microorganisms in soil and yield of hybrid Lucerne (*Medicago media*). *Pol J Environ Stud* 11: 737–744.

Okeyo-Owuor JB (1978) Insect pod borers of pigeon pea (*Cajanus cajan* (L.) Mill sp.) and their influence on developing pods and final seed yield in Kenya. MSc thesis, University of Nairobi, Kenya, 150 pp.

Okeyo-Owuor JB, Khamala CPM (1980) Insect pod borers of pigeon pea (*Cajanus cajan* (L.) Mill sp.) and their influence on developing pods and final seed yield in Kenya. *Kenya J Sci Tech* 1: 79–86.

Olupot G, Etiang J, Ssali H, Nahasirye M (2004) Sorghum yield response to kraal manure combined with mineral fertilizer in eastern Uganda. *MUARIK Bull* 7: 30–37.

Parween T, Jan S, Mahmooduzzafar, Fatma T (2012) Evaluation of oxidative stress in *Vignaradiata* L. in response to chlorpyrifos. *Int J Environ Sci Technol* 9: 605–612.

Pham CH, Min J, Gu MB (2004) Pesticide induced toxicity and stress response in bacterial cells. *Environ Contam Toxicol* 72: 380–386.

Ping-Sheng Z, Qi L (2010) Investigation on organophosphorus pesticide residue in fresh melon and legume vegetables in Huizhou. *Guangxi Agric Sci* 41: 338–340.

Prevost D, Bordeleau LM, Caaudry-Reznick S, Schulman HM, Antoun H (1987) Characterization of *Rhizobia* isolated from three legumes indigenous to the Canadian high arctic, *Astragalus alpinus, Oxytropism aydelliana, Oxytropis arctobia. Plant Soil* 98: 313–324.

Ruano A, Barcelo J, Poschericoler C (1988) Zinc phytotoxicity: effects on growth and photosynthetic pigment content of *Phaseolus vulgaris* L. during development. Anales de Edafología y Agrobiología 47: 1443–1457.

Ryu CM, Farag M, Hu CH, *et al.* (2003) Bacteria volatiles promote growth in Arabidopsis. *Proc Natl Acad Sci USA* 100: 4927–4932.

Sandermann H Jr (1992) Plant metabolism of xenobiotics. *Trends Biochem Sci* 17: 82–84.

Sanginga N, Ibewiro B, Houngnandan P, *et al.* (1996) Evaluation of symbiotic properties and nitrogen contribution of Mucunato maize grown in the derived savanna of West Africa. Plant Soil 179: 119–129.

Schroder P, Collins CJ (2002) Conjugating enzymes involved in xenobiotic metabolism of organic xenobiotics in plants. *Int J Phytorem* 4: 247–265.

Shankar PV, Shaikh NR, Vishwas PS (2012) Effect of different herbicides on the nodulation property of rhizobial isolates. *Universal J Environ Res Technol* 2: 293–299.

Sharma HC, Clement SL, Ridsdill-Smith TJ, *et al.* (2005) Insect pest management in food legumes: the future strategies. In: Kharkwal MC (ed.), *Proceedings of the Fourth International Food Legumes Research Conference*, 18–22 Oct 2005. Indian Society of Genetics and Plant Breeding, New Delhi, India, pp. 522–544.

Singh DK (2012) *Toxicology: Agriculture and Environment-Pesticide Chemistry and Toxicology*. Bentham Science Publishers.

Singh G, Wright D (1999) Effects of herbicides on nodulation, symbiotic nitrogen fixation, growth and yield of pea (*Pisum sativum*). *J Agr Sci* 133: 21–30.

Singh G, Wright D (2002) In vitro studies on the effects of herbicides on the growth of rhizobia. *Lett Appl* Microbiol 35: 12–16.

Sinha S, Mukherjee SK (2008) Cadmium-induced siderophore production by a high Cd-resistant bacterial strain relieved Cd toxicity in plants through root colonization. *Curr Microbiol* 56: 55–60.

Siti E, Cohen E, Katan J, Mordechi M (1982) Control of *Ditylenchus dipsaci* in garlic by bulb and soil treatments. *Phytoparasitica* 10: 93–100.

Smith MJ, Sholery JN, Schwyn B, Holden I, Neilands JB (1985) Rhizbactin, a structurally noval siderophore from *Rhizobium meliloti*. *J Am Chem Soc* 107: 1739–1743.

Sresen F, Kralova K, Macho V (2000) New finding about the inhibitory action of phenyl carbamates and phenyl thiocarbamates on photosynthetic apparatus. *Pestic Biochem Phys* 68: 113–118.

Tank N, Saraf M (2003) Phosphate solubilization, exopolysaccharide production and indole acetic acid secretion by rhizobacteria isolated from *Trigonellafoenum graecum*. *Indian J Microbiol* 43: 37–40.

Tank N, Saraf M (2010) Salinity resistant plant growth promoting rhizobacteria ameliorate sodium chloride stress on tomato plants. J Plant Interact 5: 51–58.

Tien TM, Gaskins MH, Hubbell DH (1979) Plant growth substances produced by *Azospirillum brasilense* and their effect on the growth of pearl millet (*Pennisetum americanum* L.). Appl Environ Microbiol 37: 1016–1024.

Vance CP, Graham PH, Allan DL (2000) Biological nitrogen fixation. Phosphorus: a critical future need. In: Pedrosa FO, Hungria M, Yates MG, Newton WE (eds), *Nitrogen Fixation: From Molecules to Crop Productivity*. Kluwer Academic Publishers, Dordrecht, pp. 506–514.

Wani PA, Khan MS, Zaidi A (2008) Chromium reducing and plant growth promoting *Mesorhizobium* improves chickpea growth in chromium amended soil. Biotechnol Lett 30: 159–163.

Xie ZP, Staehelin C, Wiemkemn A, Bolle T (1996) Ethylene responsiveness of soybean cultivars characterized by leaf senescence, chitinase induction and nodulation. *J PlantPhysiol* 149: 690–694.

Yang C, Lee C (2008) Enrichment, isolation, and characterization of 4-chlorophenol-degrading bacterium *Rhizobium* sp. 4-CP-20. *Biodegradation* 19: 329–336.

Zaidi A, Khan MS, Ahemad M, Oves M (2009) Plant growth promotion by phosphate solubilizing bacteria. *Acta Microbiol Imm H* 56: 263–284.

CHAPTER 7

Oxidative stress and antioxidant defence systems in response to pesticide stress

Fozia Bashir[1] and Sumira Jan[2]

[1]Department of Botany, Krishna College of Science and Information Technology, Bijnor, Uttar Pradesh, India

[2]Center for Research and Development (CORD), University of Kashmir, Srinagar, India

7.1 Introduction

Excessive use of synthetic fertilizers and pesticides over the past few decades has led to intensive farming and extensive crop cultivation. The consumption of fertilizers and pesticides increased significantly after World War II along with the restriction and expansion of industrialization (Ishikura, 1972; Aditya *et al.*, 1997). The increased use of fertilizers and pesticides along with denser spacing to cultivate more plants per unit area induced vigorous growth for many crop plants making them vulnerable to pest infestation.

There have been two generations of pesticides. The first comprise naturally occurring chemicals including sulphur, nicotine, mercury, arsenic and other heavy metals employed to guard crops for thousands of years; residues of these pesticides are responsible for coevolutionary development of the natural defence systems of plants. Pyrethrum is a botanical extract containing pyrethrins used since 1850, and a first-generation nonpersistent insecticide. However, it could not be employed for outdoor agricultural applications due to its thermal and photo lability (Roy & Ghosh, 1942). The second generation emerged from 1939 onwards with the development of the first synthetic pesticide, DDT, which was released to the market in 1947 (Ishikura, 1972).

There are four main types of synthetic pesticides: chlorinated hydrocarbons, organophosphates, carbamates and pyrethroids. Chlorinated hydrocarbons, such as DDT, aldrin, lindane and kepone, are broad-spectrum agents that kill insects by targeting and deactivating their nervous system. DDT was first synthesized by Zeidler (1874), but its insecticidal properties remained unknown until the 1930s and 40s, when Muller and Swiss (1940) and West and Campbell (1950) revealed the effectiveness of DDT for the control of insects and commercialized it as gesarol. This became a miracle pesticide during the World War II, when the Allied powers realized its great effectiveness against deadly diseases like malaria and typhoid (Roy & Ghosh, 1942). DDT emerged as a marvel in the fight against human epidemics and for crop protection, for example scoring highly in its efficiency in controlling the paddy stem borer, *Tryporyza oryzae* Walker, which was particularly destructive in western Japan. The enormous success of DDT encouraged the introduction of new synthetic pesticides tested for their possible effectiveness. However, subsequent decades have seen a major shift in the consumption pattern as well as the formulation of pesticides (Ishikura, 1972). Prolonged use of various pesticides like DDT resulted in bioaccumulation of residues over long periods of time (Kenaga, 1972). Crop plants growing in such contaminated soil can take up DDT, which may get translocated to the aerial parts, affecting the production and yield (Lichtenstein, 1959, 1960; Lichtenstein & Schulz, 1960, 1965; Lichtenstein *et al.*, 1965). Extensive research has been done on the uptake, translocation and effect of different pesticides, including DDT, in crop plants (both from tropical and subtropical regions) (Thakre & Saxena, 1972; Perfect *et al.*, 1979; Talekar *et al.*, 1983). Subsequent to the release of DDT in 1947, benzene hexachloride (BHC) and other chlorinated insecticides were introduced and marketed extensively (Ishikura, 1972; Roy, 2002). Organophosphates, such as

Legumes under Environmental Stress: Yield, Improvement and Adaptations, First Edition. Edited by Mohamed Mahgoub Azooz and Parvaiz Ahmad.

chlorpyrifos, quinalphos, tetraethyl pyrophosphate (TEPP) and methyl parathion, are broad-spectrum agents that kill insects by inhibiting their nerve impulses Organophosphates exhibit global accessibility through surface runoff and leaching due to their polarity and water-solublity (Hurtig, 1972). Carbamates, such as carbaryl, carbofuron and methyl carbamate, can be either broad- or narrow-spectrum agents; they also kill by deactivating the pest's nervous system and have low persistence but they are highly toxic to honey bees and therefore interfere with pollination. Although carbaryl was introduced in 1959, an extensive search for carbamate compounds was made afterwards. Goto (1970) reported that carbamates made up 16% of the total consumption of insecticides.

The growing environmental and ecological hazards caused by the continual use of these organochlorines, organophosphates and carbamate pesticides propelled the introduction of a new generation of pesticides displaying much lower persistence levels. To embark on this route, synthetic pyrethroids were introduced to the market during the 1970s and soon emerged as a shield. In the early 1960s, Elliott and his colleagues at Rothamstead Experimental Station in the UK, initiated an extensive study of the mechanisms of action and the relationship between the structures and activities of the natural pyrethroids and various synthetic analogues. This work was published in major reviews on pyrethroids by Elliott *et al.* (1973, 1978) and Soderlund *et al.* (2002). Casida and Quigstad (1998) summarized the goals of pyrethroid research as mentioned below:

- Pyrethroids should be light-stable and non-biodegradable.
- There should be target site specificity (e.g. bioresmethrin) or metabolic degradation (lower toxicity for *trans*- than for *cis*-cyclopropane- carboxylate).
- Every part of the molecule should be modified in relation to the retention of the activity.
- High insecticidal potency should be maintained with minimum toxicity to fish (e.g. the non-ester silaflufen).
- Compunds should be improved to act as effective fumigants and soil insecticides (e.g. tefluthrin).
- Potency should be optimized to allow corresponding reduction in environmental contamination and hazards.

Synthetic pyrethroids are powerful, broad-spectrum agents, and now form a diverse class of more than 1000 insecticides used to control insect pests in agriculture and stored products (Roy & Ghosh, 1942). Pyrethroids are semi-synthetic derivatives of the chrysanthemumic acids derived from *Chrysanthemum* spp., developed as insecticides that are more effective than natural pyrethrins and less toxic to mammals and birds in comparison to organochlorines and organophosphates. Deltamethrin, or (+)-*alpha*-cyano-3-phenoxybenzyl alcohol, is one of the most widely acceptable pyrethroids. Because of its high activity against a broad spectrum of insect pests, deltamethrin is widely used in agriculture and forestry (Villarini *et al.*, 1998). However, its continous application on experimental plants results in accumulation to 10 times its initial concentration in a single season, which is higher than its degradation rate (Mueller-Beilschmidt, 1990). Moreover, the excessive and continual use of synthetic pyrethroids has resulted in numerous problems such as resistance in aphids and bollworms. Leguminous plants are widely susceptible to numerous seed- and soil-borne fungi and hence their treatment with different pesticides at the seed and seedling stages is a prerequisite. Like other abiotic stresses, insecticides may cause oxidative stress in plant cells, affecting the various metabolic activities and growth components in the plants (Toscano *et al.*, 1982; Jones *et al.*, 1986).

Toxicity in pyrethroids depends much on the three-dimensional configuration of the molecule, i.e. its stereochemistry. Pyrethroids like permethrin have pairs of isomers, referred to as *cis* and *trans* isomers; each isomer has its own toxicity but usually the *cis* isomer is relatively more toxic than the *trans* isomer (Figure 7.1). Some pyrethroids have as many as eight different isomers (Mueller-Beilschmidt, 1990). Given a mixture of two isomers, toxicity depends on the ratio of the amounts of the two isomers in the formulation. Most commercial formulations have a fixed isomeric ratio.

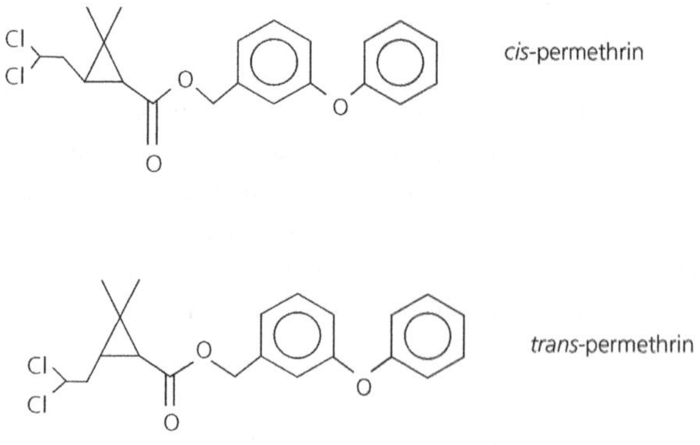

Figure 7.1 Structural formulae of *cis*- and *trans*-permethrin.

Formulations that are made up of a single isomer like deltamethrin are relatively more toxic than those formulations that are made up of two, or four to eight isomers. For example, the female rat acute oral LD_{50} of permethrin increases from 224 mg of the pyrethroid per kilogram of body weight (mg/kg) to 6000 mg/kg as the proportion of the *trans* isomer increases from 20% to 80% (Bradbury & Coats, 1989).

7.2 Fate of pesticides in plants, soil and water

Pyrethroids undergo degradation in the environment through both biotic and abiotic processes, including metabolic degradation by plants, animals and microorganisms and degradation by light via photolysis. Degradative photolysis of pyrethroids occurs in three main ways by the action of light, including: cleavage of a double bond between a carbon atom and an oxygen atom (i.e. ester cleavage); removal of halogens (chlorine, fluorine or bromine atoms), i.e. dehalogenation; and conversion of one isomeric form to another (i.e. isomerization). The main product of photolysis of pyrethroids is 3-phenoxybenzoic acid (Davies, 1985). Pyrethroid degradation in the soil is mostly by chemical and microbial action, and the rate of degradation depends on the type of pyrethroid, soil, climatic conditions, and the nature and size of microbial species present in the soil.

Insecticides are non-target specific; i.e. besides killing insect pests, they also affect the activity of beneficial microbial communities in soil and beneficial insects. Microbes present in the soil play a major role in the metabolism of both organic and inorganic constituents in soil. Accumulation of insecticidal residues in soil may have deleterious effects on the various activities of soil microbes (Edward, 1972). Excessive concentration of insecticides in the environment may shift microbial populations to temporarily favoured group(s) of microorganisms, which may overpopulate the soil (Tu & Mile, 1976).

Rhizobia exhibited varied growth patterns following pesticide treatments. Some pesticides do not affect the growth of rhizobia when applied at field rates (Mongollon *et al.*, 1986; Torralba-Redonde *et al.*, 1986; Martensson, 1992), whereas some pesticides are found to be highly toxic to rhizobia when applied in low as well as high concentrations (Bandyopadhyay *et al.*, 1979; Ruiz-Sainz *et al.*, 1985). Abu-Gharbia (1996) in his experiment observed that the growth of both free-living and symbiotic rhizobia in four leguminous species was affected by four pesticide treatments. Gaucho (an insecticide containing imidacloprid) and Vitavax-300 fungicide (carboxin and captan) when applied in higher concentrations inhibited the growth of root nodule bacteria under *in vitro* conditions (Miettinen & Echegoyen, 1996). Hancock *et al.* (1998) in an experiment on fumigated agricultural soil observed that rhizobia can oxidize methyl bromide, which suggests the importance of these microbes in decontamination and recycling of organic compounds and finally suggests their role in environmental cleanup.

Pyrethroids being highly hydrophilic can adhere to organic matter and are transported into the lipid bilayer of plant cells and get strongly adsorbed by soil particles (Briggs *et al.*, 1983; Anonymous, 1986). Once pyrethroids are absorbed in the plant body they tend to remain immobile, and their leaching into the groundwater via soil or translocation through a plant is uncommon (Anonymous, 1986). Pyrethroids mostly remain confined to the upper 5–10 cm of soil after field applications (Reed, 1983; Anonymous, 1986). But it has been observed that several principal pyrethroid degradation products like 3-phenoxybenzoic acid and dichlorovinyl acid leach readily (NRC,1986; IRPTC,1990). Pyrethroids are sometimes removed from the site of their application due to spray drift, soil erosion or volatilization (evaporation). Pyrethroids can drift from highly contaminated agricultural fields and pollute nearby surface waters; residues have been reported several months after their application. Pyrethroid runoff from a cotton field after a heavy rain shower affected various invertebrates in the nearby pond (NRC, 1986).

Blue-green algae (cyanobacteria) found within moist soils serve as excellent tools for the study of pesticide toxicity (McCann & Cullimore, 1979). Raghu and MacRae (1967) observed that sometimes pesticides in a poorly soluble state actually stimulate and enhance algal growth. Algae occasionally display positive responses to pesticides with little or no effects at low doses, and growth inhibition only at high doses. For example, impact on growth parameters like cell volume, cell number and protein content in *Chlorella* was low with 10 mg/L methyl parathion, moderate with 20 mg/L and severe with 30 mg/L (Saroja & Bose, 1982). Studies suggest that algal sensitivity shows considerable diversity to pesticides. For example, the responses of the unicellular chlorophytes *Chlorella vulgaris* and *Selenastrum capricornutum*, to three herbicides (simetryn, pretilachlor and thiobencarb) were compared (Kasai & Hatakeyama, 1993) using single species toxicity tests for population growth. The authors observed that *C. vulgaris* strains were more tolerant than *S. capricornutum* strains for these three herbicides. Kent and Currie (1995) conducted an experiment with 12 algal species exposed to fenitrothion (an organophosphorus insecticide) and observed that their tolerance may be inversely proportional to cell surface area, volume ratio, cell lipid content, and the inherent bioconcentration potential of the cell.

Some herbicides inhibit photosynthetic electron flow (Trebst *et al.*, 1991) and thus affect algal growth and development. The substituted phenyl urea herbicides applied at concentrations of between 1 and 5 mM can inhibit the Hill reaction of isolated chloroplasts by as much as 50%, due to the lipophilic nature of pesticides (Camilleri *et al.*, 1987). Hoffman (1971) found that the toxic effects of monuron and bromacil on photosynthesizing *Chlorella* were additive, indicating the probability of a common mode of action. Kirby and Sheahan (1994) compared the toxicity of the herbicides atrazine, isoproturon and mecoprop on a freshwater green flagellate alga, *Scenedesmus subspicatus*, and a macrophyte plant, *Lemna minor*. For *S. subspicatus*, atrazine and isoproturon were similar and exhibited three orders of magnitude more toxicity than mecoprop.

Karanth and Vasantharajan (1972) reported the inhibition of taproot elongation and nodulation of sun-hemp with the fungicide Dexon, which was known to inhibit the conversion of tryptophan to indole acetic acid (IAA). The inhibition of root nodulation in bean plants could be due to the inhibition of rhizobia in the soil or to metabolic changes in plants. Sen and Kapoor (1974) have shown that plants treated with carbendzim show improved growth, in the case of tomatoes and cucurbits, besides control of wilt (*Fusarium oxysposium*) and powdery mildew (*Sphaerotheca fulginea*). Benzimidazole fungicides are also known to stimulate plant growth besides controlling plant pathogens. Majumdar (1974) and Singh and Kang (1978) reported that Bavistin resulted in increased fresh or dry weight of groundnut, and the yield of pods was more than control treatments. There is considerable variation in the persistence of pesticide residues in soil, water and plant tissues. The half-life of various pyrethroids in soils varies from 1 day to 4 months. It has been observed that photolabile pyrethroids usually degrade much more quickly than photostable pyrethroids, and degradation is usually much faster in aerobic conditions in comparison to anaerobic condition (Leahey, 1985).

7.2.1 Effect of pesticides on non-target plants

Plant species that are unintentionally exposed to pesticides are called non-target plants. Any direct or indirect effects that cause a significant change in the survival, vegetative growth or reproduction of non-target plant species are called non-target plant effects. Non-target plant effects include a range of symptoms, including vegetative growth changes, plant death or altered reproductive capability; these generally result in reduced growth of non-target plants and economic losses or ecological disturbances. One of the inadvertent effects of herbicides is altered susceptibility to disease of both target and non-target plants (Altman, 1993). It is estimated that for every plant species that becomes extinct, 10--30 other non-plant organisms may also become extinct (Ware, 1993). Non-target plants growing in pesticide-contaminated environments may experience local extinction. It has been observed by various researchers that desirable species of trees, grasses and shrubs, particularly those that are highly sensitive to pesticides, may be eliminated or greatly reduced (Fink, 1972; Tomkins & Grant, 1977; Boutin *et al.*, 2000).

Hamner and Tukey (1946) studied the effects of the insecticide 2,4-D on various woody plants like juniper (*Juniperis communis* L.), elm (*Ulmus americana* L.), poplar (*Populus nigra* L. var. *Italica* Du Roi), spruce (*Picea canadensis* Mill.), willow (*Salix nigra* Marsh.), plum (*Prunus domestica* L.), pear (*Pyrus communis* L.) and pine (*Pinus strobus* L.). This study concluded that the leaves of all treated plants except juniper displayed bending, discoloration and wilt within a few days after foliar application of 2,4-D during warm weather. Powdered application of 2,4-D at 20% concentration on the cut stems of peach, poplar, pine, spruce, elm and willow elicited a typical curvature in new shoot growth within 2 days after treatment, but the degree and type of response varied in different plants. The upper leaves of 12-m (40-ft) poplar trees exhibited downward bending of the petioles within 2 days followed by progressive browning, drying and leaf fall within 1 month of treatment. The upper leaves were dead, and lower parts of the trees became twisted, chlorotic (yellowing) and brown leading to the death of the trees within 60 days after application.

7.2.2 Effects of pesticide residues in food and water

Continual and excessive use of pesticides leaves pesticide residues or their remains as a mixture of substances in food, soil, water and air. Pesticide residues include the conversion products, metabolites, reaction products and impurities. Pesticides exhibit biomagnification, where the residues of pesticides, especially insecticides, tend to increase up to lethal levels from lower to higher trophic levels in the food chain. Pesticides also undergo bioaccumulation as a characteristic feature; for example, unicellular organisms like phytoplankton accumulate pesticide in their cells at concentrations many times higher than that in the surrounding water while growing in pesticide-contaminated rivers. Zooplankton in the rivers feed on these phytoplankton and thus the residues of pesticides pass through the food chain and ultimately reach the top of the chain in higher concentrations. Fishes in polluted water accumulate DDT at concentrations more than 10,000 times that of their surroundings (Kenaga, 1971). Crop plants like potatoes and carrot grown on soil treated with organochlorine insecticides like aldrin at 0.5 kg/acre are reported to accumulate higher levels of pesticide residues than other crops grown in similar conditions.

Bioaccumulation is exhibited by both animals and plants. Chlorinated hydrocarbons have the greatest tendency for biological concentration. The indiscriminate use of pesticides can be considered a major source of pollutants with effects on the biochemistry and physiology of the plants (Kenaga, 1972). The residues of organochlorine and organophosphorus insecticides were analysed by Ahmad *et al.* (1999). About 50% of the 149 samples of different vegetables investigated were found to be contaminated by various pesticide residues, and 16% of samples contained residues above the maximum residual limit (MRL).

Plants containing high fatty acid contents rapidly take up various organochlorine and organophosphate pesticides. It has been reported by Nash *et al.* (1974) that pesticide residues in the seeds are directly related to their oil contents. Increased uptake of pesticides like DDT by tobacco plants was correlated with the high fatty acid content of this plant (Rosa & Cheng, 1973). It was suggested that the higher lipid content in oil channels of oilseeds facilitates the uptake and distribution of lipophilic chemicals like DDT. Keith (1970) also supported this theory and suggested that oils and fatty acids in the cytoplasm probably help in solubilization and dissolution of liphophilic chemicals, like DDT, within the cell. When these lipophilic chemicals bind to the lipid fraction of the cell membrane, membrane permeability and root transport are impaired. This, in turn, may affect the uptake and translocation of essential plant nutrients from soil to plants and ultimately result in inhibition of plant growth.

Pesticide monitoring programmes in California between 1982 and 1985 found permethrin (a pyrethroid) residue on cabbage, lettuce and tomatoes (Mott & Snyder, 1988). US Food and Drug Administration (FDA) monitoring programmes between 1985 and 1990 consistently found residues of various pesticides like permethrin, fenvalerate and cypermethrin in around 45 different vegetables, fruits and meat products. On average up to 6 ppm permethrin was found on leafy vegetables such as spinach, lettuce, kale, collards and turnip greens (FDA, 1990b). The maximum permethrin residue level set by the FDA on leafy vegetables is 20 ppm. To date no legal limits (tolerance levels) for pyrethroid residues on many of these products have been established.

According to FDA reports the fluvalinate residues in honey are 22 times higher than the legal limits (FDA, 1990c). It has also been reported that the pyrethroid residues (e.g. from cyfluthrin, deltamethrin, cypermethrin, fenvalerate and permethrin) remain effective for more than 10 months in grains even after milling and baking with minimal losses (Dicke *et al.*, 1988).

7.3 Pyrethroids: mode of action

DDT and pyrethroids are by far the most important insecticides.On the basis of mode of action of insecticides they are classified into two types. Type 1 includes DDT and all pyrethroids lacking a cyano-group, whereas type 2 compounds include pyrethroids that contain an α-cyano-3-phenoxybenzyl alcohol. There is ample evidence to suggest that both DDT and the pyrethroids have a common mode of action in acting on voltage-gated sodium channels. Pyrethroids prolong the period of opening of the sodium channels. Normal opening and closing should occur in less than a millisecond when an impulse passes. But pyrethroid poisoning prolongs the opening and delays the closing leading to sodium leaking out when the channel should be closed. For the type 2 pyrethroids the tail current is much more distinct and may last for minutes. The resting potential is not restored and the impulse does not pass distinctly, but becomes a train of action potentials because a lower potential rise is necessary to reach the threshold for the action potential (Stenerson, 2004). Sodium channels are pesticide specific; not only pesticides like DDT and its analogues but also poisons from plants, scorpions, sea anemones and amphibians act by binding to one of the active sites. It has also been observed that cross-resistance often occurs between DDT analogues and the pyrethroids. This type of resistance is called knockdown resistance (Kdr). Various insecticides, either naturally occurring pyrethrins or the synthetic pyrethroids, are nerve poisons.

7.4 Oxidative stress and ROS production in plants

Molecular oxygen serves as a two-edged sword due to reactive oxygen species (ROSs) in aerobic organisms (Halliwell & Gutteridge, 1989). Thus oxidative stress can simply be defined as imbalance between production and removal of ROS, leading to several physiological challenges in any cellular compartment. The hypothetical fine-tuned balance could be disturbed because of an excess accumulation of ROS or decline of antioxidants. The redox equilibrium changes when ROS levels are increased or the plant's antioxidant defence system (ADS) is depleted.

Generally, molecular oxygen (O_2) is an inert molecule in the absence of any catalysts. Higher organisms use it in the pathway of four-electron reduction, leading to production of water (H_2O). However, sometimes one-, two- and three-electron reduction of O_2 or excitation of triplet oxygen ($3O_2$) occur and cause the formation of superoxide radical (O_2^-) or hydroperoxyl radical (HO_2•), hydrogen peroxide (H_2O_2), hydroxyl radical (OH•) and singlet oxygen ($^1O_2$), respectively (Apel & Hirt, 2004). Compared to molecular oxygen, all these species are more toxic and unstable, ready to participate in unlimited chemical oxidation reactions, hence they are termed reactive oxygen species (ROS) (Bartosz, 1997).

7.4.1 Specific biochemical properties of ROS in plants

Reactive oxygen species, being derivatives of oxygen, are the most important group of free radicals in living organisms. Each one has specific properties; for example, H_2O_2 can easily diffuse through membranes and react at distant sites by comparing to charge·O_2^-. Therefore, O_2^- is significant in initiating oxidative damage in distant cellular components. On the other hand, the protonated form of O_2, the hydroperoxyl radical (HO_2^-), can cross biological membranes as a neutral molecule and by extracting protons from polyunsaturated fatty acids (PUFA) can easily propagate lipid oxidation (Gechev *et al.*, 2006). Superoxide dismutase (SOD) is an enzyme that converts O_2^- into H_2O_2. This conversion usually occurs spontaneously by dismutation of O_2^-, but the velocity of the reaction in the presence of SOD in most living organisms is greatly increased. H_2O_2 is a less reactive oxidizing agent present in micro- to millimolar ranges in plant tissues (Cheeseman, 2006; Halliwell & Gutteridge, 1999, Khan & Kour, 2007), and O_2^- and H_2O_2 by themselves are relatively harmless and removed from biological systems in the absence of any metal ions. However, their damaging effects become more explicit when they interact to form highly reactive hydroxyl radicals (OH•) in the presence of metal ions (Peixoto *et al.*, 2006; Moller *et al.*, 2007). For a historical view, towards the end of the 19th century, Fenton defined the oxidizing potential of H_2O_2 with ferrous salts (Fe^{2+}) (Fenton, 1894, 1899):

$$Fe^{2+} + H_2O_2 \rightarrow Fe^{3+} + OH + OH^- \quad (7.1)$$

The above reaction is limited by the bioavailability of ferrous ion, but with the help of reducing agents such as O_2^-, ferric ion (Fe^{3+}) can be recycled to the reduced ferrous state.

$$O_2^- + Fe^{3+} \rightarrow Fe^{2+} + O_2 \quad (7.2)$$

Forty years later, Haber and Weiss (1934) identified OH• as the oxidizing species. The sum of reactions 7.1 and 7.2 gives the formation of OH• from O_2^- and H_2O_2 catalysed by transition metal ions. Today, OH• radicals are known to be generated by the Haber–Weiss reaction (Equation 7.3) and other transition metal ions, like Cu^+ and Cu^{2+} can replace Fe^{2+} and Fe^{3+} in these reactions (Bartosz, 1997).

$$H_2O_2 + {\bullet}O_2^- \rightarrow {\bullet}OH + OH^- + O_2 \left(\text{Haber} - \text{Weiss reaction}\right) \quad (7.3)$$

The OH• radical is the most reactive product of partial oxygen reduction in biological systems and has a half-life of 1 nanosecond. As a result, it can react with and damage any type of molecule that it encounters within a cell. The important feature of ROS chemistry derives from the interconversion of one ROS into another. In this manner, the Fenton and Haber–Weiss reactions were considered to be the key events in ROS interconversions. Beside the above-mentioned reactions, ROS can interact with each other, which leads to production of a third group of ROS species that are more toxic than the primordial ones (Dat *et al.*, 2000). For example, peroxynitrite ($ONOO^-$) is produced when O_2^- reacts with nitric oxide (NO) (Bartosz, 1996; Beligni & Lamattina, 1999). Peroxynitrous acid (ONOOH) in its protonated form of $ONOO^-$ is a strong oxidizing agent and gives rise to various highly damaging radical and non-radical reactive species (Halliwell, 2006). Different ROS have very diverse properties and reactivities, their order of reactivity being: $OH\bullet > O_2^- > H_2O_2$ (Sweetlove & Moller, 2009).

Glutathione is a low molecular weight thiol and sulphur-containing tripeptide. It is also a universal redox-sensing component used as a marker of oxidative stress at the cellular level. Moreover, it plays an essential role in the detoxification of xenobiotics and also sequesters heavy metals by acting as the precursor of heavy metal-binding phytochelatins. Glutathione serves as antioxidant in cellular compartments including mitochondria, cytosol, peroxisomes, nuclei and chloroplasts. In the ascorbate-glutathione cycle (AGC), dehydroascorbate (DHA) is oxidized to ascorbate (ASC), which utilizes GSH as an electron donor. This pathway is considered to be the main pathway of free radical removal in the chloroplast stroma by equilibrating the redox status. The ASC cycle is catalysed by a set of four enzymes: ascorbate peroxidase (APX), monodehydroascorbate reductase (MDHAR), glutathione-dependent dehydroascorbate reductase (DHAR) and glutathione reductase (GR). ASC and glutathione participate in cyclic electron transfer from NADPH, which leads to reduction of H_2O_2. Senescence and plant death occur due to the progressive oxidation and degradation of glutathione and ASC pools.

7.4.2 Sites of production of ROS

The formation of ROS occurs in all cells irrespective of stress at several locations, including chloroplasts, mitochondria, plasma membranes, peroxisomes, apoplast, endoplasmic reticulum and cell walls. Reactive oxygen species are usually formed by the inevitable flow of electrons to O_2 from the electron transport activities of chloroplasts, mitochondria and plasma membranes, or as a by-product in various metabolic pathways.

7.4.2.1 Chloroplasts

The electron transport chain in photosystem I (PSI) and photosystem II (PSII) are the main sources of ROS in chloroplasts. Conditions like salinity, increased light intensity, pesticides, temperature stresses and other conditions limiting CO_2 fixation result in the enhanced production of ROS in plants. Under normal conditions, the electrons from the excited photosystem reduce NADP to NADPH, which then enters the Calvin cycle and reduces the final electron acceptor, i.e. CO_2. Electron leakage from ferredoxin to O_2 results in generation of $\bullet O_2^-$ due to decreased NADP supply and overloading of the electron transfer ch ain (ETC) under stress conditions (Elstner, 1991). This process is called the Mehler reaction:

$$2O_2 + 2\text{ferredoxin} \rightarrow 2\bullet O_2^- + 2\text{ferredoxin}$$

In PSI, the electron transport chain involves leakage of electrons to O_2 from 2Fe-2S and 4Fe-4S clusters, whereas in PSII, the plastoquinones QA and QB are electron acceptors in the ETC. Leakage of electrons from this site to O_2 results in the production of $\bullet O_2^-$ (Cleland & Grace, 1999).

The $\bullet O_2^-$ formed by O_2 reduction is a rate-limiting step. Once formed, $\bullet O_2^-$ results in the generation of more aggressive ROS. It may then either be converted to HO_2 on the internal membrane surface or to H_2O_2, by the enzymatic activity of SOD, on the external membrane surface. H_2O_2 once formed is converted to the highly dangerous OH• through the Fenton reaction.

7.4.2.2 Mitochondria

Reactive oxygen species are produced at several sites of the ETC in mitochondria. Direct reduction of oxygen to $\bullet O_2^-$ in mitochondria occurs in complex I, i.e. i.e. NADH:ubiquinone oxidoreductase (Arora *et al.*, 2002). There is reverse electron flow, i.e. electron transport from complex II to I, when NAD^+-linked substrates for complex I are limited. This reverse electron flow results in the increased production of ROS at complex I, and this flow is regulated by hydrolysis of ATP (Turrens, 2003). Complex III, the ubiquinone-cytochrome *c* reductase region of the ETC also produces $\bullet O_2^-$ from oxygen. Fully reduced ubiquinone donates an electron to cytochrome c_1 resulting in the formation of an ubisemiquinone radical, which is highly unstable and favours electron leakage to O_2 and thus finally results in the formation of $\bullet O_2^-$ (Murphy, 2009). The ETC and ATP synthases are strongly coupled in aerobic conditions but various stress factors lead to inhibition and modification of the various components, leading to over-reduction of electron carriers and, hence, increased production of ROS (Noctor *et al.*, 2007; Blokhina & Fagerstedt, 2010).

Some of the enzymes like aconitase produce ROS directly whereas some others like 1-galactono-γ-lactone dehydrogenase (GAL) furnish electrons to ETC (Andreyev *et al.*, 2005; Rasmusson*et al.*, 2008). Being the primary ROS formed by monovalent reduction in the ETC, $\bullet O_2^-$ gets quickly converted to H_2O_2 either by the MnSOD (mitochondrial form of SOD) or by APX. H_2O_2 is a relatively stable and membrane-permeable ROS; it can be further converted to the extremely active hydroxyl radical (OH•) via the Fenton reaction.

7.4.2.3 Endoplasmic reticulum

Cytochrome P450 is involved in NAD(P)H-dependent electron transport and produces $\bullet O_2^-$ in endoplasmic reticulum (Mittler, 2002). Organic substrate, RH, is first reduced by a flavoproteins after reacting with Cyt P450 to form a radical intermediate (Cyt P450 R^-). This radical intermediate can readily react with triplet oxygen as each has one unpaired electron. This oxygenated complex (Cyt P450-ROO^-) may either be reduced by cytochrome *b* or occasionally may become dissociated resulting in the production of $\bullet O_2^-$.

7.4.2.4 Peroxisomes

Peroxisomes, as a result of their essentially oxidative type of metabolism, are probably the major sites of intracellular H_2O_2 production (Del Río *et al.*, 2006). The glycolate oxidase reaction, β-oxidation of fatty acids, the enzymatic reactions of flavin oxidases, and the disproportionation of $\bullet O_2^-$ radicals are the main metabolic processes occurring in different types of peroxisomes and are responsible for the production of H_2O_2. Photorespiration occurring in peroxisomes involves the oxidation of glycolate by glycolate oxidase resulting in the production of the majority of H_2O_2 (Noctor *et al.*, 2002). In peroxisomes, like mitochondria and chloroplasts, $\bullet O_2^-$ is also produced as a consequence of their normal metabolism. Biochemical and electron spin resonance spectroscopy (ESR) methods have identified two sites of $\bullet O_2^-$ production in peroxisomes of pea leaves and watermelon cotyledons: in the organelle matrix, where xanthine oxidase (XOD) catalyses the oxidation of either xanthine or hypoxanthine to uric acid, and in the peroxisomal membranes, where and ETC made up of a flavoprotein NADH and Cyt *b* is involved in the production of $\bullet O_2^-$. Integral membrane polypeptides (PMPs) of peroxisomes having molecular masses of 18, 29 and 32 kDa were found to be involved in $\bullet O_2^-$ generation. The 18 kDa and 32 kDa PMPs use NADH as electron donor for $\bullet O_2^-$ generation while the 29 kDa PMP is capable of reducing cytochrome *c* with the help of NADPH as electron donor for $\bullet O_2^-$ generation (López-Huertas *et al.*, 1999). Of the three integral polypeptide membranes of peroxisomes, the 18 kDa PMP was the main producer of $\bullet O_2^-$. The $\bullet O_2^-$ generated by all three PMPs of peroxisomes is quickly converted into H_2O_2 via SOD.

7.4.2.5 Plasma membranes

Plasma membranes loaded with electron-transporting oxidoreductases also produce ROS. EPR spin-trapping techniques and specific dyes illustrate the production of ROS from isolated plasma membranes, especially ones from the growing and non-growing zones of hypocotyls, roots of etiolated soybean seedlings, and coleoptiles and roots of etiolated maize seedlings (Heyno *et al.*, 2011). The production of $\bullet O_2^-$ is mediated by NAD(P)H via NADPH oxidase and quinone reductase in soybean plasma membranes (Heyno *et al.*, 2011). NADPH oxidase is responsible for the transfer of electrons from cytoplasmic NADPH to O_2 resulting in $\bullet O_2^-$ generation. The $\bullet O_2^-$ formed is rapidly converted to H_2O_2 either spontaneously or enzymatically via SOD activity. NADPH oxidase has been found to play an important role in the production and accumulation of ROS in plants thriving under stress conditions (Apel & Hirt, 2004; Torres *et al.*, 2002; Kwak *et al.*, 2003).

7.4.2.6 Cell walls

Cell walls also play an active role in the production of ROS. Peroxidases associated with the cell walls are involved in the production of H_2O_2. Peroxidases associated with isolated cell walls of horseradish catalyse the production of H_2O_2 in the presence of NADH, and this reaction is enhanced by various monophenols, particularly coniferyl alcohol. NADH utilized in the production of H_2O_2 is solely provided by malate dehydrogenase (Gross, 1977). The hypersensitive response (HR) caused by the bacterium *Xanthomonas campestris* pv. *malvacearum* in cotton plant, and potassium (K) deficiency stress in *Arabidopsis* results in the generation of ROS by cell-wall-located peroxidases (Martinez *et al.*, 1998; Kim *et al.*, 2010). Diamine oxidases are also found to be involved in the production of activated oxygen in the cell wall (Elstner, 1991).

7.4.2.7 Apoplast

Enzymes located in the cell wall have been proved to be responsible for ROS production (Apel & Hirt, 2004; Heyno *et al.*, 2011). Oxalate oxidase, a cell-wall-associated enzyme (also known as germin), is responsible for H_2O_2 release. CO_2 from oxalic acid in the apoplast is also involved in H_2O_2 accumulation during interactions between different cereal species and fungi (Wojtaszek, 1997; Lane, 2002). Amine oxidase-like enzymes may provide defence responses under stress conditions in plants via production of H_2O_2 (Cona *et al.*, 2006). Heyno *et al.* (2011) observed that apoplastic OH• generation depends totally on peroxidase localized in the cell wall.

7.4.3 Role of ROS as messengers

ROS function as second messengers at very low or moderate concentrations and mediate several plant responses including stomatal closure (Neill *et al.*, 2002; Kwak *et al.*,, 2003; Yan *et al.*, 2007), programmed cell death (Bethke & Jones, 2001; Mittler, 2002) and gravitropism (Joo *et al.*, 2001), and provide tolerance to both biotic and abiotic stresses (Torres *et al.*, 2002; Miller *et al.*, 2008). Plants can sense, transduce and translate ROS signals into appropriate cellular responses via redox-sensitive enzymes, calcium mobilization, protein phosphorylation and gene expression. Xiong *et al.* (2002) provided an extensive review of ROS signalling via tyrosine phosphatase formed as result of oxidation of conserved cysteine residues. ROS can coordinate with other components such as protein phosphatases, protein kinases and transcription factors (Cheng & Song, 2006) and communicate with other signal molecules resulting in a signalling network cascade (Neill *et al.*, 2002). The balance between oxidant production and removal by the antioxidant determines the strength, lifetime and size of the ROS signalling pool. Miller *et al.* (2008) identified a signalling pathway that becomes activated in plant cells in reponse to ROS accumulation caused by abiotic stresses like temperature, salinity and osmotic stresses. This signalling pathway includes different zinc finger proteins and WRKY transcription factors.

The H_2O_2 produced as a result of abscisic acid (ABA) is an important signal in regulating stomatal closure to minimize water loss through the activation of calcium-permeable channels in the plasma membrane (Pel *et al.*, 2000). ABA-induced increases in H_2O_2 levels play an important role in ABA-induced stomatal closure up to a certain level; however, further increases in H_2O_2 levels do not favour stomata closure (Jannat *et al.*, 2011). Joo *et al.* (2001) demonstrated a role for ROS as a second messenger in mediating root gravitropism by inducing asymmetric movement of auxin within 60 min. Antioxidants like *N*-acetylcysteine, ascorbic acid and Trolox (a vitamin E analogue) that participate in ROS scavenging inhibited root gravitropism (Joo *et al.*, 2001). ROS signalling pathways interact with GAs in hormone-regulated programmed cell death in barley aluerone cells, stimulated by GA that induces ROS accumulation, whereas ABA maintains low ROS concentrations through activation of the alternative oxidase pathway and ROS scavenging system (Fath *et al.*, 2002).. An exogenous supply of H_2O_2 results in low GA signalling, which hampers germination; however, ABA signalling remains unaffected (Bahin *et al.*, 2011). Plants have developed a very complex signalling network to mediate both biotic and abiotic stress responses due to ROS synthesis. Excessive production of ROS, called oxidative burst, during plant–pathogen interactions plays an important role in signal transduction and thus ROS acts as a second messenger for the signal transmission (Klessig *et al.*, 2000; Nanda *et al.*, 2010). ROS production also plays an important role in the expression of defensive genes in tomato plants in response to wounding (Orozco-Cárdenas *et al.*, 2001).

Lignin is a protective polymer of plant cell walls that plays a significant role in protecting plants from various environmental stresses. According to Denness *et al.* (2011) ROS production causing cell wall damage regulates lignin biosynthesis in plants. ROS also play an important role in signal transduction pathways involved in responses to osmotic stress, low temperatures and heavy metals (Yuasa *et al.*, 2001; Xiong *et al.*, 2002; Yeh *et al.*, 2007). ROS also play an important role in drought conditions in plants, and it has been suggested that ROS are the signals by which plants can sense drought conditions (Yeh *et al.*, 2007).

7.4.4 ROS and oxidative damage to biomolecules

To avoid oxidative stress, production and removal of ROS should be strictly controlled. The cell is said to be in a state of 'oxidative stress' when production of ROS exceeds the capacity of the defence mechanisms to remove them. Numerous biotic and abiotic stresses such as salinity, drought, high light levels, heavy metal toxicity, pesticides and pathogens disturb the equilibrium/balance between production and scavenging of ROS. Increased levels of ROS can cause damage to biomolecules such as lipids, proteins and DNA. This imbalance disturbs various membrane properties like fluidity and ion transport, and causes loss of enzyme activity, protein cross-linking, inhibition of protein synthesis, DNA damage and ultimately death of the cell.

7.4.4.1 ROS and lipids

Increased ROS levels will enhance lipid peroxidation rates in both cellular and organellar membranes, ultimately affecting normal cellular functioning. Lipid peroxidation increases the oxidative stress by producing various lipid-derived radicals, which react with each other and damage proteins and DNA (Sharma & Dubey, 2005; Han *et al.*, 2009; Tanou *et al.*, 2009; Mishra *et al.*, 2011). Increased levels of lipid peroxidation have been used as an indicator of cellular damage due to ROS production under various biotic and abiotic stresses. It has been reported that under various environmental stresses, lipid peroxidation (degradation) is increased. Malondialdehyde (MDA) is one of the final end-products of peroxidation in phospholipids and is actively responsible for cell damage (Halliwell & Gutteridge, 1989). Double bonds between carbon atoms and ester linkages between glycerol and fatty acids are common sites of ROS attack. Fatty acids present in membrane phospholipids are very sensitive to ROS attack; a single hydroxyl ion can cause peroxidation of many polyunsaturated fatty acids because of a cyclic chain reaction. The overall process of lipid peroxidation involves three distinct stages: initiation, progression and termination.

The initial phase of lipid peroxidation involves activation of O_2, which is a rate-limiting step. The ROS $\bullet O_2^-$ and $\bullet OH$ have been found to react with methylene groups of polyunsaturated fatty acids (PUFA) forming conjugated dienes, lipid peroxy radicals and hydroperoxides (Smirnoff, 1995):

$$\mathrm{PUFA-H+X\bullet \rightarrow PUFA+X-H} \quad (7.9)$$

$$\mathrm{PUFA+O_2 \rightarrow PUFA-OO\bullet} \quad (7.10)$$

The peroxy radical formed is highly reactive and able to propagate the chain reaction:

$$\mathrm{PUFA-OO\bullet+PUFA-OOH \rightarrow PUFA-OOH+PUFA\bullet} \quad (7.11)$$

Conjugated dienes are formed when free radicals attack the hydrogens of methylene groups, separating double bonds and rearranging them with the help of metals like Fe^{2+}. Lipid hydroperoxide undergoes reductive cleavage as per the following reaction (Recknagal & Glende, 1984):

$$\begin{array}{l}\mathrm{Fe^{2+}-complex+PUFA-OOH \rightarrow}\\ \mathrm{Fe^{3+}-complex+PUFA-O\bullet}\end{array} \quad (7.12)$$

Decomposition of lipid hydroperoxide results in the formation of several reactive species like lipid alkoxyl radicals, aldehydes (malondialdehyde, acrolein and crotonaldehyde), alkanes, lipid epoxides and alcohols (Davies, 2000). PUFA-O• can initiate additional chain reactions (Buettner, 1993):

$$\mathrm{PUFA-O\bullet+PUFA-H \rightarrow PUFA-OH+PUFA\bullet} \quad (7.13)$$

Peroxidation of PUFAs by ROS can lead to chain breakage and thereby increased membrane fluidity and permeability.

7.4.4.2 ROS and proteins

ROS attack on proteins may cause a number of modifications, both direct and indirect. Modulation in protein activity through nitrosylation, carbonylation, disulphide bond formation and glutathionylation are some of the direct alterations of proteins due to ROS attack. Indirect modifications of proteins involve conjugation with breakdown products of fatty acid peroxidation, aggregation of cross-linked reaction products, peptide chain fragmentation, modification of electric charge and increased susceptibility to proteolysis (Yamauchi *et al.*, 2008). Increased concentrations of carbonylated proteins have been observed in tissues under oxidative stress, and these are often used as markers of protein oxidation (Parween *et al.*, 2011). Amino acids exhibit differential responses to ROS attack, with thiol groups and sulphur-containing amino acids being susceptible sites in any peptide chain. ROS can abstract an H atom from cysteine residues to form a thiol radical that will cross-link to a second thiol radical to form a disulphide bridge. Like pesticides, several metals like Cd, Pb and Hg have also been shown to cause the depletion of protein-bound thiol groups (Stohs & Bagchi, 1995). Methionine is converted to its methionine sulphoxide derivative with ROS (Brot & Weissbach, 1982), and tyrosine is readily cross-linked to form bityrosine products in the presence of ROS (Davies, 1987).

The irreversible oxidation of iron-sulphur centres by $\bullet O_2^-$ leads to enzyme inactivation (Gardner & Fridovich, 1991). In these cases, the metal (Fe) binds to a divalent cation-binding site on the protein. The Fe then undergoes a Fenton reaction to form a hydroxyl radical, •OH, that rapidly oxidizes an amino acid residue near the cation-binding site of the protein (Stadtman, 1986). Oxidized proteins act as a good source of substrates for proteolytic digestion. It has been reported that oxidation could predispose the protein to ubiquitination, which in turn can act as a target for proteosomal degradation (Cabiscol *et al.*, 2000). Following higher concentrations of H_2O_2 or Cd, peroxisomes exhibited increased carbonyl content. Oxidized proteins were more efficiently degraded under metal and pesticide treatment, acounting for a 20% increment in proteolytic activity (Romero-Puertas *et al.*, 2002). Further more damage causes extensive production of cross-linked and aggregated products that act as poor substrates for degradation and inhibit proteases in degrading other oxidized proteins (Grune *et al.*, 1997).

7.4.4.3 ROS and DNA

Since DNA is the cell's genetic material, any damage to the DNA can result in changes in the proteins it encodes, which may result in malfunction or complete inactivation of the encoded proteins. ROS are a major source of DNA damage (Imlay & Linn, 1988). Nuclear, mitochondrial and chloroplast DNA is susceptible to oxidative damage by ROS, resulting in deoxyribose oxidation, strand breakage, removal of nucleotides, various modifications in the organic bases of the nucleotides, and formation of DNA-protein crosslinks. In advanced stages, changes in the nucleotides of one strand can result in mismatches with the nucleotides in the other strand, resulting in subsequent mutations. Plants exposed to different environmental stresses such as salinity (Liu *et al.*, 2000), metals (Meriga *et al.*, 2004) and pesticide toxicity show enhanced DNA degradation. In DNA, both the sugars and bases can be damaged. Oxidative damage to DNA bases generally involves •OH addition to double bonds, while sugar damage mainly results from removal of hydrogen from deoxyribose (Dizdaroglu, 1993). Both purine and pyrimidine bases as well as the deoxyribose backbone react with the hydroxyl radical (Halliwell & Gutteridge, 1999). This results in the generation of various products from the DNA bases, which mainly include C8 hydroxylation of guanine to form 8-oxo-7,8-dehydro-2′-deoxyguanosine, hydroxymethyl urea, urea, thymine glycol, and ring-opened and saturated products of thymine and adenine (Tsuboi *et al.*, 1998). Of these, 8-hydroxyguanine is the most commonly observed product. O_2 reacts with guanine only, whereas H_2O_2 and $\bullet O_2^-$ do not react with bases at all (Halliwell & Aruoma, 1991; Dizdaroglu, 1993). ROS also cause various mutagenic alterations in DNA. For example, mutation arising from selective modification of G:C sites especially indicates oxidative attack on DNA by ROS. Moreover, ROS can attack DNA bases indirectly through reactive products generated by ROS attack on other macromolecules such as lipids (Fink *et al.*, 1997).

ROS cause single-strand breaks by attacking the sugars of the DNA backbone. The hydrogen atom is removed from the C4 position of deoxyribose, resulting in the formation of a deoxyribose radical that further reacts to produce DNA strand breakage (Evans *et al.*, 2004). It has been observed that under physiological conditions, neither H_2O_2 nor $\bullet O_2^-$ alone can cause *in vitro* strand breakage, but both are required. So, it was concluded that under *in vivo* conditions, toxicity associated with ROS was actually the result of a Fenton reaction. Oleinick *et al.* (1987) suggested that •OH attack on either DNA or the proteins associated with it, results in DNA-protein crosslink formation. As these DNA-protein crosslinks cannot be readily repaired, they can be lethal if replication or transcription precedes repair. Mitochondrial and chloroplast DNA are more susceptible to oxidative damage than nuclear DNA due to their lack of protective histone proteins and proximity to the ROS-producing systems (Richter, 1992). Certainly a well-developed repair system exists for damaged DNA but excessive toxicity caused by ROS leads to permanent damage to the DNA and ultimately affects the entire cell system.

7.4.4.4 ROS and lipid peroxidation

Aerobic life is threatened by the toxicity of the oxygen that is its inherent feature. Green plants, because of their oxygenic conditions and presence of ample photosensitizers and polyunsaturated fatty acids in the chloroplast envelope, are always at elevated risk of oxidative damage. It has been reported that plants can utilize only 1% of the oxygen consumed in the production of ROS like hydroxyl radical (OH•), singlet oxygen ($^1O_2$) and superoxide radical ($•O_2^-$) (Asada & Takahashi, 1987).

Lipids are most prone to oxidative damage in almost all living organisms (Zhang & Kirkham, 1996; Hung & Koa, 1997). Lipid peroxidation begins with an external oxidant, usually an oxygen-centred free radical such as OH•, $•O_2^-$ or ROO•, attacking an allylic methylene group and converting it to a new cation-centred free radical:

$$RCH = CHCH_2 + OH \rightarrow RCH = CHCH + H_2O$$

Singlet oxygen reacts with unsaturated lipids and undergoes an 'enzymatic reaction' that leads to incorporation of oxygen into the lipid chain and migration of a double bond leading to lipid peroxidation. This ensures oxidative stress as the defence mechanism in plants accompanied by the formation of various free radicals and other by-products (del-Rio *et al.*, 1996). Lipid peroxidation (LPO) results in membrane damage (Vaughan *et al.*, 1982) because the geometry of alkyl chains and the packing order in the bilayer are disrupted and altered. Membrane damage is directly proportional to lipid peroxidation – i.e. by measuring membrane damage; we can easily determine the level of lipid destruction. It has been observed that during lipid peroxidation various products including small hydrocarbons such as ketones, malondialdehyde (MDA) and other related compounds are formed from polyunsaturated precursors (Bird *et al.*, 1983; de-Vos *et al.*, 1993; Weckx & Clijsters, 1996; Reddy *et al.*, 1998). Some of these compounds react with thiobarbituric acid (TBA) to form coloured products and hence are called thiobarbituric acid reactive substances (TBARS); they can be measured by monitoring their absorption in a spectrophotometer at around 530 nm (Gray, 1978). MDA is usually used to assess the impact of adverse conditions on the organism (Bennicelli *et al.*, 1998).

7.4.5 Antioxidants

Superoxide dismutase (SOD) plays an important role in protecting photosynthesizing plants against oxidative damage (Salin, 1988; Bowler *et al.*, 1992; Foyer *et al.*, 1994). A SOD isoform found in the chloroplast has been found to act in combination with APX, MDHAR and GR (Salin & Bridges, 1981; Dalton, 1995). In plants four kinds of SOD are found: the copper-zinc type (CuZn-SOD), the manganese type (Mn-SOD), the iron type (Fe-SOD) and the nickle type (Ni-SOD). Of these, CuZn-SOD is present in chloroplasts while Fe-SOD is occasionally found in plant species (Van-Camp *et al.*, 1994). All these isoforms of SOD help in protecting cellular molecules from ROS damage (Fridovich, 1986). The mechanisms controlling and regulating the expression and activity of different isoforms of SOD are regulated by the genes and are highly complex. All these regulatory genes respond differently to different environmental signals (Sen-Gupta *et al.*, 1993).

Catalases are present in both prokaryotes and eukaryotes and are responsible for the conversion of H_2O_2 to water and oxygen. H_2O_2 generation occurs in the electron transport chain of mitochondria, PSII system in chloroplasts and peroxisomes by the action of cytochrome P450 , oxidase and dehydrogenase reactions, respectively. Once formed, H_2O_2 readily penetrates cell membranes and causes major damage to DNA by generating highly reactive hydroxyl radicals OH•, through Fenton-type reactions involving the interaction of H_2O_2 with transition metal ions such as Fe. Removal of H_2O_2 by catalases (CAT) mainly located in peroxisomes is highly energy efficient, as it does not consume cellular reducing equivalents. Different isozymes of CAT are present in plants, like CAT-1, CAT-2 and CAT-3. Among those, CAT-1 and CAT-2 are present in peroxisomes, whereas CAT-3 is present in mitochondria. In photosynthesizing plants, CAT together with SOD, forms the most efficient antioxidative machinery. Ascorbate peroxidase (APX) along with CAT helps in the detoxification of H_2O_2 and utilizes ascorbate as the electron donor for the reduction of H_2O_2 (Asada & Takahashi, 1987). Ascorbate is the most important reducing substrate for H_2O_2 in plant cells (Alscher, 1989; Smirnoff, 1993; Mehlhorn *et al.*, 1996).

Hydrogen peroxide can be detoxified by the action of APX, DHAR and GR through the ascorbate-glutathione cycle. Although it is well established that the ascorbate-glutathione cycle occurs in chloroplasts, it is now clear that the enzymes of this cycle are also found in mitochondria and peroxisomes and may represent an important antioxidant protection system against H_2O_2 generated in these organelles (Jimenez *et al.*, 1997).

Ascorbate peroxidase, or scavenging peroxidase, is a haem protein with protoporphyrin IX as its prosthetic group and a molecular size similar to the classic plant peroxidase guaiacol peroxidase (GuPX). Its primary function is rapid removal of H_2O_2 at the site of generation (Asada, 1992). APX isoenzymes are distributed in at least four distinct cell compartments: the stroma (sAPX), thylakoid membrane (tAPX), mitochondria (mAPX) and cytosol (cAPX) (Asada, 1992; Miyake & Asada, 1992; Ishikawa *et al.*, 1998). Isolation and characterization of cDNA encoding the different isoforms of APX has been done (Ishikawa *et al.*, 1998). The two chloroplastic APX (Chl APX) isoenzymes are encoded by a single gene (*ApxII*) and the mRNA undergoes alterative splicing of its two 3′ terminal exons (Yoshimura *et al.*, 1999). Diverse isoforms of APX exhibit differential responses under varied stress. The enzyme has two cytosolic forms with a purely defensive role and a membrane-bound (27 kDa) form having a functional role in addition to hydrogen peroxide scavenging.

Ascorbate can be regenerated from MDHA by the reaction catalysed by MDHAR. MDHAR are flavin nucleotide-containing enzymes found in chloroplasts and in cytosol, as well as in mitochondria and peroxisomes. They catalyse the reduction of MDHA to ascorbate by $NAD(P)H_2$. MDHA radical can also be reduced to ascorbate by photoreduced ferredoxin in the chloroplast PSI. Alternatively it can spontaneously dissociate to ascorbate and DHA, which can subsequently be reduced by another enzyme, DHA reductase, which regenerates ascorbate. DHA reductase, present in chloroplast stroma, reduces DHA to ascorbate by the ubiquitous cellular peptide, glutathione (GSH). Usually APX operates in a cycle with glutathione reductase (GR). GR uses reducing equivalents derived from glucose through the pentose phosphate pathway and NADPH to generate the reduced from of glutathione (GSH) from the oxidized disulphide form (GSSG) resulting from the action of APX.

Glutathione reductases are mainly found in chloroplast, cytoplasm and mitochondria of the plants. They range in size from around 90 to 140 kDa and usually contain two protein subunits, each with a flavin dinucleotide (FAD) at its active site. It appears that NADPH reduces the flavin nucleotide, which then transfers its electrons onto a disulphide bridge (-S-S-) in the enzyme. The two sulphydryl groups (-SH) furnished interact with GSSG and reduce it to GSH. The activity of GR suggests that the GSH/GSSG ratio in normal cells is kept high. The utilization of NADPH acts as an energy sink, which may affect indirectly the efficiency of the electron transport system. It also causes the production of a trans-thylakoidal proton gradient that is involved in control of electron transport (Foyer *et al.*, 1994). Despite the replacement of most of the old organochlorine insecticides by pyrethroids, little is known about the ecological side effects of these pesticides on the soils, overall plant growth, plant yields and the associated soil microbes in the agro-ecosystem.

7.4.6 Antioxidative defence mechanism against pesticides

Increased ROS production is a vigorous and sensitive response of plants to environmental stimuli (Jiang & Yang, 2009; Jan *et al.*, 2012). Herbicides are known to generate activated oxygen species, which are likely to contribute to the toxic effects of these herbicides (Asada & Takahashi, 1987; Halliwell, 1987). The oxidative stress induced by synthetic pyrethroid insecticide toxicity is well documented (Sayeed *et al.*, 2003; Parvez & Raisuddin, 2005). Increased lipid peroxidation rate is regarded as a general response to many stresses like heavy metals (Chaoui *et al.*, 1997; Lozano-Rodriguez *et al.*, 1997; Vanaja *et al.*, 2000), high salinity (Gharsally & Cherif, 1984; Hernandez *et al.*, 2000) and low temperature (Rodionov *et al.*, 1973). The increased concentration of TBARS in *Glycine max* at 0.20% treatment with alphamethrin and deltamethrin and 0.25% treatment with lambda-cyhalothrin insecticide suggests that these plants are highly susceptible to insecticidal stress.

Synthetic pyrethroid insecticide-treated seedlings demonstrated a significant enhancement in SOD activity, especially during the flowering stage. The relatively low activity post-flowering could be because old leaves generally contain low concentrations of antioxidants, making them more prone to enhanced oxidative injury than young leaves (Polle, 1997). It has been demonstrated that SOD plays an important role in protecting against oxidative damage in plants (Bowler *et al.*, 1992). Bashir *et al.* (2007) observed that activities of antioxidative enzymes like APX, SOD, GR, CAT and glutathione increase till the flowering stage followed by a decreasing trend during senescence of the plant. Similar observations have been recorded in the roots of Japanese radish and pea plant by Morimura *et al.* (1999) and Donahue *et al.* (1997), respectively. They found high activities of SOD (H_2O_2-generating enzyme), MDHAR, DHAR and GR (ascorbic acid (AsA)-regenerating enzymes) were associated with an increase in APX during root growth. These results suggest that the AsA-APX system may play an important role in the protection of root tissues against oxidative damage during the early stages of post-germinative growth in Japanese radish. Similar increases in the activities of APX and AsA-regenerating enzymes, MDHAR, DHAR and GR during the course of germination have also been reported for castor bean seeds (Klapheck *et al.*, 1990) and wheat seeds (Cakmak *et al.*, 1993). These observations on SOD activity find support from several earlier findings concerning the effects of herbicides (Matters & Scandalios, 1986), air pollutants (Tanaka & Sugahara, 1980; Lee and Bennett, 1982; Hernandez *et al.*, 2000), anoxic and hypoxic stress (Monk *et al.*, 1987; Biemelt *et al.* 2000), SO_2 (Madamanchi *et al.*, 1994) and salinity (Hernandez *et al.*, 1993; Gomez *et al.*, 1999; Qureshi *et al.*, 2007) on the antioxidant defence system in plants. The increase in SOD activity could be a consequence of a *de novo* synthesis of enzymatic protein (Slooten *et al.*, 1995). Methyl viologen (herbicide) treatment affects the level of chloroplast SOD as well as mitochondrial cytosolic SODs (Dodge, 1994; Van-Camp *et al.*, 1994). Iturbe-Ormaetxe *et al.* (1998) concluded that the Fe-SOD activity was being inhibited by methyl viologen, but the CuZn- and the Mn-SOD activities were not being affected. The increase in SOD activity as observed in the present study with increasing concentrations of insecticidal dose conforms well with the findings of Tsang *et al.* (1991) in *Nicotiana plumbaginifolia*.

Parween *et al.* (2011) evaluated the effect of chlorpyrifos on several metabolic and stress-related parameters of *Vigna radiata* L. Twenty-day-old seedlings were exposed to several concentrations of chlorpyrifos, ranging from 0 to 1.5 mM, through foliar spraying under field conditions. Chlorpyrifos increased the lipid peroxidation rate and proline content with a concentration of 1.5 mM by Day 20, whereas dehydroascorbate, and oxidized and total glutathione were increased by 1.5 mM by Day 10. However, a significant dose-dependent decline in relation to content of ascorbate, and reduced glutathione levels were observed at all growth stages. Among the enzymatic antioxidants, the activities of superoxide dismutase, ascorbate peroxidase and glutathione reductase were enhanced significantly by all the concentrations at Day 10. Maximum catalase activity was observed at Day 10 in controls and declined thereafter. Lipid peroxidation might be the first sign of cellular membrane damage by organophosphates (Hazarika *et al.*, 2003). The significant dose-dependent decrease in CAT activity observed in the present study at different stages of plant development is circumstantial evidence supporting the hypothesis that pesticides cause the formation of ROS (Farrington *et al.*, 1973; Scandalios, 1992, 1993; Sayeed *et al.*, 2003; Parvez & Raisuddin, 2005). This could also be due to a flux of superoxide radicals that are known to inhibit CAT activity (Kono & Fridovich, 1982). Streb *et al.* (1993) reported similar changes in CAT activity in Paraquat-treated *Secale cereale*. Decline in CAT activity is regarded as a general response to many stresses and is due to inhibition of enzyme synthesis or change in assembly of the enzyme subunits (MacRae & Fergusam, 1985; Somashekaraiah *et al.*, 1992). APX activity was relatively low during post-flowering. APX protects the cell against oxidative damage by detoxifying the toxic H_2O_2. Increase in APX activity is suggestive of its role in the detoxification of H_2O_2 under insecticide-induced oxidative stress (Morimura *et al.*, 1996). Increase in GR activity with increasing concentration of insecticides could be explained in two ways: (i) the ascorbate-glutathione cycle might be operating at a high rate in order to detoxify the ROS in these plants, or (ii) the reduced glutathione pool has to be maintained at high levels so that it does not become limiting for the synthesis of phytochelatins, the small peptides involved in the sequestration of various metal ions in the vacuoles (Cobbett, 2000; Stolt *et al.*, 2003), and in the inactivation of pesticides by conjugate formation. The non-enzymatic cellular antioxidants, like ascorbate and glutathione, undergo alterations under oxidative stress (Foyer & Halliwell, 1976; Nakano & Asada, 1981). We have noted a decrease in the ascorbate level under insecticidal stress during the present investigations. However, the decrease was more prominent during the post-flowering stage with the 0.20% dose of alphamethrin and deltamethrin and 0.25% dose of lambda-cyhalothrin. Disturbances with the GSH-independent DHAR or the structural integrity of MDHAR or the altered activity of PSI as a result of insecticidal toxicity might be the cause of depletion or decline of ascorbate (Asc). A decrease in the Asc, possibly due to shortage of a reductant to maintain the MDHAR activities, has been reported (Foyer *et al.*, 1983). The enhanced levels of GSH in *G. max* due to insecticidal toxicity suggest an active GSH participation in the detoxification of oxygen species and free radicals, directly (non-enzymatic) as well as through certain enzymes. It is assumed that GSH (Wingate *et al.*, 1988) or GSSG (Winglse & Karpinski, 1996) or a change between GSH and GSSG (Foyer *et al.*, 1997) may function as signals for activating stress-responsive gene expression in stress situations. The increase in glutathione in our study might be because synthesis of GSH is a demand-driven process, being used by the GSHS transferases in the detoxification of organic compounds and xenobiotics (May *et al.*, 1998).

7.5 Conclusion and future prospects

With introduction of new chemical formulations extensive use of pesticides has increased the risks of toxicity. Reactive oxygen species (ROS) caused by pesticides induce oxidative stress and oxidative tissue damage. This oxidative damage is a multistep process involving disturbances in the balance between the levels of oxidants/pro-oxidants and antioxidants, leading to tissue damage resulting in greater disease susceptibilities and finally to death. The mechanism(s) of pesticide-induced oxidative stress is still not wholly understood. Moreover, several other factors are considered to be linked with pesticide-induced oxidative stress in living systems and hence are central to the assessment of safety or toxicity of the pesticide concerned. Hence efforts should be made to comprehend pesticide-induced ROS in terms of monitoring alterations in the biochemical and molecular composition of different plant organs by exposing them to varying acute and sub-acute doses of pesticides. Furthermore, it is now imperative to explore plant biomolecules that could alleviate the adverse effects of ROS, including free radicals, produced by pesticide exposure. The potential of antioxidants is emphasized through application of various antioxidants that ameliorate the negative effects induced by oxidative damage.

References

Abu-Gharbia MA (1996) Growth of rhizobia, nodulation and nitrogen fixation of some legumes under pesticide treatment. *Sohag Pure Appl Sci Bull Fac Sci South Valley Univ Egypt* 12: 87–104.

Aditya CN, Banerjee H, Kole RK (1997) An appraisal of pesticide use in Indian agriculture with special reference to their consumption in West Bengal. *Sci Cult* 63: 223–228.

Ahmad N, Bugueno G, Guo L, Marolt R (1999) Determination of organochlorine and organophosphate pesticide residues in fruits, vegetables and sediments. *J Environ Sci Health* B34: 829–848.

Alscher RG (1989) Biosynthesis and antioxidant function of glutathione in plants. *Physiol Plant* 77: 457–464.

Altman J (ed.) (1993) *Pesticide Interactions in Crop Production: Beneficial and Deleterious Effects*. CRC Press, Boca Raton, FL, 579 pp.

Andreyev AY, Kushnareva YE, Starkov AA (2005) Mitochondrial metabolism of reactive oxygen species. *Biochemistry (Moscow)* 70: 200–214.

Anonymous (1986) Subcommittee on Pesticides and Industrial Organic Pesticides, Associate Committee on Scientific Criteria for Environmental Quality, National Research Council of Canada. Pyrethroids: Their effect on aquatic and terrestrial ecosystems. NRCC No. 24376. Environmental Secretariat, National Research Council of Canada, Ottawa.

Anonymous (1990a) International Registry of Potentially Toxic Chemicals, United Nations Environment Programme. Cypermethrin. *IRPTC Bulletin* 10: 24–27.

Anonymous (1990b) U.S. Food and Drug Administration. Listings of pesticides, industrial chemicals and metals data by fiscal year, origin, sample flag and industrial/product code 1985–1999. Pyrethroids. Database search (Freedom of Information Act Request F90016087).

Apel K, Hirt H (2004) Reactive oxygen species: metabolism, oxidative stress, and signal transduction. *Ann Rev Plant Bio* 55: 373–399.

Arora A, Sairam RK, Srivastava GC (2002) Oxidative stress and antioxidative system in plants. *Curr Sci* 82: 1227–1238.

Asada K (1992) Ascorbate peroxidase – a hydrogen peroxidase scavenging enzyme in plants. *Physiol Plant* 85: 235–241.

Asada K, Takahashi M (1987) Production and scavenging of active oxygen in photosynthesis. In: Kyle DJ, Osmond CJ, Arntzen CJ (eds), *Photoinhibition: Topics in Photosynthesis*. Elsevier, Amsterdam, pp. 227–287.

Bahin E, Bailly C, Sotta B, Kranner I, Corbineau F, Leymarie J (2011) Crosstalk between reactive oxygen species and hormonal signalling pathways regulates grain dormancy in barley. *Plant Cell Environ* 34: 980–993.

Bandyopadhyay S, Bhattacharjee P, Mukherjee N (1979) In vitro sensitivity of *Rhizobium* species to some fungicides and insecticides. *Pesticides* 13: 22–25.

Bartosz G (1996) Peroxynitrite: mediator of the toxic action of nitric oxide. *Acta Biochem Poi* 43: 645–660.

Bartosz G (1997) Oxidative stress in plants. *Acta Physiol Plant* 19: 47–64.

Bashir F, Mahmooduzzafar, Siddiqi TO, Iqbal M (2007) The antioxidative response system in *Glycine max* (L.) Merr exposed to deltamethrin, a synthetic pyrethroid insecticide. *Environ Pollut* 147: 94–100.

Beligni MV, Lamattina L (1999) Nitric oxide counteracts cytotoxic processes mediated by reactive oxygen species in plants. *Planta* 208: 337–344.

Bennicelli RP, Stepniewski W, Zakrzhewky DA, Ballakhnina TI, Stepriewski Z, Lipic J (1998) The effect of soil aeration on SOD activity, MDA level, pigment content and stomatal diffusive resistance in maize seedling. *Environ Exp Bot* 39: 203–211.

Bethke PC, Jones RL (2001) Cell death of barley aleurone protoplasts is mediated by reactive oxygen species. *Plant J* 25: 19–29.

Biemelt S, Keetman U, Mock HP, Grimm B (2000) Expression and activity of isoenzymes of superoxide dismutase in wheat roots in response to hypoxia and anoxia. *Plant Cell Environ* 23: 135–144.

Bird RP, Hung SO, Hadley M, Draper HH (1983) Determination of MDA in biological material by HPLC. *Anal Biochem* 128: 240–244.

Blokhina O, Fagerstedt KV (2010) Reactive oxygen species and nitric oxide in plant mitochondria: origin and redundant regulatory systems. *Physiol Plant* 138: 447–462.

Boutin C, Lee H-B, Peart ET, Batchelor PS, Maguire RJ (2000) Effects of the sulfonylurea herbicide metsulfuron methyl on growth and reproduction of five wetland and terrestrial plant species. *Environ Toxicol Chem* 19: 2532–2541.

Bowler C, Montagu MV, Inze D (1992) Superoxide dismutase and stress tolerance. *Annu Rev Plant Physiol Plant Mol Biol* 43: 83–116.

Bradbury SP, Coats JR (1989) Comparative toxicology of pyrethroid insecticides. *Rev Environ Contain Toxicol* 108: 133–177.

Briggs GG, Elliot M, Jones NF (1983) Present status and future prospects for synthetic pyrethroids. In: Takahashi N, Yoshioda H, Misato T, Matsunaka S (eds), *Pesticide Chemistry. Human Welfare and the Environment, Vol. 2 Natural Products*. Pergamon Press, New York,

Brot N, Weissbach H (1982) The biochemistry of methionine sulfoxide residues in proteins. *Trends Biochem Sci* 7: 137–139.

Brot N, Weissbach H (1983) Biochemistry and physiological role of methionine sulfoxide residues in proteins. *Arch Biochem Biophy* 223: 271–281.

Buettner GR (1993) The pecking order of free radicals and antioxidants: lipid peroxidation, α-tocopherol, and ascorbate. *Arch Biochem Biophys* 300: 535–543.

Cabiscol E, Piulats E, Echave P, Herrero E, Ros J (2000) Oxidative stress promotes specific protein damage in Saccharomyces cerevisiae. *J Biol Chem* 275: 27393–27398.

Cakmak I, Strbac D, Marschner H (1993) Activities of hydrogen peroxide scavenging enzymes in germinating wheat seeds. *J Exp Bot* 44: 127–132.

Camilleri P, Bowyer JR, Gilkerson T, Odell B, Weaver RC (1987) Structure activity relationships in the Hill inhibitory activity of substituted phenylureas. *J Agric Food Chem* 35: 479–483.

Cadenas E, Davies KJ (2000) Mitochondrial free radical generation, oxidative stress, and aging. *Free Radic Biol Med* 29: 222–230.

Casida JE, Quigstad GB (1998) Golden age of insecticide research: past, present, or future? *Ann Rev Entomol* 43: 1–6.

Chaoui A, Mazhoudi S, Ghorbal MH, El-Ferjoni E (1997) Effect of cadmium and zinc interaction on hydroponically grown bean (*Phaseolus vulgaris*). *Plant Sci* 126: 21–28.

Cheeseman JM (2006) Hydrogen peroxide concentrations in leaves under natural conditions. *J Exp Bot* 57: 2435–2444.

Cheng Y, Song C (2006) Hydrogen peroxide homeostasis and signalling in plant cells. *Science in China. Series C. Life Sciences* 49: 1–11.

Cleland RE, Grace SC (1999) Voltametric detection of superoxide production by photosystem II. *FEBS Lett* 457: 348–352.

Cobbett CS (2000) Phytochelatins and their roles in heavy metal detoxification. *Plant Physiol* 123: 825–832.

Cona A, Rea G, Angelini R, Federico R, Tavladoraki P (2006) Functions of amine oxidases in plant development and defence. *Trends Plant Sci* 11: 80–88.

Dalton DA (1995) Antioxidant defenses of plants and fungi. In: Ahmad S (ed.), *Oxidant-Induced Stress and Antioxidant Defenses in Biology*. Chapman & Hall, New York, pp. 298–355.

Dat J, Vandenabeele S, Vranová E, Van Montagu M, Inzé D, Van Breusegem F (2000) Dual action of active oxygen species during plant stress responses. *Cell Molec Life Sci* 57: 779–795.

Davies JH (1985) The pyrethroids: an historical introduction. In: Leahey JP (ed.), *The Pyrethroid Insecticides*. Taylor & Francis, London.

Davies KJ (1987) Protein damage and degradation by oxygen radicals.I. general aspects. *J Biol Chem* 262: 9895–9901.

Davies KJA (2000) Oxidative stress, antioxidant defences, and damage removal, repair, and replacement systems. *IUBMB Life* 50: 279–289.

del-Rio LAS, Palma JM, Scandalio LM, *et al.* (1996) Peroxisomes as a source of superoxide and hydrogen peroxide in stresses plants. *Biochem Soc Trans* 24: 434 –438.

Del Río LA, Sandalio LM, Corpas FJ, Palma JM, Barroso JB (2006) Reactive oxygen species and reactive nitrogen species in peroxisomes. Production, scavenging, and role in cell signaling. *Plant Phys* 141: 330–335.

Denness L, McKenna JF, Segonzac C, *et al.* (2011) Cell wall damage-induced lignin biosynthesis is regulated by a ROS- and jasmonic acid-dependent process in *Arabidopsis thaliana*. *Plant Physiol* 156: 1364–1374.

de-Vos CHR, Bookim WMY, Voojis R, Selat H, Kok LJD (1993) Effect of cadmium on fatty acid composition and peroxidation of lipids in the roots of tolerant and sensitive *Silene cucubalus*. *Plant Physiol Biochem* 17: 13–18.

Dicke W, Ocker HB, Their HP (1988) Rueckstandsanalyse von Pyrethroid Insektiziden in Getreide Mahlerzeugnissen und Brut. *Z Lebensm Unters Forsh* 186: 125–129.

Dizdaroglu M (1993) Chemistry of free radical damage to DNA and nucleoproteins. In: Halliwell B, Aruoma OI (eds), *DNA and Free Radicals*. Ellis Horwood, London, pp. 19–39.

Dodge A (1994) Herbicide action and effects on detoxication processes. In: Mullineaux P, Foyer C (eds), *Causes of Photooxidative Stress and Amelioration of Defense Systems in Plants*. CRC Press, Boca Raton, pp. 219–236.

Donahue JL, Okpodu C, Cramer CL, Grabau EA, Alscher RG (1997) Responses of antioxidants to paraquat in pea leaves. *Plant Physiol* 113: 249–257.

Edward S (1972) Insecticides. In: Coring CAI, Hamakar JW (eds), *Organic Chemicals in the Soil Environments*, vol. 2. Marcel Dekker, New York, p. 513.

Elliott M, Farnham AW, Janes NF, Needham PH, Pulman DA, Stevenson JH (1973) A photostable pyrethroid. *Nature* 246: 169–170.

Elliott M, Janes NF, Potter C (1978) The future of pyrethroids in insect control. *Annu Rev Entomol* 23: 443–469.

Elstner EF (1991) Mechanisms of oxygen activation in different compartments of plant cells. In: Pell EJ, Steffen KL (eds), *Active Oxygen/Oxidative Stress and Plant Metabolism*. American Society of Plant Physiology, Rockville, MD, pp. 13–25.

Evans MD, Dizdaroglu M, Cooke MS (2004) Oxidative DNA damage and disease: induction, repair and significance. *Mut Res* 567: 1–61.

Farrington J, Ebert M, Land E, Fletcher K (1973) Bipyridylium quaternary salts and related compounds. V. Pulse radiolysis studies of the reaction of paraquat radical with oxygen. Implications for the mode of action of bipyridyl herbicides. *Biochim Biophys Acta* 314: 372–381.

Fath A, Bethke P, Beligni V, Jones R (2002) Active oxygen and cell death in cereal aleurone cells. *J Exp Bot* 53: 1273–1282.

FDA (1990a) Listings of pesticides, industrial chemicals and metals data by fiscal year, origin, sample flag and industrial/product code 1985–1999. Pyrethroids. Database search (Freedom of Information Act Request F90016087). U.S. Food and Drug Administration.

FDA (1990b) To Test Honey for Pesticides. *San Francisco Chronicle*, August 11.

Fenton HJH (1894) Oxidation of tartaric acid in presence of iron. *J Chem Soc Trans* 65: 899–911.

Fenton HJH (1899) Oxidation of certain organic acids in the presence of ferrous salts. *Proc Chem Soc* 25: 224–228.

Fink RJ (1972). Phytotoxicity of herbicide residues in soils. *Agron J* 64: 804–805.

Fink SP, Reddy GR, Marnett LJ (1997) Mutagenicity in Escherichia coli of the major DNA adduct derived from the endogenous mutagen malondialdehyde. *Proc Nat Acad Sci USA* 94: 8652–8657.

Foyer CH, Halliwell B (1976) The presence of glutathione and glutathione reductase in chloroplasts: A proposed role in ascorbic acid metabolism. *Planta* 133: 21–25.

Foyer CH, Rowell J, Walker D (1983) Measurements of ascorbate content of spinach leaf protoplasts and protoplasts and chloroplasts during illumination. *Planta* 157: 239–244.

Foyer CH, Lenandair M, Kunert KKJ (1994) Photooxidative stress in plants. *Physiol Plant* 92: 696–717.

Foyer CH, Lopez-Delgado H, Dat JF, Scott IM (1997) Hydrogen peroxide and glutathione associated mechanism of acclimatory stress tolerance and signalling. *Physiol Plant* 100: 241–254.

Fridovich I (1986) Biological effects of the superoxide radical. *Arch Biochem Biophys* 247: 1–11.

Gardner PR, Fridovich I (1991) Superoxide sensitivity of the *Escherichia coli* 6-phosphogluconate dehydratase. *J Biol Chem* 266: 1478–1483.

Gechev TS, Van Breusegem F, Stone JM, Denev I, Laloi C (2006) Reactive oxygen species as signals that modulate plant stress responses and programmed cell death. *Bioessays* 28: 1091–1101.

Gharsally M, Cherif A (1984) Effect of sodium chloride on the sunflower plants lipid. In: Siegenthaler PA, Eichenberger W (eds), *Structure, Function and Metabolism of Plant Lipids*. Elsevier Science Publishers, Amsterdam, pp. 601–604.

Gomez JM, Hernandez JA, Jimenez A, del-Rio LA, Sevilla F (1999) Differential response of antioxidative enzymes of chloroplasts and mitochondria to long-term NaCl stress of pea plants. *Free Radical Res* 31: 11–18.

Goto S (1970) Evaluation of some pesticide residues in food. *Plant Protect J* 24: 501–505.

Gray JI (1978) Measurement of lipid peroxidation: a review. *J Am Oil Chem Soc* 55: 539–546.

Gross GG (1977) Cell wall-bound malate dehydrogenase from horseradish. *Phytochem* 16: 319–321.

Grune T, Reinheckel T, Davies KJA (1997) Degradation of oxidized proteins in mammalian cells. *FASEB J* 11: 526–534.

Haber F, Weiss J (1934) The catalytic decomposition of hydrogen peroxide by iron salts. *Proc R Soc London Ser A* 147: 332–351.

Halliwell B (1987) Oxidative damage, lipid peroxidation and antioxidant protection in chloroplast. *Chem Phys Lipids* 44: 327–340.

Halliwell B (2006) Reactive species and antioxidants. Redox biology is a fundamental theme of aerobic life. *Plant Physiol* 141: 312–322.

Halliwell B, Aruoma OI (1991) DNA damage by oxygen-derived species. Its mechanism and measurement in mammalian systems. *FEBS Lett* 281: 9–19.

Halliwell B, Gutteridge JMC (1989) *Free Radicals in Biology and Medicine*, 2nd edn. Oxford University Press, Oxford.

Halliwell B, Gutteridge JMC (1999) *Free Radicals in Biology and Medicine*, 3rd edn. Oxford University Press, Oxford.

Hamner CL, Tukey HB (1946) Herbicidal action of 2-4-dichlorophenoxyacetic acid on several shrubs, vines, and trees. *Bot Gazette* 108: 379–385.

Han C, Liu Q, Yang Y (2009) Short-term effects of experimental warming and enhanced ultraviolet-B radiation on photosynthesis and antioxidant defense of *Picea asperata* seedlings. *Plant Growth Regul* 58: 153–162.

Hancock TLC, Costello AM, Lidstrom ME, Oremland RS (1998) Strain IMB-1, a novel bacterium for the removal of methyl bromide in fumigated agricultural soils. *Appl Environ Microbiol* 64: 2899–2905.

Hazarika A, Sarkar SN, Hajare S, Kataria M, Malik JK (2003) Influence of malathion pretreatment on the toxicity of anilofos in male rats: a biochemical interaction study. *Toxic* 185: 1–8.

Hernandez JA, Corpas FJ, Gomez M, Del-Rio LA, Sevilla F (1993) Salt-induced oxidative stress mediated by activated oxygen species in pea leaf mitochondria. *Physiol Plant* 89: 103–110.

Hernandez JA, Jimenez ZA, Mullineaux PM, Sevilla F (2000) Tolerance of pea (*Pisum sativum* L.) to long-term salt stress is associated with induction of antioxidant defences. *Plant Cell Environ* 23: 853–862.

Heyno E, Mary V, Schopfer P, Krieger-Liszkay A (2011) Oxygen activation at the plasma membrane: relation between superoxide and hydroxyl radical production by isolated membranes. *Planta.* 234: 35–45.

Hoffman CE (1971) Proceedings of the 2nd International IUPAC Congress. *Pestic Chem* 5: 65–85.

Hung TC, Koa CH (1997) Lipid peroxidation in relation to senescence of maize leaves. *J Plant Physiol* 150: 283–286.

Hurtig H (1972) Long-distance transport of pesticides. In: Matsumura F, Boush GM, Misato T (eds), *Environmental Toxicity of Pesticides.* Academic Press, New York, San Francisco, London, pp. 257–276.

Imlay JA, Linn S (1988) DNA damage and oxygen radical toxicity. *Science* 240: 1302–1309.

IRPTC (1990) Cypermethrin. International Registry of Potentially Toxic Chemicals, United Nations Environment Programme. *IRPTC Bull* 10: 24–27.

Ishikawa I, Yushimura K, Sakai K, Tome M, Trareda T, Shigoekas (1998) Molecular characterization and physiological role of a glutathione bound ascorbate peroxidase from spinach. *Plant Cell Physiol* 39: 23–34.

Ishikura H (1972) Impact of pesticides use on the Japanese environment. In: Matsumura F, Boush GM, Misato T (eds), *Environmental Toxicity of Pesticides.* Academic Press, New York, San Francisco, London, pp. 1–32.

Iturbe-Ormaetxe I, Escuredo PR, Arrese-Igor C, Becana M (1998) Oxidative damage in pea plants exposed to water deficit or paraquat. *Plant Physiol* 116: 173–181.

Jan S, Parween T, Siddiqi TO, Mahmooduzzafar Iqbal M (2012) Effect of gamma radiation on morphological, biochemical and physiological aspects of plants and plant products. *Environ Rev* 20: 17–39.

Jannat R, Uraji M, Morofuji M (2011) Roles of intracellular hydrogen peroxide accumulation in abscisic acid signaling in Arabidopsis guard cells. *J Plant Physiol* 168: 1919–1926.

Jiang L, Yang H (2009) Prometryne-induced oxidative stress and impact on antioxidant enzymes in wheat. *Ecotoxicol Environ Saf* 72: 1687–1693.

Jimenez H, Hernandez JA, del Rio LA, Senilla FF (1997) Evidence for the presence of As-GSH cycle in mitochondria and peroxisomes of pea levels. *Plant Physiol* 114: 275–284.

Jones VP, Toscano NC, Johnson MW, Welter SC, Youngman RR (1986) Pesticide effect on plant physiology: integration into a pest management program. *Bull Entomol Soc Amer* 32: 103–109.

Joo JH, Yun SB, June SL (2001) Role of auxin-induced reactive oxygen species in root gravitropism. *Plant Physiol* 126: 1055–1060.

Karanth NGK, Vasantharajan VN (1972) *Symposium on Legume Inoculants: Science and Technology.* Indian National Science Academy, New Delhi.

Kasai F, Hatakeyama S (1993) Herbicide susceptibility in two green algae, *Chlorella vulgaris* and *Selenastrum capricornutum. Chemosphere* 27 : 899–904.

Kenaga EE (1971) Guidelines for environmental study of pesticides, determination of bioconcentration potential. Symposium Guidelines for Environmental Studies of Pesticides. In: *Proceedings of the 162nd National Meeting of the American Chemical Society*, Washington, DC, pp. 12–17.

Kenaga EE (1972) Factors related to bioconcentration of pesticides. In: Matsumura F, Boush GM, Misato T (eds), *Environmental Toxicity of Pesticides.* Academic Press, New York, San Francisco, London, pp. 193–228.

Keith JO, Woods J, Hunt EG (1970) *Transactions of the Thirty-fifth North American Wildlife and Natural Resources Conference.* Wildlife Management Institute, Washington DC, p. 56.

Kent RA, Currie D (1995) Predicting algal sensitivity to a pesticide stress. *Environ Toxicol Chem* 14: 983–991.

Khan SM, Kour G (2007) Sub acute oral toxicity of chlorpyriphos and protective effect of green tea extract. *Pest Biochem Physiol* 89: 118–123.

Kim MJ, Ciani S, Schachtman DP (2010) A peroxidase contributes to ros production during *Arabidopsis* root response to potassium deficiency. *Mol Plant* 3: 420–427.

Kirby MF, Sheahan DA (1994) Effects of atrazine, isoproturon and mecoprop on the macrophyte *Lemna minor* and the alga *Scenedesmus subspicatus. Bull Environ Contam Toxicol* 53: 120–126.

Klapheck S, Zimmer I, Cosse H (1990) Scavenging of hydrogen peroxide in the endosperm of *Ricinus communis* by ascorbate peroxidase. *Plant Cell Physiol* 31: 1005–1013.

Klessig DF, Durner J, Noad R (2000) Nitric oxide and salicylic acid signalling in plant defense. *Proc Nat Acad Sci USA* 97: 8849–8855.

Kono Y, Fridovich I (1982) Superoxide radical inhibits catalase. *J Biol Chem* 257: 5751–5754.

Kwak JM, Mori IC, Pei ZM (2003) NADPH oxidase AtrbohD and AtrbohF genes function in ROS-dependent ABA signaling in Arabidopsis. *EMBO J* 22: 2623–2633.

Lane BG (2002) Oxalate, germins, and higher-plant pathogens. *IUBMB Life* 53: 67–75.

Lane BG (2002) Oxalate, germins, and higher-plant pathogens. *IUBMB Life* 53: 67–75.

Leahey JP (1985) Metabolism and environmental degradation. In: Leahey JP (ed.), *The Pyrethroid Insecticides.* Taylor & Francis, London, pp. 263–342.

Lee E, Bennett J (1982) Superoxide dismutase: a possible protective enzyme against ozone injury in snap beans (*Phaseolus vulgaris*). *Plant Physiol* 69: 1444–1449.

Lichtenstein EP (1959) Absorption of some chlorinated hydrocarbon insecticides from soils into various crops. *J Agric Food Chem* 7: 430–433.

Lichtenstein EP (1960) Insecticidal residues in various crops grown in soils treated with abnormal rates of aldrin and heptachlor. *J Agric Food Chem* 8: 448–451.

Lichtenstein EP, Schulz KR (1960) Translocation of some chlorinated hydrocarbon insecticides into the aerial parts of pea plants. *J Agric Food Chem* 8: 452–456.

Lichtenstein EP, Schulz KR (1965) Residues of aldrin and heptachlor in soils and their translocation into various crops. *J Agric Food Chem* 13: 57–63.

Lichtenstein EP, Myrdal GR, Schulz KR (1965) Absorption of insecticidal residues from contaminated soils into five carrot varieties. *J Agric Food Chem* 13: 126–131.

Liu T, Van Staden J, Cress WA (2000) Salinity induced nuclear and DNA degradation in meristematic cells of soybean (*Glycine max* (L.)) roots. *Plant Growth Regul* 30: 49–54.

López-Huertas E, Corpas FJ, Sandalio LM, Del Río LA (1999) Characterization of membrane polypeptides from pea leaf peroxisomes involved in superoxide radical generation. *Biochem J* 337: 531–536.

Lozano-Rodriguez E, Hernandez LE, Boney P, Carpena-Ruiz RO (1997) Distribution of cadmium in shoot and root tissues of maize and pea plants: physiological disturbances. *J Exp Bot* 48: 123–128.

McCann AE, Cullimore DR (1979) Influence of pesticides on soil algal flora. *Residue Rev* 79: 1–31.

MacRae EA, Fergusam IB (1985) Changes in catalase activity and H_2O_2 concentration in plants in response to low temperature. *Physiol Plant* 65: 51–56.

Madamanchi NR, Donahue JV, Cramer Cl, Alscher RG, Pedersen K (1994) Differential response of Cu, Zn superoxide dismutase in two pea cultivars during a short-term exposure to sulphur dioxide. *Plant Molec Biol* 26: 95–103.

Majumdar JC (1974) Pesticide industry – some recent advances in research and development. *Pesticide Annual* 2: 29–40.

Martensson AM (1992) Effect of agrochemicals and heavy metals on fast growing rhizobia and their symbiosis with small-seeded legumes. *Soil Biol Biochem* 24: 435–445.

Martinez C, Montillet JL, Bresson E (1998) Apoplastic peroxidase generates superoxide anions in cells of cotton cotyledons undergoing the hypersensitive reaction to *Xanthomonas campestris* pv. *malvacearum* 18. *Mol Plant Microb Inter* 11: 1038–1047.

Matters GL, Scandalios JG (1986) Effect of free radical generating herbicide paraquat on the expression of superoxide dismutase (SOD) genes in maize. *Biochim Biophys Acta* 882: 29–38.

May MJ, Vernoux T, Leaver C, Montagu MV, Inze D (1998) Glutathione homeostasis in plants: implication for environmental sensing and plant development. *J Exp Bot* 49: 649–667.

Mehlhorn H, Lelandais M, Korth HG, Foyer CH (1996). Ascorbate is the natural substrate for plant peroxidases. *FEBS Lett* 378: 203–206.

Meriga B, Reddy BK, Rao KR, Reddy LA, Kishor PBK (2004) Aluminium-induced production of oxygen radicals, lipid peroxidation and DNA damage in seedlings of rice (*Oryza sativa*). *J Plant Physiol* 161: 63–68.

Miettinen P, Echegoyen PE (1996) The effects of two pesticides (Vitavax-300 and Gaucho) on rhizobia and on the nodulation of four legumes. *Agric Food Sci Finl* 5: 203–207.

Miller G, Shulaev V, Mittler R (2008) Reactive oxygen signaling and abiotic stress. *Physiol Plant* 133: 481–489.

Mishra S, Jha AB, Dubey RS (2011) Arsenite treatment induces oxidative stress, upregulates antioxidant system, and causes phytochelatin synthesis in rice seedlings. *Protoplasma* 248: 565–577.

Mittler R (2002) Oxidative stress, antioxidants and stress tolerance. *Trends Plant Sci* 7: 405–410.

Miyake C, Asada K (1992) Thylakoid bound ascorbate peroxidant in spinach chloroplast and photoreduction of its primary oxidation product MDA radicals in thylakoid plant. *Cell Physiol* 33: 541–553.

Moller IM, Jensen PE, Hansson A (2007) Oxidative modifications to cellular components in plants. *Annu Rev Plant Biol* 58: 459–481.

Mongollon T, Flores M, Cazorla M (1986) Growth determination of *Rhizobium leguminosarum* biovar trifolii under action of herbicide Amitrol. *Ann Edafol Agrobiol* 45: 631–636.

Monk LS, Fageestedt KV, Crawford RMM (1987) Superoxide dismutase as an anaerobic polypeptide key factor in recovery from oxygen deprivation in *Iris* pseudacorus. *Plant Physiol* 85: 1016–1020.

Morimura T, Ohya T, Ikawa T (1996) Presence of ascorbate-peroxidizing enzymes in roots of *Brassica campestris* L. cv. Komatsuna. *Plant Sci* 117: 55–63.

Morimura Y, Iwamoto K, Ohya T, *et al.* (1999) Light enhanced induction of ascorbate peroxidase in Japanese radish roots during postgerminative growth. *Plant Sci* 142: 123–132.

Mott L, Snyder K (1988) *Pesticide Alert*. Sierra Club Books, San Francisco, CA.

Mueller-Beilschmidt D (1990) Toxicology and environmental fate of synthetic pyrethroids. *J Pest Ref* 10: 32–37.

Muller P, Swiss P (1940) DDT das Insektizid. Dichlordiphenyltrichlorathan und seine Bedeutung 1. Birkhauser, Basel–Stuttgardt.

Murphy MP (2009) How mitochondria produce reactive oxygen species. *Biochem J* 417: 1–13.

Nakano Y, Asada K (1981) Hydrogen peroxide is scavenged by ascorbate specific peroxidase in spinach chloroplasts. *Pl Cell Physiol* 22: 867–880.

Nanda AK, Andrio E, Marino D, Pauly N, Dunand C (2010) Reactive oxygen species during plant–microorganism early interactions. *J Integr Plant Biol* 52: 195–204.

Nash RG, Beall Jr ML, Woolson EA (1974) Plant uptake of chlorinated insecticides from soils. *Agronomy J* 62: 369–372.

Neill S, Desikan R, Hancock J (2002) Hydrogen peroxide signaling. *Curr Opin Plant Biol* 5: 388–395.

Noctor G, Veljovic-Jovanovic SD, Driscoll S, Novitskaya L, Foyer CH (2002). Drought and oxidative load in the leaves of C3 plants: a predominant role for photorespiration? *Ann Bot* 89: 841–850.

Noctor G, Paepe RDe, Foyer CH (2007) Mitochondrial redox biology and homeostasis in plants. *Trends Plant Sci* 12: 125–134.

NRC (1986) Pyrethroids: Their effect on aquatic and terrestrial ecosystems. NRCC No. 24376. Subcommittee on Pesticides and Industrial Organic Pesticides, Associate Committee on Scientific Criteria for Environmental Quality, National Research Council of Canada. Environmental Secretariat, National Research Council of Canada, Ottawa.

Oleinick NL, Chiu Song-mao, Ramakrishnan N, Liang-yan Xue (1987) The formation, identification, and significance of DNA-protein cross-links in mammalian cells. *Brit J Cancer* Supp 8: 135–140.

Orozco-Cárdenas ML, Narváez-Vásquez J, Ryan CA (2001) Hydrogen peroxide acts as a second messenger for the induction of defense genes in tomato plants in response to wounding, systemin, and methyl jasmonate. *Plant Cell* 13: 179–191.

Parvez S, Raisuddin S (2005) Copper modulates in enzymatic antioxidants in the freshwater fish *Channa punctata* (Bloch) exposed to deltamethrin. *Arch Environ Contam Toxicol* 49: 371–377.

Parween T, Jan S, Mahmooduzzafar, Fatma T (2011) Alteration in nitrogen metabolism and plant growth during different developmental stages of green gram (*Vigna radiata* L.) in response to chlorpyrifos. *Acta Physiol Plant* 33: 2321–2328.

Parween T, Jan S, Mahmooduzzafar, Fatma T (2012) Evaluation of oxidative stress in *Vigna radiata* L. in response to chlorpyrifos. *Int J Environ Sci Technol* 9: 605–612.

Peixoto F, Alves-Fernandes D, Santos D, Fontaınhas-Fernandes A (2006) Toxicological effects of oxyfluorfen on oxidative stress enzymes in tilapia *Oreochromis niloticus*. *Pesticide Biochem Physiol* 85: 91–96.

Pel ZM, Murata Y, Benning G (2000) Calcium channels activated by hydrogen peroxide mediate abscisic acid signalling in guard cells. *Nature* 406: 731–734.

Perfect TJ, Cook AG, Critchley U, *et al.* (1979) The effect of DDT contamination on the productivity of a cultivated forest soil in the sub humid tropics. *J Appl Ecol* 16: 705–719.

Polle A (1997) Defence against photooxidative damage in plants. In: Scandalios JG (ed.), *Oxidative Stress and the Molecular Biology of Antioxidant Defenses*. Cold Spring Harbour Laboratory Press, New York, pp. 623–666.

Qureshi MI, Abdin MZ, Qadir S, Iqbal M (2007) Lead-induced oxidative stress and metabolic alterations in *Cassia angustifolia* Vahl. *Biol Plant* 51: 121–128.

Raghu K, MacRae IC (1967) The effect of gamma-isomer of BHC upon the microflora of submerged rice soil. I. Effect upon algae. *Can J Microbiol* 13: 173–180.

Rasmusson AG, Geisler DA, Møller IM (2008) The multiplicity of dehydrogenases in the electron transport chain of plant mitochondria. *Mitochondrion* 8: 47–60.

Recknagal RO, Glende EA (1984) Oxygen radicals in biological systems. In: Packer L (ed.), *Methods in Enzymology*, vol. 105. Academic Press, New York, pp. 331–337.

Reddy PC, Vajaranabhaian SN, Kumar MU (1998) Lipid peroxidation as a mechanism of stress tolerance in upland rice *Oryza sativa* L. *Calli Ind J Plant Physiol* 3: 68–70.

Reed WT (1983) The fate and impact of Pydrin insecticide (fenvalerate) on non target systems following Britt applications. In: Takahashi N, Yoshioda H, Misato T, Matsunaka S (eds), *Pesticide Chemistry. Human Welfare and the Environment, Vol. 2: Natural Products*. Pergamon Press, New York.

Richter C (1992) Reactive oxygen and DNA damage in mitochondria. *Mutat Res-DNAging G* 275: 249–255.

Rodionov VS, Nyuppieva CA, Zakharova LS (1973) Changes in the galacto- and phospho-lipid content of potato leaves, exposed to low temperature. *Fiziologiya Rastenii* 20: 525–531.

Romero-Puertas MC, Palma JM, Gómez M, Del Río LA, Sandalio LM (2002) Cadmium causes the oxidative modification of proteins in pea plants. *Plant Cell Environ* 25: 677–686.

Rosa N, Cheng HH (1973) Uptake of DDT by *Nicotiana tabacum*. *Can J Plant Sci* 51: 400–402.

Rosa N, Cheng HH (1974) Distribution of DDT-taC in *Nicotiana tabacum*. *Can Plant Sci* 54: 403.

Roy DN, Ghosh SM (1942) A new active constituent of pyrethrum flower. *Nature* 150: 153.

Ruiz-Sainz JE, Beringer JE, Gutierrez-Nevarro AH (1985) Effects of the fungicide captafol on the survival and symbiosis properties of *Rhizobium trifolii*. *J Appl Bacteriol* 57: 361–367.

Salin ML (1988) Toxic species and protective systems of the chloroplast. *Physiol Plant* 72: 681–689.

Salin ML, Bridges BBS (1981) Chemi-luminiscence in wounded root tissue: Evidence for peroxidase involvement. *Plant Physiol* 67: 43–46.

Saroja G, Bose S (1982) Effect of methyl parathion on the growth, cell size, pigment and protein content of *Chlorella protothecoides*. *Environ Pollut* 27: 297–308.

Sayeed I, Parvez S, Pandey S, Bin-Hafeez B, Haque R, Raisuddin S (2003) Oxidative stress biomarkers of exposure to Deltamethrin in freshwater fish, *Channa punctatus* Bloch. *Ecotoxicol Environ Saf* 56: 295–301.

Scandalios JG (1993) Oxygen stress and superoxide dismutases. *Plant Physiol* 101: 7–12.

Scandalios JG (ed.) (1992) *Molecular Biology of Free Radical Scavenging Systems*. Current Communications in Cell and Molecular Biology, vol 5. Cold Spring Harbour Laboratory Press, Cold Spring Harbor, New York.

Sen B, Kapoor JJ (1974) Field trials of systemic and non-systemic fungicides against powdery mildew of cucurbits. I. *Pesticides* 8: 43–46.

Sen-Gupta A, Webb RP, Holladay AS, Allen RD (1993) Over expression of superoxide dismutase protects plants from oxidative stress. Induction of ascorbate peroxidase in superoxide dismutase over expressing plants. *Plant Physiol* 103: 1067–1073.

Sharma P, Dubey RS (2005) Drought induces oxidative stress and enhances the activities of antioxidant enzymes in growing rice seedlings. *Plant Growth Regul* 46: 209–221.

Singh PP, Kang MS (1978) Physiological response of groundnuts to Bavistin application for the control of Tikka disease. In: *Symposium on Plant Disease Problems Udiapur*, Oct 1–3, p.13.

Slooten L, Capiiau K, VanCamp W, Van Montagu M, Sybesma C, Inze D (1995) Factors affecting the enhancement of oxidative stress tolerance in transgenic tobacco over expressing Mn-SOD in the chloroplast. *Plant Physiol* 107: 737–780.

Smirnoff N (1995) Antioxidant systems and plant response to the environment. In: Smirnoff N (ed.), *Environment and Plant Metabolism: Flexibility and Acclimation*. BIOS Scientific Publishers, Oxford, pp. 217–243.

Soderlund DM, Clark JM, Sheets LP, *et al.* (2002) Mechanisms of pyrethroid neurotoxicity: implications for cumulative risk assessment. *Toxic* 171: 3–59.

Somashekaraiah SV, Padmaji K, Prasad ARK (1992) Phytotoxicity of cadmium ions on germinating seedlings of mung bean (*Phaseolus vulgaris*): involvement of lipid peroxidation in chlorophyll degradation. *Physio Plant* 85: 85–89.

Stadtman ER (1986) Oxidation of proteins by mixed-function oxidation systems: implication in protein turnover, ageing and neutrophil function. *Trends Biochem Sci* 11: 11–12.

Stenerson J (ed.) (2004) *Chemical Pesticides, Mode of Action and Toxicology*. CRC Press, Boca Raton, FL.

Stohs SJ, Bagchi D (1995) Oxidative mechanisms in the toxicity of metal ions. *Free Rad Biol Med* 18: 321–336.

Stolt JP, Sneller FEC, Bryngelsson T, Lundborg T, Schat H (2003) Phytochelatin and cadmium accumulation in wheat. *Environ Exper Bot* 49: 21–28.

Streb P, Michael-Knauf A, Feierabend J (1993) Preferential photoinactivation of catalase and photoinhibition of photosystem II are common early symptoms under various osmotic and chemical stress conditions. *Physiol Plant* 88: 590–598.

Sweetlove LJ, Moller IM (2009) Oxidation of proteins in plants – mechanisms and consequences. In: Jacquot J-P (ed.), *Advances in Botanical Research*, Vol. 52. Academic Press, pp. 1–23.

Talekar NS, Chen JS, Rao HT (1983) Long term persistence of selected insecticides in subtropical soils, their absorption by plants. *J Econ Entomol* 76: 207–214.

Tanaka K, Sugahara K (1980) Role of superoxide dismutase in defense against SO_2 toxicity and an increase in superoxide dismutase activity with SO_2 fumigation. *Plant Cell Physiol* 21: 601–611.

Tanou G, Molassiotis A, Diamantidis G (2009) Induction of reactive oxygen species and necrotic death-like destruction in strawberry leaves by salinity. *Environ Exp Bot* 65 :270–281.

Thakre SK, Saxena SN (1972) Effects of chlorinated insecticides on plant growth and uptake of nutrients by wheat and maize. *J Ind Soc Soil Sci* 20: 45–48.

Tomkins DJ, Grant WF (1977) Effects of herbicides on species richness of two plant communities. *Ecology* 58: 398–406.

Torralba-Redonde B, Flores Rores Rodriguez M, Villar Moreno AL (1986) The amitrol action over *Rhizobium meliloti*. *Ann Edafol Agrobiol* 45: 1079–1086.

Torres MA, Dangl JL, Jones JDG (2002) Arabidopsis gp91^{phox} homologues Atrbohd and Atrbohf are required for accumulation of reactive oxygen intermediates in the plant defense response. *Proc Natl Acad Sci USA* 99: 517–522.

Toscano NC, Sances FV, Johnson MW, Lapre LF (1982) The effect of various pesticides on lettuce physiology and yield. *J Econo Entomol* 75: 738–741.

Trebst A, Kluth J, Tietjen K, Draber W (1991) Molecular analysis of photosystem II inhibition. *Pestic Chem* 14: 111–120.

Tsang EWT, Bowler C, Herouart D, *et al.* (1991) Differential regulation of superoxide dismutase dismutases in plants exposed to environmental stress. *Plant Cell* 3: 783–792.

TsuboiH, KoudaK, TakeuchiH (1998) 8-Hydroxydeoxyguanosine in urine as an index of oxidative damage to DNA in the evaluation of atopic dermatitis. *Brit J Dermatol* 138: 1033–1035.

Tu CM, Mile JRW (1976) Interactions between insecticides and soil microbes. *Res Rev* 64: 17–65.

Turrens JF (2003) Mitochondrial formation of reactive oxygen species. *J Plant Physiol* 552: 335–344.

Vanaja M, Charyawa NVN, Rao KVN (2000) Effect of cadmium on carbohydrate, nucleic acids, and phenolic content in *Stigeoclonium tenue* Kutz. *Indian J Pl Physiol* 5: 253–256.

Van-Camp W, Millekens H, Bowler C, *et al.* (1994) Elevated levels of superoxide dismutase protect transgenic plants against ozone damage. *Biotechnology* 12: 165–168.

Vaughan D, Deckok PC, Ord BG (1982) The nature and localization of superoxide dismutase in fronds of *Lemma gibba* L. and the effect of copper and zinc deficiency on its activity. *Physiol Plant* 54: 253–257.

Villarini M, Moretti M, Pasquini R, *et al.* (1998) In vitro genotoxic effects of the insecticide Deltamethrin in human peripheral blood leukocytes: DNA damage ("comet" assay) in relation to the induction of sister-chromatid exchanges and micronuclei. *Toxicology* 130: 129–139.

Ware GW (1993) *The Pesticide Book*, 5th edn. Thomson Publishers, Fresno, CA, 418 pp.

Weckx JEJ, Clijsters HMM (1996) Oxidative damage and defense mechanisms in primary leaves of *Phaseolus vulgaris* as a result of root assimilation of toxic amounts of copper. *Physiol Plant* 96: 506–512.

West TF, Campbell GA (1950) *DDT and Newer Persistent Insecticides*. Chapman & Hall Ltd., London.

Wingate VPM, Lawton MA, Lamb CJ (1988) Glutathione causes a massive and selective induction of plant defense genes. *Plant Physiol* 87: 206–210.

Winglse G, Karpinski S (1996) Differential redox regulation by glutathione and glutathione reductase and Cu-Zn SOD gene expression in *Pinus sylvestris* L. needles. *Planta* 198: 151–157.

Wojtaszek P (1997) Oxidative burst: an early plant response to pathogen infection. *Biochem J* 322: 681–692.

Xiong L, Schumaker KS, Zhu JK (2002) Cell signaling during cold, drought, and salt stress. *Plant Cell* 14: S165–S183.

Yamauchi Y, Furutera A, Seki K, Toyoda Y, Tanaka K, Sugimoto Y (2008) Malondialdehyde generated from peroxidized linolenic acid causes protein modification in heat-stressed plants. *Plant Physiol Biochem* 46: 786–793.

Yan J, Tsuichihara N, Etoh T, Iwai S (2007) Reactive oxygen species and nitric oxide are involved in ABA inhibition of stomatal opening. *Plant Cell Environ* 30: 1320–1325.

Yeh CM, Chien PS, Huang HJ (2007) Distinct signalling pathways for induction of MAP kinase activities by cadmium and copper in rice roots. *J Exper Bot* 58: 659–671.

Yoshimura K, Yabata Y, Tomoi M, Ishikawa I, Shigoeka S (1999) Alternatively spliced mRNA variants of chloroplast ascorbate peroxidase isoenzymes in spinach leaves. *Biochem J* 338: 41–48.

Yuasa T, Ichimura K, Mizoguchi T, Shinozaki K (2001) Oxidative stress activates ATMPK6, an Arabidopsis homologue of map kinase. *Plant Cell Physiol* 42: 1012–1016.

Zeidler O (1874) Verbindungen Von Chloral mit Brom-und Chlorbenzol. *Berichte* 7: 1180–1189.

Zhang J, Kirkham MB (1996) Lipid peroxidation in sorghum and sunflower seedlings as affected by ascorbic acid and propyl gallate. *J Plant Physiol* 149: 489–493.

CHAPTER 8

Legume-rhizobia symbiotic performance under abiotic stresses: Factors influencing tolerance behaviour

Haythem Mhadhbi[1], Photini V. Mylona[2] and Alexios N. Polidoros[3]

[1] *Laboratory of Legumes, Centre of Biotechnology of Borj Cedria, Hammam lif, Tunisia*

[2] *Agricultural Research Center of Northern Greece, Thermi, Greece*

[3] *Department of Genetics and Plant Breeding, School of Agriculture, Aristotle University of Thessaloniki, Greece*

8.1 Introduction

Leguminous plants (family Fabaceae) represent one of the most important food, feed and forage resources. In many developing countries, legumes are the single most important source of proteins (Roe & Kupfer, 2004). About 33% of human nutritional requirements for nitrogen come from legumes. They also play a central role in the enhancement of the natural environment since they can be fed to animals, as well as protecting the soil, stabilizing dunes and representing a source of medicinal products (Graham & Vance 2003). Due to the symbiotic association with N-fixing rhizobia, legumes sustain the production of protein-rich seeds and fodder while improving productivity of cereals and other crops in agricultural rotations (Graham & Vance, 2003). These unique characteristics make legume cultivation fundamental for sustainable agriculture, improving environmental equilibrium and quality of life. However, legume-rhizobia interaction is often compromised by adverse environmental conditions, mainly soil salinity and drought (Zahran, 1999). It is estimated that 60% of legume production in the developing world occurs under conditions of extreme drought stress (Graham & Vance, 2003). In many countries belonging to the semiarid climate, legume cultivation is in continuous regression because of the low yield, and the needs are satisfied by importation.

8.2 Symbiotic association: A specific plant-microbe interaction

The symbiotic association and the plant-pathogen interaction represent two major types of plant-microorganism interactions. In the early stages of both processes, the response of the plant is very similar. The microorganism finds the compatible host plant tissue (roots or leaves) for attachment and commencement of the invasion process. At this step, the plant defends rapidly by generation of reactive oxygen species (ROS). ROS include superoxide radical ($\bullet O_2^-$), singlet oxygen ($^1O_2$), hydrogen peroxide (H_2O_2) and the hydroxyl radical ($\bullet OH$) (Mylona & Polidoros, 2010). ROS are highly active molecules capable of oxidizing biomolecules including proteins, lipids and nucleic acids (Ahmad *et al.*, 2008, 2010, 2011, 2012, 2014, Ahmad 2014). ROS accumulation orchestrates the initiation of plant antimicrobial defences, facilitates the formation of cell wall barriers preventing further spread of the microorganism, leads to the induction of the hypersensitive response (HR), and serves as a signal in secondary signalling pathways that further activate defence responses. ROS, particularly hydrogen peroxide, are considered to be the key signals triggering plant responses to biotic and abiotic stresses (Djébali *et al.*, 2011; Ellouzi *et al.*, 2011, 2013; Mhadhbi *et al.*, 2013; Bartoli *et al.*, 2013).

In the case of symbiotic interaction a similar plant response is initially triggered. Upon infection of a legume

Legumes under Environmental Stress: Yield, Improvement and Adaptations, First Edition. Edited by Mohamed Mahgoub Azooz and Parvaiz Ahmad.

host by rhizobia, the host produces ROS, namely hydrogen peroxide and superoxide radical, at the site of infection (Santos *et al.,* 2001). This resembles the hypersensitive response of plants in incompatible plant-pathogen interactions. However, in the development of the compatible symbiotic interaction, the rhizobia are able to deal with the host defence response, and the plant adopts a different defence strategy (Mithofer, 2002). Indeed, the plant controls the invading bacteria by confining them in an organ formed for this purpose, the nodule. Failure of effective nodule formation, due to defective surface polysaccharides of rhizobia, often results in pathogenic responses (Oldroyd & Downie, 2008). Notably, the ROS level has to be controlled not only by the invading rhizobia but also by the host plant to diminish the oxidative stress effects. Both, the host plant and the rhizobia possess an arsenal of antioxidant tools, among which are the enzymes catalase, superoxide dismutase and various peroxidases (Becana *et al.,* 2000; Matamoros *et al.,* 2003). For example, *Sinorhizobium meliloti* has several antioxidant enzymes including two superoxide dismutases, SodA and SodC, and three catalases, KatA, KatB and KatC (Jamet *et al.,* 2003; Minchin *et al.,* 2008; Becana *et al.,* 2010). In mutants that lack expression of any of the three catalases or one of the two dismutases, an increased sensitivity of the bacterium to ROS is observed but not the prevention of nodulation (Jamet *et al.,* 2007; Minchin *et al.,* 2008). However, in double *katB/katC* and *katA/katC* mutants, the rhizobia lose their ability to colonize root hairs of the host plant *Medicago truncatula* (Jones *et al.,* 2007).

In addition to antioxidant defence, the efflux of ROS from the host roots is inhibited by specific rhizobial signalling molecules commonly known as Nod factors (NFs) (Shaw & Long, 2003). It is suggested that NFs suppress the activity of the ROS-generating plant system and thus allow a compatible interaction between plant and rhizobia to occur (Chang *et al.,* 2009). In this respect, host defence against oxidative stress seems also to be a crucial component for successful nodulation, since catalase deficiency induced transiently by RNAi in *Medicago truncatula* roots, resulted in reduced nodule formation, especially under osmotic stress conditions. However, the efflux of ROS from the host roots is indispensable for nodule formation and the development of a compatible plant-microbe interaction. ROS, such as hydrogen peroxide, are crucial for inter-cross-linking and strengthening the cell wall formation of infection threads. Studies have shown that inhibition of ROS production prevents root hair curling and formation of infection threads (Chang *et al.,* 2009, and references therein). Conclusively, the disparity of ROS accumulation discerned at different time points of the symbiotic interaction highlights the essential role of ROS at multiple spatiotemporal steps in the nodule formation process.

8.3 Legume-rhizobia symbiosis: A vulnerable association under osmotic stresses

Biological nitrogen fixation (BNF) by symbiotic association is a highly desirable option for restoration of marginal lands damaged by environmental stresses. However, it is well documented that BNF is vulnerable to abiotic constraints (Zahran, 1999). In addition, plant responses to abiotic stresses, such as salinity and drought, involve complex mechanisms and different pathways that are not entirely characterized. Further, understanding the legume-rhizobium symbiotic interaction under abiotic stress is even more challenging. It is generally accepted that under salinity and drought conditions, the reduced water supply is the most significant limitation on growth and crop yield (Zhu, 2001; Schleiff, 2008). The osmotic constraint induces the plant and the bacteria to decrease their internal water potential to avoid desiccation (Zahran, 1999; Tonon *et al.,* 2004). Water is a prerequisite for all biochemical activities in known life forms (Xiong & Zhu, 2002), thus any disturbance of water potential seriously affects cell metabolic activities. In nodules, responses to reduced water availability are marked by adaptive changes including growth limitation, cortical structure modification and decline of metabolic activities. Undoubtedly these changes lead to alteration of nitrogen-fixing machinery, and consequently to decline of symbiosis effectiveness.

8.4 Nodulation process and symbiotic performance variability

Nodule formation depends largely on effective development of the root hair system (Oldroyd & Downie, 2008). Water deficit adversely affects root architecture, reducing the formation of hairy roots. Additionally, disturbance of ionic balance due to water limitation impinges on the signalling interactions of host plant and rhizobia. Alterations of plant root architecture lead to the formation of atypical nodulation sites due to a rigid root cortex and deformed root hairs, causing a decline in successful infections and nodulation (Zahaf *et al.*, 2012; Zélicourt *et al.*, 2012). It is known that symbiosis establishment and functioning under stressful conditions are related to the vigour of the host legume (Zahran, 1999). The bacterial partners (i.e. rhizobia) are more tolerant than their host plants because rhizobia species are able to survive under conditions of high salt concentrations and desiccation (Jebara *et al.*, 2001; Vriezen *et al.*, 2007; Mnasri *et al.*, 2007b). Yet, the adaptation of rhizobia to water deficit is accompanied by morphological and physiological changes that decrease their infection and nodulation capacity (Zahran & Sprent, 1986; Predeepa & Ravindran, 2010).

Salinity and drought severely affect biomass production and nitrogen-fixing capacity, which are considered as essential indices of symbiosis effectiveness (Serraj *et al.*, 1999; Mhadhbi *et al.*, 2004, 2008, 2011a; Verdoy *et al.*, 2004; Mhadhbi & Aouani, 2008). Studies have shown that the nitrogenase activity and consequently the nodule N-fixing mechanism are more sensitive than plant growth to abiotic stress (Mhadhbi *et al.*, 2004, 2011a). This particular sensitivity is explained by the complexity of the symbiotic association and the high energy demand for nodule function. Energy provided by the host plant may be restricted because the salinity/drought stress causes nutrient deprivation due to a decline in water uptake and reduced function of the photosynthetic apparatus (Zahran, 1999; Lopez *et al.*, 2008). Recently, Mhadhbi *et al.* (2008, 2011a) and Pimratch *et al.* (2008) reported that more efficient symbioses were recorded in normal conditions than under osmotic stress. These results suggest that high biomass production and nitrogen-fixing capacity under osmotic stress are decreased when compared with non-stressful conditions. Nevertheless, other studies reported that bacterial strains characterized by moderate efficiency at normal conditions could enhance symbiosis tolerance under stressful conditions and vice versa (Mnasri *et al.*, 2007a; Ben Rhomdhane *et al.*, 2007; Tajini *et al.*, 2008; Mhadhbi *et al.*, 2011b). Which of the symbiotic partners plays the most significant role for achieving better tolerance under diverse conditions is the subject of a diachronic debate between microbiologists (defending the input of the bacterial partner) and plant breeders (insisting on the superiority of plant vigour). Aiming to provide novel evidence on this long-standing debate, we studied the modulation of the symbiotic effectiveness of different legume-rhizobia associations, in relation to plant genotype and rhizobial strain (Mhadhbi *et al.*, 2004, 2005, 2008, 2011a,b,c; Jebara *et al.*, 2005; Mnasri *et al.*, 2007a,b; Ben Rhomdhane *et al.*, 2007; Mhadhbi & Aouani, 2008).

8.5 Variability of symbiotic partners' input to symbiosis resilience

There is no doubt that rhizobia are more tolerant than legume host plants to environmental perturbations. Indeed, some rhizobia strains are able to survive on media containing more than 40 g/L NaCl (Mnasri *et al.*, 2007b), a salinity level higher than that of the Mediterranean Sea (36 g/L). However, the question that remains unresolved is how and through which mechanisms the microsymbiont can enhance plant growth under adverse or stressful conditions. Studies indicate that a competitive and persistent rhizobial strain is not expected to express its full capacity for nitrogen fixation if limitations constrain the host legume's vigour (Zahran, 1999). On the other hand, results showed that inoculation with tolerant and efficient rhizobial strains enhanced legume production under abiotic stress conditions (Mhadhbi *et al.*, 2004, 2008, 2011b; Mnasri *et al.*, 2007a; Ben Rhomdhane *et al.*, 2007). This provides substantial potential means of enhancing legume cultivation in marginal lands of arid and semi-arid climates. Statistical approaches have estimated the contribution of each partner as well as their interaction on the variability of symbiotic effectiveness under salt and drought stress conditions (Mhadhbi *et al.*, 2005, 2008, 2011a; Mhadhbi & Aouani, 2008). Results under controlled conditions showed that in wild legumes such as *Medicago truncatula*, the host plant genotype is the major factor contributing to the total variance of symbiotic effectiveness indices. Similar results were reported under field conditions (Fesenko *et al.*, 1994; Robinson *et al.*, 2000). The tolerance of the symbiotic association is primarily determined by the degree of host plant tolerance. The input of the bacterial partner is mostly related to its potential efficiency under stress. The prominent role of the plant genotype could be attributed to the contrasting behaviour of different *M. truncatula* lines, which imply different metabolic responses involving variations in plant tissues, nodule ionic status, nodule-cortex structure and oxygen permeability (Aydi *et al.*, 2004; Mhadhbi *et al.*, 2011a). In contrast, studies performed with legume crops, such as chickpea (*Cicer arietinum*) and common bean (*Phaseolus vulgaris*), did not discern a significant effect of the cultivar. In these experiments the bacterial partner was the key factor influencing symbiosis behaviour mainly under stressful conditions (Mhadhbi *et al.*, 2004, 2008, 2011b; Jebara *et al.*, 2005). The contribution of the rhizobial partner to the variance of symbiosis effectiveness even under osmotic stress, suggests that the selection of effective strains could enhance legume crop production especially in nitrogen-poor soils (Serraj *et al.*, 1999; Zahran 1999; Mnasri *et al.*, 2007a). Moreover, to achieve integration in agricultural practice, the adaptation of the selected rhizobial strains and their ability to compete with the local soil microflora should be considered (Mrabet *et al.*, 2005). Our experience suggests that in field conditions, native rhizobial strains nodulating chickpea (*Mezorhizobium ciceri* CMG6) and common bean (*Rhizobium etli* 12a3) were more efficient than the commercial counterparts (*M. ciceri* UPMCa7^{T}) and *R. tropici* CIAT899) (Ben Rhomdhane *et al.*, 2007; Tajini *et al.*, 2008). However, under laboratory conditions the reference strains were more or equally effective than the local ones (Mhadhbi *et al.*, 2008).

This study provided evidence for a significant contribution of the rhizobial partner to the symbiotic effectiveness of different chickpea cultivars under osmotic stress as well as under non-stressed conditions. Field trials are required to confirm our suggestion that tolerance of symbiosis to osmotic stress could be enhanced by selection of effective strains (Mhadhbi *et al.*, 2011b).

8.6 Effect of osmotic stress on nodule integrity and functioning

Salinity and drought imply secondary stresses, namely osmotic and ionic, that both affect metabolic activities (Tonon *et al.*, 2004). The primary response to metabolic disturbance is production and accumulation of ROS, resulting in oxidative stress (Zhu, 2001; Ahmad *et al.*, 2008, 2010, 2011, 2012, 2014; Ahmad 2014). In nodules, ROS are reported to have deleterious effects on tissue integrity and nodule functioning (Becana *et al.*, 2000, Hernandez-Jimenez *et al.*, 2002; Matamoros *et al.*, 2003; Mhadhbi *et al.*, 2011a). The effect of oxidative stress on nodule metabolism appears as a general slowdown of nitrogenase activity concomitant with a decrease in nodule protein content and leghaemoglobin content along with an accumulation of lipid peroxidation products (Mhadhbi *et al.*, 2008, 2011a). Such alterations indicate loss of integrity of cell membranes and perturbation in the photosynthates required for nodule performance. Nitrogenase activity is highly sensitive to oxygen concentration (Aydi *et al.*, 2004; Kratsch & Graves, 2005; Mhadhbi *et al.*, 2011a). The transport of oxygen within nodule organelles is performed by leghaemoglobin, which plays a critical role in the process of bacteroid development and cellular respiration. Leghaemoglobin oxidation (non-functional oxy-leghaemoglobin) by ROS reduces the oxygen supply to nodules (Mhadhbi *et al.*, 2011a).

Oxidative damage is not the only factor responsible for the decline of nodular activity. Indeed, as we have already reported, nodule activity is closely dependent on photosynthetic products provided by the plant. This supply is reduced under stress due to diminution of photosynthesis (decrease in quantity) as a consequence of ionic disruption. Under such circumstances the bacteroids counteract ionic disruption by accumulating osmoprotective substances such as glycine betaine (Zahran, 1999).

To overcome stressful conditions, nodules invoke various mechanisms including morphological modifications (Matamoros *et al.*, 1999; Verdoy *et al.*, 2004; Kratsch & Graves, 2005; Luqueno *et al.*, 2008) and physio-biochemical adaptations (Mhadhbi *et al.*, 2011a,b,c). Adaptive responses involve the modulation of some metabolic mechanisms to ameliorate the activity/energy ratio, and stimulation of expression of some protective molecules such as antioxidant compounds and enzymes (Hernandez-Jimenez *et al.*, 2002; Tejera *et al.*, 2004; Mhadhbi *et al.*, 2011a,c).

Correlation between antioxidant capacity and salt tolerance is reported for several food crops and model plants, including rice (Lee *et al.*, 2001), common bean (Jebara *et al.*, 2005; Mhadhbi *et al.*, 2011b), chickpea (Mhadhbi *et al.*, 2004, 2008), *Medicago truncatula* (Mhadhbi & Aouani, 2008; Mhadhbi *et al.*, 2011a,c), *Arabidopsis thaliana* (Xu *et al.*, 2008; Ellouzi *et al.*, 2011) and some halophyte species (Ellouzi *et al.*, 2011, 2013). Thus, the alleviation of oxidative damage by the use of different antioxidants and ROS scavengers can enhance plant resistance to salt stress (Mittler, 2002; Ellouzi *et al.*,

2011, 2013; Mhadhbi *et al.*, 2011a,c, 2013; Ahmad *et al.*, 2008, 2010, 2011, 2012, 2014; Ahmad, 2014).

8.7 Future prospects

Over the last decade, advances in sequencing technologies have brought an explosion of the available genomic information regarding microbial as well as plant genomes. Several genomes of microbial symbionts of legumes, the genome sequences of the model legumes *Medicago truncatula* and *Lotus japonicus*, as well as the genome sequence of the most agronomically important legume species, soybean (*Glycine max*), are now available. Gene expression analysis at the genomic level has been made possible, and transcriptome profiling offers valuable information about gene function under specific conditions. Bioinformatics tools are also used for access and analysis of the data (Li *et al.*, 2011).

Comparative genomic analyses not only enable the study of legume-specific gene families, but also provide a framework to target phenotypes with specific characteristics related to stress tolerance. Available methodologies for next generation sequencing now allow the re-sequencing of individuals with specific phenotypes so that genomic information can be translated to gain a deeper understanding of the genetic factors affecting these phenotypes.

This strategy has already provided significant information about the rhizobial partners. In bacteria there are large differences in gene content even between closely related strains, where a number of genes conserved in all strains comprise the 'core' genes, and genes that show variation between strains are 'accessory' genes; this leads to the concept of the 'pangenome', that is, the total number of genes detected not in individuals or strains, but in the species as a whole. Symbiotic diversity in *Sinorhizobium meliloti* has been studied by comparative genomic analysis between different strains, defining a set of accessory genes and regulons that may contribute to the symbiotic process (Galardini *et al.*, 2011). Stress tolerance diversity may be analysed with a similar approach in order to define genetic factors related to tolerant phenotypes. Nevertheless, advances in high-throughput sequencing may facilitate variant discovery and genotyping, but linking genotypic variants to specific phenotypes is still a major task. Future studies at the genomic and proteomic levels will further enrich our understanding of successful symbiotic associations under stress.

References

Ahmad P (2014) *Oxidative Damage to Plants: Antioxidant Networks and Signaling*. Academic Press, Elsevier.

Ahmad P, Sarwat M, Sharma S (2008) Reactive oxygen species, antioxidants and signaling in plants. *J Plant Biol* 51: 167–173.

Ahmad P, Jaleel CA, Salem MA, Nabi G, Sharma S (2010) Roles of enzymatic and non-enzymatic antioxidants in plants during abiotic stress. *Crit Rev Biotechnol* 30: 161–175.

Ahmad P, Nabi G, Ashraf M (2011) Cadmium-induced oxidative damage in mustard [*Brassica juncea* (L.) Czern. & Coss.] plants can be alleviated by salicylic acid. *South Afr J Bot* 77: 36–44.

Ahmad P, Hakeem KR, Kumar A, Ashraf M, Akram NA (2012) Salt-induced changes in photosynthetic activity and oxidative defense system of three cultivars of mustard (*Brassica juncea* L.) *Afr J Biotechnol* 11: 2694–2703.

Ahmad P, Ozturk M, Sharma S, Gucel S (2014) Effect of sodium carbonate-induced salinity-alkalinity on some key osmoprotectants, protein profile, antioxidant enzymes, and lipid peroxidation in two mulberry (*Morus alba* L.) cultivars. *J Plant Interactions* 9: 460–467.

Aydi S, Drevon JJ, Abdelly C (2004) Effect of salinity on root-nodule conductance to the oxygen diffusion in the *Medicago truncatula-Sinorhizobium meliloti* symbiosis. *Plant Physiol Biochem* 42: 833–840.

Bartoli CG, Casalongué CA, Simontacchi M, Marquez-Garcia B, Foyer CH (2013) Interactions between hormone and redox signalling pathways in the control of growth and cross tolerance to stress. *Environ Exp Bot* 94: 73–88.

Becana M, Dalton DA, Moran JF, Iturbe-Ormaetxe I, Matamoros MA, Rubio MC (2000) Reactive oxygen species and antioxidant in legume nodules. *Physiol Plant* 109: 372–381.

Becana M, Matamoros MA, Udvardi M, Dalton DA (2010) Recent insights into antioxidant defenses of legume root nodules. *New Phytol* 188: 960–976.

Ben Rhomdhane S, Tajini F, Trabelsi M, Aouani ME, Mhamdi R (2007) Competition for nodule formation between introduced strains of *Mesorhizobium ciceri* and native populations of rhizobia nodulating chickpea (*Cicer arietinum*) in Tunisia. *World J. Microbiol Biotechnol* 23: 1195–1201.

Chang C, Damiani I, Puppo A, Frendo P (2009) Redox changes during the legume-rhizobium symbiosis. *Mol Plant* 2: 370–377.

Djébali N, Mhadhbi H, Lafitte C, Dumas B, Esquerré-Tugayé MT, Aouani ME, Jacquet C (2011) Hydrogen peroxide scavenging mechanisms are components of *Medicago truncatula* partial resistance to *Aphanomyces euteiches*. *Eur J Plant Pathol* 131: 559–571.

Ellouzi H, Ben Hamed K, Fabado MAA, Muller M, Abdelly C, Bosch SM (2013) Drought and cadmium may be as effective as salinity in conferring subsequent salt stress tolerance in *Cakile maritime*. *Planta* 237: 1311–1323.

Ellouzia H, Ben Hamed K, Celab J, Bosch SM, Abdelly C (2011) Early effects of salt stress on the physiological and oxidative status of *Cakile maritima* (halophyte) and *Arabidopsis thaliana* (glycophyte). *Physiol Plant* 142: 128–143.

Fesenko AN, Provorov NA, Orlova IF, Simarov BV (1994) The role of the pea (*Pisum sativum* L.) cultivar genotype and the *Rhizobium leguminosarum* strain in the effectiveness of symbiosis. *Russ J Genet* 30: 725–729.

Galardini M, Mengoni A, Brilli M, *et al.* (2011) Exploring the symbiotic pangenome of the nitrogen-fixing bacterium *Sinorhizobium meliloti*. *BMC Genomics* 12: 235.

Graham PH, Vance CP (2003) Legumes: Importance and constraints to greater use. *Plant Physiol* 131: 872–877.

Hernandez-Jimenez MJ, Lucas MM, de Felipe MR (2002) Antioxidant defence and damage in senescing lupin nodules. *Plant Physiol Biochem* 40: 645–657.

Herouart D, Baudouin E, Frendo P, *et al.* (2002) Reactive oxygen species, nitric oxide and glutathione: a key role in the establishment of the legume:Rhizobium symbiosis. *Plant Physiol Biochem* 40: 619–624.

Jamet A, Mandon K, Puppo A, Hérouart D (2007) H_2O_2 is required for optimal establishment of the *Medicago sativa*/*Sinorhizobium meliloti* symbiosis. *J Bacteriol* 189: 8741–8745.

Jebara M, Drevon JJ, Aouani ME (2001) Effects of hydroponic culture system and NaCl on interactions between common bean lines and native rhizobia from Tunisian soils. *Agronomie* 21: 601–605.

Jebara S, Jebara M, Limam F, Aouani ME (2005) Changes in ascorbate peroxidase, catalase, guaicol peroxidase and superoxide dismutase activities in common bean (*Phaseolus vulgaris*) nodules under salt stress. *J Plant Physiol* 162: 929–936.

Kratsch HA, Graves WR (2005) Oxygen concentration affects nodule anatomy and nitrogenase activity of *Alnus maritime*. *Plant Cell Environ* 28: 688–696.

Lee DH, Kim YS, Lee CB (2001) The inductive responses of the antioxidant enzymes by salt stress in the rice (*Oryza sativa* L.). *J Plant Physiol* 158: 737–745.

Li J, Dai X, Liu T, Zhao PX (2011) LegumeIP: an integrative database for comparative genomics and transcriptomics of model legumes. *Nucleic Acids Res* 40: D1221–D1229.

Lopez M, Herrera-Cervera JA, Iribrane C, Tejera NA, Lluch C (2008). Growth and nitrogen fixation in *Lotus japonicus* and *Medicago truncatula* under NaCl stress: Nodule carbon metabolism. *J Plant Physiol* 165: 641–650.

Luqueño FF, Dendooven L, Munive A, Corlay-Chee L, Serrano-Covarrubias LM, Epinosa-Victoria D (2008) Micro-morphology of common bean (*Phaseolus vulgaris* L.) nodules undergoing senescence. *Acta Physiol Plant* 30: 545–552.

Matamoros MA, Baird LM, Escuredo PR, Dalton DA, Minchin FR, Iturbe-Ormaetxe I, Rubio MC, Moran JF, Gordon AJ, Becana M (1999) Stress-induced legume root nodule senescence, physiological, biochemical and structural alterations. *Plant Physiol* 121: 97–111.

Matamoros MA, Dalton DA, Ramos J, Clemente MR, Rubio MC, Becana M (2003) Biochemistry and molecular biology of antioxidants in the rhizobia-legume symbiosis. *Plant Physiol* 133: 449–509.

Mhadhbi H, Aouani ME (2008) Growth and nitrogen-fixing performances of *Medicago truncatula-Sinorhizobium meliloti* symbioses under salt (NaCl) stress: Micro- and macrosymbiont contribution to symbiosis tolerance. In: Abdelly C, Ozturk M, Ashraf M, Grignon C (eds), *Biosaline Agriculture and High Salinity Tolerance*. Birkhauser Verlag, pp. 91–98.

Mhadhbi H, Jebara M, Limam F, Aouani ME (2004) Rhizobial strain involvement in plant growth, nodule protein composition and antioxidant enzyme activities of chickpea-rhizobia symbioses: modulation by salt stress. *Plant Physiol Biochem* 42: 717–722.

Mhadhbi H, Jebara M, Limam F, Huguet T, Aouani ME (2005) Interaction between *Medicago truncatula* lines and *Sinorhizobium meliloti* strains for symbiotic efficiency and nodule antioxidant activities. *Physiol Plant* 124: 4–11.

Mhadhbi H, Jebara M, Zitoun A, Limam F, Aouani ME (2008) Symbiotic effectiveness and response to mannitol-mediated osmotic stress of various chickpea-rhizobia associations. *World J Microbiol Biotechnol* 24: 1027–1035.

Mhadhbi H, Chihaoui S, Mhamdi R, Mnasri B, Jebara M Mhamdi R (2011a) A highly osmotolerant rhizobial strain confers a better tolerance of nitrogen fixation and enhanced protective activities to nodules of *Phaseolus vulgaris* under drought stress. *Afr Biotechnol* 22: 4555–4563.

Mhadhbi H, Djébali N, Chihaoui S, Jebara M, Mhamdi R (2011b) Nodule senescence in *Medicago truncatula–Sinorhizobium* symbiosis under abiotic constraints: Potential mechanisms involved in maintaining nitrogen fixing capacity. *J Plant Growth Regul* 30: 480–489.

Mhadhbi H, Fotopoulos V, Mylona PV, Jebara M, Aouani ME Polidoros AN (2011c) Antioxidant gene-enzyme responses in *Medicago truncatula* genotypes with different degree of sensitivity to salinity. *Physiol Plant* 141: 201–214.

Mhadhbi H, Fotopoulos V, Mylona PV, Jebara M, Aouani ME, Polidoros AN (2013) Alternative oxidase 1 (Aox1) gene expression in roots of *Medicago truncatula* is a genotype-specific component of salt stress tolerance. *J Plant Physiol* 170: 111–114.

Minchin FR, James EK, Becana M (2008) Oxygen diffusion, production of reactive oxygen and nitrogen species and antioxidants in legume nodules. In: Dilworth MJ (ed.), *Nitrogen-fixing Leguminous Symbioses*. Springer, pp 321–362.

Mithöfer A (2002) Suppression of plant defence in rhizobia-legume symbiosis. *Trends Plant Sci* 7: 440–444.

Mittler R (2002) Oxidative stress, antioxidants and stress tolerance. *Trends Plant Sci* 7: 405–410.

Mittler R (2006) Abiotic stress, the field environment and stress combination. *Trends Plant Sci* 11: 15–19.

Mnasri B, Aouani ME, Mhamdi R (2007a) Nodulation and growth of common bean (*Phaseolus vulgaris*) under water deficiency. *Soil Biol Biochem* 39: 1744–1750.

Mnasri B, Mrabet M, Laguerre G, Aouani ME, Mhamdi R (2007b) Salt tolerant rhizobia isolated from the Tunisian Oasis that are highly effective for N_2-fixation with *Phaseolus vulgaris* constitute a novel biovar (bv. Mediterranense) of *Sinorhizobium meliloti*. *Arch Microbiol* 187: 79–85.

Mrabet M, Mhamdi R, Tajini F, Tiwari R, Trabelsi M, Aouani ME (2005) Competitiveness and symbiotic effectiveness of a *R. gallicum* strain isolated from root nodules of *Phaseolus vulgaris*. *Eur J Agron* 22: 209–216.

Mylona PV, Polidoros AN (2010) ROS regulation of antioxidant genes. In: Dutta Gupta S (ed.), *Reactive Oxygen Species and Antioxidants in Higher Plants*. CRC Press, pp. 101–127.

Oldroyd GED, Downie JA (2008) Coordinating nodule morphogenesis with rhizobial infection in legumes. *Ann Rev Plant Biol* 59: 519–546.

Pimratch S, Jogloy S, Vorasoot N, Toomsan B, Patanothai A, Holbrook CC (2008) Relationship between biomass production and nitrogen fixation under drought-stress conditions in peanut genotypes with different levels of drought resistance. *J Agron Crop Sci* 194: 15–25.

Predeepa RJ, Ravindran DA (2010) Nodule formation, distribution and symbiotic efficacy of *Vigna unguiculata* L. under different soil salinity regimes. *Emir J Food Agric* 22: 275–284.

Robinson KO, Beyene DA, van Berkum P, Knight-Mason R, Bhardwaj HL (2000) Variability in plant-microbe interaction between *Lupinus* lines and *Bradyrhizobium* strains. *Plant Sci* 159: 257–264.

Roe BA, Kupfer DM (2004) Sequencing gene rich regions of *Medicago truncatula*, a model legume. In: Hopkins A, Wang ZY, Mian R, Sledge M, Barker RE (eds), *Molecular Breeding of Forage and Turf*. Springer, pp. 333–344.

Santos R, Hérouart D, Sigaud S, Touati D, Puppo A (2001) Oxidative burst in alfalfa- *Sinorhizobium meliloti* symbiotic interaction. *Mol Plant-Microbe Interact* 14: 86–89.

Schleiff U (2008) Analysis of water supply of plants under saline soil conditions and conclusions for research on crop salt tolerance. *J Agron Crop Sci* 194: 1–8.

Serraj R, Sinclair TR, Purcell LC (1999) Symbiotic N_2 fixation response to drought. *J Exp Bot* 50: 143–155.

Shaw SL, Long SR (2003) Nod factor inhibition of reactive oxygen efflux in a host legume. *Plant Physiol* 132: 2196–2204.

Tajini F, Drevon JJ, Lamouchi L, Aouani ME, Trabelsi M (2008) Response of common bean lines to inoculation: comparison between the *Rhizobium tropici* CIAT899 and the native *Rhizobium etli* 12a3 and their persistence in Tunisian soils. *World J Microbiol Biotechnol* 24: 407–417.

Tejera NA, Campos R, Sanjuan J, Lluch C (2004) Nitrogenase and antioxidant enzyme activities in *Phaseolus vulgaris* nodules formed by *Rhizobium tropici* isogenic strains with varying tolerance to salt stress. *J Plant Physiol* 161: 329–338.

Tonon G, Kevers C, Faivre-Ranpant O, Graziani M, Gaspar T (2004) Effect of NaCl and mannitol iso-osmotic stresses on proline and free polyamine levels in embryogenic *Fraxinus angustifolia* callus. *J Plant Physiol* 161: 701–708.

Verdoy D, Lucas MM, Manrique E, Covarrubias AA, de Filipe MR, Pueyo JJ (2004) Differential organ-specific response to salt stress and water deficit in nodulated bean (*Phaseolus vulgaris*). *Plant Cell Environ* 27: 757–767.

Vriezen JAC, de Bruijn FJ, Nusslein K (2007) Responses of rhizobia to desiccation in relation to osmotic stress, oxygen, and temperature. *Appl Environ Microbiol* 73: 3451–3459.

Xiong L, Zhu JK (2002) Molecular and genetic aspects of plant responses to osmotic stress. *Plant Cell Environ* 25: 131–139.

Xu W-F, Shi W-M, Ueda A, Takabe T (2008) Mechanisms of salt tolerance in transgenic *Arabidopsis thaliana* Carring a peroxisomal ascorbate peroxidase gene from barley. *Pedosphere* 18: 486–495.

Zahaf O, Blanchet S, Zelicourt A, *et al.* (2012) Comparative transcriptomic analysis of salt adaptation in roots of contrasting *Medicago truncatula* genotypes. *Mol Plant* 5: 1068–1081.

Zahran HH (1999) Rhizobium legume symbiosis and nitrogen fixation under severe conditions and in arid climate. *Microbiol Mol Biol Rev* 63: 968–989.

Zahran HH, Sprent JI (1986) Effect of sodium chloride and polyethylene glycol on root hair infection and nodulation of *Vicia faba* L. plants by *Rhizobium leguminosarum*. *Planta* 167: 303–309.

Zélicourt A, Diet A, Marion J, *et al.* (2012) Dual involvement of a *Medicago truncatula* NAC transcription factor in root abiotic stress response and symbiotic nodule senescence. *Plant J* 70: 220–230.

Zhu JK (2001) Plant salt tolerance. *Trends Plant Sci* 6: 66–71.

CHAPTER 9

Microbial strategies for the improvement of legume production in hostile environments

Dilfuza Egamberdieva[1], Vyacheslav Shurigin[1], Subramaniam Gopalakrishnan[2] and Ram Sharma[3]

[1] *Department of Microbiology and Biotechnology, Faculty of Biology and Soil Science, National University of Uzbekistan, Tashkent, Uzbekistan*
[2] *International Crops Research Institute for the Semi-Arid Tropics (ICRISAT), Patancheru, Andhra Pradesh, India*
[3] *International Center for Agricultural Research in the Dry Areas (ICARDA), Tashkent, Uzbekistan*

9.1 Introduction

Various biotic and abiotic factors limit crop productivity, affecting nearly 1 billion people around the world (Munns & Tester, 2008; UNEP 2009). Malnutrition is recognized as the world's most serious health problem, while agricultural development is considered as the most effective sector in reducing hunger and poverty, through improvements in crop productivity (Godfray *et al.*, 2010). Crop losses due to salinity and drought are a major area of concern in coping with increasing food requirements (Egamberdiyeva *et al.*, 2007; Shanker & Venkateswarlu, 2011; Davranova *et al.*, 2013). Salinity alone affects 33% of the world's potential arable land, whereas 950 million ha of salt-affected lands occur in arid and semi-arid regions (UNEP, 2008). Major factors increasing salinity include irrigation of cultivated lands with saline water, poor cultural practices and low precipitation. The ability of plants to take up water are inhibited by salinity stress, causing ion imbalance and, in turn, resulting in a reduction of root and shoot growth (Munns, 2002).

In such situations, the presence of salt-tolerant genetic variants in a particular crop is a prerequisite for its successful cultivation (Mahmood *et al.*, 2000). Numerous studies have been reported on the management and identification of salt-tolerant crops such as cotton, wheat, maize and alfalfa (Chauhan & Singh, 2000; Soltani *et al.*, 2012). In addition, organic farming practices, such as application of farm manure, compost, biofertilizer and recycling of crop residues and/or green manures, also improve degraded soils affected by salinity (Pathma & Sakthivel, 2012; Adesemoye & Egamberdieva, 2013).

Legumes are very important crop plants for human and animal consumption, and the use of legumes in crop rotations can be a useful management strategy to increase the supply of nitrogen to non-legume plants (Lüscher *et al.*, 2011; Nyfeler *et al.*, 2011). The legume-*Rhizobium* symbiosis is known to be the most efficient system for biological nitrogen fixation (BNF) through nodulation in legume roots (Molla *et al.*, 2001). It has been estimated that some 44–66 million tons of N_2 are fixed annually by leguminous plants, providing nearly half of all N used in agriculture worldwide (Alberton *et al.*, 2006).

The response of legumes to environmental stresses such as salinity and drought varies and depends on soil type, salt toxicity and climatic factors (Cordovilla *et al.*, 1995; Serraj *et al.*, 2001; Predeepa & Ravindran, 2010; Jabborova *et al.*, 2013a). Numerous studies have shown that soil salinity inhibits legume growth and development and decreases nodulation and N_2 fixation (reviewed by Zahran, 1999; Mensah & Ihenyen, 2009; Egamberdieva *et al.*, 2013a). Grain legumes are known to be salt-sensitive crops, but differences in salt tolerance exist among genotypes (Dua, 1992; Gandour, 2002). Genotypic variation in leguminous crops for traits affecting nodulation and N_2 fixation has been found (Montealegre *et al.*, 1995). Bliss (1993) observed that the selection and breeding of common bean cultivars tolerant to salinity could improve nodulation and N_2 fixation. Thus, the selection of improved chickpea cultivars with salt-tolerant symbioses is an absolute necessity to enable its cultivation in salt-affected soils. Available reports indicate that legumes show improved

Legumes under Environmental Stress: Yield, Improvement and Adaptations, First Edition. Edited by Mohamed Mahgoub Azooz and Parvaiz Ahmad.

yield and nodulation when co-inoculated with plant-growth-promoting rhizobacteria (PGPR) under hostile environmental conditions compared to inoculation with rhizobia alone (Rokhzadi *et al.*, 2008; Yang *et al.*, 2009; Jabborova *et al.*, 2013b).

This chapter examines recent studies on the impact of salt and drought stresses on legumes and the genotypic variation among legumes for germination, seedling growth and other plant traits under hostile conditions, and the microbial technologies that can improve growth, development and symbiotic performance of legumes. In addition, some possible mechanisms of plant resistance to stress, growth stimulation and improved symbiotic performance by rhizobacteria are described.

9.2 Abiotic stresses affecting legume crop productivity

9.2.1 Plant growth and stress

Previous studies have shown that soil salinity and drought decrease rhizobial colonization, inhibit infection processes and nodule development, and reduce N_2 fixation and nitrogenase activity in legumes (Zahran & Sprent, 1986; Zahran, 1999; Kulkarni *et al.*, 2000; Serraj, 2002; Egamberdieva *et al.*, 2013b). In earlier reports the inhibition by salinity of growth, nodulation and N fixation was observed in chickpea (Singh *et al.*, 2001), common bean (Ferri *et al.*, 2000) and lentil (Golezani & Yengabad, 2012). In subsequent studies, saline soil conditions inhibited germination and seedling growth, nodulation and biomass accumulation in soybean (Essa, 2002; Li *et al.*, 2006). Similar findings were observed for soybean by Hamayun *et al.* (2010), where the plant length, biomass, chlorophyll content, number of pods, 100-seed weight and yield were all significantly reduced by salinity stress. Leaf chlorosis, leaf bleaching and necrosis were also observed as effects of salt stress on soybean (Parker *et al.*, 1987). In lentil, plant growth and nodulation were significantly reduced over the whole growing season under saline soil conditions (Van Hoorn *et al.*, 2001). The decreased nodule formation resulted in reduction of leghaemoglobin content and N_2 fixation activity (Parida & Das, 2005). Limitation of oxygen diffusion in the nodules could be the reason for inhibition of nitrogenase activity and respiration of the nodules (Serraj *et al.*, 1995). Further, salt stress affects protein synthesis, lipid metabolism and photosynthesis, and reduces the growth of roots and root hairs, thereby decreasing sites for potential rhizobial infection and further nodule development (Katerji *et al.*, 2001; Bouhmouch *et al.*, 2005). The decrease in root growth was related to endogenous levels of phytohormones such as gibberellins, abscisic acid, jasmonic acid and salicylic acid, which declined under NaCl-induced salt stress (Debez *et al.*, 2001).

Several reports have indicated that germination and seedling growth of chickpea are reduced in saline soils, with responses varying according to cultivars (Gandour, 2002; Al-Mutawa, 2003). Krouma (2009) evaluated the growth, nodulation, nitrogen fixation and ionic repartition in two chickpea varieties, and found that the salt-tolerant cultivar was able to protect its photosynthetic and symbiotic apparatus against the toxic $Na^+$and Cl^- ions.

We have also observed a significant effect of salinity (salt concentration) on germination, and on shoot and root length of chickpea genotypes (Table 9.1). The 29 genotypes differed significantly for germination and shoot and root length. There was a significant genotype × salinity interaction on germination and shoot and root length. According to Almansouri *et al.* (2001), seed germination is usually the most critical stage in seedling establishment. Acccording to Sadiki and Rabih (2001), chickpea is a salt-sensitive species, and conditions of 25 mM NaCl resulted in a 71% reduction in growth. We have also observed that most chickpea genotypes were salt sensitive, with germination capacity decreasing with increasing salinity. The present result agrees with the work of Gandour (2002) and Vadez *et al.* (2007), who observed decreases in percentage germination and seedling emergence of chickpea with increases in salinity. Atak *et al.* (2006) and Neamatollahi *et al.* (2009) pointed out that higher salinity may reduce germination due to higher osmotic pressures. The seeds of six chickpea genotypes, namely Jahongir, Uzbekiston-32, Lazzat, Zimistoni, Flip 1-22 and Flip 1-31, showed better germination (40–45%) than other chickpea genotypes. According to Tejovathi *et al.* (1988) the ability of a seed to germinate under salt stress indicates that it has genetic potential for salt tolerance. The 29 genotypes of chickpea differed in their response to different salinity levels (Table 9.2).

The reduction in seed germination rate (at 10 days after sowing), as compared to the respective controls, was less than 25% for Sino, Flip 1-01, Flip 1-04, Flip 1-05, Flip 1-19, Flip 03-27c and Flip 06-155c. Seeds of

Table 9.1 Germination[a] of various soybean genotypes 2, 6 and 10 days after sowing in Petri dishes at different concentrations of NaCl.

Time / Salinity (dS/m)	2 days			6 days			10 days		
	0	5.0	10.0	0	5.0	10.0	0	5.0	10.0
Variety									
Jahongir	5.3	4.8	2.5	12.0	11.5	4.0	14.5	12.5	8.5
Uzbekistan-32	11.0	6.0	2.8	15.3	13.5	4.3	16.5	14.3	8.0
Lazzat	5.8	5.0	4.5	10.3	6.0	6.3	15.3	10.0	8.3
Zimistoni	10.3	9.0	3.8	12.8	13.3	4.0	18.3	15.0	9.0
Xalima	4.3	6.0	0.1	9.5	9.3	2.5	16.8	15.0	6.5
Miroz	4.0	2.3	0.5	7.5	4.8	2.3	17.3	11.3	6.3
Muqtadir	4.5	5.5	3.3	7.0	7.3	4.5	9.0	8.3	5.3
Xisor-32	8.8	8.0	3.8	9.0	8.0	5.8	9.3	8.8	6.5
Sino	3.5	2.8	1.0	7.0	5.3	2.3	8.0	7.0	3.3
Flip 1-01	7.5	2.8	1.3	10.0	4.8	3.0	13.8	10.5	4.0
Flip 1-04	3.8	2.8	0.8	7.5	4.3	2.3	14.0	11.0	4.8
Flip 1-05	3.0	0.8	0.3	5.5	4.3	1.0	12.3	9.5	2.8
Flip 1-19	7.8	4.8	0.3	12.8	10.5	2.5	19.0	16.3	5.0
Flip 1-21	7.5	5.0	3.8	11.0	8.3	5.5	19.3	15.8	7.5
Flip 1-22	12.3	8.8	2.0	14.8	11.3	3.0	18.8	15.0	8.3
Flip 1-29	9.0	7.8	1.5	12.0	9.0	2.3	17.5	15.3	5.3
Flip 1-31	2.8	2.8	0.8	6.5	5.3	3.8	15.0	11.3	8.5
Flip 1-33	13.8	4.5	1.8	17.0	12.3	4.3	19.3	17.0	6.3
Flip 03-102c	6.5	6.0	3.5	8.3	7.8	4.3	10.3	9.3	5.5
CIEW-45	6.0	7.0	6.0	9.0	8.8	8.8	10.3	9.3	8.8
Flip 05-69 c	5.3	3.5	5.0	7.5	7.0	5.0	9.3	8.8	6.3
Flip 03-74c	5.0	5.0	4.5	7.5	6.5	6.3	9.0	8.0	6.8
Flip 06-102c	4.8	4.5	4.0	6.5	7.5	6.3	9.5	9.0	7.0
Flip 06-66	5.5	6.0	5.0	8.5	7.0	6.3	10.3	9.5	7.8
Flip 05-65	5.0	6.5	5.8	7.0	8.3	6.5	9.8	10.0	7.3
Flip 06-124c	6.3	7.0	6.0	8.8	7.8	6.0	9.8	9.0	7.0
Flip 06-80c	8.0	7.8	6.5	9.0	8.5	7.3	9.5	9.3	8.8
Flip 03-27c	7.5	6.5	1.3	8.0	7.5	2.8	9.8	8.8	3.0
Flip 06-155c	1.5	5.5	1.8	2.8	7.3	4.3	3.0	7.5	2.0
LSD (5%)	1.4	2.0	1.4	1.4	1.8	1.4	0.6	1.8	1.4
Mean	6.4	5.3	2.9	9.3	8.0	4.4	12.9	11.1	6.3
CV (%)	14.0	23.2	29.1	12.1	14.0	19.1	7.7	10.1	13.23

[a]Number of germinated seeds.

CV, coefficient of variation; LSD, least significant difference.

Table 9.2 Mean squares from analysis of variance for various plant traits of 29 genotypes of chickpea at three concentrations of NaCl.

Source	df	Germination	Shoot length	Root length
Replication	2	1.7	1.9	3.1
Concentration	2	1018.4**	1219.3**	841.9**
Genotype	28	66.3**	63.9**	15.4**
Genotype × concentration	56	16.6**	29.8**	19.3**
Error	181	1.0	1.3	1.0

**Significant at $P=0.01$.

Jahongir, Uzbekistan-32, Lazzat, Zimistoni, Flip 1-22, Flip 1-31, CIEW-45 and Flip 06-80c showed better germination (40–45%) at 10 dS/m salt concentration than other chickpea genotypes (Table 9.1). Seed germination was decreased slightly with increasing salinity levels for Xisor-32, CIEW-45, Flip 06-66, Flip 06-124c, Flip 06-80c and Flip 03-27c. In contrast, significant differences between NaCl treatments were observed for seed germination ($P<0.05$) at 5 and 10 dS/m.

In our work the increase in salt concentration reduced shoot and root length in all genotypes, and it seemed to reduce the availability of the nutrients required for the growth and development of the plants. Growth response of chickpea to NaCl treatment (0, 5 and 10 dS/m) varied with the genotype, as shown by shoot and root lengths. The influence was more pronounced at 10 dS/m salinity level (Table 9.3). Genotypes Flip 1-04, Flip 1-05, Flip 1-22, Flip 06-66, Flip 03-27c, Flip 06-155c and Sino were

Table 9.3 Effect of salinity on shoot and root length of chickpea genotypes (seedlings were grown in a gnotobiotic sand system for 3 weeks).

	Root length (cm)			Shoot length (cm)		
	Salinity (dS/m)					
Genotype	0	5.0	10.0	0	5.0	10.0
Jahongir	16.7	16.2	15.8	11.2	10.4	9.4
Uzbekistan-32	14.3	13.2	13.9	10.5	14.0	9.9
Lazzat	15.1	15.2	12.1	15.8	11.9	10.5
Zimistoni	18.4	14.9	9.3	22.6	13.3	7.6
Xalima	17.0	9.4	10.6	15.4	13.3	10.5
Miroz	18.6	18.5	13.6	14.3	12.5	8.8
Muqtadir	15.6	14.3	11.6	10.4	9.6	7.7
Xisor-32	16.2	13.4	10.5	17.6	12.8	7.8
Sino	19.0	14.9	0.0	19.1	13.3	0.0
Flip 1-01	18.4	15.3	12.4	16.7	12.2	8.8
Flip 1-04	16.4	14.0	11.7	11.7	11.8	10.5
Flip 1-05	19.6	15.9	6.3	11.3	13.3	9.0
Flip 1-19	16.2	16.7	11.6	12.9	13.0	10.0
Flip 1-21	19.8	17.1	10.9	15.5	13.8	10.2
Flip 1-22	7.7	6.1	5.7	12.4	10.6	8.3
Flip 1-29	17.9	16.4	10.0	15.5	13.8	10.2
Flip 1-31	21.2	19.7	14.4	13.9	12.1	10.0
Flip 1-33	13.0	11.6	10.9	14.5	11.5	8.5
Flip 03-102c	16.3	16.1	11.3	13.2	11.4	8.9
CIEW-45	19.6	16.9	10.8	13.5	11.3	8.8
Flip 05-69c	22.5	16.9	11.4	13.1	11.3	9.0
Flip 03-74c	14.2	13.2	9.9	14.5	12.2	10.2
Flip 06-102c	22.0	11.0	10.5	14.5	12.0	9.8
Flip 06-66	12.6	11.5	6.8	14.1	11.8	8.6
Flip 05-65	11.4	9.6	8.1	12.2	9.8	9.2
Flip 06-124c	18.1	15.8	10.4	11.9	11.6	9.3
Flip 06-80c	22.2	15.6	14.2	18.7	13.3	8.7
Flip 03-27c	16.1	14.3	0.0	16.7	13.2	0.0
Flip 06-155c	24.7	5.4	0.0	14.5	11.2	0.0
LSD (5%)	2.4	1.6	1.5	1.8	1.8	1.1
Mean	17.7	14.2	10.0	14.4	12.2	8.0
CV(%)	8.4	6.9	9.3	0	9.0	1.8

CV, coefficient of variation; LSD, least significant difference.

found to be salt-sensitive whereas genotypes Jahongir, Uzbekistan-32, Xalima, Miroz, Xisor 32 and Flip 06-102 were found to be salt-tolerant at 10 dS/m salinity level. There was a significant interaction between salt and shoot and root growth ($P<0.05$) (Table 9.2).

However, other genotypes remained intermediate in their salt tolerance with respect to seedling shoot and root length. This would suggest the possibility of exploiting genotypic variation in chickpea for tolerance to specific concentrations of salts. The present observations are in line with earlier reports in bean (Kaymakanova, 2009), ground nut (Mensah *et al.*, 2006) and chickpea (Al-Mutawa, 2003), where increased salinity also led to decreased radicle lengths.

Salinity also affects uptake by plants of nutrients such as phosphorus, nitrogen and potassium (Egamberdiyeva & Hoflich 2004). According to Van Hoorn *et al.* (2001) salinity strongly reduced soil nitrogen content, probably through inhibiting nitrogen fixation and soil biological activity responsible for transformation of organic nitrogen. Salehi *et al.* (2008) reported that salt stress reduced plant dry weight, nitrogen content and the number of active nodules in alfalfa cultivars.

9.2.2 Rhizobia-legume symbiosis and stress

It has been reported that the colonization and infection of root hairs by rhizobial cells is sensitive to environmental stresses (Zahran, 1999; Räsänen *et al.*, 2003). Several environmental stresses, including salinity, drought, extreme temperature and nutrient deficiency, are known to decrease survival and proliferation of rhizobia in the rhizosphere, and inhibit the infection process leading to symbiotic association with their plant host (Biswas *et al.*, 2008; Ali *et al.*, 2009). Salt inhibits the absorption of Ca, which reduces the growth of roots, root tips and root hairs, thereby decreasing sites for potential rhizobial infection and further nodule development (Bouhmouch *et al.*, 2005). The number of rhizobial cells was found to be reduced in the root of soybean, common bean and chickpea grown under salt stress conditions (Zahran & Sprent, 1986; Bouhmouch *et al.*, 2005).

We have also observed that increased salt content decreased the ability of *Rhizobium galegae* sv. *officinalis* cells to colonize goat's rue roots (Egamberdieva *et al.*, 2013b). Similar observations were made about the decrease of root colonization of liquorice (*Glycyrrhiza uralensis*) by *Mesorhizobium* sp. under salt stress, where a salt concentration of 100 mM NaCl totally inhibited bacterial survival (Figure 9.1).

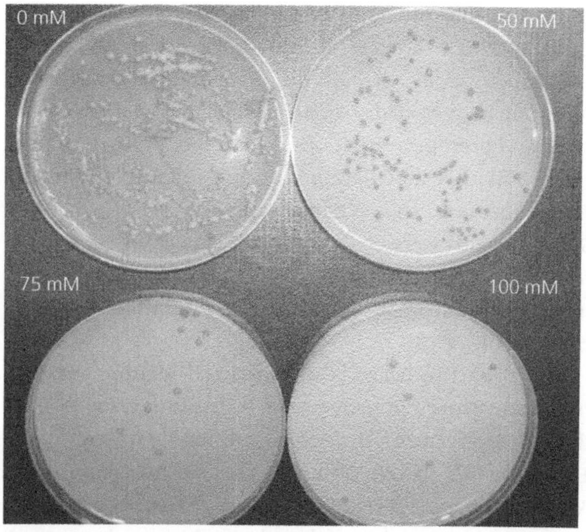

Figure 9.1 The effect of NaCl concentration on the colonization of *Mezorhizobium* sp. in the rhizosphere of *Glycyrrhiza uralensis*.

However, the species vary in tolerance to environmental stresses (Sridhar *et al.*, 2005; Wei *et al.*, 2008; Biswas *et al.*, 2008). The rhizobial strains from various grain legumes may tolerate 100–300 mM concentrations of NaCl (Predeepa & Ravindran, 2010). Leguminous plants growing in saline environments require both the rhizobia and the host to be tolerant to salt. The salt tolerance of rhizobia is important for improved symbiotic performance of legumes under stress conditions, where they may enhance the nodulation and nitrogen fixation ability of plants (Shamseldin & Werner, 2005; Ali *et al.*, 2009). Salt-tolerant strains of rhizobia improved the salt tolerance of host plants (Zou *et al.*, 1995; Hashem *et al.*, 1998; Shamseldin & Werner, 2005). Zahran (1999) reported that rhizobia use distinct mechanisms for osmotic adaptation to salt stress. Rüberg *et al.* (2003) reported the accumulation of low-molecular-weight organic solutes (osmolytes) by rhizobia to equilibrate internal and external osmotic concentrations under salt stress.

9.3 Improving legume yield by inoculation with rhizobacteria

The utilization of root-associated bacteria that interact with plants to mitigate the effects of various stresses opens a novel, inexpensive and advanced technology for combating the problems of salinity (Egamberdiyeva & Hoflich, 2003; Berg *et al.*, 2010; Pliego *et al.*, 2011; Lugtenberg *et al.*, 2013). In our previous work, we reported increased dry matter yield, nodulation, seed yield and protein content of soybean seeds grown under saline soil after inoculation with salt-tolerant strains of *Bradyrhizobium* spp. (Egamberdiyeva *et al.*, 2004). Similar results were observed for chickpea, where shoot, root length and dry matter increased after inoculation of PGPR strain under stressful soil conditions (Mishra *et al.*, 2010).

The tripartite bacterial-mycorrhizal-legume symbiosis also showed improvement in legume growth and development under stressful conditions. For example, Estévez *et al.* (2009) observed that co-inoculation of *Rhizobium tropici* CIAT899 with *Chryseobacterium balustinum* Aur9 improved growth and symbiotic performance of salt-stressed soybeans compared with the single inoculation (CIAT899). Molla *et al.* (2001) also showed that total root length, root number, dry matter, root hair development, number of nodules and shoot dry matter of soybean were significantly increased by *Azospirillum lipoferum* and *Bradyrhizobium japonicum*. Dardanelli *et al.* (2008) also observed improved rhizobia-legume symbioses under saline conditions by co-inoculation of common bean with *Rhizobium* and *Azospirillum* strains.

Dual inoculation of *Rhizobium* with *Azotobacter* showed increased nodulation and plant growth of faba bean under drought conditions (Dashadi *et al.*, 2011). Han and Lee (2005) also observed alleviation of induced stress in plants and improvement in root and shoot growth of soybean by dual inoculation with *Bacillus subtilis* and *Bradyrhizobium japonicum*.

There are also several reports on the improvement of nodulation and plant growth of legumes by phosphate-solubilizing bacteria. For example, Rosas *et al.* (2006) reported on the positive effect of phosphate-solubilizing bacteria *Bradyrhizobium japonicum* and *Pseudomonas putida* on the root and shoot growth of soybean. Similar results were observed by Kumar and Chandra (2008), who showed that phosphate-solubilizing bacteria improved the symbiotic performance of introduced rhizobia in field-grown lentils. In another study, *Pseudomonas* spp. inoculated together with rhizobia significantly increased the number of pods per plant, number of seeds per pod per plant and seed yield per hectare of soybean (Argaw, 2012). Phosphate-solubilizing bacteria (PSB) are known to help plants to acquire more phosphorus from soil, thus stimulating P uptake by plants and also improving nodulation and nitrogen fixation (Elkoca *et al.*, 2008). El-Azouni (2008) observed significant increases of dry matter, N and P uptake and yield of soybean grown in Egyptian soil inoculated with the phosphate-solubilizing fungi *Aspergillus niger* and *Penicillium italicum*. In another study, the sulphur-oxidizing bacteria *Thiobacillus* sp. stimulated nodule number, plant biomass and yield of groundnut, and increased soil available S (Anandham *et al.*, 2007).

Rabie and Almadini (2005) observed that the dual inoculation of faba bean (*Vicia faba*) with nitrogen-fixing bacteria (NFB) *Azospirillum brasilense* and the arbuscular mycorrhizal (AM) fungus *Glomus clarum* increased salt tolerance of plants and improved growth and development under saline conditions. In their study significant effects of inoculation were reported in the plants for salinity tolerance, mycorrhizal dependency, phosphorus level, phosphatase enzymes, nodule number, nitrogen uptake, protein content and nitrogenase enzymes.

A greenhouse experiment demonstrated that the salt tolerance of goat's rue was clearly improved when the plant was inoculated with its own specific symbiont *Rhizobium galegae* sv. *officinalis* and *Pseudomonas extremorientalis* TSAU20 (Egamberdieva *et al.*, 2013). Dual inoculation with *Rhizobium* and plant growth-promoting rhizobacteria (PGPR) *Pseudomonas* could also enhance formation of nodules on soybean grown in salinized potting soil. In recent studies, Aamir *et al.* (2013) demonstrated that combined inoculation of mung bean enhanced nodulation, grain yield (up to 30%) and protein content (up to 48%) compared to an individually inoculated control.

The survival of introduced PGPR strains in the roots of legumes is important, as bacteria may deliver biologically active compounds directly to the rhizosphere of plants. We have investigated the effect of salinity on the colonization of two selected plant-growth-promoting bacteria *P. extremorientalis* TSAU20 and *P. chlororaphis* TSAU13 in the rhizosphere of common bean, using a gnotobiotic sand system, and determined their survival on the plant roots in pot experiments under salinated soil conditions (Egamberdieva, 2011). Results showed that bacterial strains were able to colonize the rhizosphere of bean at the highest salinity (12.5 dS/m) and were also able to survive in the root of bean grown in saline soil.

9.4 Biomechanisms regulating growth and development

Although the mechanisms playing in the PGPR stimulation of plant growth are not yet well understood, it is likely that PGPR can promote plant growth through several different mechanisms, which might function in tandem or separately, and depend on the species of plant and bacteria, the environment, and the nature and degree of stress (Ashraf *et al.*, 2004; Adesemoye & Egamberdieva, 2013; Berg *et al.*, 2013). Mechanisms by which bacteria are able to promote plant growth and prevent physiological plant disorders caused by salinity include production of phytohormones like indole acetic acid (IAA), gibberellic acid and cytokinins (Mishra *et al.*, 2010), solubilization of phosphates and micronutrients (Medeot *et al.*, 2010), production of 1-aminocyclopropane-1-carboxylate (ACC) deaminase to reduce the level of stress ethylene in the roots (Dey *et al.*, 2004) and symbiotic nitrogen fixation (Ardakani *et al.*, 2009). It has also been reported that PGPR strains can produce exopolysaccharides (EPSs), which may bind Na and decrease the content of Na available for plant uptake. Dardanelli *et al.* (2008) suggested that inoculation with *Azospirillum* spp. combined with rhizobia in common bean induced the synthesis of flavonoids by roots.

9.4.1 Phytohormone production

Phytohormones play an important role in plant physiology, and regulate many aspects of plant development, including the differentiation of vascular tissues, elongation growth, apical dominance, lateral root initiation, and stress responses and adaptation (Sharma *et al.*, 2005; Egamberdieva, 2009; Javid *et al.*, 2011). The exogenous application of auxins to alfalfa (Gruodien & Zvironaite, 1971), groundnut (Srinivasan & Gopal, 1977) and mung bean (Hayat *et al.*, 2008) promoted plant growth and nodulation. Abiotic stresses inhibit phytohormone synthesis in plants, resulting in decreased plant growth and development. According to Figueiredo *et al.* (2008) drought stress causes a change in the balance of plant hormones, like cytokinin, zeatin, IAA and gibberellins in common bean. Phytohormones produced by root-associated bacteria will be taken up by plant cells, and can stimulate plant cell proliferation; this mechanism might be responsible for the enlarged root system and increased number of infection sites prior to nodulation, especially under stressed conditions (Tanimoto, 2005; Tilak *et al.*, 2006; Jabborova *et al.*, 2013a). Root-associated bacteria synthesize and release phytohormones such as auxins and gibberellins as secondary metabolites because of the rich supplies of substrates exuded from the roots (Egamberdiyeva & Hoflich, 2002; Tsavkelova *et al.*, 2007; Shahab *et al.*, 2009). Nutman (1977) reported that the IAA-producing abilities of *Rhizobium* strains have a positive influence on plant growth and also play an important role in the formation and development of root nodules. In another study Sridevi and Mallaiah (2007) reported IAA production by 26 *Rhizobium* strains isolated from the root nodules of the green manure crop, *Sesbania sesban*.

The comparative effects of auxin (IAA) and IAA-producing bacteria on the amelioration of salt stress on seedling growth of soybean under saline conditions were studied by Jabborova *et al.* (2013b). They observed that seedling growth of soybean was inhibited at 100 mM NaCl concentration and salt stress reduced the length of root by up to 56%. The plant growth regulator IAA did reverse the growth-inhibiting effect of salt stress to a certain extent in both shoot and root, whereas the IAA-producing bacterial strain *Pseudomonas putida* TSAU1 significantly increased seedling root growth by up to 29% in non-salinated conditions and by up to 86% at 100 mM NaCl compared to control plants. This study suggests that the application of low concentrations of auxin (IAA) or IAA-producing bacteria may improve the growth of soybean seedlings, which could enhance the tolerance of plants to soil salinity.

9.4.2 ACC (1-aminocyclopropane-1-carboxylate) deaminase

Ethylene is one of the endogenous hormones that play important roles in plant growth and development. Its production by plants is also considered as one of the stress responses, and is closely associated with various stress factors such as salinity, drought, metal toxicity and nutrient deficiency (Lynch & Brown, 1997; Schmidt, 2001; Li *et al.*, 2009). It has been reported that in leguminous plants, extra ethylene production inhibits nodulation by rhizobia (Hirsch & Fang, 1994). The level of ethylene in stressed plants can be reduced by ACC deaminase enzyme, which can cleave the ethylene precursor ACC to α-ketobutyrate and ammonium (Hontzeas *et al.*, 2005; Glick *et al.*, 2007). It has been reported that many root-colonizing PGPR have the ability to produce ACC deaminase and may enhance the survival of seedlings under stressed conditions (Glick *et al.*, 1998). We have also observed in our previous work that *Pseudomonas trivialis* 3Re27 was able to utilize ACC as an N source, indicating the presence of ACC deaminase, and thereby increase the salt tolerance of goat's rue under salinated soil conditions (Egamberdieva *et al.*, 2013b). Similar results were observed by Shaharoona *et al.* (2006), who reported that co-inoculation of *Bradyrhizobium* with PGPR isolates possessing ACC deaminase activity enhanced nodulation in mung bean compared with inoculation with *Bradyrhizobium* alone. In another study Ma *et al.* (2003) reported that ACC deaminase-producing *R. leguminosarum* could lower ethylene production in pea roots and improve nodulation. Similar results were observed for lentil (Shahzad *et al.*, 2010; Zahir *et al.*, 2011) and chickpea (Aslam *et al.*, 2010). Arshad *et al.* (2008) studied the effect of ACC deaminase-producing PGPR strains on plant growth of pea under water stress conditions, and found that plant inoculation with PGPR decreased the effects of water stress and improved plant growth and yield. PGPR strains that produce ACC deaminase showed positive effects on plant growth of chickpea (Roopa *et al.*, 2012) and lentil (Zafar-ul-Hye *et al.*, 2013), resulting in increased number of nodules, root and shoot growth, and yield of plant under stressful conditions.

9.4.3 Cell wall degrading enzymes

The complete erosion of the plant cell wall through which the bacterial symbiont penetrates to establish an intracellular endosymbiotic relationship with the host is a key event of the infection process (Robledo *et al.*, 2008). Cellulase enzyme may help rhizobia penetrate more easily into intercellular spaces of root cells, which may result in development of more nodules. Sindhu and Dadarwal (2001) explained the improvement of nodulation and symbiotic performance of rhizobia in legumes through production of hydrolytic enzymes such as cellulases by root-colonizing *Pseudomonas* strains. In their study, co-inoculations of five cellulase-producing *Pseudomonas* strains with *Mesorhizobium* increased the number of nodules and especially nodule biomass in chickpea. Robledo *et al.* (2008) showed that both the cell-bound cellulase enzyme from *R. leguminosarum* bv. *trifolii* and the purified enzyme could erode cell walls at the tip of the root hair of the host, white clover, making a localized hole of sufficient size to allow rhizobial cell penetration. Co-inoculation of the cellulase-producing strain *P. trivialis* 3Re27 with *R. galegae* HAMBI 540 significantly increased nodulation and nitrogen content of fodder galega, whereas cellulase-negative *P. extremorientalis* TSAU20 showed no significant stimulation (Egamberdieva *et al.*, 2010).

9.5 Conclusions and future prospects

The negative effect of abiotic factors such as salinity on plant growth, development and yield has been described by many studies. Agricultural biotechnology, particularly the use of rhizobia or consortia of plant-beneficial microbes, can be an effective approach for enhancing a plant's tolerance to adverse environmental stresses, increasing legume productivity and the supply of biologically fixed N at low cost under stressed conditions. Several mechanisms of action are used by PGPR to alleviate salt stress, improve symbiotic performance of legumes and stimulate plant growth; however, more detailed studies are needed on the induction of salt stress tolerance at the plant tissue, cell and molecular levels. Recent studies have demonstrated that tripartite bacterial-legume symbioses represent the best microbial strategy for arid and saline regions, and are of great interest as the subject of future research. Revelations about the mechanisms of PGPR action on improving legume symbioses open new doors for improving the efficacy of microbial strategies under harsh environments.

References

Aamir M, Aslam A, Khan MY, *et al.* (2013) Co-inoculation with *Rhizobium* and plant growth promoting rhizobacteria (PGPR) for inducing salinity tolerance in mung bean under field condition of semi-arid climate. *Asian J Agri Biol* 1: 17–22.

Adesemoye AO, Egamberdieva D (2013) Beneficial effects of plant growth promoting rhizobacteria on improved crop production: the prospects for developing economies. In: Maheshwari DK, Saraf M, Aeron A (eds), *Bacteria in Agrobiology: Crop Productivity*. Springer-Verlag, Berlin, Heidelberg, pp. 45–63.

Anandham R, Sridar R, Nalayini P, Poonguzhali S, Madhaiyan M, Tongminsa (2007) Potential for plant growth promotion in groundnut (*Arachis hypogaea* L.) cv. ALR-2 by co-inoculation of sulfur-oxidizing bacteria and *Rhizobium*. *Microb Res* 162: 139–153.

Alberton O, Kaschuk G, Hungria M (2006) Sampling effects on the assessment of genetic diversity of rhizobia associated with soybean and common bean. *Soil Biol Bioch* 38: 1298–1307.

Ali B, Hayat S, Fariduddin Q, Ahmad A (2009) Nickel: essentiality, toxicity and tolerance in plants. In: Ali B, Hayat S, Ahmad A (eds), *Nickel in Relation to Plants*. Narosa Publishing House, India, pp. 69–88.

Almansouri M, Kinet JM, Lutts S (2001) Effect of salt and osmotic stresses on germination in durum wheat (*Triticum durum* Desf.). *Plant Soil* 231: 243–254.

Al-Mutawa MM (2003) Effect of salinity on germination and seedling growth of chickpea (*Cicer arietinum* L.) genotypes. *Int J Agro Biol* 5: 227–229.

Ardakani MR, Pietsch G, Moghaddam A, Raza A, Friedel JK (2009) Response of root properties to tripartite symbiosis between lucerne (*Medicago sativa* L.), rhizobia and mycorrhiza under dry organic farming conditions. *Am J Agr Biol Sci* 4: 266–277.

Argaw A (2012) Evaluation of co-inoculation of *Bradyrhizobium japonicum* and phosphate solubilizing *Pseudomonas* spp. effect on soybean (*Glycine max* L. (Merr.)) in Assossa Area. *J Agr Sci Tech* 14: 213–224.

Arshad M, Shaharoona B, Mahmood T (2008) Inoculation with plant growth promoting rhizobacteria containing ACC-deaminase partially eliminates the effects of water stress on growth, yield and ripening of *Pisum sativum* L. *Pedosphere* 18: 611–620.

Ashraf M, Berge SH, Mahmood OT (2004) Inoculating wheat seedlings with exopolysaccharide producing bacteria restricts sodium uptake and stimulates plant growth under salt stress. *Biol Fert Soils* 40: 57–162.

Aslam M, Ahmad HK, Himayatullah, *et al.* (2010) Nodulation, grain yield and grain protein contents as affected by rhizobium inoculation and fertilizer placement in chickpea cultivar bittle. *Sarhad J Agr* 26: 467–474.

Atak M, Kaya MD, Kaya G, Çıkılı Y, Çiftçi CY (2006) Effects of NaCl on the germination, seedling growth and water uptake of triticale. *Turk J Agr For* 30: 39–47.

Berg G, Egamberdieva D, Lugtenberg B, Hagemann M (2010) Symbiotic plant–microbe interactions: Stress protection, plant growth promotion and biocontrol by *Stenotrophomonas*. In: Seckbach J, Grube M (eds), *Symbioses and Stress, Cellular Origin, Life in Extreme Habitats and Astrobiology*, Vol. 17(4). Springer-Verlag, pp. 445–460.

Berg G, Alavi M, Schmidt CS, *et al.* (2013) Biocontrol and osmoprotection for plants under saline conditions. In: de Bruijn FJ (ed.), *Molecular Microbial Ecology of the Rhizosphere*. Wiley-Blackwell.

Biswas S, Das RH, Sharma GL (2008) Isolation and characterization of a novel cross-infective rhizobial from *Sesbania aculeata* (Dhaincha). *Curr Microb* 56: 48–54.

Bliss FA (1993) Breeding common bean for improved biological nitrogen fixation. *Plant Soil* 152: 71–77.

Bouhmouch I, Souad-Mouhsine B, Brhada F, Aurag J (2005) Influence of host cultivars and *Rhizobium* species on the growth and symbiotic performance of *Phaseolus vulgaris* under salt stress. *J Plant Physiol* 162: 1103–1113.

Chauhan MP, Singh IS (2000) Variability estimates and identifying chick pea genotypes for yield and yield attributes in salt affected soil. *Legume Res* 23: 199–200.

Cordovilla MP, Ocana A, Ligero F, Lluch C (1995) Salinity effects on growth analysis and nutrient composition in four grain legumes-*Rhizobium* symbiosis. *J Plant Nutr* 18: 1595–1609.

Dardanelli M, Fernández de Córdoba F, Espuny M, *et al.* (2008) Effect of *Azospirillum brasilense* co-inoculated with *Rhizobium* on *Phaseolus vulgaris* flavonoids and Nod factor production under salt stress. *Soil Biol Bioch* 40: 2713–2721.

Dashadi M, Khosravi H, Moezzi A, Nadian H, Heidari M, Radjabi R (2011) Co-inoculation of *Rhizobium* and *Azotobacter* on growth of faba bean under water deficit conditions. *Amer-Eur J Agr Env Sci* 11: 314–319 .

Davranova N, Egamberdieva D, Ismatov Z, Wirth S (2013) Impact of crop management practice on soil microbial populations in a semi arid soil of Uzbekistan. *Soil Water J* 2: 921–927.

Debez A, Chaibi W, Bouzid S (2001) Effect du NaCl et de regulatoeurs de croissance sur la germination d' *Atriplex halimus* L. *Cahiers Agr* 10: 135–138.

Dey R, Pal KK, Bhatt DM, Chauhan SM (2004) Growth promotion and yield enhancement of peanut (*Arachis hypogaea* L.) by application of plant growth-promoting rhizobacteria. *Microb Res* 159: 371–394.

Dua RP (1992) Differential response of chickpea (*Cicer arietinum* L.) genotypes to salinity. *J Agric Sci* 119: 367–371.

Egamberdieva D (2009) Alleviation of salt stress by plant growth regulators and IAA producing bacteria in wheat. *Acta Phys Plant* 31: 861–864.

Egamberdieva D (2011) Survival of *Pseudomonas extremorientalis* TSAU20 and *P. chlororaphis* TSAU13 in the rhizosphere of common bean (*Phaseolus vulgaris*) under saline conditions. *Plant Soil Env* 57: 122–127.

Egamberdiyeva D, Hoflich G (2002) Root colonization and growth promotion of winter wheat and pea by *Cellulomonas* spp. at different temperatures. *J Plant Growth Reg* 38: 219–224.

Egamberdiyeva D, Hoflich G (2003) Influence of growth promoting bacteria on the growth of wheat at different soils and temperatures. *Soil Biol Bioch* 35: 973–978

Egamberdiyeva D, Hoflich G (2004) Importance of plant growth promoting bacteria on growth and nutrient uptake of cotton and pea in semi-arid region Uzbekistan. *J Arid Env* 56: 293–301.

Egamberdiyeva D, Qarshieva D., Davranov K (2004) The use of *Bradyrhizobium japonicum* to enhance growth and yield of soybean varieties in Uzbekistan conditions. *Plant Growth Reg* 23: 54–57.

Egamberdieva D, Gafurova L, Islam KR (2007) Salinity effects on irrigated soil chemical and biological properties in the Syr Darya basin of Uzbekistan. In: Lal R, Sulaimanov M, Stewart B, Hansen D, Doraiswamy P (eds), *Climate Change and Terrestrial C Sequestration in Central Asia*. Taylor-Francis, New York, pp. 147–162.

Egamberdieva D, Berg G, Lindstrom K, Rasanen L (2010) Root colonizing *Pseudomonas* spp. improve growth and symbiosis performance of fodder galega (*Galega orientalis* LAM) grown in potting soil. *Eur J Soil Biol* 46: 269–272.

Egamberdieva D, Jabborova D, Wirth S (2013a) Alleviation of salt stress in legumes by co-inoculation with *Pseudomonas* and *Rhizobium*. In: Arora NK (ed.), *Plant Microbe Symbiosis – Fundamentals and Advances*. Springer India, pp. 292–299.

Egamberdieva D, Berg G, Lindström K, Räsänen LA (2013b) Alleviation of salt stress of symbiotic *Galega officinalis* L. (goat's rue) by co-inoculation of rhizobium with root colonizing *Pseudomonas*. *Plant Soil* 369: 453–465.

El-Azouni I (2008) Effect of phosphate solubilizing fungi on growth and nutrition uptake of soybean (*Glycine max* L.). *Plants J Appl Sci Res* 4: 592–598.

Elkoca E, Kantar F, Sahin F (2008) Influence of nitrogen fixing and phosphorus solubilizing bacteria on the nodulation, plant growth, and yield of chickpea. *J Plant Nutr* 31: 157–171.

Essa TA (2002) Effect of salinity stress on growth and nutrient composition of three soybean (*Glycine max* (L.) Merrill) cultivars. *J Agron Crop Sci* 188: 86–93.

Estévez J, Dardanelli MS, Megías M, Rodriguez-Navarro DN (2009) Symbiotic performance of common bean and soybean co-inoculated with rhizobia and *Chryseobacterium balustinum* Aur9 under moderate saline conditions. *Symbiosis* 49: 29–36.

Ferri A, Lluch C, Ocana A (2000) Effect of salt stress on carbon metabolism and bacteroid respiration in root nodules of common bean (*Phaseolus vulgaris* L.). *Plant Biol* 2: 396–402.

Figueiredo MVB, Martinez CR, Burity HA, Chanway CP (2008) Plant growth promoting rhizobacteria for improving nodulation and nitrogen fixation in the common bean (*Phaseolus vulgaris* L.). *World J Micr Biotech* 24: 1187–1193.

Gandour G (2002) Effect of salinity on development and production of chickpea genotypes. PhD thesis, Aleppo University, Syria.

Glick BR, Penrose DM, Li JA (1998) Model for the lowering of plant ethylene concentrations by plant growth promoting bacteria. *J Theor Biol* 190: 63–68.

Glick BR, Cheng Z, Czarny J, Duan J (2007) Promotion of plant growth by ACC deaminase producing soil bacteria. *Eur J Plant Pathol* 119: 329–339.

Godfray HCJ, Beddington JR, Crute IR, *et al.* (2010) Food security: the challenge of feeding 9 billion people. *Science* 327: 812–818.

Golezani KG, Yengabad FM (2012) Physiological responses of lentil (*Lens culinaris* Medik) to salinity. *Int J Agr Crop Sci* 4: 1531–1535.

Gruodien J, Zvironaite V (1971) Effect of IAA on growth and synthesis of N compounds in Lucerne. *Luk TSR Aukstuja Mosklo Darbai Biologia* 17: 77–87.

Hamayun M, Khan SA, Shinwari ZK, Khan AK, Ahmad N, Lee IJ (2010) Effect of polyethylene glycol induced drought stress on physio-hormonal attributes of soybean. *Pak J Bot* 42: 977–986.

Han HS, Lee D (2005) Physiological responses of soybean – inoculation of *Bradyrhizobium japonicum* with PGPR in saline soil conditions. *Res J Agr Biol Sci* 1: 216–221.

Hashem FM, Swelim DM, Kuykendall LD, Mohamed AI, Abdel-Wahab SM, Hegazi NI (1998) Identification and characterization of salt- and thermo-tolerant *Leucaena*-nodulating *Rhizobium* strains. *Biol Fert Soils* 27: 335–341.

Hayat R, Ali S, Ijaz SS, Chatha TH, Siddique MT (2008) Estimation of N_2-fixation of mung bean and mash bean through xylem ureide technique under rainfed conditions. *Pak J Bot* 40: 723–734.

Hirsch AM, Fang Y (1994) Plant hormones and nodulation: what's the connection? *Plant Mol Biol* 26: 5–9.

Hontzeas N, Richardson AO, Belimov A, Safronova V, Abu-Omar MM, Glick BR (2005) Evidence for horizontal transfer of 1-aminocyclopropane-1-carboxylate deaminase genes. *Appl Env Microb* 71: 7556–7558.

Jabborova D, Qodirova D, Egamberdieva D (2013a) Improvement of seedling establishment of soybean using IAA and IAA producing bacteria under saline conditions. *Soil Water J* 2: 531–539.

Jabborova D, Egamberdieva D, Räsänen L, Liao H (2013b) Salt tolerant *Pseudomanas* strain improved growth, nodulation and nutrient uptake of soybean grown under hydroponic salt stress condition. In: *XVII International Plant Nutrition Colloquium and Boron Satellite Meeting Proceedings Book*. Sabanci University, Istanbul, pp. 260–261.

Javid MG, Sorooshzadeh A, Moradi F, Modares Sanavy AM, Allahdadi I (2011) The role of phytohormones in alleviating salt stress in crop plants. *Aust J Crop Sci* 5: 726–734.

Katerji N, Van Hoorn JW, Hamdy A, Mastrorelli M (2001) Salt tolerance of crops according to three classification methods and examination of some hypotheses about salt tolerance. *Agr Water Man* 47: 1–8.

Kaymakanova M (2009) Effect of salinity on germination and seed physiology in bean (*Phaseolus vulgaris* L.). *Biotech Equip* 23: 326–329.

Krouma A (2009) Physiological and nutritional responses of chickpea (*Cicer arietinum* L) to salinity. *J Agric For* 33: 503–512.

Kulkarni S, Sarangnath PA, Salakinkop SP, Goddi AV (2000) Response of chickpea to rock phosphate and phosphate solubilizers in typic chromustert. *Legume Res* 23: 21–24.

Kumar R, Chandra R (2008) Influence of PGPR and PSB on *Rhizobium leguminosarum* bv. *viciae* strain competition and symbiotic performance in lentil. *World J Agr Sci* 4 (Suppl 3): 297–301.

Li H, Wong WS, Zhu L, Guo HW, Ecker J, Li N (2009) Phosphoproteomic analysis of ethylene-regulated protein phosphorylation in etiolated seedlings of *Arabidopsis* mutant ein2 using two-dimensional separations coupled with a hybrid quadrupole time-of-flight mass spectrometer. *Proteomics* 9: 1646–1661.

Li X, An P, Inanaga S, Eneji AE, Tanabe K (2006) Salinity and defoliation effects on soybean growth. *J Plant Nutr* 29: 1499–1508.

Lugtenberg B, Malfanova N, Kamilova F, Berg G (2013) Plant growth promotion by microbes. In: de Bruijn FJ (ed.), *Molecular Microbial Ecology of the Rhizosphere*. Wiley-Blackwell, Hoboken, pp. 561–573.

Lüscher A, Soussana JF, Huguenin-Elie O (2011) Role and impacts of legumes in grasslands for high productivity and N gain from symbiotic N_2 fixation. In: Lemaire G, Hodgson J, Chabbi A (eds), *Grassland Productivity and Ecosystem Services*. CAB International, Wallingford, UK, pp. 101–107.

Lynch JP, Brown KM (1997) Ethylene and plant responses to nutritional stress. *Physiol Plant* 100: 613–619.

Ma JH, Yao JL, Cohen D, Morris B (1998) Ethylene inhibitors enhance in vitro root formation from apple shoot cultures. *Plant Cell Rep* 17: 211–214.

Mahmood IA, Nawaz S, Aslam M (2000) Screening of rice (*Oryza sativa* L.) genotypes against salinity. *Int J Agric Biol* 2: 147–150.

Medeot DB, Paulicci NS, Albornoz AI, *et al.* (2010) Plant growth promoting rhizobacteria improving the legume-rhizobia symbiosis. In: Khan MS, Zaidi A, Musarrat J (eds), *Microbes for Legume Improvement*. Springer-Verlag, New York, pp. 473–494.

Mensah JK, Ihenyen J (2009) Effects of salinity on germination, seedling establishment and yield of three genotypes of mung bean (*Vigna mungo* L. Hepper) in Edo State, Nigeria. *Nigerian Ann Nat Sci* 8: 17–24.

Mensah JK, Akomeah PA, Ikhajiagbe B, Ekpekurede, EO (2006) Effects of salinity on germination, growth and yield of five groundnut genotypes. *Afr J Biotech* 5: 1973–1979.

Mishra M, Kumar U, Mishra PK, Prakash V (2010) Efficiency of plant growth promoting rhizobacteria for the enhancement of *Cicer arietinum* L. growth and germination under salinity. *Adv Biol Res* 4: 92–96.

Molla AH, Shamsuddin ZH, Halimi MS, Morziah M, Puteh AB (2001) Potential for enhancement of root growth and nodulation of soybean coinoculated with *Azospirillum* and *Bradyrhizobium* in laboratory systems. *Soil Biol Bioch* 33: 457–463.

Montealegre C, Graham PH, Kipe-Nolt JA (1995) Preference in the nodulation of *Phaseolus vulgaris* cultivar RAB39. *Can J Micr* 41: 992–999.

Munns R (2002) Comparative physiology of salt and water stress. *Plant Cell Env* 20: 239–250.

Munns R, Tester M (2008) Mechanisms of salinity tolerance. *Annu Rev Plant Biol* 59: 651–681.

Neamatollahi E, Bannayan M, Souhani Darban A, Ghanbari A (2009) Does hydro and osmo-priming improve fennel (*Foeniculum vulgare*) seed germination and seedling growth? *Notulae Botanicae Horti Agrobotanici Cluj-Napoca* 37: 190–194.

Nutman PS (1977) Study frameworks for symbiotic nitrogen fixation. In: Newton WE, Postgate JR, Rodriguez-Barrueco C (eds), *Recent Developments in Nitrogen Fixation*. Academic Press, London, pp. 443–448.

Nyfeler D, Huguenin-Elie O, Suter M, Frossard E, Lüscher A (2011) Grass-legume mixtures can yield more nitrogen than legume pure stands due to mutual stimulation of nitrogen uptake from symbiotic and nonsymbiotic sources. *Agri Ecos Env* 140: 155–163.

Parida SK, Das AB (2005) Salt tolerance and salinity effects on plants. *Ecotox Environ Safe* 60: 324–349.

Pathma J, Sakthivel N (2012) Microbial diversity of vermicompost bacteria that exhibit useful agricultural traits and waste management potential. *SpringerPlus* 1: 26.

Parker MB, Gaines TP, Hook JE, Gascho GJ, Maw BW (1987) Chloride and water stress effects on soybean in pot culture. *J Plant Nutr* 10: 517–538.

Pliego C, Kamilova F, Lugtenberg B (2011) Plant growth-promoting bacteria: fundamentals and exploitation. In: Maheshwari DK (ed.), *Bacteria in Agrobiology: Crop Ecosystems*. Springer, pp. 295–343.

Predeepa RJ, Ravindran DA (2010) Nodule formation, distribution and symbiotic efficacy of *Vigna unguiculata* L. under different soil salinity regimes. *Emir J Food Agr* 22: 275–284.

Rabie GH, Almadini AM (2005) Role of bioinoculants in development of salt-tolerance of *Vicia faba* plants under salinity stress. *Afr J Biotech* 4: 210–222.

Räsänen LA, Saijets S, Jokinen K, Lindström K (2003) Evaluation of the roles of two compatible solutes, glycine betaine and trehalose for the *Acacia senegal-Sinorhizobium* symbiosis exposed to drought stress. *Plant Soil* 260: 237–251.

Robledo M, Jiménez-Zurdo JI, Velázquez E, *et al.* (2008) *Rhizobium* cellulase CelC2 is essential for primary symbiotic infection of legume host roots. *Proc Natl Acad Sci USA* 105: 7064–7069.

Rokhzadi A, Asgharzadeh A, Darvish F, Nour-Mohammadi G, Majidi E (2008) Influence of plant growth promoting rhizobacteria on dry matter accumulation of chickpea (*Cicer arietinum* L) under field conditions. *J Agr Env Sci* 3 (Suppl. 2): 253–257.

Roopa B, Maya C, Makari HK (2012) Effect of different PGPR strain along with rhizobium on nodulation and chick pea productivity. *Asian J Exp Biol Sci* 3: 424–426.

Rosas BS, Javier AA, Marisa R, Correa NS (2006) Phosphate solubilizing *Pseudomonas putida* can influence the rhizobia legume symbiosis. *Soil Biol Biochem* 38: 3502–3505.

Rüberg S, Tian ZX, Krol E, *et al.* (2003) Construction and validation of a *Sinorhizobium meliloti* whole genome DNA microarray: genome wide profiling of osmoadaptative gene expression. *J Biotech* 106: 255–268.

Sadiki M, Rabih K (2001) Selection of chickpea (*Cicer arietimum*) for yield and symbiotic nitrogen fixation ability under salt stress. *Agronomie* 21: 659–666.

Salehi M, Salehi F, Poustini K, Heidari-Sharifabad H (2008) The effect of salinity on the nitrogen fixation in four cultivars of *Medicago sativa* L. in the seedling emergence stage. *Res J Agr Biol Sci* 4: 413–415.

Schmidt W (2001) From faith to fate: ethylene signaling in morphogenic responses to P and Fe deficiency. *J Plant Nutr Soil Sci* 164: 147–154.

Shahab S, Nuzhat A, Nasreen SK (2009) Indole acetic acid production and enhanced plant growth promotion by indigenous PSBs. *Afr J Agr Res* 4: 1312–1316.

Shahzad SM, Khalid A, Arshad M, Tahir J, Mahmood T (2010) Improving nodulation, growth and yield of *Cicer arietinum* L. through bacterial ACC-deaminase induced changes in root architecture. *Eur J Soil Biol* 46: 342–347.

Shanker AK, Venkateswarlu B (2011) *Abiotic Stress in Plants – Mechanisms and Adaptations*. In Tech Publisher, Janeza Tridne Rijeka, Croatia, 428 pp.

Sharma N, Abrams SR, Waterer DR (2005) Uptake, movement, activity, and persistence of an abscisic acid analog (80 acetylene ABA methyl ester) in marigold and tomato. *J Plant Growth Regul* 24: 28–35.

Serraj R (2002) Response of symbiotic nitrogen fixation to drought and salinity stresses. *Phys Mol Biol* 8: 77–86.

Serraj R, Fleurat-Lessard P, Benoit J, Drevon JJ (1995) Structural changes in the inner-cortex cells of soybean root-nodules are induced by short-term exposure to high salt or oxygen concentrations. *Plant Cell Environ* 18: 455–462.

Serraj RH, Vasquez-Diaz G, Hernandez, Drevon JJ (2001) Genotypic difference in response of nitrogenase activity (C_2H_2 reduction) to salinity and oxygen in common bean. *Agronomie* 21: 645–650.

Shaharoona B, Arshad M, Zahir ZA (2006) Effect of plant growth promoting rhizobacteria containing ACC-deaminase on maize (*Zea mays* L.) growth under axenic conditions and on nodulation in mung bean (*Vigna radiata* L.). *Lett Appl Microbiol* 42: 155–159.

Shamseldin A, Werner D (2005) High salt and high pH tolerance of new isolated *Rhizobium etli* strains from Egyptian soils. *Curr Microb* 50: 11–16.

Singh AK, Singh RA, Sharma SG (2001) Salt stress induced changes in certain organic metabolites during seedling growth of chickpea. *Legume Res* 24: 11–15.

Sindhu SS, Dadarwal KR (2001) Chitinolytic and cellulolytic *Pseudomonas* sp. antagonistic to fungal pathogens enhances nodulation by *Mesorhizobium* sp. *cicer* in chickpea. *Microb Res* 156: 353–358.

Soltani A, Khodarahmpour Z, Jafari AA, Nakhjavan S (2012) Selection of alfalfa (*Medicago sativa* L.) cultivars for salt stress tolerance using germination indices. *Afr J Biotech* 11: 7899–7905.

Sridevi M, Mallaiah KV (2007) Bioproduction of indole acetic acid by *Rhizobium* strains isolated from root nodules of green manure crop *Sesbania sesban* (L.) Merr. *Iran J Biotech* 5: 3.

Sridhar KR, Arun AB, Narula N, Deubel A, Merbech W (2005) Patterns of sole-carbon-utilization by fast growing coastal and dune rhizobia of the Southwest coast of India. *Eng Life Sci* 5: 425–430.

Son TTN, Diep CN, Giang TTM (2006) Effect of *Bradyrhizobia* and phosphate solubilizing bacteria application on soybean in rotational system in the Mekong delta. *Omonrice* 14: 48–57.

Srinivasan PS, Gopal KS (1977) Effect of plantofix and NAA formulation on groundnut var TMU-7. *Curr Sci* 46: 119–120.

Tanimoto E (2005) Regulation of root growth by plant hormones: Roles for auxin and gibberellin. *Crit Rev Plant Sci* 24: 249–265.

Tejovathi G, Khadeer MA, Anwer SY (1988) Studies on certain enzymes in salt tolerant and sensitive genotypes of sunflower. *Indian J Bot* 11: 113–117.

Tilak KVB, Ranganayaki N, Manoharachari C (2006) Synergistic effects of plant-growth promoting rhizobacteria and Rhizobium on nodulation and nitrogen fixation by pigeonpea (*Cajanus cajan*). *Europ J Soil Sci* 57: 67–71.

Tsavkelova EA, Cherdyntseva TA, Klimova S, Shestakov AI, Botina SG, Netrusov AI (2007) Orchid-associated bacteria produce indole-3-acetic acid, promote seed germination, and increase their microbial yield in response to exogenous auxin. *Arch Micr* 188: 655–664.

UNEP (2008) In dead water. Merging of climate change with pollution, over-harvest, and infestations in the world's fishing grounds. UNEP/GRID-Arendal, Arendal, Norway. Available at: http://www.grida.no/_res/site/file/publications/InDeadWater_LR.pdf [accessed 28 June 2014]

UNEP (2009) The environmental food crisis: The environment's role in averting future food crises. A UNEP rapid response assessment. Nellemann C, MacDevette M, Manders T, *et al.* (eds). United Nations Environment Programme/GRID-Arendal. Available at: http://www.grida.no

Upadhyay SK, Singh JS, Singh D P (2011) Exopolysaccharide-producing plant growth-promoting rhizobacteria under salinity condition. *Pedosphere* 2: 214–222.

Vadez V, Krishnamurthy L, Serraj R, *et al.* (2007) Large variation in salinity tolerance in chickpea is explained by differences in sensitivity at the reproductive stage. *Field Crops Res* 104: 123–129.

Van Hoorn JW, Katerji N, Hamdy A, Mastrorilli M (2001) Effect of salinity on yield and nitrogen uptake of four grain legumes and on biological nitrogen contribution from the soil. *Agr Water Man* 51: 87–98.

Wei GH, Yang XY, Zhang ZX, Yang YZ, Lindstrom K (2008) Strain *Mesorhizobium* sp. CCNWGX035; a stress tolerant isolate from *Glycyrrhiza glabra* displaying a wide host range of nodulation. *Pedosphere* 18: 102–112.

Yang J, Kloepper JW, Ryu C-M (2009) Rhizosphere bacteria help plants tolerate abiotic stress. *Trends Plant Sci* 14: 1–4.

Zafar-ul-Hye M, Ahmad M, Shahzad SM (2013) Synergistic effect of rhizobia and plant growth promoting rhizobacteria on the growth and nodulation of lentil seedlings under axenic conditions. *Soil Env* 32: 79–86.

Zahir ZA, Munir A, Asghar HN, Shaharoona B, Arshad M (2008) Effectiveness of rhizobacteria containing ACC deaminase for growth promotion of peas (*Pisum sativum*) under drought conditions. *J Micr Biot* 18: 958–963.

Zahir ZA, Zafar-ul-Hye M, Sajjad S, Naveed M (2011) Comparative effectiveness of *Pseudomonas* and *Serratia* sp. containing ACC-deaminase for coinoculation with *Rhizobium leguminosarum* to improve growth, nodulation, and yield of lentil. *Biol Fert Soils* 47: 457–465.

Zahran HH (1999) Rhizobium-legume symbiosis and nitrogen fixation under severe conditions and in an arid climate. *Microbiol Mol Biol Rev* 63: 968–989.

Zahran HH, Sprent JI (1986) Effects of sodium chloride and polyethylene glycol on root-hair infection and nodulation of *Vicia faba* L. plants by *Rhizobium leguminosarum*. *Planta* 167: 303–309.

Zou N, Dort PJ, Marcar NE (1995) Interaction of salinity and rhizobial strains on growth and N_2 fixation by *Acacia ampliceps*. *Soil Biol Bioch* 27: 409–413.

CHAPTER 10

Role of abscisic acid in legumes under abiotic stress

Analia Llanes, Genoveva Devinar and Virginia Luna

Laboratorio de Fisiología Vegetal, Departamento de Ciencias Naturales, Facultad de Ciencias Exactas, Físico-Químicas y Naturales, Universidad Nacional de Río Cuarto, Río Cuarto, Argentina

10.1 Introduction

Plants are sessile organisms and have to regulate their growth and development to allow them to respond to numerous external stimuli and constantly changing living conditions. These conditions include a complex set of biotic and abiotic environmental stresses, and plants must be able to response with compensatory mechanisms to limit or repair the impact of such stresses (Peleg & Blumwald, 2011). As the world's human population continues to expand (increasing from about 6 billion people in 2000 to a projected 10 billion in 2050), so has the need to grow more food, which requires an increase in the area of land under cultivation as well as in land productivity measured as yield per hectare. In this regard, the UN Food and Agriculture Organization (FAO) has indicated that the average world crop yield will need to reach 5 t/ha from its present 3 t/ha. Furthermore, less than 10% of the arable lands in the world may be free from adverse abiotic conditions (Dudal, 1976). Environmental concerns about the great amounts of chemical fertilizers needed and the increasing cost of fossil fuels as well as the emergence of biofuels competing with food production, may further complicate the situation.

In addition, changing climatic conditions, with increased temperatures and evapotranspiration losses, and decreasing rainfall, will have a particularly negative effect on the soils and productivity of many developing regions (Rijsberman, 2006; Lobell *et al.*, 2008). In fact, the global agricultural systems will be drastically affected by the anticipated increase of 3–5°C in average surface temperatures in the next 50–100 years; such climate predictions also show an increased frequency of drought, flood and heatwaves (Mittler & Blumwald, 2010). However, some of the negative impact of climatic changes on food production could be alleviated by adaptation of crop plants and by relatively inexpensive changes, such as shifting planting dates or switching from an existing crop variety.

One of the main types of crops worldwide are legumes. These plants have important impacts on farming, and also on animal and human nutrition and health (Graham & Vance, 2003). Legumes represent the third largest group of angiosperms and are the second largest group of food and feed crops grown globally. Around 12–15% of available arable land is cultivated for legumes, which account for more than 25% of the world's primary crop production; annual global production of grain legumes is 247 million tonnes (Jensen *et al.*, 2007). Among the more important leguminous food and feed crop species are soybean, pea, clover, chickpea, alfalfa and mung bean. Nitrogen fixation allows legumes to have tissue with a high content of protein; thus, legumes are used in agricultural practice to amend deficiencies of nitrogen in soil.

Adaptation of legumes will determine the future severity of the effects of climatic change. Furthermore, some relatively inexpensive agricultural practices, such as shifting planting dates or switching from an existing crop variety, may moderate the negative effects of climatic change. Several environmental factors such

Legumes under Environmental Stress: Yield, Improvement and Adaptations, First Edition. Edited by Mohamed Mahgoub Azooz and Parvaiz Ahmad.

as drought, salinity, mineral toxicity and deficiencies, and high or low temperatures affect plant growth, development and final yield performance of a crop. Salinity and drought exert a negative effect mainly by disrupting the ionic and osmotic homeostasis of the cell, while low temperatures exert mechanical constraints. It has been proposed that to know the specific responses of plants to multiple adverse environmental factors, it is crucial to impose the stresses simultaneously and treat each set of environmental conditions as an entirely new stress (Mittler, 2006).

In general, abiotic stress may elicit a wide range of cellular adaptive responses that plants have evolved. These responses include accumulation of compatible solutes, induction of stress proteins and acceleration of scavenging systems for reactive oxygen species (Zhu, 2002). One of the most important water tolerance responses is the leaf abscission induced by water stress (Hamdia, 2008; Chalk & Alves, 2010). Plants can 'sense' changes in environmental conditions by generating non-heritable modifications reflecting the morphological and physiological plasticity of the plant, or by generating heritable modifications. These responses are mediated by plant growth regulators (phytohormones), compounds derived from plant biosynthetic pathways that can act either at the site of synthesis or, following their transport, elsewhere in the plant (Javid *et al.*, 2011). These regulators are essential for adaptation to abiotic stresses by mediating physiological adaptive responses. The phytohormone abscisic acid (ABA) plays a central role in regulating physiological responses.

The rapid accumulation of ABA in response to stress and the regulation of a wide range of adaptive responses define this phytohormone as 'the stress hormone' helping the plant's survival under stresses. Early investigations demonstrated that ABA acts as a long-distance water stress signal in sensing impending soil drying (Davies & Zhang, 1991). ABA synthesis is a common and fast response of plants to abiotic stresses, triggering ABA-inducible gene expression (Yamaguchi-Shinozaki & Shinozaki, 2006) and causing stomatal closure, thereby reducing water loss via transpiration and eventually decreasing cellular growth (Wilkinson & Davies, 2010). In water-stressed plants, ABA provokes the rapid closure of stomata a few minutes after perception of stress, triggering a signalling cascade in guard cells that leads to reduction of water loss by transpiration (Schroeder *et al.*, 2001). In addition to these short-term effects, modulation of the transcript levels of hundreds of genes has been reported for *Arabidopsis thaliana* subjected to salt (Sottosanto *et al.*, 2004), drought and heat (Rizhsky *et al.*, 2004) or cold stress (Provart *et al.*, 2003; Vogel *et al.*, 2005). In drying soil, the roots of plants synthesize ABA and it is transported from dehydrated roots to the xylem and thence to regulate stomatal opening and leaf growth in the shoots (Zhang *et al.*, 1987; Zhang & Davies, 1990a, 1990b). pH and ionic conditions in the xylem modify this mechanism (Trejo & Davies, 1991; Bacon *et al.*, 1998; Wilkinson *et al.*, 1998). In fact, Hartung *et al.* (2002) have shown that pH changes play a central role in the ABA redistribution in leaf tissues and control the stomata when no significant changes in ABA levels are detected in the xylem. On the other hand, ABA is required to regulate the effects of nitrate on root branching, signalling through part of the ABA response pathway comprising the transcription factors ABA INSENSITIVE (ABI) of *Arabidopsis*, ABI4 and ABI5. It has been proposed that ABI1, ABI2 and ABI3 are not required for the lateral root response, while these ABA signalling genes play a central role in seed germination. This differential participation of ABI genes indicates a branched pathway (Signora *et al.*, 2001). Interestingly, although auxin has been thought to be the primary regulator of lateral root formation, the ABA-induced inhibition of lateral root development appears to be auxin dependent (De Smet, 2003). Thus, an ABA role in regulating root architecture is to mediate nutritional signals in an auxin-independent manner.

In legumes, it was shown that ABA was rapidly transported from roots to shoots and accumulated in the apoplast of guard cells of leaves responding to drought (Jia *et al.*, 1996). Additionally, ABA is produced in lupins (*Lupinus cosentinii* and *L. angustifolius*) subjected to soil water deficit, and accumulation of ABA in the leaf is correlated with a reduction in stomatal conductance (Henson, 1981; Gallardo *et al.*, 1994). Thus, numerous studies suggest a relevant role for this phytohormone in legumes, and that knowledge of ABA responses would help to improve crop productivity.

In this chapter we highlight the latest progress in determining the role of ABA in the responses of legumes to abiotic conditions. We then discuss recent advances in engineering of ABA hormone-associated genes aimed at ameliorating environmental stresses.

10.2 Effect of abiotic stress on ABA biosynthesis, catabolism and transport

In higher plants, the synthesis of ABA follows an 'indirect' pathway via cleavage of a C_{40} carotenoid precursor, followed by a two-step conversion of the intermediate xanthoxin to ABA via ABA aldehyde. The epoxidation of zeaxanthin and antheraxanthin to violaxanthin occurs in the plastids. This is the first key step in ABA biosynthesis pathway and it is catalysed by zeaxanthin epoxidase (ZEP). Molecular identification of this enzyme in tobacco was reported by Marin *et al.* (1996). The violaxanthin is converted to 9-*cis*-epoxycarotenoid through structural modifications. Epoxycarotenoid 9-*cis*-neoxanthin undergoes an oxidative cleavage by 9-*cis*-epoxycarotenoid dioxygenase (NCED) resulting in a C_{15} intermediate called xanthoxin (Schwartz *et al.*, 1997). This reaction, which is the first relevant step in the ABA biosynthetic pathway, is followed by the export of xanthoxin to the cytosol. Here, xanthoxin is converted to ABA-aldehyde and finally to free ABA through two reactions. The *AtABA2* gene encodes a short-chain alcohol dehydrogenase/reductase (SDR) that catalyses the first reaction step in ABA aldehyde generation. The enzyme ABA aldehyde oxidase (AAO) catalyses the final step in ABA biosynthesis (Rook *et al.*, 2001; Gonzalez-Guzman *et al.*, 2002, Cheng *et al.*, 2002).

The biosynthesis of ABA can be regulated directly or indirectly by certain environmental signals. However, the environmental signals that dramatically activate ABA biosynthesis are drought and salt stress. Under these abiotic stresses there is an increase in ABA levels resulting mainly from enhanced *de novo* biosynthesis. Both drought and salt stress induce ABA biosynthesis by controlling the transcriptional regulation of genes encoding enzymes of the ABA biosynthetic pathway (Nambara & Marion-Poll, 2005). Therefore, it is essential to understand how ABA biosynthesis is regulated by transcriptional regulation of ABA biosynthetic genes and by the specific activities of ABA biosynthesis enzymes.

The catabolism of ABA is divided into two pathways, hydroxylation and glucose conjugation. There are three different methyl groups on the carbon skeleton of ABA (C-7, C-8 and C-9), but the C-8 position is the predominant position for the hydroxylation reaction. These three forms of ABA hydroxylate still have substantial biological activity (Zhou *et al.*, 2004), but the hydroxylation reaction triggers further steps of inactivation (Zeevaart & Creelman, 1988; Cutler & Krochko, 1999). ABA conjugation and deconjugation are clearly dynamic processes that play a key role in controlling the amount of biologically active ABA under both non-stressful and stressful conditions (Verslues & Zhu 2007). In the conjugation pathway, ABA is conjugated with glucose by ABA-glucosyltransferase (ABAGTase) (Figure 10.1). It has been reported that other conjugates with the hydroxyl groups of ABA and its hydroxylated catabolites could be physiologically inactive and accumulate in vacuoles (Bray & Zeevaart, 1985; Lehmann & Glund, 1986). However, recently a role for the ABA conjugate, abscisic acid glucose ester (ABA-GE), was proposed: it could be involved in long-distance ABA transport (Hartung *et al.*, 2002; Wilkinson & Davies, 2002). ABA-GE had previously been thought to be a physiologically inactive by-product stored in the vacuoles (Zeevaart & Creelman, 1983). ABA-GE cannot migrate passively through plasma membranes, and the mechanisms underlying the transport of conjugates of ABA remain incompletely understood, as do the functions of conjugated forms of ABA (Schachtman & Goodger, 2008). However, ABA-GE was reported as an allelopathic substance of *Citrus junos* (Kato *et al.*, 2003), and soil in agricultural fields contains higher ABA-GE content (up to 30 nM) than ABA. It has been hypothesized that ABA-GE is taken up by the root (Saito *et al.*, 2004). The ABA-GE glucosidase activity is increased by salt stress and is inhibited competitively by ABA-GE or zeatin riboside. Nevertheless, recently Llanes *et al.* (2014) proposed that ABA and ABA-GE from roots and leaves work together to create and intensify a specific stress signal. This mechanism had not been previously reported for legumes. ABA-GE was the major metabolite found in the roots of the legume *P. strombulifera* under salt stress. Both the roots and leaves of Na_2SO_4-treated plants at −1.9 and −2.6 MPa showed high amounts of ABA-GE and free ABA, suggesting the presence of a system for transport of ABA-GE from roots to leaves. These plants showed a high glucosidase activity in the roots and mainly in the leaves, suggesting that ABA-GE could be a reservoir of free ABA in roots as well as an ABA-transport form from roots to leaves in this species. In NaCl-treated *P. strombulifera* ABA-GE peaked in roots at 12 and 24 h, whereas in leaves there were no differences from the controls. Thus, roots could act as a reservoir of free ABA (Llanes *et al.*, 2014). However, few

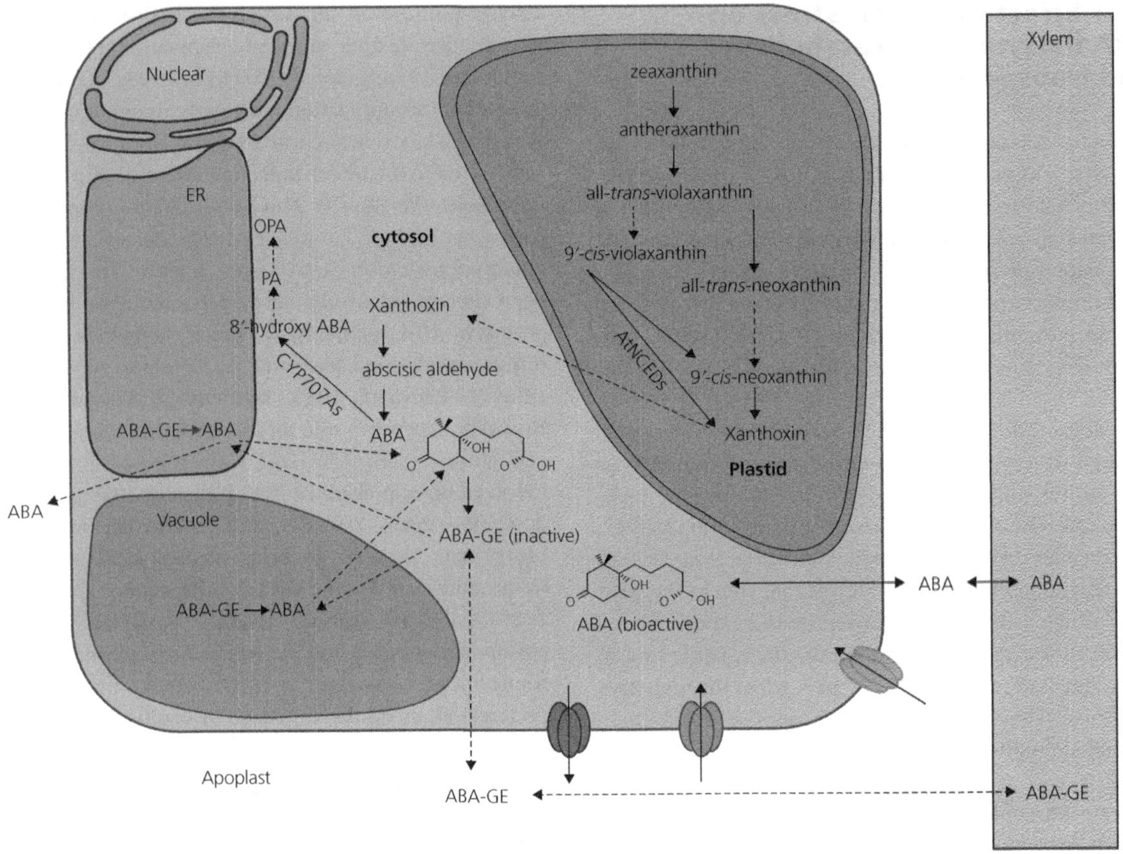

Figure 10.1 Abscisic acid (ABA) biosynthesis and catabolism pathways. The steps from zeaxanthin to xanthoxin in *de novo* ABA synthesis occurring in plastids are shown. Xanthoxin moves from the plastids to the cytoplasm and is converted to ABA. In the catabolic pathways, ABA is inactivated through either oxidation or conjugation. Hydroxylation as well as hydrolysis of ABA-GE is shown with the corresponding enzymes. Postulated pathways are shown by broken lines and the confirmed pathways are shown as solid lines. ER, endoplasmic reticulum; PA, phaseic acid; DPA, dihydrophaseic acid; ABA-GE, ABA glucosyl ester. Adapted from Xu *et al.* (2013).

investigations have studied the ABA-GE role in legumes, and the molecular mechanism underlying the transport of ABA or its conjugates remains unclear.

Recently, some investigations have intensively studied the specific enzymes of the ABA 8′-hydroxylation reaction in plants. These enzymes are members of the cytochrome P450 (CYP) superfamily, named CYP707A. Kushito *et al.* (2004) isolated four genes of the CYP707A superfamily that encode ABA 8′-hydroxylase in *Arabidopsis*. Further, Vallabhaneni and Wurtzel (2010) identified three genes of this superfamily from bean. These investigations established that the genes of CYP707A superfamily are mainly regulated at the transcriptional level and control ABA levels under adverse conditions. Zheng *et al.* (2012) identified and analysed the function of the CYP707A gene family isolated from soybean under dehydration and salt stresses. The construction of a cyp707a2T-DNA knock-down mutant showed overexpressed soybean GmCYP707A1 and was a single mutant that reported a significant phenotype related to seed germination under ABA treatment (Kushiro *et al.*, 2004). A series of experiments analysing the germination rate indicated that GmCYP707A1a possibly encodes an ABA 8′-hydroxylase due to the complementation of overexpression of GmCYP707A1a to AtCYP707A2. Alteration of ABA sensitivity by the loss of function of AtCYP707A2 was reflected by primary root growth, formation of lateral roots and germination of seeds. Expression of different members of the GmCYP707A family, *A1a, A1b, A2b, A2c, A3a, A3b, A4a, A4b* and *A5*, both in roots and in leaves, was induced by drought stress, whereas transcription of *A2a, A2b, A3b, A4a* and *A5* in roots, and only *A3b* and *A5* in leaves, was stimulated by rehydration. In addition, short- or long-term salt stress induced expression of all soybean CYP707A genes.

ABA catabolism might be affected by expression of the soybean root-specific protein kinase, GmWNK1, interacting with GmCYP707A1. Thus, the second effect of the GmWNK1 on ABA could be related to metabolism (Wang *et al.*, 2010). Four homologues of CYP707A (CYP707A 1–4) with major functions in ABA catabolism were isolated from *Arabidopsis* (Kushiro *et al.*, 2004; Nambara & Marion-Poll, 2005). Sensitivity to ABA or related stresses was increased with mutations of these genes (Okamoto *et al.*, 2006; Umezawa *et al.*, 2006). The mechanism of how the root-specific protein kinase acts on ABA or affects ABA dependence under abiotic stress remains unclear. In the legume *Prosopis strombulifera* low ABA hydroxylate levels were detected under different salinity conditions (Llanes *et al.*, 2014). Control plants had a low dihydrophaseic acid (DPA) content and usually no detectable phaseic acid (PA), indicating that PA was rapidly metabolized. Rapid metabolism of ABA is essential to prevent accumulation of PA, and to regulate precisely stomatal conductance (Nambara & Marion-Poll 2005). This is consistent with a high content of DPA in the leaves of *P. strombulifera* plants treated with Na_2SO_4 (−2.6 MPa), which also showed active production of free ABA and ABA-GE. The biological activity of PA and DPA appears to be very low. Previous studies showed that different ABA-binding proteins in aleurone cells of barley are unable to bind to PA, suggesting that this metabolite is inactive in at least some physiological processes (Zhang *et al.*, 2001).

10.3 Perception of ABA in legumes under abiotic stress

The cell needs to perceive ABA and transduce the signal to cellular components for ABA function. Progress in understanding ABA perception and signal transduction has been achieved recently (Cutler *et al.*, 2010; Raghavendra *et al.*, 2010; Umezawa *et al.*, 2010; Weiner *et al.*, 2010). Most of the studies of ABA receptors have been done in *Arabidopsis* with a few reports in other plants (Jia *et al.*, 2011; Kim *et al.*, 2012). Recently, ABA receptors and group A 2C-type protein phosphatases were shown to govern the ABA signalling pathway. Moreover, recent studies have suggested that there are interactions between the major ABA signalling pathway and other signalling factors in stress responses and seed and seedling development (Nakashima & Yamaguchi-Shinozaki, 2013). Control of the expression of ABA signalling factors may improve tolerance to environmental stresses.

The PYR/PYL/RCAR family (hereafter referred to as PYLs) is known to include ABA receptors. Ge *et al.* (2010) reported the characterization of PYL homologues (GmPYLs) in soybean. Twenty-three putative GmPYLs can be found from the soybean genome sequence and categorized into three subgroups. GmPYLs interact with AtABI1 and two GmPP2Cs in diverse ways. Many of the subgroup I GmPYLs interact with PP2Cs in an ABA-dependent manner, whereas most of the subgroup II and III GmPYLs bind to PP2Cs in an ABA-independent manner. The subgroup III GmPYL23, which cannot interact with any of the tested PP2Cs, differs from other GmPYLs. The CL2/gate domain is crucial for GmPYL–PP2C interaction, and a mutation in the conserved proline (P109S) abolishes the interaction between GmPYL1 and AtABI1. Furthermore, GmPYL1 interacts with AtABI1 in an ABA-dependent manner in plant cells. Three GmPYLs differentially inhibit AtABI1 and GmPP2C1 in an ABA-dependent or -enhanced manner *in vitro*. In addition, ectopically expressed GmPYL1 partially restores ABA sensitivity of the *Arabidopsis* triple mutant pyr1/pyl1/pyl4. Thus, soybean GmPYLs are ABA receptors that function by interacting and inhibiting PP2Cs.

It has been demonstrated that a protein phosphorylation reaction in ABA signalling involves the participation of several protein kinases (Verslues & Zhu, 2007; Wang *et al.*, 2010). The guard cell-specific protein kinase AAPK from *Vicia faba* and its orthologue OST1/SnRK2.6 in *Arabidopsis* are involved in ABA signalling (Yoshida *et al.*, 2002; Rodriguez *et al.*, 2010). Expression of the soybean GmWNK1 is involved in ABA signalling through activation of SPAK and OSR1 triggering a MAPK cascade or different co-transporters regulating ABA-dependent osmotic stresses in legumes (Wang *et al.*, 2010)

10.4 ABA mediating whole-legume responses to abiotic stress

Generally, the action of ABA has been related to processes of inhibition, although there is recent evidence of its presence in developing tissues where it may have a promoting action (Finkelstein & Rock, 2002; Sansberro *et al.*, 2004; Peng *et al.*, 2006). It was shown that endogenous ABA levels were increased in both flowers and pods of soybean by drought stress, and it was related to a reduction in pod set. Liu *et al.* (2003) suggested that drought induced an increase in ABA levels causing pod abortion. However, direct cause and effect could not be shown, and evidence to support this hypothesis is not clear. Nevertheless, Liu *et al.* (2004) demonstrated that different stresses modified endogenous ABA content in pods from plants, and that this response was correlated with pod set. ABA treatments were applied simultaneously with both well-watered and drought stressed conditions to the soybean plants to test if ABA could alleviate the effect of drought stress on pod set. Results showed that pod set was correlated negatively with pod ABA content when the latter was above a threshold value; this conforms with early reports in wheat (Westgate *et al.*, 1996). In addition, ABA treatments reduced pod set in plants under well-watered conditions, results that are in agreement with data for wheat (Wang *et al.*, 2001) and maize (Mambelli & Setter, 1998).These findings suggest an ABA role in controlling pod set in legumes. However, it was also reported that ABA treatments had a positive effect on pod set in plants under drought-stressed conditions. These contrasting effects of ABA treatments on pod set in soybean plants in both well-watered and drought stressed conditions are still unclear. However, one reason could be that ABA affects pod set directly by altering processes within the ovary (i.e. cell division) or indirectly by influencing the availability of photosynthates. A similar mechanism had been proposed when ABA applied in wheat provokede seed abortion (Waters *et al.*, 1984). Liu *et al.* (2004) demonstrated that the effect of ABA treatments on pod set was due partially to its effect on photosynthesis. Plants in the early stages of drought stress showed induction of closure of stomata when ABA was applied. Also observed was a slowing in development of water deficits that resulted in a high leaf water potential. Furthermore, several reports, have shown that a high leaf water potential is essential for seed set in wheat and pod set in soybean (Dembinska *et al.*, 1992; Liu *et al.*, 2004).

ABA also has important functions in vegetative growth under abiotic stresses, as was observed in *Prosopis strombulifera*, which showed that endogenous ABA was affected by salt type and concentration and relative humidity (Devinar *et al.*, 2013). It was demonstrated that when three stressing factors are applied simultaneously they interact bringing about a response that is different to that triggered by any of them individually. At low humidity (30%) ABA accumulation seemed to be triggered by the conjunction of low humidity plus salinity, especially in leaves of Na_2SO_4-treated plants from −1 to −2.6 MPa. These high ABA levels are correlated with increasing toxicity symptoms caused by this salt, and the maintenance of primary root elongation and shoot inhibition. Thereby, low humidity amplified the effects of salinity in plant growth and ABA response; thus, ABA has a role not only as a stress signal but also as a growth regulator governing vegetative growth under different relative humidity and salinity conditions (Devinar *et al.*, 2013). Similar results were obtained previously in this species, where the highest levels of ABA and ABA-GE, and the highest ABA-GE glucosidase activity were observed in Na_2SO_4-treated plants. Therefore, ABA may act as a triggering signal for several adaptive mechanisms for survival of plants in this extremely stressing situation. However, the stress imposed by the presence of sulphate anion in the culture medium blocked the activity of ABA, stomata remained open, and high transpiration values were recorded (Llanes *et al.*, 2014).

Moreover, it is well known that ABA accumulation triggers H_2O_2 production as a signal mediating stress tolerance responses (Xing *et al.*, 2008). Several studies have demonstrated that ABA accumulation occurs prior to H_2O_2 appearance and that both molecules are involved in stomatal closure as well as in stomatal opening inhibition (Lü *et al.*, 2013). In this regard, Furlán *et al.* (2012) evaluated the time-course of production of both compounds, ABA and H_2O_2, in peanut plants exposed to drought stress. In these plants, significant ABA accumulation occurred by 12 h in polyethylene glycol (PEG)-treated plants, whereas this increment was less pronounced at 24 and 48 h compared with the non-treated plants. On the other hand, a significant increase in H_2O_2 content occurred after 24 h. These authors set sample times at 12 and 24 h to address the question of whether the increase in H_2O_2 production was related to ABA accumulation in peanut leaves exposed to water stress. After pretreatment with 5 mM tungstate, an ABA biosynthesis inhibitor, the increase in ABA levels induced by water stress was inhibited. At the same time, this pretreatment suppressed the increase in H_2O_2 generation in leaves exposed to water stress at 12 h. Treatment with 100 μM ABA replaced the ABA partially inhibited by tungstate at 24 h and prevented the reduction in H_2O_2 generation. A possible explanation for this increase of H_2O_2 content could be that some of the ROS sources, which include electron transport, many non-enzymatic and enzymatic reactions (Foyer & Noctor, 2000; Asada, 1999), photorespiration and the Mehler reaction (Cruz de Carvalho, 2008), might be ABA-independent. Similar results were found in the tropical legume *Stylosanthes guianensis* (Zhou *et al.*, 2005). Therefore, ABA accumulation is the first signal that triggers H_2O_2 generation during the initial phase, but its subsequent production is partially ABA-independent (Furlán *et al.*, 2012).

In addition, ABA sprayed on the leaves of soybean benefits dry matter accumulation favouring vegetative growth of the plants grown in field conditions (Travaglia *et al.*, 2009). For this cultivar and at the dose used, no restraint of shoot growth was observed as had been previously demonstrated (Sloger & Caldwell, 1970). Similarly, field-grown wheat plants under water stress that had been treated with ABA showed higher shoot biomass accumulation (Travaglia *et al.*, 2010). This ABA effect has also been seen in investigations carried out under controlled conditions with different levels of water deficit, where plants sprayed with ABA showed greater growth than those with lower endogenous ABA levels. These studies suggest that ABA is an important regulator controlling water content of both cells and whole-plant, likely due to enhanced turgor that allows optimal expansion of plant cells (Finkelstein & Rock 2002; Sansberro *et al.*, 2004).

According to He and Cramer (1996), ABA application on *Brassica* can improve salt and drought tolerance, and it was proposed that an adequate concentration of ABA in plants under abiotic stress may lead to growth stimulation. In this regard, Khadri *et al.* (2007) demonstrated that in both salt-sensitive and salt-tolerant *Phaseolus vulgaris* plants, in symbiosis with *Rhizobium tropici* CIAT899 strain, shoot and root growth decreased under salt treatment. In addition, these authors showed that nitrogenase activity and ureide levels in nodules and nitrogen percentage in shoot tissues were also decreased. However, when these plants were pretreated with ABA growth parameters increased, nodule weight of salt-sensitive plants was normalized and nitrogenase activity was restored or less inhibited in both salt-sensitive and salt-tolerant cultivars. Besides, ABA treatment in bean plants seems to limit Na^+ translocation to shoots, resulting in the maintenance of a high K/Na ratio. Therefore, these results support the role ABA in the mitigation of salt stress in common bean plants (Khadri *et al.*, 2007).

Some studies support a role for ABA in legume root development. ABA acts in regulating lateral root formation in higher plants, but the mechanism of this response is still not clear. Enhancement of the number of lateral roots by ABA has been previously reported (Trewavas & Jones, 1991). Liang and Harris (2005) have proposed that ABA inhibits lateral root development in non-leguminous species, but has a positive effect in leguminous species. The increasing density of lateral roots provoked by ABA can be a positive effect to obtain a higher number of root nodules by the plant. Therefore, ABA is a key regulator in the responses of plants to abiotic stresses such as drought, osmotic stress, salinity and chilling. Lateral root formation coordinated by ABA provides one of the first links between environmental conditions and control of root architecture (Finkelstein *et al.*, 2002). Besides, there is a direct correlation between the response of lateral roots of legumes and non-legumes to ABA and their ability to accommodate nitrogen-fixing bacteria in root nodules. Thus, the characterization of genes involved in the ABA signalling cascade of legume lateral roots may help to elucidate the changes in development and physiology of the root necessary to accommodate a rhizobial endosymbiont.

10.5 ABA regulation of leaf expansion under abiotic stress

Numerous studies have reported that ABA is involved in the regulation of leaf expansion under salt and drought stress in legumes (Hartung *et al.*, 1999; Cramer & Quarrie, 2002; Thaler & Bostock, 2004). Previous experiments with *Phaseolus vulgaris* demonstrated a direct relationship between stomatal conductance and leaf or xylem ABA levels under salinity (Sibole *et al.*,

2000). However, Sibole *et al.* (2003) showed that in *Medicago* species there was no correlation between leaf growth and leaf ABA. Root ABA accumulation was not significantly different in either species under salt stress, and this growth regulator only increased in leaves and xylem of salt-treated *Medicago citrina*. The increase in leaf ABA was correlated with decreased leaf water content, which suggests that ABA could be involved in osmotic adjustment by lowering water potential. In this study it was proposed that the higher xylem ABA accumulation found in salt-treated *M. citrina* could be correlated with its retranslocation from the leaves via the phloem to xylem (Wolf *et al.*, 1990). Correspondingly, the lack of response in leaf ABA content in *M. arborea* plants could be related to the observed increase in leaf water content with NaCl (Hartung *et al.*, 2002).

Cabot *et al.* (2009) observed a decrease in ABA content in *Phaseolus vulgaris* pretreated with fluoridone under salt stress. These plants had lower Na^+ uptake and higher leaf Na^+ exclusion capacity in comparison with non-pretreated plants. Moreover, leaf Na^+ exclusion was decreased by ABA and Na^+ uptake showed an increment. Leaf ABA increased in NaCl- and KCl-treated plants, and this ABA content caused a decrease in transpiration rates, whereas supplemental Ca^{2+} and higher strength nutrient solution caused a decrease in leaf ABA and leaf Na^+. These results evidenced the osmotic component of salt associated with non-ion-specific increase in ABA. The higher ABA levels observed in *Phaseolus vulgaris* plants may be correlated with higher leaf Na^+ concentrations due to lower Na^+ exclusion or increased root–shoot Na^+ translocation. However, in the legume *P. strombulifera* an anionic component of salt was observed. The high ABA levels found in Na_2SO_4-treated plants at 48 days of culture may be correlated with a failure in ion compartmentalization by these plants when compared with NaCl-treated plants. In fact, NaCl-treated plants accumulated levels of Na^+ in their leaves similar to those of Na_2SO_4-treated plants but they successfully accomplished ion compartmentalization and osmoregulation with direct consequences for their growth. Therefore, production of high levels of ABA in leaves under Na_2SO_4 treatment may be considered as the stress signal to protect this legume against dehydration and ion toxicity caused by the sulphate anion as well as a regulating factor controlling vegetative growth (Devinar *et al.*, 2013).

10.6 ABA as a regulator of nodulation under abiotic stress

Abscisic acid is a negative regulator of nodulation, as was first described by Phillips (1971) in *Pisum sativum*. In fact, Cho and Harper (1993) demonstrated that ABA application to roots reduced nodule number and isoflavonoid levels in the wild-type as well as in the super-nodulating mutant NOD1-3 in *Glycine max* (soybean). Indeed, all phases of soybean nodulation (nodule initiation, nodule development, and nitrogen fixation) were inhibited by exogenous ABA (Bano & Harper, 2002), suggesting general physiological toxicity or suppression. Furthermore, *Lotus japonicus* nodule number decreased after ABA application to roots but increased with abamine, an inhibitor of ABA biosynthesis (Suzuki *et al.*, 2004). ABA levels in wild-type soybean and the super-nodulating mutant nts382 were investigated to elucidate the possible role of ABA in nodule development (Carroll *et al.*, 1985a,b; Searle *et al.*, 2003; Nontachaiyapoom *et al.*, 2007). In wild-type roots, ABA levels were significantly higher than in nts382, and increased further upon inoculation with *Bradyrhizobium*, but only in the wild-type (Caba *et al.*, 2000). Endogenous ABA also increased in the shoot at the onset of nodule regulation, but again, only in the wild-type and not in nts382. The ABA/cytokinin ratio has been positively correlated with nodule suppression and autoregulation, with the root ratio always being higher in wild-type soybean compared to the super-nodulating nts382mutant (Bano *et al.*, 2002). Therefore, it was suggested that ABA may be a root-induced inhibitor, decreasing nodule numbers in wild-type but not in super-nodulating mutants. Additionally, the mutant latd, a nodulation-defective mutant of *Medicago truncatula*, produced nodules that were immature and non-nitrogen fixing. This mutant is deficient in the growth of three types of meristems – those of the primary root, lateral root and nodules. Organization and function of both primary and lateral root meristems, but not nodulation, were restored by ABA application. The mutant was later found to have reduced sensitivity to ABA (Liang *et al.*, 2007). Another mutagenesis analysis was performed in *Lotus japonicus* MG-20, and the ABA-insensitive mutant Beyma was isolated. This mutant failed to show any nodulation sensitivity to ABA, and had a slow-growing (50–70% of wild-type

MG-20) and near-constitutive wilty phenotype associated with its inability to regulate stomatal opening (Biswas *et al.*, 2009). In the absence of exogenous ABA, Beyma did not show alteration of nodule number control, and the number and patterning of nodules formed in the mutant were similar to MG-20. As MG-20 showed that application of ABA on one side of the split root inhibited nodulation locally but not systemically, it was proposed that ABA is not directly involved in autoregulation of nodulation. Similarly, ABA insensitivity as well as a hypernodulating phenotype was exhibited by *Medicago truncatula* seedlings overexpressing *ABI*-1 gene from *Arabidopsis*, which suppresses the ABA signalling pathway (Ding *et al.*, 2009). In addition, the mutant enhanced nitrogen fixation 1 (enf1) showed lower endogenous ABA concentration in comparison with wild-type *Lotus japonicus* seedlings. This mutant enf1 not only showed increased root nodule number but also enhanced nitrogen fixation (Tominaga *et al.*, 2009). Ding *et al.* (2008) suggested that the effect of ABA inhibition on nodulation was linked to the abolition of both Nod factor-induced calcium spiking and gene expression to coordinate Nod factor and cytokinin (CK) signaling during the regulation of nodulation. It has been observed that cortical responses leading to nodule primordia formation are mostly the function of CKs and auxin signalling (Oldroyd & Downie, 2008). Besides, nodule initiation is associated with suppression of lateral root formation (Nutman, 1948), probably following changes in ABA/CK ratios and their opposing effects on the inception of cell division associated with nodules and lateral root formation.

In addition, González *et al.* (2001) proposed that ABA could be involved in nodule senescence by controlling the development of the plant root system in response to nitrogen nutrition (Gawronska *et al.*, 2003) and its implication in sugar- and nitrate-sensing pathways in plants (Puppo *et al.*, 2005).

Under saline conditions, Salah *et al.* (2013) demonstrated an inverse relationship between ABA in nodules and the growth-related parameters of *Medicago ciliaris*, and with symbiotic nitrogen fixation. In contrast, ABA levels in leaves and roots were positively correlated with both growth and symbiotic nitrogen fixation under salt stress, suggesting complex hormonal interactions between different organs in relation to the response to abiotic stress and to the symbiotic relationship, which requires further investigation.

10.7 ABA and assimilate accumulation under abiotic stress

Abiotic stress can modify ABA accumulation, regulating assimilate partitioning between source and different sink tissues (Hartig & Beck, 2006; Pérez-Alfocea *et al.*, 2010). ABA can also trigger changes in carbohydrate concentrations and enzyme activities involved in carbohydrate biosynthesis (Kobashi *et al.*, 1999; Taji *et al.*, 2002). Blöchl *et al.* (2005) observed an accumulation of raffinose and stachyose in ABA-treated somatic embryos of alfalfa; these compounds have been proposed to act generally as compatible solutes or osmoprotectants to allow osmotic adjustment of plant cells exposed to abiotic stress. Oligosaccharides have been considered as factors involved in glass formation in cells and may act against deterioration of macromolecular structures during desiccation (Horbowicz & Obendorf, 1994). In drought-stressed lupin seeds, oligosaccharides were the predominant fraction (75–77%) of all soluble carbohydrates (Zalewski *et al.*, 2001), and raffinose showed significant increases during water deficit in *Populus euphratica* (Brosche *et al.*, 2005). Carbohydrates induced by drought conditions and ABA treatments may serve not only as osmoprotectants and free radical scavengers, but also as phloem-mobile anti-senescence metabolites (Wattana & Monica, 1999).

In legumes, the accumulation of starch is generally associated with storage protein accumulation (Wobus & Weber, 1999). However, in *Vicia narbonensis* a reduction in starch synthesis due to antisense expression of ADP-glucose pyrophosphorylase was accompanied by increased total protein accumulation, possibly reflecting an increased period of seed-filling or delayed maturation (Weber *et al.*, 2000). Besides, one set of genes, named *LEAFY COTYLEDON* (*LEC*) and *FUS3*, is involved in the regulation of a variety of embryonic stages and interact genetically with ABA-dependent regulators to suppress vivipary at mid-embryogenesis (Raz *et al.*, 2002), and with several of the ABI loci to mediate ABA sensitivity at seed maturity (Finkelstein & Gampala, 2002). *Arabidopsis* mutants of these regulators – i.e. *fusca 3* (*fus3*), *leafy cotyledon 1* (*lec1*) and *leafy cotyledon 2* (*lec2*) – have excessive starch accumulation but almost no storage protein or lipid accumulation (Harada, 2001). The *DE-ETIOLATED 1* (*DET1*)/*FUS2*, *FUS3* and *LEC1* loci are also negative regulators of some light responses that are enhanced by sugars, such as anthocyanin accumulation, such that the (derepressed) mutant seeds are abnormally pigmented. This effect is greatly enhanced by some of the *abi* mutations. Additional interactions between ABA and light signalling during seed development are reflected in the dependence on DET1, FUS3 and LEC1 for adequate accumulation of ABI3 protein (Gampala *et al.*, 2001). Taken together, these results suggest that the regulators ABI3, ABI4, ABI5, DET1, FUS3 and LEC1 may all participate in the ABA signalling cascade under different environmental conditions. These regulators may also participate in sugar accumulation during seed development. Furthermore, exogenous ABA applied to grains leads to the accumulation of carbohydrates, thus accelerating the filling process. In addition, Grappin *et al.* (2000) reported that the application of fluridone reduced the activity of enzymes associated with starch synthesis. In soybean experiments, where stress was of short duration, the application of ABA immediately increased carbohydrate amounts at the flowering period in shoots of the ABA-treated plants as compared to controls; the difference in shoots disappeared at harvest because of increased carbohydrate remobilization (21%) to the seed in the ABA-treated plants (Travaglia *et al.*, 2009). This effect has been also reported in rice (Yang *et al.*, 2004) and by Moreno (2009) who found that ABA stimulates carbon translocation from source to sink. Thus, these studies confirm the participation of ABA in promoting source-to-sink transport of assimilates during the stage of seed filling. Additionally, treatment with ABA did not affect the quality of seeds, an important characteristic since seed quality is one of the key aspects for farming success. On the other hand, ABA mediates amino acid and proline accumulation. Khadri *et al.* (2006) showed that pretreatment of plants with ABA prevented or reduced amino acid and proline production under salt stress. In this regard, Trotel-Aziz *et al.* (2000) have proposed that proline accumulated under adverse environmental conditions is rapidly reduced when those conditions are alleviated. The results of Khadri *et al.* (2006) suggest that in common bean ABA exogenously applied has a central function in the plant's adaptive response to salt stress.

10.8 ABA mediating the expression of abiotic stress-responsive genes

ABA can modify plant cells' transcriptome by upregulating or downregulating several genes, termed ABA-responsive genes. Such genes encode transcription factors and enzymes, hydrophilic proteins, etc. (Yamaguchi-Shinozaki & Shinozaki, 2006). ABA-responsive genes or their products are believed to orchestrate various cellular and physiological responses to ABA. Consistent with this idea, numerous studies have illustrated the ABA-responsive transcriptome from some species (Cuming *et al.*, 2007; Richardt *et al.*, 2010). In this regard, Seki *et al.* (2002a) identified 299 ABA-responsive genes, 155 of which were upregulated under drought stress conditions. ABA treatment of guard cells revealed 64 differentially expressing genes (Leonhardt *et al.*, 2004). Using microarray techniques, Seki *et al.* (2002b), Kreps *et al.* (2002) and Kawaguchi *et al.* (2004) identified 27 genes that were induced under water deficit stress; they were classified into six functional categories: metabolism, transport, signalling, transcription, hydrophilic proteins and unknown proteins (Bray, 2004). Therefore, the activation of regulatory components for stress perception leads to the production of effector molecules that are directly involved in alleviating stress. Regulatory components are represented by transcription factors and signal transduction components such as kinases and phosphatases (Yamaguchi-Shinozaki & Shinozaki, 2006). There is a group of stress-responsive genes whose expression is ABA-independent; these encode the DREB1 (cold) and DREB2 (salt, dehydration) families of proteins (Liu *et al.*, 1998; Nakashima *et al.*, 2000). The ABA-dependent pathway is mainly associated with a subfamily of basic leucine zipper (bZIP) class transcription factors called ABRE-binding factors, or ABFs (Choi *et al.*, 2000; Uno *et al.*, 2000). Other transcription factors involved in the abiotic stress response have been reported (Yamaguchi-Shinozaki & Shinozaki, 2006).

Transcriptome analysis under water-deficit stress has also been performed in various other species, like *Populus trichocarpa* and *Populus euphratica*, maize, chickpea and wheat (Brosche *et al.*, 2005; Andjelkovic & Thompson, 2006; Mantri *et al.*, 2007; Talame *et al.*, 2007). In the model legume *Lotus japonicus*, a gene encoding for beta-glucosidase, which hydrolyses glycosides of abscisic acid to generate free ABA, was highly induced during water deficit (Betti *et al.*, 2012). Overexpression of this enzyme in *Arabidopsis* enhanced drought and salinity tolerance (Xu *et al.*, 2012). Different genes encoding for transcription factors (TFs) were also highly modulated in *Lotus japonicum* under drought stress (Betti *et al.*, 2012). Interestingly, while a NAM, ATAF1/2 and CUC2 (NAC) domain TF orthologue to *Arabidopsis* NAC47 was highly induced, RAD-like 5 (AtRL5), a gene related to the myeloblastosis (MYB) family of TFs, was the most repressed under drought stress in *L. japonicum*. RAD-like TFs are a subfamily of the MYB factors. Members of the RAD-like family of TFs are involved in floral development in *Arabidopsis* (Baxter *et al.*, 2007). However, the exact role of AtRL5 is unknown.

Ge *et al.* (2010) demonstrated that cell structure, protein synthesis, energy and secondary metabolism genes were significantly affected by the alkaline stress treatment in soybean leaf. The responses of soybean to $NaHCO_3$ highlight specific transcription factors, such as the C2C2-CO-like, MYB-related, WRKY, GARP-G2-like and ZIM families. Co-expression genes were involved in regulation of transcription factors, including AP2-EREBP, bHLH, MYB/MYB-related, C2C2-CO-like, C2C2-DOF, C2C2, C3H and GARP-G2-like transcription factors. In this study, it was revealed that ABA-mediated signalling participates in signal transduction in the alkaline stress response.

Numerous studies of legume transcriptomes are being carried out, but a comprehensive analysis of ABA mediation of stress-responsive genes during various stages of the legume life cycle under different abiotic stresses is not yet available.

10.9 Concluding remarks and future prospects

Several environmental factors, including drought, salinity, nutrient availability and temperature, adversely affect plant growth and development of a crop. The need for a change in focus in plant abiotic stress research has become apparent. This chapter summarizes our current understanding of the role of ABA as a key phytohormone controlling different processes during plant life cycles under abiotic conditions. In this regard, recent progress in plant molecular biology has shown that a large group of genes are involved in plant stress responses. Although not all of these gene expressions are ABA-dependent, it is certain that ABA plays a central role as modulator of many morpho-physiological responses that help legumes to survive abiotic stresses. Both the production and catabolism of ABA are delicately stress-regulated. Indeed, such regulation indicates that ABA can act as a stress signal in legumes when abiotic stress modifies their growth.

The challenge for plant researchers will be to investigate multiple signalling components in stress tolerance in agronomically important legumes, thus improving the yield potential, particularly in areas with adverse environmental conditions.

References

Andjelkovic V, Thompson R (2006) Changes in gene expression in maize kernel in response to water and salt stress. *Plant Cell Rep* 25: 71–79.

Asada K (1999) The water-water cycle in chloroplasts: scavenging of active oxygen and dissipation of excess photons. *Annu Rev Plant Biol* 50: 601–639.

Bacon MA, Wilkinson S, Davies WJ (1998) pH-regulated leaf cell expansion in droughted plants is abscisic acid dependent. *Plant Physiol* 118: 1507–1515.

Bano A, Harper JE (2002) Plant growth regulators and phloem exudates modulate root nodulation of soybean. *Funct Plant Biol* 29: 1299–1307.

Bano A, Harper JE, Auge RM, Neuman DS (2002) Changes in phytohormone levels following inoculation of two soybean lines differing in nodulation. *Funct Plant Biol* 29: 965–974.

Baxter CJ, Redestig H, Schauer N, et al. (2007) The metabolic response of heterotrophic Arabidopsis cells to oxidative stress. *Plant Physiol* 143: 312–325.

Betti M, Pérez-Delgado C, García-Calderón M, Díaz P, Monza J, Márquez A (2012) Cellular stress following water deprivation in the model legume *Lotus japonicus*. *Cell* 1: 1089–1106.

Biswas DK, Xu H, Li YG, *et al.* (2008) Genotypic differences in leaf biochemical, physiological and growth responses to ozone in 20 winter wheat cultivars released over the past 60 years. *Global Change Biol* 14: 46–59.

Blochl A, Grenier-de MG, Sourdioux M, Peterbarer T, Richter A (2005) Induction of raffinose oligosaccharide biosynthesis by abscisic acid in somatic embryos of alfalfa (*Medicago sativa* L.). *Plant Sci* 168: 1075–-1082.

Bray EA (2004) Genes commonly regulated by water deficit stress in Arabidopsis thaliana. *J Exp Bot* 55: 2331–2341.

Bray EA, Zeevaart JAD (1985) The compartmentation of abscisic acid and β-D-glucopyranosyl abscisate in mesophyll cells. *Plant Physiol* 79: 719–722.

Bright J, Desikan R, Hancock JT, Weir IS, Neill SJ. (2006) ABA-induced NO generation and stomatal closure in *Arabidopsis* are dependent on H_2O_2 synthesis. *Plant J* 45: 113–122.

Brosche M, Vinocur B, Alatalo ER, *et al.* (2005) Gene expression and metabolite profiling of *Populus euphratica* growing in the Negev desert. *Genome Biol* 6: R101.

Caba JM, Centeno ML, Fernandez B, Gresshoff PM, Ligero F (2000) Inoculation and nitrate alter phytohormone levels in soybean roots: differences between a super nodulating mutant and the wild type. *Planta* 211: 98–104.

Cabot C, Sibole JV, Barcelo J, Poschenrieder C (2009) Abscisic acid decreases leaf Na^+ exclusion in salt treated *Phaseolus vulgaris* L. *J Plant Growth Regul* 28: 187–192.

Carroll BJ, McNeil DL, Gresshoff PM (1985a) Isolation and properties of soybean [*Glycine max* (L.) Merr.] mutants that nodulate in the presence of high nitrate concentrations. *Proc Natl Acad Sci USA* 82: 4162–4166.

Carroll BJ, McNeil DL, Gresshoff PM (1985b) A supernodulation and nitrate-tolerant symbiotic (nts) soybean mutant. *Plant Physiol* 78: 34–40.

Chalk PM, Alves BJR (2010) Integrated effects of abiotic stresses on inoculants performance, legume growth and symbiotic dependence estimated by N dilution. *Plant Soil* 328: 1–16.

Cheng WH, Endo A, Zhou L, *et al.* (2002) A unique short-chain dehydrogenase/reductase in *Arabidopsis* glucose signaling and abscisic acid biosynthesis and functions. *Plant Cell* 14: 2723–2743

Cho MJ, Harper JE (1993) Effect of abscisic acid application on root isoflavonoid concentration and nodulation of wild type and nodulation-mutant soybean plants. *Plant Soil* 152: 145–149

Choi H, Hong J, Ha J, Kang J, Kim SY (2000) ABFs, a family of ABAresponsive element binding factors. *J Biol Chem* 275: 1723–1730.

Cramer GR, Quarrie SA (2002) Abscisic acid is correlated with the leaf growth inhibition of four genotypes of maize differing in their response to salinity. *Funct Plant Biol* 29: 111–115.

Cruz De Carvalho MH (2008) Drought stress and reactive oxygen species: production, scavenging and signaling. *Plant Signaling Behav* 3: 156–165.

Cuming AC, Cho SH, Kamisugi Y, Craham H, Quatrano RS (2007) Microarray analysis of transcriptional responses to abscisic acid, osmotic, salt, and drought stress in the moss, *Physcomitrella patens*. *New Phytol* 176: 275–287.

Cutler AJ, Krochko JE (1999) Formation and breakdown of ABA. *Trends Plant Sci* 4: 472–478.

Cutler S, Rodriguez P, Finkelstein R, Abrams S (2010) Abscisic acid: emergence of a core signaling network. *Annu Rev Plant Biol* 61: 651–679.

Davies WJ, Zhang JH (1991) Root signals and the regulation of growth and development of plants in drying soil. *Annu Rev Plant Physiol Plant Mol Biol* 42: 55–76.

Dembinska O, Lalonde S, Saini H S (1992) Evidence against the regulation of grain set by spikelet abscisic acid levels in water-stressed wheat. *Plant Physiol* 100: 1599–1602.

De Smet I (2003) An abscisic acid-sensitive checkpoint in lateral root development of *Arabidopsis*. *Plant J* 33: 543–555.

Devinar G, Llanes A, Masciarelli O, Luna V (2013) Different relative humidity conditions combined with chloride and sulfate salinity treatments modify abscisic acid and salicylic acid levels in the halophyte Prosopis strombulifera. *Plant Growth Regul* 70: 247–256.

Ding X, Richter T, Chen M, *et al.* (2009) A rice kinase-protein interaction map. *Plant Physiol* 149: 1478–1492.

Ding Y, Kalo P, Yendrek C, *et al.* (2008) Abscisic acid coordinates Nod factor and cytokinin signaling during the regulation of nodulation in *Medicago truncatula*. *Plant Cell* 20: 2681–2695.

Dische Z (1962) General color reactions. *Method Carbohydrate Chem* 1: 478–492.

Dudal R (1976) *Inventory of Major Soils of the World with Special Reference to Mineral Stress*. Cornell University Agricultural Experiment Station, Ithaca, NY, pp. 89–96.

Finkelstein RR, Rock CD (2002) Abscisic acid biosynthesis and signaling. In: Somerville CR, Meyerowitz, EM (eds), *The Arabidopsis Book*. American Society of Plant Biologists, Rockville, MD, pp. 1–52.

Finkelstein RR, Gampala SS, Rock CD (2002) Abscisic acid signalling in seeds and seedlings. *Plant Cell* 14: 15–45.

Foyer CH, Noctor G (2000) Oxygen processing in photosynthesis: regulation and signalling. *New Phytol* 146: 359–388.

Furlan A, Llanes A, Luna V, Castro S (2012) Physiological and biochemical responses to drought stress and subsequent rehydration in the symbiotic association *Peanut Bradyrhizobium* sp. *ISRN Agronomy* ID 318083: 1–8.

Gallardo M, Turner NC, Ludwig C (1994) Water relations, gas exchange and abscisic acid content of *Lupinus cosentinii* leaves in response to drying different proportions of the root system. *J Exp Bot* 45: 909–918.

Gampala SS, Finkelstein RR, Sun SS, Rock CD (2001) ABA INSENSITIVE-5 interacts with ABA signaling effectors in rice protoplasts. *J Biol Chem* 9: 1–9.

Gawronska H, Deji A, Sakakibara H, Sugiyama T (2003) Hormone-mediated nitrogen signaling in plants: implication of participation of abscissic acid in negative regulation of cytokinin-inducible expression of maize response regulator. *Plant Physiol Biochem* 41: 605–610.

Ge Y, Li Y, Zhu YM, *et al.* (2010) Global transcriptome profiling of wild soybean (Glycine soja) roots under $NaHCO_3$ treatment. *BMC Plant Biol* 10: 153.

González EM, Gálvez L, Arrese-Igor C (2001) Abscisic acid induces a decline in nitrogen fixation that involves leghemoglobin, but is independent of sucrose synthase activity. *J Exp Bot* 52: 285–293.

Gonzalez-Guzman M, Apostolova N, Belles JM, Barrero JM, Piqueras P (2002) The short-chain alcohol dehydrogenase ABA2 catalyzes the conversion of xanthoxin to abscisic aldehyde. *Plant Cell* 14: 1833–1846.

Graham PH, Vance CP (2003) Legumes: importance and constraints to greater use. *Plant Physiol* 131: 872–877.

Grappin P, Bouinot D, Sotta B, Miginiac E, Jullien M (2000) Control of seed dormancy in Nicotiana plumbaginifolia: Post-imbibition abscisic acid synthesis imposes dormancy maintenance. *Planta* 210: 279–285.

Hamdia MA (2008) Review on the physiology of senescence and abscission. *Curr Topic Plant Biol* 9: 45–67.

Hartig K, Beck E (2006) Crosstalk between auxin, cytokinins, and sugars in the plant cell cycle. *Plant Biol* 8: 389–396.

Hartung W, Peuke AD, Davies WJ (1999). Abscisic acid – a hormonal long-distance stress signal in plants under drought and salt stress. In: Pessarakli M (ed.), *Handbook of Crop Stress*, 2nd edn. Marcel Dekker Inc., New York, pp. 731–747.

Hartung W, Sauter A, Hose E (2002) Abscisic acid in the xylem: where does it come from, where does it go to? *J Exp Bot* 53: 27–32.

He T, Cramer GR. (1996) Abscisic acid concentrations are correlated with leaf area reductions in two salt-stressed rapid-cycling *Brassica* species. *Plant Soil* 179: 25–33.

Henson IE (1981) Abscisic acid and after-effects of water stress in pearl millet (*Pennisetum americanum* (L.) Leeke). *Plant Sci Lett* 21: 129–135.

Holbrook NM, Shashidhar VR, James RA, Munns R (2002) Stomatal control in tomato with ABA-deficient roots: response of grafted plants to soil drying. *J Exp Bot* 53: 1503–1514.

Horbowicz M, Obendorf RL (1994) Seed desiccation tolerance and storability: dependence on flatulence-producing oligosaccharides and cyclitols. *Seed Sci Res* 4: 385–405.

Javid M, Sorooshzadeh A, Moradi F, Ali S, Modarres M (2011) The role of phytohormones in alleviating salt stress in crop plants. *Aust J Crop Sci* 5: 726–734.

Jensen E, Crozat Y, Schmidtke K, Carrouee B (2007) The role of grain legumes in future European crop rotations. In: Schneider A, Craig J (eds), *Grain Legumes and the Environment: How to Assess Benefits and Impacts*. European Association for Grain Legume Research, Paris, pp. 79–81.

Jia HF, Chai YM, Li CL, *et al.* (2011) Abscisic acid plays an important role in the regulation of strawberry fruit ripening. *Plant Physiol* 157: 188–199.

Jia W, Zhang J, Zhang DP (1996) Metabolism of xylem-delivered ABA in relation to ABA flux and concentration in leaves of maize and *Commelina communis*. *J Exp Bot* 47: 1085–1091.

Kawaguchi R, Girke T, Bray EA, Bailey-Serres JN (2004) Differential mRNA translation contributes to gene regulation under non-stress and dehydration stress conditions in Arabidopsis thaliana. *The Plant J* 38: 823–839.

Kato Noguchi H, Tanaka Y (2003) Allelopathic potential of citrus fruit peel and abscisic acid-glucose ester. *Plant Growth Regul* 40: 117–120.

Khadri M, Tejera NA, Lluch C (2007) Sodium chloride-ABA interaction in two common bean (*Phaseolus vulgaris*) cultivars differing in salinity tolerance. *Environ Exp Bot* 60: 211–218.

Kim JM, To TK, Seki M (2012) An epigenetic integrator: New insights into genome regulation, environmental stress responses and developmental controls by HISTONE DEACETYLASE 6. *Plant Cell Physiol* 53: 794–800.

Kobashi K, Gemma H, Iwahori S (1999) Sugar accumulation in peach fruit as affected by abscisic acid (ABA) treatment in relation to some sugar metabolizing enzymes. *J Jap Soc Hortic Sci* 68: 465-470.

Kreps JA, Wu Y, Chang H-S, Zhu T, Wang X, Harper JF (2002) Transcriptome changes for Arabidopsis in response to salt, osmotic, and cold stress. *Plant Physiol* 230: 2129–2141.

Kushiro T, Okamoto M, Nakabayashi K, et al. (2004) The *Arabidopsis* cytochrome P450 CYP707A encodes ABA 80-hydroxylases: key enzymes in ABA catabolism. *EMBO J* 23: 1647–1656.

Lehmann H, Glund K (1986) Abscisic acid metabolism-vascular/extravascular distribution of metabolites. *Planta* 168: 559–562.

Leonhardt N, Kwak JM, Robert N, Waner D, Leonhardt G, Schroeder JI (2004) Microarray expression analyses of Arabidopsis guard cells and isolation of a recessive abscisic acid hypersensitive protein phosphatase 2C mutant. *Plant Cell* 16: 596–615.

Liang Y, Harris J (2005) Response of root branching to abscisic acid is correlated with nodule formation both in legumes and nonlegumes. *Am J Bot* 10: 1675–1683.

Liang Y, Mitchell DM, Harris JM (2007) Abscisic acid rescues the root meristem defects of the *Medicago truncatula* latd mutant. *Develop Biol* 304: 297–307.

Liu F, Andersen MN, Jensen CR (2003) Loss of pod set caused by drought stress is associated with water status and ABA content of reproductive structures in soybean. *Funct Plant Biol* 30: 271–280.

Liu F, Andersen MN, Jensen CR (2004) Root signal controls pod growth in drought-stressed soybean during the critical, abortion-sensitive phase of pod development. *Field Crops Res* 85: 159–166.

Liu Q, Kasuga M, Sakuma Y, *et al.* (1998) Two transcription factors, DREB1 and DREB2, with an EREBP/AP2 DNA binding domain separate two cellular signal transduction pathways in drought- and low-temperature-responsive gene expression, respectively, in Arabidopsis. *Plant Cell* 10: 1391–1406.

Llanes A, Masciarelli O, Ordoñez R, Isla M I, Luna V (2014) Differential growth responses to sodium salts involve different ABA catabolism and transport in the halophyte *Prosopis strombulifera*. *Biol Plant* 58: 80–88.

Lobell DB, Burke MB, Tebaldi C, Mastrandrea MD, Falcon WP, Naylor RL (2008) Priorizing climate change adaptation needs for food security in 2030. *Nature* 319: 607–610.

Lü D, Wei W, Chen M (2013). ATHK1 acts downstream of hydrogen peroxide to mediate ABA signaling through regulation of calcium channel activity in *Arabidopsis* guard cells. *Chinese Sci Bull* 58: 336–343.

Mambelli S, Setter TL (1998) Inhibition of maize endosperm cell division and endoreduplication by exogenously applied abscisic acid. *Physiol Plant* 104: 266–272.

Mantri N, Ford R, Coram T, Pang E (2007) Transcriptional profiling of chickpea genes differentially regulated in response to high-salinity, cold and drought. *BMC Genomics* 8: 303.

Marin E, Nussaume L, Quesada A, Gonneau M, Sotta B (1996) Molecular identification of zeaxanthin epoxidase of *Nicotiana plumbaginifolia*, a gene involved in abscisic acid biosynthesis and corresponding to the ABA locus of *Arabidopsis thaliana*. *EMBO J* 15: 2331–2342.

Mittler R (2006) Abiotic stress, the field environment and stress combination. *Trends Plant Sci* 11: 15–19.

Mittler R, Blumwald E (2010) Genetic engineering for modern agriculture: challenges and perspectives. *Annu Rev Plant Biol* 61: 443–462.

Moreno LP (2009) Respuesta de las plantas al estrés por déficit hídrico. Una revisión. *Agron Colomb* 27: 179–191.

Nakashima K, Yamaguchi-Shinozaki K (2013) ABA signaling in stress-response and seed development. *Plant Cell Rep* 7: 959–970.

Nakashima K, Shinwari ZK, Sakuma Y, Seki M, *et al.* (2000) Organization and expression of two Arabidopsis DREB2 genes encoding DRE-binding proteins involved in dehydration- and high-salinity-responsive gene expression. *Plant Mol Biol* 42: 657–665.

Nambara E, Marion-Poll A (2005) Abscisic acid biosynthesis and catabolism. *Annu Rev Plant Biol* 56: 165–185.

Nontachaiyapoom S, Scott PT, Men AE, Kinkema M, Schenk PM, Gresshoff PM (2007) Promoters of orthologous *Glycine max* and *Lotus japonicus* nodulation auto-regulation genes interchangeably drive phloem-specific expression in transgenic plants. *Mol Plant Microbe Interact* 20: 769–780.

Nutman PS (1948) Physiological studies on nodule formation. I. The relation between nodulation and lateral root formation in red clover. *Ann Bot* 12: 81–96.

Okamoto M, Kuwahara A, Kushiro T, Kamiya Y, Koshiba T, Nambara E. (2006) Physiological roles of ABA 8′-hydroxylase family, key enzymes in ABA catabolism during seed development through to postgermination growth in *Arabidopsis thaliana*. *Plant Cell Physiol* 47: S147.

Oldroyd GE, Downie JA (2008) Coordinating nodule morphogenesis with rhizobial infection in legumes. *Annu Rev Plant Biol* 59: 519–546.

Peleg Z, Blumwald E (2011) Hormone balance and abiotic stress tolerance in crop plants. *Curr Opin Plant Biol* 14: 290–295.

Peng YB, Zou C, Wang DH, Gong HQ, Xu ZH, Bai SN (2006) Preferential localization of abscisic acid in primordial and nursing cells of reproductive organs of *Arabidopsis* and cucumber. *New Phytol* 170: 459–466.

Pérez-Alfocea F, Albacete A, Ghanem ME, Dodd IC (2020) Hormonal regulation of source–sink relations to maintain crop productivity under salinity: a case study of root-to-shoot signalling in tomato. *Funct Plant Biol* 37: 592–603.

Phillips DA (1971) Abscisic acid inhibition of root nodule initiation in *Pisum sativum*. *Planta* 100: 181–190.

Provart NJ, Gil P, Chen W, *et al.* (2003) Gene expression phenotypes of *Arabidopsis* associated with sensitivity to low- temperatures. *Plant Physiol* 132: 893–906.

Puppo A, Groten K, Bastian F, *et al.* (2005) Legume nodule senescence: roles for redox and hormone signalling in the orchestration of the natural aging process. *New Phytol* 165: 683–701.

Raghavendra AS, Gonugunta VK, Christmann A, Grill E (2010) ABA perception and signalling. *Trends Plant Sci* 15: 395–401.

Raz V, Bergervoet JH, Koornneef M (2001) Sequential steps for developmental arrest in Arabidopsis seeds. *Development* 128: 243–252.

Richardt S, Timmerhaus G, Lang D, Qudeimat E, Correa LG, Reski R (2010) Microarray analysis of the moss *Physcomitrella patens* reveals evolutionarily conserved transcriptional regulation of salt stress and abscisic acid signalling. *Plant Mol Biol* 72: 27–45.

Rijsberman FR (2006) Water scarcity: Fact or fiction? *Agric Water Manage* 80: 5–22.

Rizhsky L, Liang H, Shuman J, Shulaev V, Davletova S, Mittler R (2004) When defense pathways collide, the response of *Arabidopsis* to a combination of drought and heat stress. *Plant Physiol* 134: 1683–1696.

Rodriguez MC, Petersen M, Mundy J (2010) Mitogen-activated protein kinase signaling in plants. *Annu Rev Plant Biol* 61: 621–649.

Rook F, Corke F, Card R, Munz G, Smith C (2001) Impaired sucrose-induction mutants reveal the modulation of sugar-induced starch biosynthetic gene expression by abscisic acid signaling. *Plant J* 26: 421–433.

Saito S, Hirai N, Matsumoto C, Ohigashi H, Ohta D. (2004) *Arabidopsis* CYP707As encode (+)-abscisic acid 8-hydroxylase, a key enzyme in the oxidative catabolism of abscisic acid. *Plant Physiol* 134: 1439–1449.

Salah B, Albacete I, Messedi A, *et al.* (2013) Hormonal responses of nodulated *Medicago ciliaris* lines differing in salt tolerance. *Environ Exp Bot* 86: 35–46.

Sansberro P, Mroginski L, Bottini R (2004) Abscisic acid promotes growth of *Ilex paraguariensis* plants by alleviating diurnal water stress. *Plant Growth Regul* 42: 105–111.

Schachtman D, Goodger J (2008) Chemical root to shoot signaling under drought. *Trends Plant Sci* 13: 281–287.

Schroeder JI, Allen GJ, Hugouvieux V, Kwak JM, Waner D (2001) Guard cell signal transduction. *Annu Rev Plant Physiol Plant Mol Biol* 52: 627–658.

Schwartz S, Tan B, Gage D, Zeevaart J, McCarty D. (1997) Specific oxidative cleavage of carotenoids by VP14 of maize. *Science* 276: 1872–1874.

Searle IR, Men AE, Laniya TS, *et al.* (2003) Long-distance signaling in nodulation directed by a CLAVATA1-like receptor kinase. *Science* 299: 109–112.

Seki M, Narusaka M, Ishida J, *et al.* (2002a) Monitoring the expression pattern of around 7000 Arabidopsis genes under ABA treatments using a full-length cDNA microarray. *Funct Integr Genomics* 2: 282–291.

Seki M, Narusaka M, Kamiya A, *et al.* (2002b) Monitoring the expression profiles of 7000 Arabidopsis genes under drought, cold and high-salinity stresses using a full-length cDNA microarray. *Plant J* 31: 279–292.

Sibole JV, Montero E, Cabot C, Poschenrieder Ch, Barcelo ÂJ (2000) Relationship between carbon partitioning and Na^+, Cl^- and ABA allocation in fruits of salt-stressed bean. *J Plant Physiol* 157: 637–642.

Sibole JV, Cabot C, Poschenrieder C, Barcelo AJ (2003). Efficient leaf ion partitioning, an overriding condition for abscisic acid-controlled stomatal and leaf growth responses to NaCl salinization in two legumes. *J Exp Bot* 54: 2111–2119.

Signora L, De Smet I, Foyer CH, Zhang H. (2001) ABA plays a central role in mediating the regulatory effects of nitrate on root branching in *Arabidopsis*. *Plant J* 28: 655–662.

Sloger C, Caldwell B (1970) Response of cultivars of soybean to synthetic abscisic acid. *Plant Physiol* 46: 634–635.

Sottosanto JB, Gelli A, Blumwald E (2004) DNA array analyses of *Arabidopsis thaliana* lacking a vacuolar Na^+/H^+ antiporter: impact of AtNHX1 on gene expression. *Plant J* 40: 752–771.

Suzuki A, Akune M, Kogiso M, *et al.* (2004) Control of nodule number by the phytohormone abscisic acid in the roots of two leguminous species. *Plant Cell Physiol* 45: 914–922.

Taji T, Ohsumi C, Iuchil S, *et al.* (2002) Important roles of drought-and cold-inducible genes for galactinol synthase in stress tolerance in Arabidopsis thaliana. *Plant J* 29: 417–426.

Talame V, Ozturk NZ, Bohnert HJ, Tuberosa R (2007) Barley transcript profiles under dehydration shock and drought stress treatments: a comparative analysis. *J Exp Bot* 58: 229–240.

Thaler SJ, Bostock RM (2004) Interactions between abscisic acid mediated responses and plant resistance to pathogens and insects. *Ecology* 85: 48–58.

Tominaga A, Nagata M, Futsuki K, *et al.* (2009) Enhanced nodulation and nitrogen fixation in the abscisic acid low-sensitive mutant enhanced nitrogen fixation of *Lotus japonica*. *Plant Physiol* 151: 1965–1976.

Travaglia C, Reinoso H, Bottini R (2009) Application of abscisic acid promotes yield in field-cultured soybean by enhancing production of carbohydrates and their allocation in seed. *Crop Pasture Sci* 60: 1131–1136.

Travaglia C, Reinoso H, Cohen A, Luna C, Tommasino E, Castillo C, Bottini R (2010) Exogenous ABA increases yield in field grown wheat with moderate water restriction. *J Plant Growth Regul* 29: 366–374.

Trejo CL, Davies WJ (1991) Drought-induced closure of *Phaseolus vulgaris* L. stomata precedes leaf water deficit and any increase in xylem ABA concentration. *J Exp Bot* 42: 1507–1516.

Trewavas A, Jones H (1991) An assessment of the role of ABA in plant development. In: Davies WJ, Jones H (eds), *Abscisic Acid: Physiology and Biochemistry*. Bios Scientific Publishers, Oxford, UK, pp. 169–188.

Trotel-Aziz P, Niogret MF, Larher F (2000) Proline level is partly under the control of abscisic acid in canola leaf discs during recovery from hyper-osmotic stress. *Physiol Plant* 110: 376–383.

Umezawa T, Okamoto M, Kushiro T, *et al.* (2006) CYP707A3, a major ABA 8′-hydroxylase involved in dehydration and rehydration response in *Arabidopsis thaliana*. *Plant J* 46: 171–182.

Umezawa T, Nakashima K, Miyakawa T, *et al.* (2010) Molecular basis of the core regulatory network in ABA responses: sensing, signaling and transport. *Plant Cell Physiol* 51: 1821–1839.

Uno Y, Furihata T, Abe H, Yoshida R, Shinozaki K, Yamaguchi-Shinozaki K (2000) Arabidopsis basic leucine zipper transcription factors involved in an abscisic acid-dependent signal transduction pathway under drought and high-salinity conditions. *Proc Natl Acad Sci USA* 97: 11632–11637.

Vallabhaneni R, Wurtzel E (2010) From epoxycarotenoids to ABA: the role of ABA 8′-hydroxylases in drought-stressed maize roots. *Arch Bioch Biophys* 504: 112–117.

Verslues PE, Zhu JK (2007) New developments in abscisic acid perception and metabolism. *Curr Opin Plant Biol* 10: 447–452.

Vogel JT, Zarka DG, Van Buskirk HA, Fowler SG, Thomashow MF (2005) Roles of the CBF2 and ZAT12 transcription factors in configuring the low-temperature transcriptome of *Arabidopsis*. *Plant J* 41: 195–211.

Wang YX, Suo HC, Zheng Y, *et al.* (2010) The soybean root-specific protein kinase GmWNK1 regulates stress-responsive ABA signaling on the root system architecture. *Plant J* 64: 230–242.

Waters SP, Martin P, Lee BT (1984) The influence of sucrose and abscisic acid on the determination of grain number in wheat. *Exp Bot* 35: 829–840.

Wattana P, Monica AM (1999) Water deficit on raffinose family oligosaccharide metabolism in coleus. *Plant Physiol* 121: 987–993.

Weber H, Rolletschek H, Heim U, Golombek S, Gubatz S, Wobus U (2000) Antisense-inhibition of ADP-glucose pyrophosphorylase in developing seeds of *Vicia narbonensis* moderately decreases starch but increases protein content and affects seed maturation. *Plant J* 24: 33–43.

Weiner JJ, Peterson FC, Volkman BF, Cutler SR (2010) Structural and functional insights into core ABA signaling. *Curr Opin Plant Biol* 13: 495–502.

Westgate ME, Passioura JB, Munns R (1996) Water status and ABA content of floral organs in drought-stressed wheat. *Aus J Plant Physiol* 23: 763–772.

Wilkinson S, Davies WJ. (2002) ABA-based chemical signaling: the coordination of responses to stress in plants. *Plant Cell Environ* 25: 195–210.

Wilkinson S, Davies WJ. (2010) Drought, ozone, ABA and ethylene: new insights from cell to plant to community. *Plant, Cell Environ* 33:510–525.

Wilkinson S, Corlett JE, Oger L, Davies WJ (1998) Effects of xylem pH on transpiration from wild-type and flacca tomato leaves. A vital role for abscisic acid in preventing excessive water loss even from well watered plants. *Plant Physiol* 117: 703–710.

Wilkinson S, Clephan AL, Davies WJ (2001) Rapid low temperature-induced stomatal closure occurs in cold-tolerant *Commelina communis* L. leaves but not in cold-sensitive *Nicotiana rutica* L. leaves, via a mechanism that involves apoplastic calcium but not abscisic acid. *Plant Physiol* 126: 1566–1578.

Wobus U, Weber H (1999) Sugars as signal molecules in plant seed development. *J Biol Chem* 380: 937–944.

Wolf O, Jeschke WD, Hartung W (1990) Long-distance transport of abscisic-acid in NaCl-treated intact plants of *Lupinus albus*. *J Exp Bot* 41: 593–600.

Xing Y, Jia W, Zhang J (2008) AtMKK1 mediates ABA-induced CAT1 expression and H_2O_2 production via AtMPK6-coupled signaling in *Arabidopsis*. *Plant J* 54: 440–451.

Xu ZY, Lee KH, Dong T, *et al.* (2012) A vacuolar beta-glucosidase homolog that possesses glucose-conjugated abscisic acid hydrolyzing activity plays an important role in osmotic stress responses inArabidopsis . *Plant Cell* 24: 2184–2199.

Xu Z, Kim D, Hwang I (2013) ABA homeostasis and signaling involving multiple subcellular compartments and multiple receptors. *Plant Cell Rep* 32: 807–813.

Yamaguchi-Shinozaki K, Shinozaki K (2006) Transcriptional regulatory networks in cellular responses and tolerance to dehydration and cold stresses. *Annu Rev Plant Biol* 57: 781–803.

Yang J, Zhang J, Wang Z, Xu G, Zhu Q (2004) Activities of key enzymes in sucrose-to-starch conversion in wheat grains subjected to water deficit during grain filling. *Plant Physiol* 135: 1621–1629.

Yoshida S, Ito M, Nishida I, Watanabe A (2002) Identification of a novel gene HYS1/CPR5 that has a repressive role in the induction of leaf senescence and pathogen-defense responses in *Arabidopsis thaliana*. *Plant J* 29: 427–437.

Zalewski K, Lahuta L, Horbowicz M (2001) The effect of soil drought on the composition of carbohydrates in yellow lupin seeds and triticale kernels. *Acta Physiol Plant* 23: 73–78.

Zeevaart JAD, Creelman RA (1988) Metabolism and physiology of abscisic acid. *Annu Rev Plant Physiol Plant Mol Biol* 39: 439–473.

Zhang J, Davies WJ (1990a) Changes in the concentration of ABA in xylem sap as a function of changing soil water status will account for changes in leaf conductance. *Plant Cell Environ* 13: 277–285.

Zhang J, Davies WJ (1990b) Does ABA in the xylem control the rate of leaf growth in soil-dried maize and sunflower plants. *J Exp Bot* 41: 1125–1132.

Zhang J, Schurr U, Davies WJ (1987) Control of stomatal behavior by abscisic acid which apparently originates in roots. *J Exp Bot* 38: 1174–1181.

Zhang X, Zhang L, Dong F, Gao J, Galbraith DW, Song C (2001). Hydrogen peroxide is involved in abscisic acid-induced stomatal closure in *Vicia faba*. *Plant Physiol* 126: 1438–1448.

Zheng Y, Huang Y, Xian W, Wang J, Liao H (2012) Identification and expression analysis of the Glycine max CYP707A gene family in response to drought and salt stresses. *Ann Bot* 110: 743–756.

Zhou R, Squires T M, Ambrose S J, Abrams SR, Ross ARS, Cutler AJ (2003) Rapid extraction of abscisic acid and its metabolites for liquid chromatography-tandem mass spectrometry. *J Chromatogr A* 1010: 75–85.

Zhou R, Cutler AJ, Ambrose SJ, Galka MM, Nelson KM (2004) A new abscisic acid catabolic pathway. *Plant Physiol* 134: 361–369.

Zhu JK (2002) Salt and drought stress signal transduction in plants. *Annu Rev Plant Physiol Plant Mol Biol* 53: 247–273.

Zou JT, Abrams GD, Barton DL, Taylor DC, Pomeroy MK (1995) Induction of lipid and oleosin biosynthesis by (+)-abscisic acid and its metabolites in microspore derived embryos of *Brassica napus* L. cv Reston. Biological responses in the presence of 8-hydroxyabscisic acid. *Plant Physiol* 108: 563–571.

CHAPTER 11

Exogenous application of phytoprotectants in legumes against environmental stress

Mirza Hasanuzzaman[1,2], Kamrun Nahar[1,3], Md. Mahabub Alam[1], Shakeel Ahmad[4] and Masayuki Fujita[1]

[1] *Laboratory of Plant Stress Responses, Department of Applied Biological Science, Kagawa University, Kagawa, Japan*

[2] *Department of Agronomy, Faculty of Agriculture, Sher-e-Bangla Agricultural University, Dhaka, Bangladesh*

[3] *Department of Agricultural Botany, Faculty of Agriculture, Sher-e-Bangla Agricultural University, Dhaka, Bangladesh*

[4] *Department of Agronomy, Bahauddin Zakariya University, Multan, Pakistan*

11.1 Introduction

The groups of plants included in the family Fabaceae (or Leguminosae; McNeill *et al.*, 2011) are commonly known as the legumes. Legumes are the third largest land plant family in terms of number of species, and include 730 genera and over 19,400 species (for updates see Stevens, 2014) of trees, shrubs and perennial or annual herbaceous plants. Legumes are essentially distributed worldwide except for Antarctica and the high Arctic (Stevens, 2014). Legumes rank third after cereals and oilseeds in world production (Popelka *et al.*, 2004) and account for 27% of the world's primary crop production (Graham & Vance, 2003). Legumes are utilized as important sources of protein for human beings, and constitute 33% of the dietary protein nitrogen (N) needs of humans (Vance *et al.*, 2000). Apart from their use as human foods, legumes have multiple uses like forage, grain, flowers, pharmaceuticals, industrial, fallow or green manure, timber and aesthetic purposes, which make the legumes one of the most commercially important plant groups.

Abiotic stresses such as salinity, drought, waterlogging, high temperature (HT), low temperature, toxic metals, ozone (O_3), ultraviolet (UV) radiation, nutrient deficiency, and so on are common natural or anthropogenic stresses responsible for reducing the productivity of crop plants by over 50% (Hasanuzzaman *et al.*, 2012a). More than 90% of arable lands or crop-growing areas experience one or more environmental stresses (Dita *et al.*, 2006). Legumes are a large group of plants distributed worldwide and having diverse uses in different continents; hence they experience a wide range of environments some of which might not be favourable for growth and development. Abiotic stresses may affect root growth, interrupt symbiotic interactions, or impede vegetative or reproductive development, and thus limit productivity (Van Hoorn *et al.*, 2001; Sharma & Lavanya, 2002). These stresses also lead to oxidative stress due to excessive generation of reactive oxygen species (ROS) derived from the reduction of molecular oxygen (O_2). ROS include some free radicals such as superoxide ($O_2{\bullet}^-$), hydroxyl radical (•OH), alkoxyl (RO•) and peroxyl (ROO•), as well as non-radical products like hydrogen peroxide (H_2O_2) and singlet oxygen ($^1O_2$) (Gill & Tuteja, 2010). Stress signals induce the synthesis of several metabolites, including phytohormones and osmolytes, and changes in endogenous macromolecules are often involved in plant stress adaptation. Phytohormones also activate cellular responses, including cell death, to diverse stress situations in plants (Eyidogan *et al.*, 2012).

The antioxidant defense system of plants is constituted chiefly of antioxidant enzymes and non-enzymatic antioxidant components and distributed in different cell organelles, such as chloroplasts, mitochondria and peroxisomes, or in the apoplast, which are coordinated to act together (Hasanuzzaman *et al.*, 2012a,b). Non-enzymatic antioxidants may include ascorbic acid (AsA),

Legumes under Environmental Stress: Yield, Improvement and Adaptations, First Edition. Edited by Mohamed Mahgoub Azooz and Parvaiz Ahmad.

glutathione (GSH), phenolic compounds, alkaloids, non-protein amino acids and α-tocopherols. Antioxidant enzymes are superoxide dismutase (SOD), catalase (CAT), ascorbate peroxidase (APX), glutathione reductase (GR), monodehydroascorbate reductase (MDHAR), dehydroascorbate reductase (DHAR), glutathione peroxidase (GPX), guaicol peroxidase (GOPX), peroxidase (POD) and glutathione-*S*-transferase (GST) (Hasanuzzaman *et al.*, 2012a). Polyamines (PAs), aliphatic amines and organic polycations have been recognized as regulators of plant metabolism for over 100 years, for their roles in cell proliferation, somatic embryogenesis, seed germination, growth, morphogenesis, differentiation, reproductive development, fruit set and programmed cell death (Kusano *et al.*, 2008; Hussain *et al.*, 2011). Osmolytes or osmoprotectants, like proline (Pro) and glycine betaine (GB), are organic metabolites that play potential protective roles against stresses (Ashraf & Foolad, 2007; Chen & Murata, 2008). Different signalling molecules have emerged in connection with redox-regulated or defense-related gene expression involved in establishing plant stress tolerance (Sung & Hong, 2010). Some trace elements have been identified as not only promoting growth and development of plants but also increasing resistance to certain abiotic stresses (Feng *et al.*, 2013). The roles of phytohormones, amino acids, antioxidants, signalling molecules and trace elements in overcoming stress effects on plants or in developing abiotic stress resistance in plant species have been studied by many research groups. But the complexities of the species-specific and stress-specific mechanisms in plants mean that much more research is needed.

This chapter presents a comprehensive review regarding the responses of legumes to common environmental stresses, and the roles of different groups of compounds or molecules like phytohormones, amino acids and derivatives, antioxidants, signalling molecules and trace elements, in mitigating the effects of stress in legumes.

11.2 Importance of legumes

Legumes are plants belonging to the family Fabaceae (formerly known as Leguminosae) that bear their seeds in pods and contain nodules on their root system. The latter property makes them different from other crops because they not only provide food and feeds but also can act as potent agents to enrich soil fertility. These and other benefits of legume crops mean that more than 200 species of legumes are known to be cultivated worldwide, including: herbs (e.g. *Lens culinaris, Pisum sativum, Phaseolus vulgaris*), shrubs (*Gliricidia sepium, Calliandra calothyrsus, Sesbania sesban* and *S. grandiflora)* and trees (*Sesbania grandiflora, Moringa oleifera, Leucaena leucocephala*). Some of the major pulses are presented in Figure 11.1.

Globally, legume crops (pulses) are the third most important group of crops after cereals and oilseeds. Developing countries are the major source of pulses, contributing 74% of total production (Khedar *et al.*, 2008). Pulses play an important role in human diets as well as enriching soil fertility, and hence are a major contributor to food security and sustainable agriculture. Nearly 70% of pulses are used for food, mostly in developing countries; 25% are used for feed, mostly in Europe, the Commonwealth of Independent States and Australia; the remaining 5% go for seed and other purposes (Khedar *et al.*, 2008). Being a rich source of protein, minerals, vitamins, crude fibre and several amino acids required for human health, pulses are considered as a health food and offer security against protein malnutrition to millions of people in the developing world. The nutritional compositions of some major pulses are depicted in Table 11.1.

Pulses are a 'wonderful gift' of nature for all living beings, for the soil, and for natural resources and the environment. Every plant of a pulse crop may be called a 'mini N factory' because with the help of *Rhizobium* bacteria, pulses fix atmospheric N_2 and so improve soil fertility (Figure 11.2). Legumes are possibly the most important cultivated crops capable of fixing carbon and N. In many areas, legumes are cultivated as cover crops after the harvest of main crops, first, because they provide economic benefit and, second, they increase the fertility by fixing atmospheric N. Pulse crops can fix 30–200 kg N ha^{-1} $year^{-1}$ (Khedar *et al.*, 2008; Table 11.2). Hence, pulse crops sustain the productivity of the soil and make the soil economically more sound and environmentlly more friendly.

Figure 11.1 Major pulse crops grown in South Asia.

Table 11.1 Nutritional composition (per 100g of edible parts) of the major pulses.

Crop	Energy (kcal)	Protein (g)	Carbohydrate (g)	Fat (g)	Fibre (g)	Ash (g)	Ca (mg)	P (mg)	Fe (mg)
Lentil (*Lens culinaris*)	383	26.9	67.8	0.8	4.3	3.2	71	331	7.7
Grasspea (*Lathyrus sativus*)	379	29.9	65.2	1.2	8.0	3.6	–	447	10.9
Chickpea (*Cicer arietinum*)	360	17.1	60.9	5.3	3.9	–	202	312	4.6
Mung bean (*Vigna radiata*)	334	24.0	56.7	1.3	4.1	3.9	124	326	4.4
Black gram (*Vigna mungo*)	385	23.5	59.6	1.8	4.9	3.8	123	390	9.4
Field pea (*Pisum sativum*)	315	25.6	56.5	1.1	4.5	4.1	75	331	6.6
Cowpea (*Vigna unguiculata*)	323	24.1	54.5	1.0	3.8	–	77	414	8.6
Field bean (*Phaseolus vulgaris*)	346	22.9	60.6	1.7	4.8	6.2	260	425	12.4
Pigeon pea (*Cajanus cajan*)	383	21.9	72.7	1.5	8.1	4.2	179	316	16.6
Soybean (*Glycine max*)	446	36.5	30.2	19.9	9.3	–	377	704	15.7

Source: Khedar *et al.* (2008).

Table 11.2 Amount of nitrogen fixation in soil by different pulse crops.

Pulse crops	Nitrogen fixed in soil (kg/ha/year)
Lentil (*Lens culinaris*)	60–147
Grasspea (*Lathyrus sativus*)	61
Chickpea (*Cicer arietinum*)	41–134
Mung bean (*Vigna radiata*)	30–74
Black gram (*Vigna mungo*)	30–74
Field pea (*Pisum sativum*)	30–125
Cowpea (*Vigna unguiculata*)	198
Field bean (*Phaseolus vulgaris*)	30
Pigeon pea (*Cajanus cajan*)	31–97

Source: Khedar *et al.* (2008).

11.3 Legume responses to environmental stresses

Abiotic stresses are the greatest constraint for crop production worldwide and account for significant yield reductions. Crop plants, being sessile organisms, encounter unavoidable abiotic stresses during their life cycles, including salinity, drought, extreme temperatures, metal toxicity, flooding, UV-B radiation, O_3 and so on, which all pose a serious challenge to plant growth, metabolism and productivity (Hasanuzzaman *et al.*, 2012a, 2013a,b,c,d, 2014a; Hasanuzzaman & Fujita, 2012a, 2013). Although some legumes are moderately tolerant to certain stresses like drought and HT, most are sensitive to various abiotic stresses. However, the responses of legumes vary greatly depending on the plant species, genotype, stress intensity and duration as well as management practices (Sultana *et al.*, 2014). Legumes show various kinds of response under environmental adversities, which result in substantial yield reductions (Table 11.3).

11.3.1 Salinity

Salt stress is one of the most devastating abiotic stresses; it can severely affect agricultural productivity in various ways because it can cause both osmotic stress and ionic toxicity (Hasanuzzaman *et al.*, 2009, 2013a,b). Most cultivated legumes are sensitive to salt stress at all growth stages (Sultana *et al.*, 2014). Compared to cereals, the yield of legumes is affected more at the same level of salinity and hence legumes are considered to be more salt sensitive than cereals (Katerji *et al.*, 2011). This is mainly due to their indeterminate growth cycle, which makes the reproductive stage longer than that of cereals (Katerji *et al.*,, 2005a,b). Apart from the reduction in growth and yield, symbiotic N fixation in leguminous crops is also negatively affected by salinity. In many reports salinity was found to be responsible for reducing the survival rate and replication of the rhizobia and subsequent nodulation (Katerji *et al.*, 2011).

Sidari *et al.* (2007) conducted an experiment with four lentil (*Lens culinaris*) genotypes, which were treated with different levels of salt (50, 100, 150 and 200 mM NaCl), and observed that germination percentage, growth parameters and the relative water content (RWC) decreased with the increase in NaCl concentration. Decreases in plant growth due to elevated levels of salt have been reported in many legume crops, including *V. aconitifolia* (Mathur *et al.*, 2006), *V. unguiculata* (Taffouo *et al.*, 2009) and *V. mungo* (Kapoor & Srivastava, 2010).

Table 11.3 Yield reduction due to different abiotic stress in legumes.

Plant species	Stress treatment and duration	Yield reduction (%)	References
Glycine max	7 dS/m, entire growth period	Seed yield: 40%	Papiernik (2005)
Pisum sativum	6.25 dS/m, 20 DAS	Total yield: 75%	Hussein *et al.* (2006)
Vigna radiata	7.82 dS/m, 3 months	No. of pods: 14%	Hossain *et al.* (2008)
V. radiata	12 dS/m, throughout the growth stages	Seed yield: 67%	Ahmed (2009)
V. mungo	500 mM NaCl, germination stage	Grain yield: 33%	Mensah & Ihenyen (2009)
V. unguiculata	50 mM NaCl, pretreatment	Grain yield: 60%	Taffouo *et al.* (2009)
Arachis hypogaea	12.50 dS/m, throughout the growing period	Yield: 43%	Hammad *et al.* (2010)
G. max	140 mM NaCl	Grain yield: 85%	Hamayun *et al.* (2010b)
G. max	9 dS/m NaCl, throughout the growing period	Grain yield: 78% Oil yield: 77%	Ghassemi-Golezani & Taifeh-Noori (2011)
Phaseolus vulgaris	150 mM NaCl, 48 h	Pod yield: 42%	Rady (2011)
Vicia faba	8.97 dS/m NaCl, 2 weeks	Seed dry weight: 65%	Hellal *et al.* (2012)
P. sativum	150 mM NaCl, 3-day interval during experimental period	Green seed yield: 86%	Howladar & Rady (2012)
Lens culinaris	12 dS/m NaCl, throughout the growing period	Grain yield: 57%	Ghassemi-Golezani & Mahmoodi-Yengabad (2012)
L. culinaris	6 dS/m NaCl	Seed yield: 20%	Islam *et al.* (2012)
P. vulgaris	12 dS/m NaCl, regular irrigation	Grain yield: 53%	Ghassemi-Golezani *et al.* (2012)
V. faba	5 dS/m NaCl, entire growth period	No. of seeds: 65% No. of pods: 51%	Kardoni (2013)
L. culinaris	130 mm evaporation from class A pan after seedling establishment	Grain yield: 58%	Alami-Milani *et al.* (2013)
G. max	Drought by 30–40% field capacity	Seed yield: 39%	Li *et al.* (2013)
V. radiata	25 mg/kg Cd	100-seed weight: 61% Seed yield: 79%	Wahid & Ghani (2008)
V. radiata	12 mg kg^{-1} Cd & at 3 times during different growth stages	No. of seeds plant^{-1}: 95%	Ghani (2010)
G. max	200 mg/kg Cd	Yield: 55%	Abdo *et al.* (2012)
Cicer arietinum	Chilling (5 °C) during seed development stage	Seed weight: 43%	Nayyar *et al.* (2005)
C. arietinum	Chilling (2.3/11.7 °C day/night), 1 month	Seed yield: 38%	Nayyar *et al.* (2007)
G. max	Chilling (3 °C)	No. of pods: 29%	Tacarindua *et al.* (2013)
G. max	Waterlogging, 15 days	Seed dry weight: 58%	Youn *et al.* (2008)
G. max	Waterlogging at pod filling stage, 192 h	Yield: 20%	Rhine *et al.* (2010)
V. radiata	Water logging at vegetative stage, 9 days	Yield: 52%	Kumar *et al.* (2013)
G. max	150 nL/L O_3	Yield: 37%	Betzelberger *et al.* (2010)
G. max	38–120 nL L^{-1} O_3	Yield: 37–39%	Betzelberger *et al.* (2012)
G. max	13 kJ/m^2/day UV-B radiation, 7 h/day	Yield: 40%	Liu *et al.* (2013)
P. sativum	+7.1 kJ/m^2 UV-B radiation	Yield:19%	Agrawal & Mishra (2009)

In contrast, Dantus *et al.* (2005) found that moderate salinity led to increases in plant height in *V. unguiculata*. Lentil is one of the most tolerant legumes to high salinity, as reported by Chattopadhyay *et al.* (2011). They observed that lentil plants exposed to high concentrations of salt (500 mM NaCl) did not show any visible signs of wilting until 36 h of stress. Common bean (*P. vulgaris*) seedlings showed marked decline in shoot and root lengths and plant dry mass, and chlorophyll (chl) and carotenoid contents when exposed to 100 mM NaCl (Rady *et al.*, 2013). Salt stress caused sharp increases in $O_2{\bullet}^-$, H_2O_2 and malondialdehyde (MDA) content, which indicated the sensitivity of bean to salinity. Thus sensitivity of legumes to salinity not only depends on the plant species but also on the genotypes. Tavakkoli *et al.* (2012) compared genotypic variation in faba bean (*Vicia faba*) in response to two levels of salt stress (75 and 150 mM, 49 days). Results revealed that salt stress

Figure 11.2 Root system of *Sesbania aculeata* containing nodules, organs formed in leguminous plants that help to fix N in soil. They are a product of successful interactions between the host plant and the soil bacteria, *Rhizobium* spp.

greatly affected the yield and there was a highly negative correlation between salt concentration and yield. The effects of salt stress on growth and metabolism in *V. faba* was also studied by Qados (2011). Five lentil cultivars varying in their salt tolerance were grown with seawater irrigation (electrical conductivity (EC) 0.9–12 dS/m) by Natsheh *et al.* (2012). They observed that seawater of salinity 3 and 6 dS/m did not cause any negative effects on germination and growth, whereas at higher salinity (EC 12 dS/m) the germination rate was reduced and the mean germination time was greater in all cultivars tested. Chickpea (*C. arietinum*) is highly sensitive to salinity. Haileselasie and Teferii (2012) reported that increased concentration of salt had a negative impact on the germination and growth of chickpea by reducing its germination, water uptake and length of root and shoot. In a laboratory experiment, Ashagre *et al.* (2013) observed the effect of different salt concentrations (50, 100, 150 and 200 mM NaCl) on germination and seedling vigour of two chickpea cultivars, viz. Worku and Shasho. Regardless of the salt concentration, salinity markedly decreased the germination rate and seedling vigour of chickpea. However, a higher germination percentage was recorded in Shasho than in Worku, which indicated its better tolerance to salinity. The dry weight and length of radical and plumule were reduced due to salt stress in dose-dependent ways, but the effects were less in salt-tolerant Shasho than in Worku. In soybean, Queiroz *et al.* (2012) observed a drastic alteration in growth and metabolism when plants were exposed to elevated levels of salt. Soybean receiving 200 mM NaCl exhibited 34% and 62% reductions in shoot length and fresh weight, respectively. Interestingly, upon salt treatment no significant differences were observed in the water content of the aerial parts, whereas this increased significantly in the roots. Ashraf *et al.* (2013) investigated three mung bean (*V. radiata*) cultivars in response to salt stress. They observed marked decreases in plant height, shoot relative water content, photosynthetic pigments, endogenous levels of K^+ and K^+/Na^+ ratios, and elevated levels of H_2O_2, MDA, Na^+ and Cl^-. These parameters differed greatly depending on the cultivars. The most tolerant cultivar (NCM-209) exhibited the smallest decline in K^+, K^+/Na^+ ratio and photosynthetic pigments as well as lower levels of H_2O_2 and MDA. The variations in the endogenous H_2O_2 and MDA under salt stress were modulated by the activities of the antioxidant enzymes, which were greatly dependent on cultivars. Yield reduction due to salt stress differed significantly among the mung bean cultivars; because of its greater tolerance the cultivar NCM-209 showed the highest yield. Rasool *et al.* (2013) studied the relative performance of eight chickpea cultivars at different levels of salinity (25, 50, 75 and 100 mM NaCl). Salinity caused great reductions in fresh and dry weights of root and shoot in dose-dependent ways. The oxidative damage was also higher in the plants exposed to salinity as measured by increased levels of lipid peroxidation. Salinity also reduced the chl content. Although salt stress increased the activities of antioxidant enzymes in all cultivars, the salt-tolerant genotypes (SKUA-06 and SKUA-07) showed higher activities than others. Rasool *et al.* (2013) also grew bean seedlings under different concentrations of salt (60, 120 and 240 mM NaCl) and observed that growth, osmotic potential, chl content and protein content all substantially decreased due to salt stress.

11.3.2 Drought

Among the abiotic stresses, drought is the most complex and devastating on a global scale, and has pronounced adverse effects on the growth and development of crop plants. Drought stress usually leads to reductions in crop yield, which can result from many drought-induced morphological, physiological and metabolic changes in the plants (Hasanuzzaman & Fujita 2011; Hasanuzzaman *et al.*, 2014a). Since drought is closely associated with both plants and soil conditions, improving grain legume productivity in drought-prone environments is considered not only in terms of increased grain yield, but also as improvements to the physical, chemical and biological characteristics of the soil (Subbarao *et al.*, 1995). In fact different legume crops behave differently under drought stress, whereas only a few can tolerate moderate levels of salt stress. Numerous research findings have indicated a strong negative correlation between water shortage in soil and legume growth and yield. Among the growth stages, the reproductive stage is considered the most sensitive to water stress (Katerji *et al.*, 2011). It is noteworthy that drought stress also considerably impairs the symbiotic N fixation in leguminous crops (Katerji *et al.*, 2011).

Zare *et al.* (2012) reported that grain yield of *V. radiata* was decreased by 23% after imposition of drought compared with non-stress conditions; this was attributed to reductions in pod weight, number of pods per plant and plant dry weight under drought stress. While studying lentil and grass pea, Talukdar (2013) observed that plant growth traits and seed yield components were reduced significantly in both crops with increasing duration of water stress. However, lentil was more sensitive to drought compared to grass pea. Drought stress caused decreases in leaf RWC, K^+/Na^+ ratio, chl *a*, chl *a*/*b* ratio, stomatal conductance and net photosynthetic rate. Nodulation of both crops was drastically reduced due to drought stress. Seed yield differences among genotypes under drought stress conditions have been reported in *P. vulgaris* (Terán & Singh, 2002a; Beebe *et al.*, 2008) besides associated decreases in yield attributes such as number of pods per plant, number of seeds per plant, number of seeds per pod and 100-seed weight (Terán & Singh, 2002b; Beebe *et al.*, 2008, 2010). Under drought stress (non-irrigated conditions) *P. vulgaris* showed 65%, 34%, 29% and 12% reductions in seed yield, seed number per square metre, pod number per square metre and 100-seed weight compared to irrigated conditions (Assefa *et al.*, 2013). However, there was great variation among the genotypes tested for drought stress response. Importantly, pod harvest index, which was reported as a selection criterion to improve drought resistance in white pea bean, was greater in drought-tolerant genotypes. Under laboratory conditions, polyethylene glycol (PEG) is widely used to impose drought stress in plants. In an experiment, Muscolo *et al.* (2014) investigated four lentil genotypes (Castelluccio, Eston, Pantelleria and Ustica) under different levels of drought (10, 15, 18 and 21% PEG-6000). Drought stress caused reductions in germination percentage, root length, tissue water content, α- and β-amylase and α-glucosidase activities, and osmolyte content in dose-dependent manners. The drought-induced damage was variable depending on the genotypes – drought-tolerant Eston and Castelluccio genotypes exhibited higher accumulations of Pro and soluble sugars as well as greater activities of antioxidant enzymes.

11.3.3 Toxic metals

The rapid increase in human population together with speedy industrialization is causing serious environmental problems, including the production and release of considerable amounts of toxic metals into the environment (Sarma, 2011). Excessive uptake of toxic metals by plants may occur, and these elements may participate in some physiological and biochemical reactions that can harm the normal growth of the plant by disturbing absorption, translocation or synthetic processes (Hasanuzzaman & Fujita, 2012a,b). Most of the cultivated legumes are sensitive to toxic metals; however, many legumes have been found as natural pioneers in contaminated sites undergoing revegetation, prompting their use for remediation purposes (Dashti *et al.*, 2009; Khan *et al.*, 2009). For instance, *Lupinus albus* L., *Mimosa caesalpiniaefolia*, *Erythrina speciosa* and *Schizolobium parahyba* have been proposed as potential species to remediate sites, leaving aside considerations of growth or biomass production in soils contaminated with metals like zinc (Zn) and lead (Pb) (Pastor *et al.*, 2003; de Souza *et al.*, 2012). Heavy metals also hamper the legume-bacteria symbiosis. Talano *et al.* (2013) observed considerable reduction in germination and growth of soybean plants at early stages of growth when they were exposed to 10 μM arsenate or arsenite. In addition, arsenic (As) contamination also reduced the number of effective nodules, even after inoculation with *Bradyrhizobium japonicum* – a metalloid-tolerant microorganism. It was also observed that roots accumulated more arsenic (250 times higher) than shoots (Talano *et al.*, 2013).

Ergün and Öncel (2010) investigated the toxic impacts on growth of *L. esculenta* of cadmium (Cd) and Zn over a concentration gradient of 250 mg/kg. Increased levels of Cd caused a huge decrease in root and shoot growth as well as their dry weight. Both Cd and Zn at 250 mg/kg resulted in increased Pro content, but this decreased upon exposure to 500 mg/kg of these metals (Ergün & Öncel, 2010). Cokkizgin and Cokkizgin (2012) also investigated the effects of Pb stress on germination of four lentil genotypes. They observed that germination percentage, vigour index, plumule length and radicle length were decreased after Pb exposure in dose-dependent ways. For instance, the germination percentage was decreased from 98.5% to 32.8% upon exposure to 4.5 mM $PbCl_2$. These elevated level of Pb also decreased the vigour index to 175.4, compared with 1027.4 in control plants. However, all of these parameters varied greatly within genotypes. Fatoba *et al.* (2012) determined the comparative response of both *Arachis hypogaea* and *G. max* grown in soil with different levels of Cd and Pb (10, 20, 30 and 40 mg/kg). These metals at high concentrations (40 mg/kg) caused drastic reductions in germination, growth and productivity. Cadmium treatment at 10 mg/kg concentration produced a significant ($P<0.05$) effect on seed germination in *A. hypogaea* and *G. max*. Increasing the concentration to 30 mg/kg reduced the numbers of leaves, leaf area, number of pods per plant and number of seeds per plant (Fatoba *et al.*, 2012). However, further increasing the concentration of Cd to 40 mg/kg drastically reduced all the growth indices significantly compared to the controls. In *G. max* a lead concentration of 20 mg/kg greatly affected the leaf area, number of pods per plant and number of seeds per plant, while a significant effect was recorded in *A. hypogaea* under a regime of 10 mg/kg in number of leaves, height, leaf area and number of seeds per plant (Fatoba *et al.*, 2012). Both Cd and Pb at 30 mg/kg resulted in a great reduction in the leaf sizes in *G. max* and *A. hypogaea*. Different toxic metals also cause oxidative stress in plants, which is a general phenomenon for both redox-active or non-active metals. As reported by Saidi *et al.* (2013), exposure of *P. vulgaris* to 20 µM Cd resulted in significant decreases in chl and carotenoid concentrations, with concomitant enhancement of electrolyte leakage, H_2O_2 content and accumulation of MDA. Plants exposed to Cd also showed impairment of antioxidant metabolism. While studying *L. culinaris*, Janas *et al.* (2010) observed a sharp reduction of growth and increased lipid peroxidation and H_2O_2 accumulation when the plants were exposed to copper (Cu) at variable concentrations (0.05, 0.1 and 0.5 µM).

11.3.4 High temperature (HT)

One of the predicted negative effects of global warming on plants is the damage due to HT on plant growth, development and yield. The ever-increasing threat of climate change including very HT might lead to catastrophic declines in crop productivity and result in massive food insecurity (Bita & Gerats, 2013). High-temperature stress results in detrimental effects on legumes, but the adverse temperature range varies for different kinds of legume. The maximum threshold seed zone temperature for the emergence of *V. unguiculata* is about 37 °C. Among the growth stages, the reproductive phase of plants is highly sensitive to HT stress, and plant fertility is considerably reduced with increases in temperatures. Pre-anthesis heat stress in *P. vulgaris* caused anomalous pollen and anther development (Porch & Jahn, 2001). Heat stress caused abscission and abortion of flowers, young pods and developing seeds in soybean. Lower pollen viability at 37/27°C (day/night) reduced pod setting in soybean (Kitano *et al.*, 2006). The heat stress tolerance of four *A. hypogaea* cultivars was investigated; the HT stress (28/22 or 38/22°C from 21 to 90 days after planting) reduced total dry weight by 20 to 35%, and the seed harvest index by 0 to 65% (Craufurd *et al.*, 2002). *Cicer arietinum* exposed to 35/16°C (day/night for 10 days) during flower and pod development stages reduced the number of pods per plant by 53% and seed yield by 48% (Gan *et al.*, 2004). High-temperature stress at pre-anthesis may reduce pollen viability and germinability, and these factors were investigated as common reasons for poor fruit set in legumes including common bean and peanut (Gross & Kigel, 1994; Kozai *et al.*, 2004). Heat stress reduced the chl *a* content of *L. esculentum* leaves by 10–32%, chl *b* by 10% and chl *a*/*b* by 5%, which was the reason for the reduction of photosynthesis (Malgorzata *et al.*, 2009). In *P. vulgaris* phenology, partitioning, plant-water relations, and shoot growth and extension were all hampered by heat stress (Koini *et al.*, 2009). At normal temperatures, seeds of *Cassia tora* exhibited 92% germination. But at 40, 50 and 60 °C (for 10 days) the germination percentages were decreased to 85, 63 and 32%, respectively (Pant *et al.*, 2012). Shoot growth in hydroponically grown *P. aureus* seedlings was reduced at 40/30 and 45/35 °C by 18% and 34%, respectively. Root growth at those temperatures was reduced by 13% and 23%, respectively (Kumar *et al.*, 2011). Heat stresses significanty altered photosynthesis, reproductive development and yield. *Glycine max* exposed to HT stress (38/28 °C, day/night) for 14 days during the flowering stage showed a 20% decrease in photosynthesis and 16% decrease in stomatal conductance; a decreased total chl content (18%), chl *a* content (7%) and chl *a*/*b* ratio (3%); increased thickness of the palisade and spongy mesophyll layers of leaf tissue (Tan *et al.*, 2011). High temperatures inhibit flower differentiation and development, and reduce the size of the ovaries of leguminous flowers (Johkan *et al.*, 2011). As in other plant species, HTs also cause oxidative stress by altering the antioxidant system. In *G. max* HT (38/28 °C, day/night) stress resulted in oxidative stress and damaged plasma membrane, and distorted mitochondrial membranes, cristae and matrix (Tan *et al.*, 2011). Heat stress (45/40 °C) created oxidative stress in *C. arietinum* plants together with altered activities of antioxidant enzymes SOD, CAT, APX, GR and concentrations of the non-enzymatic antioxidants AsA and GSH (Kaushal *et al.*, 2011). Exposure of *V. mungo* seeds to 50 °C for 10, 20 and 30 min significantly reduced germination and vigour index (Piramila *et al.*, 2012).

11.3.5 Chilling

Studies with some legume species (*Melilotus albus, Medicago lupulina, Lotus corniculatus, Trifolium pretense* and *T. repens*) with different combination of temperature treatments revealed that low temperatures negatively affected the water uptake and germination under unfavourable conditions. Seeds remained impermeable and dormant until the next favourable season (Van Assche *et al.*, 2003). Depending upon the sensitivity of varieties to chilling, low temperature (<10 °C) at flowering has adverse effects on *C. arietinum* L. production, which may cause 15–20% yield losses (Chaturvedi *et al.*, 2009). Different legumes like field pea, faba bean, lentil and chickpea were very sensitive to chilling and freezing temperatures. The reproductive phases, viz. flowering, early pod formation and seed filling, are most sensitive to low temperatures. Low temperatures at the reproductive stage may cause flower abortion, poor pod set and impaired pod filling, and drastic reductions of yield and quality (Maqbool *et al.*, 2010). In *C. arietinum* L. biochemical parameters were adversely affected at low temperatures (5 and 10 °C); these included increased electrolyte leakage, total soluble sugars and total free amino acids. The chilling-tolerant genotypes showed better performance compared to sensitive genotypes (Chohan & Raina, 2011). Chilling temperatures (2 and 4 °C) were responsible for causing oxidative stress in relation to the activities of downregulated antioxidant enzymes in non-acclimated *C. arietinum*. However, in acclimated plants the activities of SOD, APX, GR and POD increased, which enhanced their chilling tolerance (Turan & Ekmekçi, 2011). Low temperatures (5 and 10 °C) affected morphological and reproductive attributes of *C. arietinum* genotypes. Morphological characters, viz. plant height, number of branches, number of leaves, pollen viability percentage, were better in chilling-tolerant genotypes compared to sensitive genotypes (Chohan & Raina, 2011). *Stylosanthes guianensis*, an important forage legume in tropical and subtropical regions, is sensitive to chilling stress. The physiological mechanism of chilling resistance, and the physiological impact of chilling on antioxidants and photosynthesis in chilling-resistant mutants and the parent cultivar CIAT184 were analysed. Chilling temperature reduced net photosynthetic rate, the maximum photochemical efficiency, quantum efficiency of photosystem II (PSII) photochemistry, photochemical quenching efficiency and excitation energy allocated to PSII photochemistry. Compared to the parent plant, these values were higher in mutants (Lu *et al.*, 2013). At low temperatures (6 °C) for up to 4 days the *S. guianensis* mutants had higher activities of SOD, CAT and APX, and higher contents of AsA and GSH compared to its parental line. These improved antioxidant systems provided protection for photosynthesis against chill-induced oxidative damage and increased chill resistance in mutants (Lu *et al.*, 2013).

11.3.6 Flooding

Waterlogging or flooding is damaging for some legumes such as pigeon pea and soybean. The performance of two pigeon pea (*C. cajan*) genotypes, ICP 301 (tolerant) and Pusa 207 (susceptible), was examined under waterlogged conditions. Waterlogging for 4 to 6 days caused yellowing and senescence of leaves, and decreased leaf area, dry matter, relative water content and chl content of leaves. The more severe consequences were observed in the susceptible Pusa 207 than in ICP 301 (Kumutha *et al.*, 2009). Waterlogging-induced declines in relative water content, membrane stability index and chl content, and increased chl *a*/*b* ratio, were greater in the susceptible Pusa Baisakhi mung bean (*V. radiata*) cultivar than in a T44 (tolerant) cultivar (Sairam *et al.*, 2009). Soybean plants were subjected to flooded growing conditions at different growth stages: five trifoliate vegetative stages, full-bloom flowering stage and pod fill reproductive stage. Plants were able to withstand flooding for 48–96 h without crop injury. Flooding at the trifoliate vegetative growth stage entailed the least amount of yield loss. The greatest yield losses occurred from flooding at the pod fill reproductive stage – 20–39% less compared to non-flooded checks (Rhine *et al.*, 2010). Flooding stress significantly reduced photosynthetic rate and leaf water potential in mung bean. This stress also damaged lateral roots, to which the plant responded by forming numerous adventitious roots (Islam *et al.*, 2010). Nodulation of legumes is an important factor that is affected by flooding. Upon waterlogging, the roots have no regular root hairs and the water-stressed legumes develop numerous lateral roots. The bases of these lateral roots form abundant nodules, as observed in *Sesbania rostrata* (Capoen *et al.*, 2010). Waterlogging (for 4 and 6 days) is also responsible for oxidative stress in the pigeon pea (*C. cajan*) cultivars ICP 301 (tolerant) and Pusa 207 (susceptible), as indicated by generation of $O_2^{\bullet -}$ and H_2O_2. Oxidative stress decreased the membrane stability index in roots and leaves, and increased thiobarbituric acid-reactive substances (TBRS) over control plants. Injury indices were higher in the susceptible cultivar. With increased oxidative stress, the activities of antioxidant enzymes were also increased, such as SOD, APX, GR and CAT. These were higher in ICP 301, and were correlated with lower oxidative damage and enhanced tolerance compared to Pusa 207 (Kumutha *et al.*, 2009). Waterlogging stress (3 and 7 days) in soybean resulted in increased lipid peroxidation and H_2O_2 levels in roots. Besides these, some proteins have been identified in stressed plants that are anaerobically induced, waterlogging-responsive proteins; these serve for signal transduction, programmed cell death, RNA processing, redox homeostasis and metabolism of energy (Alam *et al.*, 2010). A gradual increase in water levels exacerbated defoliation in *Lotus tenuis*. Flooding consistently increased plant height, probably as a response to cope with flood conditions at the expense of food reserves (Striker *et al.*, 2011).

11.3.7 Ozone

Tropospheric O_3 is a major threat to crop productivity and forest ecosystems (Heagle, 1989). It can also be generated from anthropogenic activity like fossil fuel combustion and biomass burning, and from biogenic emissions (Kesselmeier & Staudt, 1999), and is formed by photochemical oxidation of primary pollutants such as N oxides, hydrocarbons and carbon monoxide (Lelieveld & Crutzen, 1990). Ozone is responsible for various injuries in plants. Leaf chlorosis, necrosis and mottling were observed in *V. faba* and *P. sativum* L. upon exposure to O_3. Reducing sugars, non-reducing sugars, total sugars, fats, total protein and dissolved protein all decreased due to O_3 exposure. The decreases were greater with higher O_3 concentrations, and *V. faba* L. was more sensitive to ozone than *P. sativum* L. (Almohisen & Alyemani, 2013). Ozone enters through stomata of leaves and is rapidly broken down into different ROS, which cause oxidative damage (Yan *et al.*, 2010). Studies of the effect of O_3 on 38 different accessions of the model legume, *M. truncatula*, proved that O_3 causes substantial physiological disorder including modification of antioxidant defence systems. Low stomatal conductance and reduced transpiration rate were characteristic features of O_3 stress, and increased ROS production and lipid peroxidation were evident. Ozone increased the levels of reduced ascorbate and total ascorbate, total GSH and reduced GSH; this stress also significantly raised the concentrations of oxidized forms of ascorbate and glutathione (Puckette *et al.*, 2007). Ozone fumigation of *Trifolium pratense* and *T. repens* leaves induced elevated oxidative stress with a decrease of total glutathione content, mainly due to the strong increase in glutathione disulphide (GSSG) and decrease in GSH (Scebba *et al.*, 2003).

11.3.8 Ultraviolet radiation

Solar UV-B irradiance has increased due to stratospheric ozone depletion, which has impacted adversely on living organisms including plants. Most research has shown that higher intensities and longer duration of UV radiation are harmful for legumes, whereas some research findings also showed that low-intensity, short-duration exposure to UV radiation might have some benefits. Natural levels of UV-B radiation (UV-B: 280–315 nm) reduced soybean growth and yield, which was correlated with CAT and APX activity and epidermal transmittance (Mazza *et al.*, 2013). Exposure to enhanced levels of UV-B decreased total biomass accumulation, harvest index, root:shoot ratio and yield in *V. radiata* L. cultivars, and *Malviya janpriya* was more responsive to UV-B than *M. jyoti* (Agrawal *et al.*, 2006). The impact of UV-B radiation (UV-B: 280 to ~320 nm) on growth and biochemical and antioxidant enzyme activity was studied in seedlings of the green manure plant *Indigofera tinctoria* L. UV-B radiation for 4 to 8 days significantly reduced leaf area, shoot length, and fresh and dry weight of seedlings. It decreased chl and carotenoid contents but increased flavonoid and anthocyanin contents. These seedlings were burdened with high levels of ROS produced under UV-B stress that also altered antioxidant and enzymatic systems as evidenced by decreased CAT and polyphenol oxidase activities, and increased peroxidase, SOD and phenylalanine ammonia-lyase activities (Ravindran *et al.*, 2010). After the fifth day of exposure to UV-B radiation (7 and 9 kJ/m^2 daily doses), shoot heights of *P. sativum* were decreased by 58 and 59%, and leaf area was decreased by 52 and 63% compared to controls. Shoot dry mass was reduced by 55% under 9 kJ/m^2 of UV-B. Significant decreases of leaf stoma counts, chl *a* and chl *b* contents, chl *a*/*b* ratio and carotenoid content were also noted under UV-B stress (Juozaityte *et al.*, 2008). Increased shoot weight and length, root length and weight, leaf area and number of nodules were observed due to UV-C treatment of seeds of groundnut and mung bean (10, 15, 30 and 60 min) and the reason was a reduction in root-infecting fungi. The same UV-C treatment also resulted in increments in total chl and carbohydrate contents (Siddiqui *et al.*, 2011). Short-term (10, 20 and 30 min) exposure to UV-B (200–400 nm) radiation enhanced salinity tolerance in *V. radiata* (Hamid *et al.*, 2012a).

11.4 Application of phytoprotectants for enhancing stress tolerance

Although environmental stresses are considered the major cause of crop losses in pulses, and most of them are sensitive to environmental extremes, the mechanisms of stress tolerance are not yet fully understood and attempts to improve yield under stress conditions by plant breeding have been largely unsuccessful, primarily due to the multigenic origin of the adaptive responses. Exploring suitable ameliorants or stress alleviants is one of the tasks of plant biologists. In recent decades, exogenous protectants such as osmoprotectants, plant hormones, antioxidants, signalling molecules, polyamines and trace elements have been found to be effective in mitigating stress-induced damage in plants. These phytoprotectants can enhance plant growth and yield as well as conferring stress tolerance (Table 11.4).

11.4.1 Phytohormones

Phytohormones are chemical compounds that are produced in one part of a plant and transported to exert an effect on another part. They are critical for plant growth, development and stress signalling, or defense signal networking, and influence physiological and biochemical processes in both synergistic and antagonistic ways. The adaptive responses of crops under abiotic stresses are mainly controlled by phytohormones. So, phytohormone homeostasis and signalling are essential for improving plant performance under stressful environments (Eyidogan *et al.*, 2012).

11.4.1.1 Abscisic acid

Abscisic acid (ABA) acts as an endogenous messenger in a plant's perception and adaptive responses to abiotic stress. It is a chief regulator of plants' water status; it can specifically target guard cells to induce stomatal closure, and signal systemically for adjustment in case of severe water shortage (Swamy & Smith, 1999; Mahajan & Tuteja, 2005). The signals generated by ABA are vital throughout a plant's life cycle to control developmental processes during normal or favourable growth conditions as well as under stressful conditions. Stress-responsive genes can be expressed either through an ABA-dependent or ABA-independent pathway (Chinnusamy *et al.*, 2004). That is why ABA is known as the plant stress hormone (Swamy & Smith, 1999; Mahajan & Tuteja 2005).

Drought stress was imposed at the 50% flowering stage of mung bean plant, and reduced root and shoot fresh and dry weights. Drought decreased pink bacteroid tissue diameter and number of root nodules. Exogenous ABA (10^{-6} M) treatments alleviated the damage to a great extent and improved drought tolerance. The tolerance mechanisms provided by ABA were increase in chl content and RWC of leaves; increase in Pro, protein and sugar contents; and enhanced activities of antioxidant enzymes of leaves. All these contributed to better performance of the plants and better survival of *Rhizobium* in soil following ABA application (Farooq & Bano, 2006). Abscisic acid (100 μM) alleviated the harmful effects of drought stress in soybean plant (*G. max* L. cv. Giza 22) by regulating endogenous phytohormones and polyamine contents. Total soluble protein and the protein banding pattern of soybean was also improved by ABA (Hassanein *et al.*, 2009). *Vicia faba* L. seedlings exposed to water-deficient growing media showed reduced plant height and root length, root and shoot fresh weights, and root and shoot dry weights compared with controls. This stress also reduced total soluble carbohydrates, chl *a* and *b* contents, and

Table 11.4 Protective effects of phytoprotectants on the growth and physiology of legumes grown under environmental stresses.

Plant species	Stress treatment and duration	Treatment with phytoprotectants	Major protective effects	References
Pisum sativum	70 mM NaCl, 25 d	25–100 mM SA, 7 days × 2 sprays	• Reduced plant growth • Imbalanced antioxidant metabolism	Barba-Espín *et al.* (2011)
Vigna sinensis	25, 50, 100, and 150 mM NaCl, 45 days	0.05 mg/L brassinolide, 2 sprays	• Increased fresh weight, dry weight and length of root and shoot • Increased antioxidant enzymes' activities • Decreased lipid peroxidation	El-Mashad & Mohamed (2011)
Glycine max	100 mM NaCl, 13 days	0.5, 1.0 and 5.0 mM GA_3, 13 days	• Increased plant height and fresh weight • Maintained the hormonal balance	Hamayun *et al.* (2010a)
G. max	60 mM NaCl, 2 weeks	20 and 30 mM MeJA, 24 h	• Increased plant growth • Increased chl content, leaf photosynthetic rate, leaf transpiration	Yoon *et al.* (2009)
Cicer arietinum	25 mM NaCl, 2, 4 and 6 days	0.2 and 1 mM SNP, 2, 4 and 6 days	• Increased RWC • Decreased relative membrane injury	Sheokand *et al.* (2010)
Phaseolus vulgaris	100 mM NaCl, 48 h	25, 50, 100 mM AsA as a nutrient solution	• Increase chl content • Decreased MDA content	Dolatabadian & Jouneghani (2009)
P. mungo	Seed soaked in 25, 50, 100 and 200 mM NaCl, 2 h	15 mg/L IAA foliar application	• Increase shoot and root length, fresh and dry weight, leaf area, chl content, soluble sugar • Decrease Pro content	Devi *et al.* (2012)
G. max	50 mM NaCl	Soaked in 400 mg/L AsA for 4 h before sowing	• Increased shoot length, germination percentages, total dry weight and protein content • Increased POD and CAT activities • Decreased SOD activity	Dehghan *et al.* (2011)
Trigonella foenum-graecum	50, 100 mM NaCl during germination and seedling stagedevelopment	Seed soaked in 100 mg/L AsA, 24 h	• Increased germination percentage, shoot length, seedling vigour, chl content	Behairy *et al.* (2012)
G. max	EC = 11.1 dS/m during experiment period	2.5, 5, 7.5 and 10 kg/ha GB, six foliar sprays at flowering stages	• Increased number of lateral branches, number of pods per plant, number of grains per pod, 1000-grain weight	Rezaei *et al.* (2012)
G. max	Water stressed −90 to − 100 kPa during experiment period	1.1 kg/ha boric acid, flowering and seed-filling stages	• Increased grain weight, nodule mass, protein content	Bellaloui *et al.* (2013)
G. max	12% PEG, 14 days	200 mg/L Si, 2 weeks	• Increased shoot length, shoot dry weight	Hamayun *et al.* (2010b)
V. unguiculata	−1.97 to −2.01 MPa, 7 days	5 mM SA foliar spray during water-deficit stress period	• Increased plant biomass, vitamin B_{12} at vegetative and reproductive stages • Increased chl content, NR activity at vegetative stages	Umebese & Bankole (2013)
V. unguiculata	Mild water deficit, 14 days	Se 0.2 mg/L, 2 times	• Decreased MDA, GSH content • Increased AsA, RWC content, leaf/plant, plant height • Increased activities of CAT and GPX	Ajiboso & Adenuga (2012)

Table 11.4 (Cont'd)

Plant species	Stress treatment and duration	Treatment with phytoprotectants	Major protective effects	References
Vicia faba	Water stress, 30 days	20 mM Pro, 30 days	• Increased root length, plant height, root and shoot fresh weight, root and shoot dry weight, leaf number, leaf area, chl content • Decreased MDA content • Increased CAT and POD activities	Ali *et al.* (2013)
P. aureus	10 μM As, 10 days	5 μM Se, 10 days	• Increased chl content • Increased SOD, APX, CAT, GR and GST activities • Decreased MDA, H_2O_2 content	Malik *et al.* (2012)
V. radiata	9 μM Cd, 72 h	10 μM ABA, 12 h	• Increased root number, root fresh weight, • Increased POD activity and GSH content • Decreased activities of SOD and CAT	Li *et al.* (2014)
V. radiata	10.0 mM $AlCl_3$, 7 days	10^{-8} M of 24-epibrassinolide, foliar spray	• Increased shoot and root length, root and shoot fresh weight, root and shoot dry weight • Increased RWC, water use efficiency • Increased chl, stomatal conductivity, internal CO_2 concentration • Increased carbonic anhydrase activity and photosynthetic rate	Ali *et al.* (2008)
P. aureus	50 μM Cd, 3 days	100 μM SA, 16 h	• Decreased TBARS content, relative electrolyte leakage, H_2O_2 level, $O_2{\bullet}^-$ content • Increased SOD and APX activities • Decreased CAT activity	Zhang *et al.* (2011)
G. max	45 μM Hg(II) every week as a solution during 4 weeks	60 μM Se(IV), 4 times	• Forms a high molecular weight entity containing Se and Hg in plants	Yathavakilla & Caruso (2007)
V. radiata	0.2 mg/L Cd, 10 days	6.3, 7.3, and 8.3 mg/L Zn, 3 times	• Increased plant height, fresh weight, number of root nodules, chl content • Increased activities of NR, GOGAT, GDH and protein content	Kumari *et al.* (2011)
P. aureus	1, 25, 50, 100, 250 mg/L As and Cd, 5 days	Seed soaked in 25 mg/kg Zn	• Increased seed germination, germination index and seedling vigour index • Increased metal-helating activity	Patel *et al.* (2013)
Cajanus cajan	5 mM Cu, 12 days	1 μM JA, 12 days	• Increased germination (%), root and shoot length, total chl, carotenoids • Decreased MDA and Pro contents • Decreased SOD and POD activities	Poonam *et al.* (2013)
V. radiata	Chilling (5 °C), 10 h)	1 mM ABA, 24 h	• Decreased levels of MDA and H_2O_2 • Increased total chl content and carbohydrates • Increased CAT, POD and APX activities	Saleh (2007)
C. arietinum	Chilling (2 °C), 24 h)	500, 1000 μM SA	• Increased RWC, chl content • Decreased electrolyte leakage	Imami *et al.* (2011)

ABA, abscisic acid; AsA, ascorbic acid; CAT, catalase; chl, chlorophyll; DAS, days after sowing; EC, electrical conductivity; GA_3, gibberellic acid; GB, glycine betaine; GDH, glutamate dehydrogenase; GOGAT, glutamate synthase; GPX, glutathione peroxidase; GR, glutathione reductase; GSH, glutathione; GST, glutathione-*S*-transferase; IAA, indole acetic acid; JA, jasmonic acid; MDA, malondialdehyde; MeJA, methyl jasmonic acid; NR, nitrate reductase; PEG, polyethylene glycol; POD, peroxidase; Pro, proline; RWC, relative water content; SA, salicyclic acid; SNP, sodium nitroprusside; SOD, superoxide dismutase; TBARS, thiobarbituric acid-reactive substances.

activities of peroxidases (POX) and CAT, whereas water stress increased MDA and Pro levels. Compared to water-stressed seedlings alone, when seedlings were supplemented with 0.1 mM ABA under water stress they showed improved growth parameters, increased antioxidant enzyme activities and decreased MDA and Pro levels (Ali *et al.*, 2013). Increased leaf and xylem ABA concentrations were observed in salt-treated (1, 50, 100 or 200 mM NaCl; 30 days) *Medicago citrina*, which showed higher ion (Na^+/Cl^-) compartmentation and gas exchange parameters compared to *M. arborea* (Sibole *et al.*, 2003). High-temperature stress (35/30, 40/35 and 45/40 °C day/night) resulted in oxidative stress in *C. arietinum*. Exogenous ABA (2.5 μM) regulated endogenous levels of ABA and osmolytes, reducing oxidative damage by decreasing MDA and H_2O_2 levels. Treatment with ABA also improved the growth of *C. arietinum* under HT stress compared to no treatment (Kumar *et al.*, 2012). Chilling stress (5 °C for 5 or 10 h) induced significant increases in lipid peroxidation, membrane leakage and H_2O_2 level, as well as decreased activities of CAT, POD and APX in *V. radiata* seedlings. Total chl, total carbohydrate, protein and Pro levels were also decreased by chilling stress. Exogenous ABA (0.5 and 1 mM) treatments ameliorated the chilling injuries by enhancing antioxidant enzymes that reduced lipid peroxidation, membrane leakage and the H_2O_2 level. Increaes in total chl, carbohydrates, protein content and Pro levels were other beneficial effects of ABA treatment (Saleh, 2007). Under $CdCl_2$ (1, 3, 5, 7 and 9 μM) stress the adventitious root development of *V. radiata* seedlings was inhibited. On the other hand, ABA (1, 5, 10 and 15 μM) enhanced the number and fresh weight of adventitious roots by ameliorating the adverse effects. Exogenous ABA increased indole acetic acid (IAA) oxidase and phenol levels to a greater extent than the controls. Cadmium increased oxidative stress by significantly reducing the activities of SOD, APX, POD and IAA oxidase, as well as GSH level, but increased AsA content. But ABA upregulated antioxidative defence systems towards the alleviation of wounding and Cd-induced oxidative stress (Li *et al.*, 2014).

11.4.1.2 Salicylic acid

The roles of salicylic acid (SA) in seed germination, growth, flowering and fruit yield are long documented (Klessig & Malamy, 1994). Its role in fundamental physiological processes like ion uptake and transport (Harper & Balke, 1981), photosynthetic rate, stomatal conductance, transpiration and glycolysis has also been examined (Khan *et al.*, 2003). The presence of SA in a wide range of plant species and its role in basic plant processes have made SA a potent phytohormone in responses to environmental stresses (Raskin, 1992; Yalpani *et al.*, 1994; Senaratna *et al.*, 2000; Alam *et al.*, 2013; Hasanuzzaman *et al.*, 2014b,c).

Salt stress (67, 134, 202 and 270 mM NaCl) hampered different growth parameters of *V. radiata*. But seed priming with SA (75, 150 and 300 mg/L) alleviated salt stress effects and improved germination rate, vigour, root length, shoot length and dry weight (Entesari *et al.*, 2012). The physiological performace of two mung bean cultivars, one salt-tolerant (Pusa Vishal) and one salt-sensitive (T44), was examined under salt stress (50 mM NaCl) alone or in combination with SA (0.5 mM). The application of SA increased N and sulphur (S) assimilation, GSH content, and activity of APX and GR. The SA addition with salinity also restricted Na^+ and Cl^- contents in leaf, and maintained greater efficiencies of PSII, photosynthetic N-use and water relations. In response to SA under salt stress, Pusa Vishal performed better than T44 (Nazar *et al.*, 2011). The positive effects of SA have also been documented in developing salt tolerance in faba bean (Azooz, 2009). The ameliorative impact of salicylic acid (SA) on *V. unguiculata* L. during water deficit stress was investigated. Plants were subjected to water deficit stress (7 days) at the vegetative stage and reproductive stage, and leaf water potentials (ψw) of −1.9 MPa and −2.01 MPa were maintained, respectively. Stress reduced almost all parameters studied. Foliar applications of SA (3 and 5 mM) increased leaf water potential by 27%, chl content by 94%, plant biomass by 75%, nitrate reductase (NR) activity by 7% and Pro content by 38% in the vegetative stage. Similar but minor improvements were observed in the reproductive stage. Stress increased the levels of Pro and the antioxidant vitamin B_{12} in the vegetative stage and vitamin C in the reproductive stage; SA further increased osmolyte Pro and antioxidants, which maintained better water potential and growth under stress (Umebese & Bankole, 2013). Seed imbibition following soil drenching with different levels of SA (0.05, 0.1, 0.5, 1.0 and 5.0 mM) or acetyl SA improved the tolerance of *P. vulgaris* against HT stress (54 °C, 3 h), chilling stress (0 °C) and drought stress (withholding water for 7 days) as evidenced by increased plant height, vigour and leaf number, compared to untreated stressed plants (Senaratna *et al.*, 2000). Chilling stress (2 °C) imposed on *C. arietinum* L. resulted in several negative impacts on growth and physiological parameters. But application of 500 and 1000 mM of SA reduced chilling stress effects and improved plant height, fresh weight, chl content, relative water content and Pro content, and also reduced electrolyte leakage compared to chilling stress alone (Imami *et al.*, 2011). Cadmium stress (50, 100 µM $CdCl_2$; for 3 and 6 days) reduced the growth of *P. aureus* and *V. sativa* and significantly increased the production of H_2O_2 and $O_2{\bullet}^-$. Seed-soaking pretreatment with 100 µM SA for 16 h decreased Cd-induced production of H_2O_2 and $O_2{\bullet}^-$ by enhancing activities of symplastic and apoplastic SOD, symplastic APX and apoplastic CAT under Cd stress. The SA also improved the growth of these two species under Cd stress (Zhang *et al.*, 2011). Zhou *et al.* (2009) reported that 0.2 mM SA alleviated mercury (10 µM Hg^{2+}) toxicity, protecting *M. sativa* roots from oxidative damage by increasing activities of NADH oxidase, APX and POD, and by accumulatiion of AsA, GSH and Pro.

11.4.1.3 Gibberellic acid

Gibberellic acids (GAs) constitute a group of phytohormones in higher plants mediating many developmental processes. Its roles in seed germination, vegetative growth, flowering induction and fruit development have been studied extensively (Sun & Gubler, 2004). The results of previous studies have also proved that GAs have vital roles in plant responses to biotic and abiotic stress conditions (Wigoda *et al.*, 2006; Ko *et al.*, 2007; Robert-Seilaniantz *et al.*, 2007; Navarro *et al.*, 2008).

Hamayun *et al.* (2010a) reported that exogenous GA_3 application not only enhanced the growth of *G. max* but also mitigated the adverse effects of salt stress. The salt stress mitigation effects of GA_3 application were in regulating the level of phytohormones like bioactive GAs (GA_1 and GA_4) and increasing jasmonic acid (JA), and decreasing endogenous ABA and SA contents (Hamayun *et al.*, 2010a). Pre-soaking the seeds in GA_3 (200 mg/L) was beneficial in minimizing the deleterious effects of salinity (100, 200 and 300 mM NaCl) in mung bean plants. The NaCl caused drastic reductions in shoot and root lengths and their fresh and dry weights, in number of lateral roots and leaves, and in total area of leaves. Salinity also reduced pigment contents, photosynthetic activity, contents of reducing sugars and sucrose, total soluble-N, total-N and protein-N contents, and nucleic acid content. The rates of respiration, O_2 uptake and CO_2 evolution were also reduced, while Na^+ and Cl^- contents increased and K^+, Ca^{2+} and Mg^{2+} levels decreased. The exogenous GA improved growth and physiological attributes and decreased Na^+ and Cl^- contents in root and shoot (Haleem & Mohammed, 2007). Five levels of GA (0, 100, 200, 300 and 400 mg/L) were applied as a foliar spray at 3 and 6 weeks after sowing to assess the yield and yield traits of different groundnut cultivars in wet and dry growing seasons. Compared to the untreated controls, GA increased the yield and yield components. The highest pod, kernel and haulm yields were obtained at 100 mg/L GA application both in wet and dry seasons (Yakubu *et al.*, 2013). Reduced irrigation frequency (at 2-, 4-, 6- and 8-day intervals) in mung bean drastically reduced plant height, internode length, number of leaves per plant, leaf area per plant, leaf area index, number of inflorescences per plant, pod length, pod number per plant, seed number per pod, biomass yield and seed yield per plant. Application of GA_3 (100 and 200 mg/L) with water stress improved plants' performance regarding these parameters (compared to stress treatment alone), where the combination of 200 mg/L GA_3 and 2-day irrigation interval showed the best performance amongst all treatments (Abdel & Al-Rawi, 2011). Addition of GA_3 also improved germination and seedling growth of *C. arietinum* L. cv. PBG-1 under drought stress induced by PEG (Kaur *et al.*, 1998). Gradual increase of temperature 30 °C, 40 °C reduced the α-amylase activity in seedlings of different genotypes of *V. radiata*. The reduction was greatest at a lethal temperature of 50 °C. But application of GA_3 (100 µM) alleviated the effect of heat stress by increasing α-amylase activity (Mansoor & Naqvi, 2012) and it was mentioned that elevated α-amylase activity in roots, shoots and cotyledons was positively correlated with the progress of seedling growth of mung bean plant. Cadmium (20 mg/kg $CdCl_2$) and Pb (100 mg/kg lead acetate) reduced the contents of chls, soluble carbohydrates and proteins in *V. faba* and also in *L. albus*. Under Cd and Pb stress bean and lupin plants supplemented with 100 ppm GA showed differential responses. Gibberellic acid supplementation in broad bean enhanced the activities of CAT and POD under Pb stress, and amylase activity under Cd stress. In lupin, GA supplementation enhanced amylase, CAT and POD activities under Cd stress, and improved CAT and amylase activities under Pb stress.

Nonetheless, GA has beneficial effects on both species under Cd or Pb stresses (Sharaf *et al.*, 2009).

11.4.1.4 Naphthalene acetic acid

Naphthalene acetic acid (NAA) is a synthetic plant hormone of the auxin family. It is often used in embyogenesis and *in vitro* plant regeneration, and for the vegetative propagation of plants from stem and leaf cuttings. Application of NAA in plant tissue culture is also common (Navalón *et al.*, 1997). Exogenous NAA application has been shown to improve fruit development, quality and total yield (Stern *et al.*, 2007). Although the information is sparse, NAA plays effective roles in countering different abiotic stresses (Aslam *et al.*, 2010; Kusuma & Begum, 2013).

The response of *C. arietinum* to NAA (4.5% a.i.) was studied under control and reduced moisture levels. Stress with 50, 65, 80 and 95% available soil moisture depletion levels was examined with NAA application and without NAA. With depleted water levels NAA could maintain seed yield and yield components in chickpeas, indicated by more pods per plant and seeds per pod, and greater 100-seed weight, biological yield and seed yield, compared to those of chickpeas grown in a drought environment without NAA. Ttreatment with NAA was able to maintain enhanced performance up to an 80% reduction in water levels (Aslam *et al.*, 2010). Drought stress had several adverse effects on *G. max* var. Nannong 99-6, including reductions in the rate of biomass production, leaf area, leaf relative water content, chl content, photosynthesis rate, stomatal conductance to CO_2, intercellular CO_2 concentration and transpiration rate. Pretreatment with NAA alleviated those adverse effects, and improved physiological adaptation capability and drought resistance capacity (Peng *et al.*, 2011). With a view to develop salt-stress-tolerant plants of *A. hypogaea* Linn (JL-24), *in vitro* plant regeneration was performed under Murashige and Skoog (MS) regeneration medium with a combination of 2.5 mg/L benzylaminopurine (BAP) and 0.2 mg/L NAA. The NAA-supplemented medium was good for the development of healthy plants subjected to different regimes of salt stress; they were found to be tolerant up to a maximum concentration of 1.0 mg/L NaCl (Kusuma & Begum, 2013).

11.4.1.5 Indole acetic acid

Indole acetic acid (IAA) is a hormone of auxin group that stimulates cell elongation and lateral root formation. It has vital roles in the absorption of minerals and in plant growth. Exogenously applied IAA has been documented to promote growth as well as to alleviate adverse effects of different abiotic stresses in different plant species (Chakrabarti & Mukherji, 2003; Ali *et al.*, 2007; Egamberdieva, 2009).

Salt stress (NaCl at 25, 50, 100 and 200 mM) in *Phaseolus mungo* caused significant reductions in germination, seedling growth, root and shoot length, fresh weight and leaf area. Reductions in protein content and NR activity were documented at all levels of stresses, but Pro level, free amino acids, and POD and CAT activities were increased by salt stresses. Exogenous IAA enhanced salt stress tolerance by improving those antioxidant defence processes. Foliar application of IAA (15 mg/L) to the NaCl-stressed plants increased the percentage of germination, shoot and root length, fresh weight, total chl content and dry weight, and improved yield. IAA also decreased levels of Pro, free amino acids and total soluble sugar levels, increased NR activity and modulated CAT activity (Devi *et al.*, 2012). Drought stress (10% PEG) exerted several adverse effects on the enzymatic activities of germinated *C. arietinum* cv. PBG-1 seedlings. Treatment with IAA (3 μM) enhanced the activities of different metabolic enzymes including amylase, acid invertase, alkaline invertase, sucrose synthase and sucrose phosphate synthase, improved and carbohydrate composition. Germination was better in IAA-supplemented seeds under drought treatment, compared to seeds under drought stress alone (Kaur *et al.*, 2000). Gangwar *et al.* (2011) reported that when *P. sativum* L. cv. Azad P-1 seedlings were exposed to different levels of manganese (Mn 50, 100 and 250 μM) their antioxidant system components were disrupted, including the non-enzymatic and enzymatic antioxidants, thereby increasing the oxidative stress (i.e. increased H_2O_2 and MDA levels). Indole acetic acid (10 and100 μM) application alleviated Mn toxicity symptoms by promoting SOD, APX, GR and DHAR activities; the concentrations of AsA and GSH; and the ratio of their reduced to oxidized forms. Thus the oxidative stress was eliminated (Gangwar *et al.*, 2011). Cadmium exposure inhibited the germination of *C. arietinum* by hampering various attributes. But soaking the seeds in 4-Cl-IAA (4-chloroindole-3-acetic acid) improved the RWC, moisture content, and activities of NR and POD. The combined effects resulted in improvement of the rate of germination (Ali *et al.*, 2007). Reduced root and shoot growth, chl, N, phosphorus (P), total sugar, and soluble protein under Cd stress were ameliorated by 4-Cl-IAA (10–6 M) in *V. radiata* seedlings (Mazid *et al.*, 2010).

11.4.1.6 Ethylene

Ethylene is a gaseous phytohormone involved in fruit ripening, but also with roles in senescence, and leaf and fruit abscission. Its participation in germination, seedling morphogenesis, root emergence, root hair elongation, and promotion of flowering is also important. The stress response and ethylene regulation has attracted the attention of researchers (Ligero *et al.*, 1987; Lee & LaRue, 1992). Like other hormones, ethylene also has antagonistic and synergistic effects.

When *M. sativa* and *Cassia corymbosa* plants were exposed to drought stress the ethylene level significantly increased. In the case of long-term (8–12 day) severe drought stress (−0.4 MPa) ethylene production was related to senescence and abscission (Berumen-padilla *et al.*, 2009). Mung bean (*V. radiata*) and pea (*P. sativum*) pre-exposed to ethylene were more resistant to O_3. The resistance of ethylene-pretreated plants was due to suppression of H_2O_2 through stimulation of APX activity (Mehlhorn, 1990). Ethylene diurea (EDU) treatment led to a significant increase in biomass and yield of *P. vulgaris* plants exposed to O_3 (Brunschon-Harti *et al.*, 1995a). The antioxidative roles of EDU as a free radical scavenger or in reducing oxidative damage were studied in common bean and snap bean (Kostca-Rick & Manning, 1993; Brunschon-Harti *et al.*, 1995b; Lee *et al.*, 1997). Roles for ethylene in mediating adaptive responses to P deficiency stress in common bean, soybean (*G. max*) and pea (*P. sativum*) involved morphological changes in root systems, gravitropism, aerenchyma formation and root hair development (Bonser *et al.*, 1996). Ethylene generally plays a negative role in legume symbiosis by restricting infection. Ethylene production significantly increased in *Medicago sativa* and *P. sativum* roots infected by rhizobia, whereas exogenous ethylene decreased the number of nodules (Lee & LaRue, 1992). Exogenous ethylene did not decrease the number of rhizobial infections in root, but most infections were blocked in the basal epidermal cell or in the outer cortical cells (Lee & LaRue, 1992). Ethylene and ROS act as signal molecules in Nod factor-triggered nodule formation. In *S. rostrata*, ethylene and ROS act downstream from the Nod factors and play roles in the generation of infection pockets and initiation of nodule primordia (D'Haeze *et al.*, 2003).

11.4.1.7 Jasmonic acid

Jasmonic acid (JA) is a growth regulator that influences plant developmental processes. By acting as signal molecule, it influences physiological and metabolic regulation in responses to stress (Poonam *et al.*, 2013). JA has both synergistic and antagonistic effects on plant developmental processes, including seed germination, root growth, fertility, fruit ripening, senescence and stomatal closure. JA or its methyl esters are omnipresent in plants and play vital roles under stress conditions (Wasternack & Hause, 2002; Cheong & Choi, 2003; Alam *et al.*, 2014a).

Methyl jasmonate (MeJA) (20 and 30 µM) counteracted the negative effects of salinity (60 mM NaCl, 2 weeks) stress in *G. max* seedlings. Improved plant growth, chl content, leaf photosynthetic rate, leaf transpiration rate and Pro content were recorded in MeJA-treated plants. JA also reduced the effects of salt on seed carbohydrates, lipids, proteins, N, P and K (Yoon *et al.*, 2009). Treating *P. sativum* with 10 mM JA improved photosynthesis, RWC and protein content in conditions of salt stress (30 mM NaCl, 4 and 7 days) (Fedina & Tsonev, 1997). Exogenously applied JA (100 µM) increased drought tolerance in *G. max*; JA supplementation increased the endogenous JA, ABA, ethylene and cytokinin contents in drought-affected plants. Enhancements of total PA, putresceine (Put), spermidine (Spd) and spermine (Spm) also correlated with adaptive responses and safeguard mechanisms of soybean plants coping with drought stress (Hassanein *et al.*, 2009). Improvement of drought tolerance in *A. hypogaea* plants was enabled by JA's role as a growth substance (El-Meleigy *et al.*, 1999). The *C. cajan* plants were subjected to Cu^{2+} (5 mM) stress that resulted in destruction of photosynthetic pigments and antioxidant systems. Exogenous JA supplementation with Cu^{2+} neutralized the deleterious effects of Cu^{2+} on seedling growth, manifest by the accumulation of osmolyte Pro, improved photosynthetic pigments, and enhanced antioxidant enzymes SOD and POD activities, thus contributing to Cu^{2+} tolerance (Poonam *et al.*, 2013). Supplementation by MeJA (100 µM) prevented petiole abscission in $AgNO_3$-treated *V. radiata* plants. In contrast, $AgNO_3$-treated plants were characterized by heavy petiole abscission and it was suggested that MeJA was involved in ethylene-independent abscission. In a recent study, Alam *et al.* (2014a) showed that exogenous spraying with JA (0.5 mM) might be an effective agent for eliciting short-term drought tolerance in different *Brassica* species although the responses are different. These species especially appeared to respond differently with respect to antioxidant and glyoxalase enzyme activities under short-term drought stress (15% PEG, 48 h). JA reduced the oxidative stress in all cases and improved physiological adaptation in most of the species under drought stress; the best response to JA was observed with *Brassica juncea* under drought stress. This species showed fewer deleterious effects than all the other species under drought stress, which reflects its natural drought tolerance capacity.

11.4.1.8 Brassinosteroids

Research with higher plants suggests that brassinosteroids (BR) are present in most or all parts of the plant and play crucial roles in many processes like vegetative growth, floral initiation, and development of flowers and fruits (Hayat & Ahmad, 2003; Sasse, 2003). Brassinosteroid metabolism has long been recognized in altering plants' responses to abiotic stresses, and brassinosteroids confer stress tolerance via activation of different mechanisms (Bajguz & Hayat, 2009).

Salt stress exerted several adverse effects on growth, yield and physiological attributes of *P. sativum* cv. Climax. Brassinosteroids alleviated those stress effects in diverse ways. Compared to the untreated controls, seed treatment with 24-epibrassinolide (10 μM) significantly enhanced the fresh and dry biomass, seedling height, nodule number and nodule dry biomass of *P. sativum* L. cv. Climax. Photosynthesis rate, stomatal conductance, total chl and Pro contents were maintained, and activities of antioxidant enzymes like SOD, POD and CAT were enhanced. Activities of nitrate metabolism enzymes like NR were also improved by 24-epibrassinolide treatment. All these might help in achieving higher seed yield, seed number and seed protein contents (Shahid *et al.*, 2011). Salinity stress (50–400 mM) inhibited nitrate uptake and metabolism in *C. cajan*. Levels of proteins, amino acids, nitrate, NR of roots and the composition of xylem sap amino acids were reduced by salt stress. But exogenous 24-epibrassinolide (10^{-7} M and 0.5×10^{-9} M) increased nitrate uptake and elevated the activity of NR together with achieving greater levels of free amino acids and soluble proteins in roots of plants under saline conditions (Dalio *et al.*, 2013). HT or NaCl stresses, individually or in combination, increased electrolyte leakage and lipid peroxidation, and decreased the membrane stability index and leaf water potential in *V. radiata*. High temperature and NaCl led to a significant decline in growth, photosynthetic parameters and maximum quantum yield of PSII. 28-Homobrassinolide improved photosynthesis, chl fluorescence and the antioxidant system so that the oxidative damage was alleviated, partly due to improvement in photosynthetic parameters (Hayat *et al.*, 2010). Spraying with BRs (25 and 50 mg/L) increased vegetative growth, total yield, quality of pods and total phenolic acids in the pod of *P. vulgaris* under HT stress of 34.7 to 35.2 °C (El-Bassiony *et al.*, 2012). Gradual increases of Cd toxicity (0, 50, 100 or 150 mM) decreased plant fresh and dry mass, number of nodules, and their fresh and dry mass in *C. arietinum* cv. Uday. This stress also decreased leghaemoglobin, N and carbohydrate content in the nodules. Increased Cd levels also gradually reduced the leaf chl content, and NR and carbonic anhydrase activities. These adverse effects were overcome by spraying with 0.01 mM 28-homobrassinolide; this spray also regulated the content of Pro and the activities of CAT, POD and SOD towards Cd tolerance (Hasan *et al.*, 2008). Exposure of mung bean plants to Al (1 and 10 mM) in the nutrient medium caused a sharp reduction in length, and fresh and dry mass of root and shoot. Aluminium decreased the activity of carbonic anhydrase, RWC, water use efficiency, chl content and the rate of photosynthesis. The activities of antioxidant enzymes (CAT, POD, SOD) in shoot and root and the Pro content were increased by Al. Spraying with 10^{-8} M 24-epibrassinolide or 28-homobrassinolide improved antioxidant enzyme activities, Pro content, growth and physiological parameters in the absence or presense of Al (Ali *et al.*, 2008).

11.4.2 Proline and glycine betaine

The accumulation of osmolytes such as Pro is a well-known adaptive mechanism of plants under osmotic stress, and it has been suggested that Pro accumulation can serve as a selection criterion for the tolerance of most species to stress conditions (Ashraf & Foolad, 2007). Glycine betaine (GB) is another small water-soluble organic molecule that can potentially play a protective role against osmotic stress by protecting plant cells through osmotic adjustment, protein stabilization, photosynthetic apparatus protection and reduction of ROS (Ashraf & Foolad, 2007).

Hossain *et al.* (2010) investigated the effect of exogenously applied Pro and GB in defending oxidative stress in *V. radiata* seedlings exposed to Cd. Cadmium stress (1 mM $CdCl_2$, 48 h) caused a significant increase in GSH and GSSG contents, while the AsA content decreased significantly with a sharp increase in H_2O_2 and MDA contents. The activities of APX, GST and GPX increased in response to Cd stress, while the activities of CAT, MDHAR, DHAR and GR were sharply decreased. Exogenous application of Pro (5 mM) and GB (5 mM) resulted in an increase in GSH and AsA contents, maintenance of a high GSH/GSSG ratio and increase in the activities of APX, DHAR, MDHAR, GR, GST, GPX and CAT compared to control and mostly also Cd-stressed plants, with a concomitant decrease in GSSG, H_2O_2 and MDA contents. These authors suggested that both Pro and GB had a protective action against Cd-induced oxidative stress in *V. radiata* seedlings. Hossain *et al.* (2011) studied the effect of exogenous Pro and GB in mitigating salt stress in *V. radiata* seedlings. Subjecting the seedlings to 48 h of salt stress (200 mM NaCl) resulted in sharp increases in lipid peroxidation and H_2O_2 content. Salt stress altered antioxidant metabolism, which was revealed by increases in GSH and GSSG and a decrease in AsA content. But the antioxidant enzymes responded differently under salt stress. The activities of MDHAR, DHAR and CAT decreased markedly upon exposure to salt stress, whereas the activities of GR, GPX and GST increased in response to salt stress. However, these activities were not sufficient to protect against salt stress-induced damage. The application of exogenous Pro and GB caused significant increments of the activities in salt-stressed seedlings, which resulted in increased levels of AsA and GSH and reduced the oxidative damage as indicated by decreased levels of MDA and H_2O_2. The results of this study (Hossain *et al.*, 2011) prompted the suggestion that upregulation of antioxidant defence by Pro and GB could confer salt stress tolerance in *V. radiata*. Rezaei *et al.* (2012) grew soybean plants supplemented with exogenous GB (2.5, 5, 7.5 and 10 kg ha^{-1}) in saline soil (11.1 dS/m). The uptake of Na^+ was decreased with the addition of GB in salt-stressed plants. Glycine betaine application also increased the number of lateral branches and pods per plant, while no difference was observed in seed number per pod. Exogenous GB was also able to increase the 1000-grain weight in salt-stressed plants. However, the impact of GB on 1000-grain weight was cultivar-, dose- and time-dependent. When *V. faba* was grown under water-deficit conditions, plant height and root length, root fresh weight, shoot fresh weight, root dry weight, shoot dry weight, total soluble carbohydrates, chl *a* and *b*, and activity of POD and CAT were all strongly depressed with a concomitant increase in MDA content (Ali *et al.*, 2013). However, application of Pro improved all growth characteristics by improving Pro, total soluble carbohydrate (TSC) and photosynthetic pigments as well as the activities of CAT and POD. These results indicated that application of Pro helped the plant to restore the physiological processes altered by water stress (Ali *et al.*, 2013). Taie *et al.* (2013) found physiological, biochemical and anatomical attributes were enhanced in seawater-treated *Vicia faba* plants when the seeds were pretreated with Pro. When 16-day-old plants were irrigated with seawater for 50 days they exhibited reductions in all growth parameters (plant height, number of leaves and shoot dry weight per plant), photosynthetic pigments (chl *a*, chl *b* and carotenoids), total carbohydrate, contents of P, Ca^{2+}, K^+ and K:Na ratio compared to controls. Increasing seawater stress also led to increases in total phenolics, free amino acids, Pro and soluble carbohydrate as well as values of N, Na^+, and Cl^-. On the other hand, the plants grown from Pro-pretreated seeds showed salt tolerance reflected by improved growth parameters, photosynthetic pigments, soluble carbohydrate and total carbohydrate and decreased Na^+ and Cl^- (Taie *et al.*, 2013). The beneficial effect of foliar applied Pro (20 mM) was found to protect against Cd stress in *C. arietinum* as reported by Hayat *et al.* (2013). Foliar treatment of Pro increased the plant growth. Exogenous Pro also increased the photosynthetic attributes and yield characteristics in the Cd-stressed plants. Moreover, the activities of antioxidative enzymes were also increased when the plants were supplemented with Pro.

11.4.3 Polyamines

Polyamines (PAs) are low-molecular-mass aliphatic amines and organic polycations found as three different forms, viz. diamine putrescine (Put), triamine spermidine (Spd) and tetraamine spermine (Spm), which are involved in various processes such as cell proliferation, somatic embryogenesis, seed germination, growth, morphogenesis, differentiation, development of flowers and fruits and programmed cell death (Hasanuzzaman *et al.*, 2014d).

In *P. vulgaris*, exogenous Put treatment (10 μM) improved the germination and growth of plants grown in saline soil (1% NaCl) (Zeid, 2004). The germination percentage of Put-treated seeds increased under salinity stress, which was associated with the activation of amylase and protease enzymes during germination. The rise of the K^+/Na^+ ratio in shoots of Put-treated seedlings under NaCl-induced stress suggested that Put treatment may partially ameliorate the adverse effects of sodium and chloride ions. In *P. vulgaris* Put treatment improved the growth of salt-stressed seedlings due to partial inhibition of amylase and protease enzymes, and increased the total content of nucleic acids and photosynthetic pigments. The same author also observed changes in protein banding patterns that suggested that a defence-response gene could be activated by Put treatment (Zeid, 2004). Jiménez-Bremont *et al.* (2006) suggested that PA could alleviate salt stress in *P. vulgaris* by prompting Pro accumulation. Since the metabolic pathways involved in Pro and PA biosynthesis require glutamate, ornithine and arginine as precursors, these pathways are probably well coordinated. They also observed differential responses of PA in salt-sensitive and -tolerant cultivars. When sensitive plants were subjected to salt stress (150 mM) in the presence of PAs, the content of Pro decreased in relation to controls with PAs and NaCl alone, whereas such a decrease did not occur in resistant plants (Jiménez-Bremont *et al.*, 2006). In *V. sinensis*, Alsokari (2011) observed the protective effect of Spm on some physiological aspects under salt stress (50% seawater from the Red Sea). Salt stress negatively affected most of the physiological and biochemical parameters in plants. Plants irrigated with seawater exhibited reductions in chl *a*, chl *b*, carotenoids, carbohydrates, protein, and Spd and Spm levels as well as amylase activity. Salt stress also decreased different yield attributes and seed yield of bean. In contrast, salt caused increases in Pro, K^+, Na^+ and Put concentrations, and in POD activity. Importantly, exogenous Spm application (0.3 mM) mitigated the deleterious effects of salinity stress on growth and yield by enhancing the physiological and yield attributes of plants (Alsokari, 2011). Exogenous Put could also alleviate HT stress (affected by late sowing) in *P. vulgaris* (Ibrahim *et al.*, 2012). Upon exposure to HT, the plants exhibited significant reductions in height, branches per plant, number of leaves per plant, fresh weight and dry matter of plants and content of chl. However, foliar spray of Put (200 mg/L) alleviated the harmful effects of HT stress in the late-sown plants, which indicated the protective role of PA in heat stress tolerance (Ibrahim *et al.*, 2012). Radhakrishnan and Lee (2013) observed that exogenous Spm-induced changes in antioxidant metabolism and ABA levels could ameliorate the effects of osmotic stress in soybean. When soybean plants were exposed to osmotic stress (9% PEG) at the flowering stage, the fresh weight of pods and seeds and the protein content were decreased significantly. However, exogenous application of Spm (0.1 mM) induced the enhancement of pod and seed fresh weight and protein content under osmotic stress conditions. Osmotic stress also reduced the enzymatic activities of antioxidant enzymes in the pod, which resulted in higher lipid peroxidation (MDA). In contrast, exogenous Spm application could upregulate the activities of antioxidant enzymes and maintain a lower level of the MDA, which indicated the role of Spm in antioxidant defence in soybean (Radhakrishnan & Lee, 2013).

11.4.4 Antioxidants

11.4.4.1 Glutathione

Glutathione (GSH) is a water-soluble non-protein thiol compound that serves a broad range of biochemical functions in almost all cell organelles (Foyer & Noctor, 2003). GSH can protect plants from different abiotic stresses due to its dynamic biochemical properties like high water solubility, relative stability, capacity to form mercaptide bonds with metals and reactivity with selected electrophiles (Gill *et al.*, 2013a) (Figure 11.3). It is a vital component of the antioxidant defence system, playing crucial roles in cellular redox homeostasis and stress signalling (Wu *et al.*, 2011; Chen *et al.*, 2012). Glutathione has a special role in the elimination of metal toxicity due to its role in the activity of phytochelatin (PC) synthesis (Figure 11.3).

Pretreatment of *Vigna cutjang* with L-cysteine and GSH produced resistance against water deficit stress by modulating the H_2O_2 and ascorbic acid contents. Water stress (−0.5, −1.0, −1.5 MPa) increased glycolate oxidase activity and H_2O_2 and Pro contents in *V. cutjang* seedlings, and resulted in decreased leaf water potential and relative water content, catalase activity, and ascorbic acid and protein contents. Pretreatment with L-cysteine and reduced glutathione (10^{-3} M) decreased glycolate oxidase activity, H_2O_2 content, ascorbic acid oxidase activity and Pro content and also slightly improved the water status of leaves stressed (−1.0 MPa) for 2 days (Mukharjee & Choudhuri, 1983). It was proposed that *Phaseolus vulgaris* seedlings prevented damage due to HT stress (38–39 °C, 2 h followed by exposure to 45–48 °C, 8 h) by enhancing GSH and AsA contents, and the activities of β-amylase, acid phosphatase and guaiacol-specific peroxidase (Babu & Devaraj, 2008). Mung bean (*V. radiata*

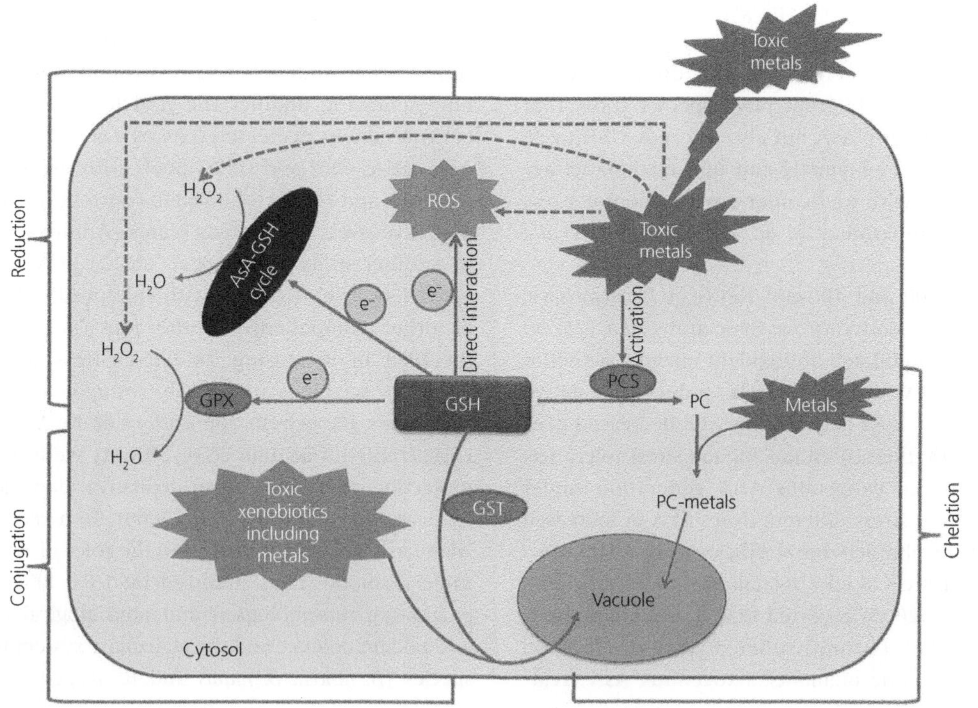

Figure 11.3 Glutathione-mediated toxic metal and xenobiotic detoxification in plant cells. AsA, ascorbic acid; GPX, glutathione peroxidase; GSH, glutathione; GST, glutathione-*S*-transferase; PC, phytochelatin, PCS, phytochelatin synthase; ROS, reactive oxygen species.

cv. Binamoog-1) seedlings were exposed to salt stress (NaCl, 200 mM), which disrupted the antioxidant system components resulting in oxidative stress. Exogenous GSH reduced the oxidative stress, marked by reduced MDA and H_2O_2 levels. GSH application decreased oxidative stress by increasing activities of antioxidant enzymes, contents of GSH and AsA, and the GSH/GSSG ratio, and decreasing GSSG content (Nahar *et al.*, 2012). Drought stress imposed on *V. radiata* cv. Binamoog-1 seedlings increased the oxidative stress; deleterious effects were more severe with 48 h duration, compared to 24 h duration. However, exogenous co-application of GSH with drought stress enhanced the antioxidant system, which included enhanced activities of APX, MDHAR, DHAR, GPX and CAT. Thus, GSH reduced the oxidative damage by reducing lipoxygenase activity and MDA and H_2O_2 levels under drought stress. Exogenous GSH also enhanced the activities of glyoxalase enzymes, which are required for methylglyoxal detoxification (Nahar *et al.*, 2013). GSH has also proved to be a potent protectant against heavy metal stress. Seedlings of mung bean (*V. radiata* cv. Binamoog-1) showed damage due to cadmium ($CdCl_2$, 1 mM) toxicity. Protection from oxidative damage induced by Cd was achieved by application of GSH (1 mM). The exogenous GSH improved the endogenous GSH level and GSH/GSSG ratio, decreased GSSG, and increased the activities of antioxidant enzymes MDHAR, DHAR and GR. The reduced oxidative stress due to exogenous GSH was also partly achieved by the enhanced glyoxalase system as evidenced by increased Gly I and Gly II activities (Nahar *et al.*, 2012).

11.4.4.2 Ascorbic acid

Ascorbic acid (AsA) is one of the key components of the antioxidant defence system, acting as a redox buffer in plant cells. It plays significant roles in plant growth, metabolism, development and stress responses (Noctor & Foyer, 1998; Anjum *et al.*, 2012a,b, 2014; Alam *et al.*, 2014b). AsA protects the plasma membrane against oxidative damage and is a precursor for some plant metabolites (Smirnoff, 1996; Noctor & Foyer, 1998), helping in the regeneration of α- tocopherol and zeaxanthin, and in PSII activity (Smirnoff, 1996). It has roles in controlling cell cycle progression (Potters *et al.*, 2002) and root growth and development (Liso *et al.*, 2004). The protective effects of AsA under different stress conditions in different plant species show that the levels of not only AsA but also the AsA-GSH cycle components, both enzymatic and non-enzymatic, are often correlated with and confer environmental stress tolerance (Hasanuzzaman *et al.*, 2012c; Nahar *et al.*, 2012, 2013).

Salt stress (100 and 400 mM NaCl) in *P. vulgaris* cv. Naz enhanced oxidative stress. Exogenous AsA (25, 50 and 100 mM) modulated antioxidant enzyme activities (SOD, CAT, APX, GR and POD), reduced oxidative stress, prevented lipid peroxidation and decreased generation of MDA, thereby enhancing salt stress tolerance. AsA also reduced exogenous ABA generation under NaCl. Under salt stress different doses of AsA increased the maximum photochemical efficiency of PSII, compared to NaCl stress alone (Dolatabadian & Jouneghani, 2009). Beltagi (2008) reported that *C. arietinum* plants treated with NaCl (40 mM) suffered adverse effects on their photosynthetic pigments. Exogenous AsA treatment improved the chl *a* content and chl stability index. Exogenously applied AsA (400 mg/L, 4 h) counteracted the adverse effects of salt stress (12.5 and 50 mM NaCl), improved antioxidant enzyme activities (CAT, POX, SOD) and improved germination and growth performance of *G. max* seedlings (Dehghan *et al.*, 2011). Water stress (−1 MPa) reduced N fixation and carbon flux in nodules of *P. sativum*, but 5 mM AsA with water stress (−1 MPa) increased the nodule AsA + dehydroascorbate (DHA) pool compared to water-stressed nodules without AsA, modulated guaiacol peroxidase activity in leaves and nodules, restored nodule carbon and N enzymes, and prevented the decline in N fixation and the reduction of carbon flux in nodules (Groten *et al.*, 2006). The N_2-fixing legume nodules of *G. max* are vulnerable to oxidative damage caused by leghaemoglobin, which produces ROS through spontaneous autoxidation. Treatment of *G. max* with excess AsA with irrigation increased nitrogenase activity, nodule leghaemoglobin content, and activity of APX. The concentration of lipid peroxides, an indicator of oxidative damage and onset of senescence, was reduced by exogenous AsA. Thus exogenous AsA decreased oxidative stress and improved capacity to fix N_2 (Bashor & Dalton, 1999). Nodule development and senescence in *P. vulgaris* is partly controlled by endogenous AsA or its associated enzymes. *P. vulgaris* treated with salt (150 mM NaCl), Cd (100 µM $CdCl_2$) or oxidative stress (10 mM H_2O_2) showed that under these stress conditions or in naturally senescing nodules the AsA level and associated enzyme activity decreased (Loscos *et al.*, 2008). Balanced AsA, DHA, GSH and GSSG pools with higher activities of DHAR and MDHAR helped in controlling Cd-induced oxidative stress in *V. radiata* plants (Anjum *et al.*, 2011). In another study, Nahar *et al.* (2012) proved that elevated AsA levels together with enhanced concentration of other components of the AsA-GSH cycle were involved in increasing oxidative stress tolerance in *V. radiata* seedlings under salt, drought and Cd stress conditions. Exogenous applications of the AsA precursor L-galactono-1,4-lactone (Gal, 50 mM) were effective in protecting thylakoids from oxidative damage caused by paraquat (Tambussi *et al.*, 2004). In a recent report, Alam *et al.* (2014b) investigated the roles of AsA (1 mM) under osmotic stress (induced by 15% PEG-6000) by examining morphological and physiological attributes, antioxidant defence and the glyoxalase system in *Brassica* species (*B. juncea, B. napus* and *B. campestris*). Osmotic stress reduced fresh and dry weights, leaf RWC and chl content; and increased Pro, MDA and H_2O_2 contents and LOX activity. The ascorbate content in *B. napus, B. campestris* and *B. juncea* decreased, increased and remained unaltered, respectively. The dehydroascorbate content increased only in *B. napus*, and the AsA/DHA ratio was reduced by osmotic stress in all species except *B. juncea*. Osmotic stress increased GSH content only in *B. juncea*, increased GSSG content and decreased the GSH/GSSG ratio in all species. Osmotic stress increased activities of APX (except in *B. napus*), GR (except in *B. napus*), GST (except in *B. juncea*) and GPX, and decreased the activities of CAT (in all species) and of MDHAR (only in *B. campestris*). Osmotic stress decreased glyoxalase I (Gly I) and increased glyoxalase II (Gly II) activity. Addition of AsA in combination with PEG improved fresh weight, RWC and chl content; and decreased Pro, MDA and H_2O_2 levels. It improved levels of AsA-GSH cycle components, and improved the activities of all antioxidant and glyoxalase enzymes in most cases. So, exogenous AsA improved physiological adaptation and alleviated oxidative damage under osmotic stress by improving the antioxidant and glyoxalase systems in all species, with *B. napus* showing greater antioxidant capacity. The performance of *B. juncea* was also better in response to exogenous AsA addition under osmotic stress (Alam *et al.*, 2014b).

11.4.4.3 Tocopherols

Tocopherols are exclusively located in green tissues such as algae and some cyanobacteria. Their presence in plastid or thylakoid membranes protects against photosynthesis-derived ROS (Munné-Bosch, 2005; Hasanuzzaman *et al.*, 2014e). The tocopherols in cooperation with other antioxidants take part in controlling ROS, and maintaining the redox state (Munné-Bosch 2005, 2007). Tocopherol levels within plants change according to stages of growth and development or in response to environmental stress. Tocopherol synthesis is linked to levels of different stress hormones like JA, SA and ABA, and thus tocopherol is supposed to have roles in stress signalling (Falk *et al.*, 2002; Sandorf & Holländer-Czytko, 2002).

Pisum sativum plants subjected to drought (leaf water potential of −1.3 MPa) showed remarkable elevation (67%) of α-tocopherol (Moran *et al.*, 1994). Chilling stress (10 °C) increased lipid peroxidation or MDA and H_2O_2 accumulation in *M. sativa* leaves. However, the redox properties of α-tocopherol played roles in sequestering free radicals. α-Tocopherol also gave antioxidant protection as higher α-tocopherol levels were correlated with increased activities of SOD, CAT, APX and GR (Bafeel & Ibrahim, 2008). Studies with different lines of *G. max* showed that in comparison with normal growth temperatures, HT stress resulted in a several-fold increase in α-tocopherol content, whereas the same temperature had no significant effect on total tocopherols. When those same soybean cultivars were exposed to drought stress, a similar increase in α-tocopherol but no change in total tocopherol was observed (Britz *et al.*, 2008). Different levels of salt tress (25, 50, 100, 200 mM) increased the anthocyanin, ascorbic acid and α-tocopherol contents in soybean seedlings, but the increases in these metabolites were shown to be beneficial only in 25 mM NaCl stress (Mohammed *et al.*, 2012). Exogenous application of the α-tocopherol analogue 6-hydroxy-2,5,7,8-tetramethylchroman-2-carboxylic acid (Trolox), showed effective results in protecting thylakoids from oxidative damage caused by paraquat (Tambussi *et al.*, 2004). Bean (*P. vulgaris*) seedlings were exposed to different heavy metals ($PbCl_2$, $CuCl_2$, $CdCl_2$ and $HgCl_2$) for 10 days. All heavy metal treatments increased α-tocopherol, AsA and Pro contents. The extent of the increase of these metabolites, from highest to lowest, was: Hg > Cd > Cu > Pb. The increase in α-tocopherol together with other antioxidants and metabolites is thought to be due to enhancement of the plants' self-defence mechanism under heavy metal stress conditions (Zengin & Munzuroglu, 2005).

11.4.5 Nitric oxide

Recently NO has emerged as an important signalling molecule and antioxidant. NO triggers expression of many kinds of redox-regulated (defence-related) genes, directly or indirectly, to establish plant stress tolerance. Several recent reports indicate that the application of exogenous NO donors confers tolerance to various abiotic stresses (Hasanuzzaman *et al.*, 2010a, 2011a, 2012c, 2013e; Hasanuzzaman & Fujita, 2013; Gill *et al.*, 2013b). Nitric oxide has a protective function against oxidative stress mediated by (i) reaction with lipid radicals, which stops the propagation of lipid oxidation; (ii) scavenging $O_2{\bullet}^-$ and formation of peroxynitrite ($ONOO^-$), which can be neutralized by other cellular processes; (iii) activation of antioxidant enzymes (SOD, CAT, APX, GPX, GR, POD, etc.); and (iv) functioning as a signalling molecule in the cascade of events leading to changes in gene expression. These mechanisms together confer enhanced antioxidant protection against oxidative stress (Hasanuzzaman *et al.*, 2013e) (Figure 11.4). Many experiments have been conducted to investigate the effect of NO on plant stress tolerance using sodium nitroprusside (SNP) as NO donor.

In *C. arietinum* plants, Sheokand *et al.* (2008) showed the antioxidant effects of NO (0.2 mM SNP) under salt stress (100 mM NaCl, 48 h). In their study, salt stress caused a decrease in RWC and increases in electrolyte leakage and lipid peroxidation. Interestingly, an exogenous NO donor (SNP) could completely ameliorate the toxic effects of salt stress on electrolyte leakage and lipid peroxidation and partially prevent a decline in RWC. Salt stress also activated the antioxidant defence system by increasing the activities of POD, CAT, SOD and APX without altering the activities of GR and DHAR. However, NO donors further increased the activities of these enzymes under salt stress. In their experiment, Moussa and Mohamed (2011) investigated the effects of seed pretreatment with an NO donor (10 μM SNP) on drought tolerance in *P. sativum* seedlings. Drought stress (PEG-6000, −0.5 MPa) caused a decrease in the activities of APX, GPX and CAT enzymes and overproduction of $O_2{\bullet}^-$ in pea leaves, which in turn caused exacerbation of lipid peroxidation and depression of photosynthesis. On the other hand, NO-pretreated plants increased the

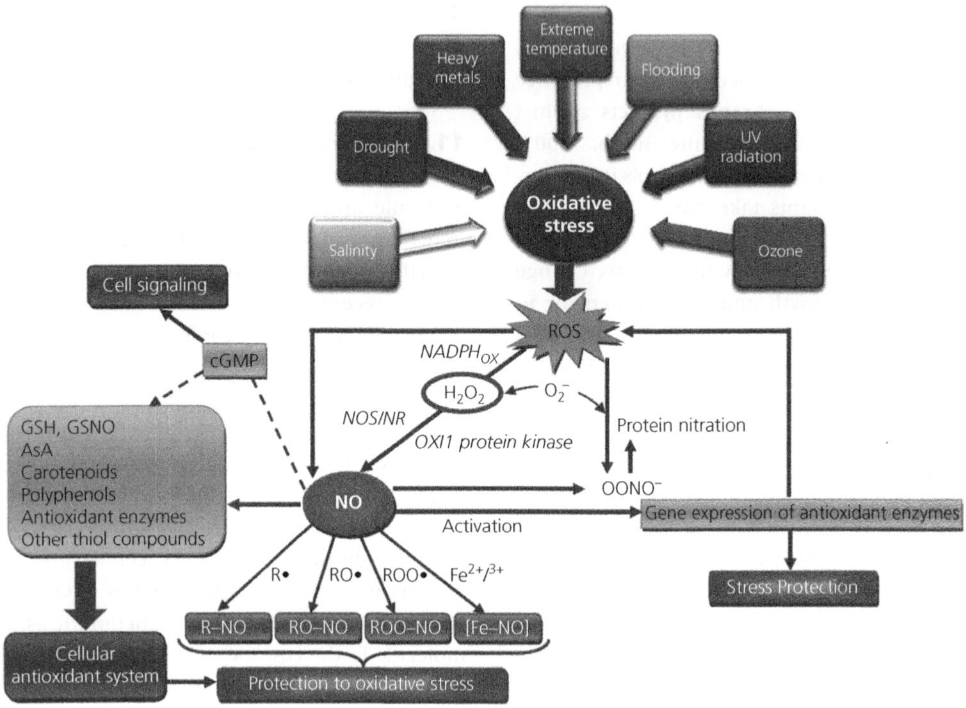

Figure 11.4 Possible mechanisms of NO-induced oxidative stress protection. Adapted from Hasanuzzaman *et al.* (2013e), and reproduced with permission of Springer Science+Business Media.

enzyme activities and decreased $O_2{\bullet}^-$ production, and hence inhibited lipid peroxidation. Levels of H_2O_2, Pro and Evan blue uptake in seedlings pretreated with SNP were markedly lower than under drought stress, indicating the operation of the antioxidant system in them. Moreover, seedlings arising from H_2O_2 or SNP pretreatment enhanced the membrane stability, as revealed from greatly reduced MDA content (Moussa & Mohamed, 2011). Kumari *et al.* (2010) studied the effect of foliar applied NO in alleviating the toxic effects of short-term and long-term Cd stress on growth, oxidative metabolism and Cd accumulation in *C. arietinum*. Exogenous SNP decreased the Cd-induced ion leakage and lipid peroxidation in chickpea. Cadmium treatment for a short term resulted in a higher induction of the CAT, POD, APX, GR and SOD activities as compared to long-term Cd stress. However, upon NO supplementation these activities were upregulated further, associated with a higher redox status of AsA and GSH. With long-term and short-term Cd treatment, roughly 50% and 12% reductions in seed yield were observed in chickpea. However, the yield reduction was lower when the plants were supplemented with an NO donor. In *P. radiata*, presoaking of leaf discs with NO donor (150 μM SNP) provided better protection against HT stress-induced damage (Yang *et al.*, 2006). Compared with heat-shocked (45 °C, 90 min) leaf discs, the levels of thiobarbituric acid-reactive substances (TBARS) and H_2O_2 and electrolyte leakage were reduced, while maximal quantum yield of PSII (*Fv*/*Fm*) was significantly increased by SNP presoaking. The activities of antioxidant enzymes (POD, CAT and SOD) suppressed by heat shock were all restored by SNP presoaking. The effect of NO was confirmed by the opposite results obtained when the leaf discs were soaked in bovine haemoglobin (a powerful NO scavenger). These authors concluded that exogenous NO could prevent HT-induced damage by promoting leaf photochemical activity, cell membrane integrity and upregulation of the activities of antioxidant enzymes (Yang *et al.*, 2006). In *P. vulgaris*, NO supplementation caused incomplete alleviation of Ni toxicity by NO supplementation, as reported by Mihailovic and Drazic (2011). Four-day exposure of the seedlings to 0.2 mM $NiCl_2$ resulted in a marked increase in soluble protein, Pro and SOD activity, whereas POD and CAT activities were decreased. Nickel toxicity also reduced the contents of K, Ca, Mg, Mn and Zn, while the contents of Cu and Fe were increased. However, when the seedlings were supplemented and grown under Ni stress, soluble protein and Pro contents decreased. Nitric oxide supplementation also increased the SOD activity and eliminated the Ni-induced inhibition of POD and CAT activities. Although exogenous NO had no effect on the impact of Ni on K, Cu, Fe and Mn concentrations, it did improve the Ca and Zn concentrations in Ni-treated seedlings. These results suggested that exogenous NO efficiently attenuates oxidative stress in *P. vulgaris*, but could not prevent Ni-induced ion leakage. Cai *et al.* (2011) reported that exogenously NO could protect soybean plants from Al-induced oxidative stress due to the enhancement of the antioxidant defence induced by the NO donor. Soybean plants exposed to Al (50 μM $AlCl_3$) showed a significant reduction in root elongation and a slight reduction in chl content, which was associated with higher Al content in roots. Aluminum treatment also resulted in slight increases in SOD and CAT activities, while POD activity decreased with a concomitant increase in MDA content. On the other hand, the seedlings pretreated with NO donor (0.25 mM SNP) showed improvement in root elongation and chl content, upregulated the activities of the antioxidant enzymes and reduced lipid peroxidation. The protective effects of SNP were reversed by the addition of an NO scavenger – 50μM cPTIO, 2-(4-carboxy-2-phenyl)-4,4,5,5-tetramethylimidazoline-1-oxyl-3-oxide – which confirmed the obvious role of NO in Al stress in soybean (Cai *et al.*, 2011). When chickpea plants were sprayed with NO donor (150 and 300 μM SNP) they performed well under chilling stress, with improvements in plant height and leaf area, decreased electrolyte leakage, and improved yield components and final yield of the crop (Chohan *et al.*, 2012). They also observed that the effects were different in chill-sensitive and chill-tolerant genotypes, and that the lower concentration of SNP (150 μM) was more effective than the higher concentration (300 μM).

11.4.6 Micronutrients

Although the main role of micronutrients is to make plants healthy and ensure that their life cycle runs smoothly, the management of plant micronutrients is very helpful in developing plants' tolerance to several abiotic stresses. Better use of micronutrients can effectively alleviate stress-induced damage by improving several defence systems.

In *V. radiata*, foliar application of Zn provided better protection against water stress (Thalooth *et al.*, 2006). When the bean plants were subjected to lack of irrigation, whether at vegetative, flowering or pod formation stages, significant reductions were seen in all growth parameters, yield components and photosynthetic pigment content compared with unstressed plants. However, foliar application of Zn in water-stressed plants improved growth parameters, yield and yield components (Thalooth *et al.*, 2006). The protective effect of Zn and B in mitigating salt stress in *V. radiata* was reported by Arora *et al.* (2012). Different levels of salinity (10, 20, 40, 60, 80, 100 and 200 mM NaCl) resulted in decreases in seed germination as well as in growth parameters such as root length, shoot length, number of branches per plant and dry weight of *V. radiata* plants in dose-dependent ways. However, when different combinations of Zn (1 to 10 μM) and B (1 to 5 μM) were applied, the negative effects of salt stress were partially mitigated and provided improved plant growth. Concentrations of 4 μM Zn and 3 μM B were found to be optimal in mitigating salt stress in *V. radiata* (Arora *et al.*, 2012). The protective effect of Zn was also reported in Cd-stressed *V. radiata* (Kumari *et al.*, 2011). When *V. radiata* plants were exposed to Cd (0.2 mg/L), there were significant reductions in growth parameters like plant height, fresh weight of plants, number of root nodules, and chl *a* and *b* contents, and in biochemical parameters like NR activity, glutamine oxoglutarate aminotransferase activity, and protein content. However, these growth and biochemical parameters were increased by combined treatment with Zn and Cd compared to Cd treatment alone (Kumari *et al.*, 2011). Patel *et al.* (2013) reported that exogenous application of Zn on *P. aureus* could partially restore seedling growth and reduce oxidative stress during seed development at low concentrations of the heavy metals Cd and As (25 and 50 mg/kg), while the Zn could not protect the seedlings from higher concentrations of Cd and As. The Zn-induced alleviation of Cd and As toxicity was associated with an enhanced level of free Pro and improved metal-chelating activity as well as the higher antioxidant defense system in *P. aureus* seedlings (Patel *et al.*, 2013). Bellaloui (2011) reported that soybean plants grown under waterlogged conditions and supplemented with foliar application of boron (B) improved their seed protein, oil, fatty acid and nitrogen metabolism compared with plants that did not receive B. In a subsequent study, Bellaloui and his group (2013) reported that foliar B application could improve the seed composition, cell wall B content, and seed $\delta^{15}N$ and $\delta^{13}C$ isotopes in water-stressed soybean plants. The boron-induced increases in $\delta^{15}N$ and $\delta^{13}C$ isotopes under water stress suggests a role of B in regulation of N fixation, which is of great importance for legumes.

11.4.7 Trace elements

11.4.7.1 Silicon

Many reports have appeared on silicon (Si) research in the last few decades, but this element continues to be an anomaly – an underappreciated element and in most cases one not considered essential for plant function. Numerous research reports have provided evidence for the notion that Si may play a vital role in conferring tolerance to adverse environmental factors in plants. Hence, Si is considered as a beneficial element for plants growing under stressful conditions (Hasanuzzaman *et al.*, 2014c).

Wang and Han (2007) studied the mechanisms of Si-mediated salt stress tolerance in alfalfa (*M. sativa*), reporting that exogenous Si could alter the ion distribution in different plant parts. They found that salt stress (NaCl 120 mM) affected not only the macronutrient distribution but also the micronutrient distribution in alfalfa plants, which was reflected by a drastic increase in Na^+ content, a slight increase in Zn^{2+} content and significant decreases in K^+, Ca^{2+}, Mg^{2+} and Cu^{2+}. Compared with the NaCl treatment, the added Si significantly decreased Na^+ content in the roots, but notably increased K^+ content in the shoots and leaves of the highly salt-tolerant Zhongmu No. 1 cultivar. Applying Si to both sensitive and tolerant cultivars under NaCl stress did not significantly affect the Fe^{3+}, Mg^{2+} and Zn^{2+} contents in the roots, shoots and leaves of Defor or in the roots and shoots of Zhongmu No. 1, but increased the Ca^{2+} content in the roots of Zhongmu No. 1 and the Mn^{2+} content in the shoots and leaves of both cultivars, while it decreased Ca^{2+} and Cu^{2+} contents of the shoots and leaves of both cultivars under salt stress. These well-integrated ion distribution systems in plants due to Si application conferred better salt tolerance in alfalfa (Wang & Han, 2007). Zuccarini (2008) reported that exogenous Si could enhance the photosynthesis, water relations and nutrient uptake of *P. vulgaris* under NaCl stress. This study showed that salt stress (30 and 60 mM NaCl) significantly reduced growth, stomatal conductance and net photosynthetic rate, and increased Na^+ and Cl^- contents. However, Si application (1.5 mM) enhanced growth of salt-stressed plants and reduced Na^+ content. Exogenous Si also improved the photosynthesis gas-exchange parameters and net photosynthesis (Zuccarini, 2008). While studying soybean, Hamayun *et al.* (2010b) observed that both salinity (100 mM NaCl) and drought stresses (12% PEG) decreased shoot length, and plant fresh and dry weights. However, these adverse effects of salinity and drought on plant growth were alleviated by adding Si (100 and 200 mg/L) to salt- and drought-stressed treatments. During salt and drought stress, exogenous Si also improved physio-hormonal attributes of soybean contributing to the mitigation of the adverse effects of these stresses (Hamayun *et al.*, 2010b). In alfalfa (*M. sativa*), Wang *et al.* (2011) revealed the role of Si in enhancing antioxidant defence in plants exposed to salt stress. They grew two alfalfa cultivars (highly salt-tolerant Zhongmu No. 1 and poorly salt-tolerant Defor) under saline conditions (120 mM NaCl). After 15 days of salt stress, the plants exhibited alterations of their antioxidant metabolism in the roots and shoots of both cultivars. However, applying Si (1 mM) to both cultivars under NaCl stress significantly increased the antioxidant enzymes' activities in different plant parts, except for SOD. These results suggested that Si could alter the activity of antioxidant enzymes of plants to improve their salt tolerance. Responses of two widely cultivated pulse crops, *V. radiata* and *V. unguiculuta*, were studied under drought stress (100, 75, 50 and 25% soil moisture regimes) by Hamid *et al.* (2012b). Both crops showed reductions in carbohydrates and protein upon exposure to drought. However, Si application (20, 40 or 60 mg/kg) significantly increased total carbohydrate and protein contents in those plants under drought conditions. However, the effect of Si on plant metabolism depends upon the applied concentration and the plant species (Hamid *et al.*, 2012b). In their experiment, Parande *et al.* (2013) observed that salt stress (1–5 dS/m) caused significant reductions in number of seeds per pod, number of fertile pods and dry matter in *P. vulgaris*. However, application of Si (0.5, 1 or 2 mM) significantly increased the 100-seed weight and yield of beans under saline environments. However, the effects of both salt and Si greatly depended on the doses of application. It was evident that application of 1 mM Si could give the best protection against salinity (Parande *et al.*, 2013). In salt-sensitive *V. faba*, high salinity (100 mM NaCl) caused a sharp increase in Na^+ concentration in the leaf apoplast, which was significantly ameliorated by the addition of Si. Salinity-induced decrease in shoot growth also recovered upon addition of Si (Shahzad *et al.*, 2013). In *G. max*, the addition of Si to salt-stressed plants substantially alleviated the adverse effects of NaCl on growth, as it enhanced endogenous GA_3, while reducing the levels of ABA and Pro (Lee *et al.*, 2010).

11.4.7.2 Selenium

Although selenium (Se) has been studied widely in plants and its beneficial role is well known, it is still unresolved whether Se is an *essential* micronutrient for plants. In many plant species, Se exerts a positive effect on plant growth and productivity at low concentrations (Feng *et al.*, 2013). However, the precise mechanisms underlying the beneficial role of Se in plants have not been clearly elucidated yet. Some plant species supplemented with Se have shown enhanced resistance to certain abiotic stresses (Hasanuzzaman *et al.*, 2010b, 2011b, 2014c; Hasanuzzaman & Fujita, 2011, 2012b). Recently, Feng *et al.* (2013) reviewed the possible mechanisms responsible for the production and quenching of ROS after Se addition to stressed plants. They proposed that Se can regulate the ROS levels in stressed plants through three pathways: (i) by stimulating the spontaneous dismutation of $O_2{\bullet}^-$ into H_2O_2; (ii) by a direct reaction between Se-containing compounds and ROS; and (iii) by the regulation of antioxidative enzymes. Another possible mechanism is that Se may affect the assembly of photosynthetic complexes to regulate the levels of ROS (Feng *et al.*, 2013).

Ajiboso and Adenuga (2012) investigated the influence of Se on *V. unguiculata* and observed that exogenous Se could alter some biochemical parameters of the plants under water deficit conditions. Upon exposure to water deficit for 14 days, *V. unguiculata* showed a 28.6% reduction in RWC. Water deficit also caused drastic increases in MDA levels and antioxidant enzyme activities. On the other hand, Se-supplemented plants showed a higher RWC, reduced lipid peroxidation, further increase in the activities of antioxidant enzymes such as CAT and GPX, and a higher level of AsA. Exogenous Se also enhanced some growth parameters like plant height and number of leaves per plant under water deficit conditions. Based on the study, Se was recommended as an antioxidant chemical to be used by legume growers under conditions of short-term drought (Ajiboso & Adenuga, 2012). Selenium-induced mitigation of As stress has been reported by Malik *et al.* (2012). When plants were subjected to different levels of As (2.5, 5.0 and 10 μM), they exhibited drastic reductions in shoot and root growth and chl content. Arsenic stress also caused a sharp increase in electrolyte leakage, and MDA and H_2O_2 contents. When As-treated plants were supplemented with Se (5 μM) they showed improvement in growth indicating an antagonistic interaction between the two elements. The Se-treated plants also experienced less damage to membranes, chl and cellular viability induced by As. Moreover, the oxidative damage due to As was reduced with Se application, which could be related to the upregulation of antioxidant enzymes such as SOD, CAT, APX, GR and GST, as well as enhanced synthesis of water-soluble antioxidants like AsA and GSH (Malik *et al.*, 2012). Exogenous Se also increased the metallothioneins (MTs) and thiols under elevated As, which rendered the plants more tolerant to As. The beneficial effect of Se on Pb-stressed *Vicia faba* L. *minor* roots was also reported by Mroczek-Zdyrska and Wójcik (2012). When *V. faba* roots were exposed to 50 μM Pb, they exhibited damage due to increased accumulation of Pb. The added Pb increased the production of $O_2{\bullet}^-$, lipid peroxidation (MDA content) and thiol content. The antioxidant metabolism was also altered due to Pb stress. On the other hand, supplementation of the plants with Se (1.5 and 6 μM) alleviated Pb toxicity, which was accompanied by decreased $O_2{\bullet}^-$ production in the apical parts of roots, and increased thiol content and GPX activity. However, the higher Se concentration intensified MDA and thiol accumulation and GPX activity in Pb-treated plant roots. At the lower concentration, Se improved cell viability, whereas at the higher concentration it was pro-oxidant and enhanced lipid peroxidation and cell membrane injury (Mroczek-Zdyrska & Wójcik, 2012).

11.5 Conclusion and future perspectives

Legumes are important for human nutrition and for maintaining soil fertility. However, there are several constraints on enhancing the productivity of pulse crops, including environmental adversities. The situation is becoming more serious due to the unpredictability of environmental conditions and global climate change and variability. Some legumes are very sensitive to environmental stresses and hence finding ways to improve stress tolerance is a most important task for plant biologists. Numerous research results have shown the benefits of using exogenous phytoprotectants in mitigating stress-induced damage in legumes. However, due to the multigenic nature of abiotic stress tolerance the task of developing stress-tolerant varieties is challenging. Coordinated approaches involving traditional plant breeding, along with molecular approaches, should be followed to identify and cultivate stress-tolerant varieties. However, effective enrichment of legumes with phytoprotectants via soil or foliar sprays may play a protective role in stress mitigation. However, it may be challenging due to their high cost and difficulties in application. Notably, findings from experimental field research studies are still inadequate. Field studies that confirm the ability of exogenous applications of phytoprotectants to mitigate abiotic stress will point to potential options for their extended practical application for increasing the productivity of legume crops.

11.6 Acknowledgements

We wish to thank Mr Anisur Rahman, Laboratory of Plant Stress Responses, Faculty of Agriculture, Kagawa University, Japan, for providing several supporting articles and suggestions for improving the manuscript. We are also very grateful to Mr Hasanuzzaman and Jubayer-Al-Mahmud, Faculty of Agriculture, Sher-e-Bangla Agricultural University, Dhaka-1207, Bangladesh, for their critical reading of the manuscript draft.

References

Abdel CG, Al-Rawi IMT (2011) Response of mungbean (*Vigna radiata* L., Wilczek) to gibberellic acid (GA_3) rates and varying irrigation frequencies. *Int J Biosci* 1: 85–92.

Abdo FA, Nassar DMA, Gomaa EF, Nassar RMA (2012) Minimizing the harmful effects of cadmium on vegetative growth, leaf anatomy, yield and physiological characteristics of soybean plant [*Glycine max* (L.) Merrill] by foliar spray with active yeast extract or with garlic cloves extract. *Res J Agric Biol Sci* 8: 24–35.

Agrawal SB, Mishra S (2009) Effects of supplemental ultraviolet-B and cadmium on growth, antioxidants and yield of *Pisum sativum* L. *Ecotoxicol Environ Saf* 72: 610–618.

Agrawal SB, Rathore D, Singh A (2006) Combined effects of enhanced ultraviolet-B radiation and mineral nutrients on growth, biomass accumulation and yield characteristics of two cultivars of *Vigna radiata* L. *J Environ Biol* 27, 55–60.

Ahmed S (2009) Effect of soil salinity on the yield and yield components of mungbean. *Pak J Bot* 41: 263–268.

Ajiboso SO, Adenuga GA (2012) The influence of zinc and selenium on some biochemical responses of *Vigna unguiculata* and *Zea mays* to water deficit condition and rehydration. *Biokemistri* 24: 108–115.

Alam I, Lee DG, Kim KH, *et al.* (2010) Proteome analysis of soybean roots under waterlogging stress at an early vegetative stage. *J Biosci* 35: 49–62.

Alam MM, Hasanuzzaman M, Nahar K, Fujita M (2013) Exogenous salicylic acid ameliorates short-term drought stress in mustard (*Brassica juncea* L.) seedlings by upregulating the antioxidant defense and glyoxalase system. *Aust J Crop Sci* 7: 1053–1063.

Alam MM, Nahar K, Hasanuzzaman M, Fujita M (2014a) Exogenous jasmonic acid modulates the physiology, antioxidant defense and glyoxalase system in imparting drought stress tolerance in different *Brassica* species. *Plant Biotechnol Rep* 8: 279–293.

Alam MM, Nahar K, Hasanuzzaman M, Fujita M (2014b) Exogenous ascorbic acid-induced alleviation of osmotic stress in three Brassica species through enhancement of growth and physiological parameters, antioxidant defense and glyoxalase system. *Biol Plant* (in press).

Alami-Milani M, Amini R, Mohammadinasab AD, Shafaghkal vanegh J, Asgharzade A, Emaratpardaz J (2013) Yield and yield components of lentil (*Lens culinaris* Medick.) affected by drought stress and mulch. *Intl J Agri Crop Sci* 5: 1228–1231.

Ali B, Rani I, Hayat S, Ahmad A (2007) Effect of 4-Cl-indole acetic acid on seed germination of *Cicer arietinum* exposed to cadmium. *Acta Bot Croat* 66: 57–65.

Ali B, Hasan SA, Hayat S, *et al.* (2008) A role for brassinosteroids in the amelioration of aluminium stress through antioxidant system in mung bean (*Vigna radiata* L. Wilczek). *Environ Exp Bot* 62: 153–159.

Ali HM, Siddiqui MH, Al-Whaibi MH, Basalah MO, Sakran AM, El-Zaidy M (2013) Effect of proline and abscisic acid on the growth and physiological performance of faba bean under water stress. *Pak J Bot* 45: 933–940.

Almohisen IAA, Alyemeni MN (2013) Effect of ozone on the quality of two legume crops (*Vicia faba* and *Pisum sativum*) around Riyadh city. *Life Sci J* 10: 2297–2302.

Alsokari SS (2011) Synergistic effect of kinetin and spermine on some physiological aspects of seawater stressed *Vigna sinensis* plants. *Saudi J Biol Sci* 18: 37–44.

Anjum NA, Umar S, Iqbal M, Khan NA (2011) Cadmium causes oxidative stress in mung bean by affecting the antioxidant enzyme system and ascorbate-glutathione cycle metabolism. *Russ J Plant Physiol* 58: 92–99.

Anjum NA, Ahmad I, Mohmood I, *et al.* (2012a) Modulation of glutathione and its related enzymes in plants' responses to toxic metals and metalloids – a review. *Environ Exp Bot* 75: 307–324.

Anjum NA, Umar S, Ahmad A (2012b) *Oxidative Stress in Plants: Causes, Consequences and Tolerance*. IK International Publishing House, New Delhi.

Anjum NA, Gill SS, Gill G, *et al.* (2014) Metal/metalloid stress tolerance in plants: role of ascorbate, its redox couple, and associated enzymes. *Protoplasma* doi: 10.1007/s00709-014-0636-x.

Arora AS, Umer S, Mishra SN (2012) Boron and zinc response on growth in *Vigna radiata* L. Wilczek var. Pusa Vishal under salinity. *Int J Plant Anim Environ Sci* 2: 131–138.

Ashagre H, Hamza IA, Fasika E, Temesgen F (2013) Effect of salinity stress on germination and seedling vigour of chickpea (*Cicer arietinum* L.) cultivars. *Acad J Agric Res* 1: 161–166.

Ashraf M, Foolad MR (2007) Roles of glycine betaine and proline in improving plant abiotic stress resistance. *Environ Exp Bot* 59: 206–216.

Ashraf MA, Rasool M, Alio Q, Haider MZ, Noman A, Azeem M (2013) Salt-induced perturbation in growth, physiological attributes, activities of antioxidant enzymes and organic solutes in mungbean (*Vigna radiata* L.) cultivars differing in salinity tolerance. *Arch Agron Soil Sci* 59: 1695–1712.

Aslam M, Ahmad A, Saguu AG, *et al.* (2010) Effect of plant growth regulator (NAA) and available soil moisture on yield and yield components of chickpea. *Sarhad J Agric* 26: 325–335.

Assefa T, Beebe SE, Rao IM, *et al.* (2013) Pod harvest index as a selection criterion to improve drought resistance in white pea bean. *Field Crops Res* 148: 24–33.

Azooz MM (2009) Salt stress mitigation by seed priming with salicylic acid in two faba bean genotypes differing in salt tolerance. *Int J Agric Biol* 11: 343–350.

Babu R, Devaraj VR (2008) High temperature and salt stress response in French bean (*Phaseolus vulgaris*). *Aust J Crop Sci* 2: 40–48.

Bafeel SO, Ibrahim MM (2008) Antioxidants and accumulation of α-tocopherol induce chilling tolerance in *Medicago sativa*. *Int J Agri Biol* 10: 593–598.

Bajguz A, Hayat S (2009) Effects of brassinosteroids on the plant responses to environmental stresses. *Plant Physiol Biochem* 47: 1–8.

Barba-Espín G, Clemente-Moreno MJ, Álvarez S, García-Legaz MF, Hernández JA, Díaz-Vivancos P (2011) Salicylic acid negatively affects the response to salt stress in pea plants. *Plant Biol* 13: 909–917.

Bashor CJ, Dalton DA (1999) Effects of exogenous application and stem infusion of ascorbate on soybean (*Glycine max*) root nodules. *New Phytol* 142: 19–26.

Beebe SE, Rao IM, Cajiao C, Grajales M (2008) Selection for drought resistance in common bean also improves yield in phosphorus limited and favorable environments. *Crop Sci* 48: 582–592.

Beebe SE, Rao IM, Blair M, Acosta-Gallegos JA (2010) Phenotyping common beans for adaptation to drought. In: Ribaut JM, Monneveux P (eds), *Drought Phenotyping in Crops: From Theory to Practice. Generation Challenge Program Special Issue on Phenotyping*. Generation Challenge Program of CGIAR, Mexico City, Mexico, pp. 311–334.

Behairy RT, El-Danasoury M, Craker L (2012) Impact of ascorbic acid on seed germination, seedling growth, and enzyme activity of salt-stressed Fenugreek. *J Medic Active Plants* 1: 106–113.

Bellaloui N (2011) Effect of water stress and foliar boron application on seed protein, oil, fatty acids, and nitrogen metabolism in soybean. *Am J Plant Sci* 2: 692–701.

Bellaloui N, Hu Y, Mengistu A, Kassem MA, Abel CA (2013) Effects of foliar boron application on seed composition, cell wall boron, and seed δ15N and δ13C isotopes in water-stressed soybean plants. *Front Plant Sci* 4: 1–12; doi: 10.3389/fpls.2013.00270.

Beltagi MS (2008) Exogenous ascorbic acid (vitamin C) induced anabolic changes for salt tolerance in chick pea (*Cicer arietinum* L.) plants. *Afr J Plant Sci* 2: 118–123.

Berumen-padilla S, Lownds NK, Mexal JG (2009) *Ethylene and Drought Tolerance in Selected Legumes. Ethylene Production in Cassia Corymbosa and Alfalfa Cultivars as a Stress Response and its Role in Drought Tolerance*. Lambert Academic Publishing.

Betzelberger AM, Gillespie KM, McGrath JM, Koester RP, Nelson RL, Ainsworth EA (2010) Effects of chronic elevated ozone concentration on antioxidant capacity, photosynthesis and seed yield of 10 soybean cultivars. *Plant Cell Environ* 33: 1569–1581.

Betzelberger AM, Yendrek CR, Sun J, *et al.* (2012) Ozone exposure response for U.S. soybean cultivars: Linear reductions in photosynthetic potential, biomass, and yield. *Plant Physiol* 160: 1827–1839.

Bita CE, Gerats T (2013) Plant tolerance to high temperature in a changing environment: scientific fundamentals and production of heat stress-tolerant crops. *Front Plant Sci* 4: 273.

Bonser AM, Lynch J, Snapp S (1996). Effect of phosphorus deficiency on growth angle of basal roots in *Phaseolus vulgaris*. *New Phytol* 132: 281–288.

Britz SJ, Kremer DF, Kenworthy WJ (2008) Tocopherols in soybean seeds: Genetic variation and environmental effects in field-grown crops. *J Am Oil Chem Soc* 85: 931–936.

Brunschon-Harti S, Andreas F, Jager HJ (1995a) Influence of ozone and ethylene diurea (EDU) on growth and yield of beans (*Phaseolus vulgaris* L.) in open top field chambers. *Environ Pollut* 90: 89–94.

Brunschon-Harti S, Fangmeier A, Jage HJ (1995b) Effects of ethylene diurea and ozone on the antioxidative systems in beans (*Phaseolus vulgaris* L.). *Environ Pollut* 90: 95–103.

Cai MZ, Zhang SN, Wang FM, Wang N, Xu SY (2011) Protective effect of exogenously applied nitric oxide on aluminum-induced oxidative stress in soybean plants. *Russ J Plant Physiol* 58: 791–798.

Capoen W, Goormachtig S, Holsters M (2010) Water-tolerant legume nodulation. *J Exp Bot* 61: 1251–1255.

Chakrabarti N, Mukherji S (2003) Effect of phytohormone pretreatment on nitrogen metabolism in *Vigna radiata* under salt stress. *Biol Plant* 46: 63–66.

Chattopadhyay A, Subba P, Pandey A, *et al.* (2011) Analysis of the grasspea proteome and identification of stress-responsive proteins upon exposure to high salinity, low temperature, and abscisic acid treatment. *Phytochemistry* 72: 1293–1307.

Chaturvedi SK, Mishra DK, Vyas P, Neelu M (2009) Breeding for cold tolerance in chickpea. *J Trends Biosci* 2: 1–6.

Chen JH, Jiang HW, Hsieh EJ, *et al.* (2012) Drought and salt stress tolerance of an *Arabidopsis* glutathione *S*-transferase U17 knockout mutant are attributed to the combined effect of glutathione and abscisic acid. *Plant Physiol* 158: 340–351.

Chen THH, Murata N (2008) Glycinebetaine: an effective protectant against abiotic stress in plants. *Trend Plant Sci* 13: 499–505.

Cheong JJ, Choi YD (2003) Methyl jasmonate as a vital substance in plants. *Trends Genet* 19: 409–413.

Chinnusamy V, Schumaker K, Zhu JK (2004) Molecular genetic perspectives on cross-talk and specificity in abiotic stress signaling in plants. *J Exp Bot* 55: 225–236.

Chohan A, Raina SK (2011) Comparative studies on morphological and biochemical characters of chickpea genotypes under chilling stress. *J Environ Biol* 32: 189–194.

Chohan A, Parmar U, Raina SK (2012) Effect of sodium nitroprusside on morphological characters under chill stress in chickpea (*Cicer arietinum* L.). *J Env Biol* 33: 695–698.

Cokkizgin A, Cokkizgin H (2012) Effects of lead ($PbCl_2$) stress on germination of lentil (*Lens culinaris* Medic.) lines. *Afr J Biotechnol* 9: 8608–8612.

Craufurd PQ, Prasad PVV, Summerfield RJ (2002) Dry matter production and rate of change of harvest index at high temperature in peanut. *Crop Sci* 42: 46–151.

Dalio RJD, Pinheiro HP, Sodek L, Haddad CRB (2013) 24-epibrassinolide restores nitrogen metabolism of pigeon pea under saline stress. *Bot Stud* 54: 9.

Dantus BF, Ribeiro L, Aragao CA (2005) Physiological response of cowpea seeds to salinity stress. *Rev Bras Sementes* 27: 144–148.

Dashti N, Khanafer M, El-Nemr I, Sorkhoh N, Ali N, Radwan, S. (2009) The potential of oil-utilizing bacterial consortia associated with legume root nodules for cleaning oily soils. *Chemosphere* 74: 1354–1359.

Dehghan G, Rezazadeh L, Habibi G (2011) Exogenous ascorbate improves antioxidant defense system and induces salinity tolerance in soybean seedlings. *Acta Biol Szeged* 55: 261–264.

de Souza SCR, López de Andrade A, Anjos de Souza L, Schiavinato MA (2012) Lead tolerance and phytoremediation potential of Brazilian leguminous tree species at the seedling stage. *J Environ Manage* 15: 299–307.

Devi RG, Pandiyarajan V, Gurusaravanan P (2012) Alleviating effect of IAA on salt stressed *Phaseolus mungo* (L.) with reference to growth and biochemical characteristics. *Recent Res Sci Technol* 4: 22–24.

D'Haeze W, Rycke RD, Mathis R, *et al.* (2003) Reactive oxygen species and ethylene play a positive role in lateral root base nodulation of a semiaquatic legume. *Proc Natl Acad Sci USA* 100: 11789–11794.

Dita MA, Rispail N, Prats E, Rubiales D, Singh KB (2006) Biotechnology approaches to overcome biotic and abiotic stress constraints in legumes. *Euphytica* 147: 1–24.

Dolatabadian A, Jouneghani RS (2009) Impact of exogenous ascorbic acid on antioxidant activity and some physiological traits of common bean subjected to salinity stress. *Not Bot Hort Agrobot Cluj* 37: 165–172.

Egamberdieva D (2009) Alleviation of salt stress by plant growth regulators and IAA producing bacteria in wheat. *Acta Physiol Plant* 31: 861–864.

El-Bassiony AM, Ghoname AA, El-Awadi ME, Fawzy ZF, Gruda N (2012) Ameliorative effects of brassinosteroids on growth and productivity of snap beans grown under high temperature. *Gesunde Pflanzen* 64: 175–182.

El-Mashad AAA, Mohamed HI (2011) Brassinolide alleviates salt stress and increases antioxidant activity of cowpea plants (*Vigna sinensis*). *Protoplasma* 249: 625–635.

El-Meleigy EA, Hassanein RA, Kader DZA (1999) Improvement of drought tolerance in *Araches hypogaea* L. plant by some growth substances I. Growth and productivity. *Bull Fac Sci Assiut Univ* 28: 159–185.

Entesari M, Sharif-Zadeh F, Zare S, Farhangfar M, Dashtaki M (2012) Effect of seed priming on mung been (*Vigna radiata*) cultivars with salicylic acid and potassium nitrate under salinity stress. *Int J Agric Res Rev* 2: 926–932.

Ergün N, Öncel I (2010) Effects of cadmium and zinc on growth and some biochemical parameters of lentil seedlings (*Lens esculenta* L.). *Fen Bilimleri Dergisi* 31: 63–72.

Eyidogan F, Oz MT, Yucel M, Oktem HA (2012) Signal transduction of phytohormones under abiotic stresses. In: Khan NA, Nazar R, Iqbal N, Anjum NA (eds), *Phytohormones and Abiotic Stress Tolerance in Plants*. Springer, pp. 1–48.

Falk J, Krauss N, Dähnhardt D, Krupinska K (2002) The senescence associated gene of barley encoding 4-hydroxyphenylpyruvate dioxygenase is expressed during oxidative stress. *J Plant Physiol* 159: 1245–1253.

Farooq U, Bano A (2006) Effect of abscisic acid and chlorocholine chloride on nodulation and biochemical content of *Vigna radiata* L. under water stress. *Pak J Bot* 38: 1511–1518.

Fatoba PO, Ogunkunle CO, Salihu BZ (2012) Toxic effects of cadmium (Cd) and lead (Pb) on growth and productivity of *Arachis hypogaea* (L.) and *Glycine max* (L.). *J Asian Sci Res* 2: 254–259.

Fedina IS, Tsonev TD (1997) Effect of pretreatment with methyl jasmonate on the response of *Pisum sativum* to salt stress. *J Plant Physiol* 151: 735–740.

Feng R, Wei C, Tu S (2013) The roles of selenium in protecting plants against abiotic stresses. *Environ Exp Bot* 87: 58–68.

Foyer CH, Noctor G (2003) Redox sensing and signaling associated with reactive oxygen in chloroplasts, peroxisomes and mitochondria. *Physiol Plant* 119: 355–364.

Gan Y, Wang J, Angadi SV, Mcdonald CL (2004) Response of chickpea to short periods of high temperature and water stress at different developmental stages. In: 4th International Crop Science Congress, Brisbane, Australia, 26 September–1 October 2004.

Gangwar S, Singh VP, Prasad SM, Maurya JN (2011) Differential responses of pea seedlings to indole acetic acid under manganese toxicity. *Acta Physiol Plant* 33: 451–462.

Ghani A (2010) Effect of cadmium toxicity on the growth and yield components of mungbean [*Vigna radiata* (L.) Wilczek]. *World Appl Sci J* 8: (Special Issue of Biotechnology & Genetic Engineering), pp. 26–29.

Ghassemi-Golezani K, Mahmoodi-Yengabad F (2012) Physiological responses of lentil (*Lens culinaris* Medik.) to salinity. *Intl J Agri Crop Sci* 4: 1531–1535.

Ghassemi-Golezani K, Taifeh-Noori M (2011) Soybean performance under salinity stress. In Tzi-Bun Ng (ed.), *Soybean – Biochemistry Chemistry and Physiology*. InTech, pp. 631–642.

Ghassemi-Golezani K, Nikpour-Rashidabad N, Zehtab-Salmasi S (2012) Effect of salinity on yield and yield components of pinto bean cultivars. *Int J Plant Anim Environ Sci* 2: 47–51.

Gill SS, Tuteja N (2010) Reactive oxygen species and antioxidant machinery in abiotic stress tolerance in crop plants. *Plant Physiol Biochem* 48: 909–930.

Gill SS, Anjum NA, Hasanuzzaman M, Gill R, Trivedi DK, Ahmad I, Pereira E, Tuteja N (2013a) Glutathione reductase and glutathione: a boon in disguise for plant abiotic stress defense operations. *Plant Physiol Biochem* 70: 204–212.

Gill SS, Hasanuzzaman M, Nahar K, Macovei A, Tuteja NN (2013b) Importance of nitric oxide in cadmium stress tolerance in crop plants. *Plant Physiol Biochem* 63: 254–261.

Graham PH, Vance CP (2003) Legumes: importance and constraints to greater use. *Plant Physiol* 131: 872–877.

Gross Y, Kigel J (1994) Differential sensitivity to high temperature of stages in the reproductive development of common bean (*Phaseolus vulgaris* L.). *Field Crops Res* 36: 201–212.

Groten K, Dutilleul C, van Heerden PDR, *et al.* (2006) Redox regulation of peroxiredoxin and proteinases by ascorbate and thiols during pea root nodule senescence. *FEBS Lett* 580: 1269–1276.

Haileselasie TH, Teferii G (2012) The effect of salinity stress on germination of chickpea (*Cicer arietinum* L.) land race of Tigray. *Curr Res J Biol Sci* 4: 578–583.

Haleem A, Mohammed MA (2007) Physiological aspects of mungbean plant (*Vigna radiata* L. Wilczek) in response to salt stress and gibberellic acid treatment. *Res J Agric Biol Sci* 3: 200–213.

Hamayun M, Khan SA, Khan AL, *et al.* (2010a) Effect of salt stress on growth attributes and endogenous growth hormones of soybean cultivar Hwangkeumkong. *Pak J Bot* 42: 3103–3112.

Hamayun M, Sohn EY, Khan SA, Shinwari ZK, Khan AL, Lee IJ (2010b) Silicon alleviates the adverse effects of salinity and drought stress on growth and endogenous plant growth hormones of soybean (*Glycine max* L.). *Pak J Bot* 42: 1713–1722.

Hamid N, Hussain H, Jawaid F, Rahman A (2012a) Short term exposure of UV-B radiation enhances salinity tolerance in *Vigna radiata*. *Pak J Bot* 44: 841–844.

Hamid N, Naz B, Rehman A (2012b) Effect of exogenous application of silicon with drought stress on protein and carbohydrate contents of edible beans (*Vigna radiata* and *Vigna unguiculata*). *Pak J Chem* 2: 99–105.

Hammad SAR, Shaban KA, Tantawy MF (2010) Studies on salinity tolerance of two peanut cultivars in relation to growth, leaf water content. Some chemical aspects and yield. *J Appl Sci Res* 6: 1517–1526.

Harper JR, Balke NE (1981) Characterization of the inhibition of K^+ absorption in oats roots by salicylic acid. *Plant Physiol* 68: 1349–1353.

Hasan SA, Hayat S, Ali B, Ahmad A (2008) 28-Homobrassinolide protects chickpea (*Cicer arietinum*) from cadmium toxicity by stimulating antioxidants. *Environ Pollut* 151: 60–66.

Hasanuzzaman M, Fujita M (2011) Selenium pretreatment upregulates the antioxidant defense and methylglyoxal detoxification system and confers enhanced tolerance to drought stress in rapeseed seedlings. *Biol Trace Elem Res* 143: 1758–1776.

Hasanuzzaman M, Fujita M (2012a) Heavy metals in the environment: Current status, toxic effects on plants and possible phytoremediation. In: Anjum NA, Pereira MA, Ahmad I, Duarte AC, Umar S, Khan NA (eds), *Phytotechnologies: Remediation of Environmental Contaminants*. CRC Press, Boca Raton, FL, pp. 7–73.

Hasanuzzaman M, Fujita M (2012b) Selenium and plants' health: the physiological role of selenium. In: Aomori C, Hokkaido M (eds), *Selenium: Sources, Functions and Health Effects*. Nova Science Publishers, New York, pp. 101–122.

Hasanuzzaman M, Fujita M (2013) Exogenous sodium nitroprusside alleviates arsenic-induced oxidative stress in wheat seedlings by enhancing antioxidant defense and glyoxalase system. *Ecotoxicology* 22: 584–596.

Hasanuzzaman M, Fujita M, Islam MN, Ahamed KU, Nahar K (2009) Performance of four irrigated rice varieties under different levels of salinity stress. *Int J Integr Biol* 6: 85–90.

Hasanuzzaman M, Hossain MA, Fujita M (2010a) Physiological and biochemical mechanisms of nitric oxide induced abiotic stress tolerance in plants. *Am J Plant Physiol* 5: 295–324.

Hasanuzzaman M, Hossain MA, Fujita M (2010b) Selenium in higher plants: Physiological role, antioxidant metabolism and abiotic stress tolerance. *J Plant Sci* 5: 354–375.

Hasanuzzaman M, Hossain MA, Fujita M (2011a) Nitric oxide modulates antioxidant defense and the methylglyoxal detoxification system and reduces salinity-induced damage of wheat seedlings. *Plant Biotechnol Rep* 5: 353–365.

Hasanuzzaman M, Hossain MA, Fujita M (2011b) Selenium-induced up-regulation of the antioxidant defense and methylglyoxal detoxification system reduces salinity-induced damage in rapeseed seedlings. *Biol Trace Elem Res* 143: 1704–1721.

Hasanuzzaman M, Hossain MA, da Silva JAT, Fujita M (2012a) Plant responses and tolerance to abiotic oxidative stress:

Antioxidant defense is a key factor. In: Bandi V, Shanker AK, Shanker C, Mandapaka M (eds), *Crop Stress and its Management: Perspectives and Strategies*. Springer, Berlin, pp. 261–316.

Hasanuzzaman M, Hossain MA, Fujita M (2012b) Exogenous selenium pretreatment protects rapeseed seedlings from cadmium-induced oxidative stress by upregulating the antioxidant defense and methylglyoxal detoxification systems. *Biol Trace Elem Res* 149: 248–261.

Hasanuzzaman M, Nahar K, Alam MM, Fujita M (2012c) Exogenous nitric oxide alleviates high temperature induced oxidative stress in wheat (*Triticum aestivum*) seedlings by modulating the antioxidant defense and glyoxalase system. *Aust J Crop Sci* 6: 1314–1323.

Hasanuzzaman M, Nahar K, Fujita M (2013a) Plant response to salt stress and role of exogenous protectants to mitigate salt-induced damages. In: Ahmad P, Azooz MM, Prasad MNV (eds), *Ecophysiology and Responses of Plants Under Salt Stress*. Springer, New York, pp. 25 – 87.

Hasanuzzaman M, Nahar K, Fujita M, *et al.* (2013b) Enhancing plant productivity under salt stress – Relevance of polyomics. In: Ahmad P, Azooz MM, Prasad MNV (eds), *Salt Stress in Plants: Omics, Signaling and Responses*. Springer, Berlin, pp. 113–156.

Hasanuzzaman M, Nahar K, Fujita M (2013c) Extreme temperatures, oxidative stress and antioxidant defense in plants. In: Vahdati K, Leslie C (eds), *Abiotic Stress*. InTech, Rijeka, doi: 10.5772/54833.

Hasanuzzaman M, Nahar K, Alam MM, Roychowdhury R, Fujita M (2013d) Physiological, biochemical, and molecular mechanisms of heat stress tolerance in plants. *Int J Mol Sci* 14: 9643–9684.

Hasanuzzaman M, Gill SS, Fujita M (2013e) Physiological role of nitric oxide in plants grown under adverse environmental conditions. In: Tuteja N, Gill SS (eds), *Plant Acclimation to Environmental Stress*. Springer, New York, pp. 269 – 322.

Hasanuzzaman M, Nahar K, Gill SS, Fujita M (2014a) Drought stress responses in plants, oxidative stress and antioxidant defense. In: Tuteja N, Gill SS (eds), *Climate Change and Plant Abiotic Stress Tolerance, Vol. 1: Biological Techniques*, 1st edn. Wiley-VCH Verlag GmbH & Co. Weinheim, Germany, pp. 209–250.

Hasanuzzaman M, Alam MM, Nahar K, Jubayer-Al-Mahmud, Ahamed KU, Fujita M (2014b) Exogenous salicylic acid alleviates salt stress-induced oxidative damage in *Brassica napus* by enhancing the antioxidant defense and glyoxalase systems. *Aust J Crop Sci* 8: 631–648.

Hasanuzzaman M, Nahar K, Fujita M (2014c) Silicon and selenium: Two vital trace elements in conferring abiotic stress tolerance to plants. In: Parvaiz A, Rasool S (eds), *Emerging Technologies and Management of Crop Stress Tolerance, Vol. 1: Biological Techniques*. Academic Press, New York, pp. 375–420.

Hasanuzzaman M, Nahar K, Fujita M (2014d) Regulatory role of polyamines in abiotic stress tolerance in plants. In: Anjum NA, Gill SS (eds), *Plant Adaptation to Environmental Change: Significance of Amino Acids and their Derivatives*. Centre for Agricultural Bioscience International (CABI), Wallingford, Oxon, pp. 157–193.

Hasanuzzaman M, Nahar K, Fujita M (2014e) Role of tocopherol (vitamin E) in plants: Abiotic stress tolerance and beyond. In: Parvaiz A, Rasool S (eds), *Emerging Technologies and Management of Crop Stress Tolerance, Vol. 2: A Sustainable Approach*. Academic Press, New York, pp. 267–289.

Hassanein RA, Hassanein AA, El-din AB, Salama M, Hashem HA (2009) Role of jasmonic acid and abscisic acid treatments in alleviating the adverse effects of drought stress and regulating trypsin inhibitor production in soybean plant. *Aust J Basic Appl Sci* 3: 904–919.

Hayat S, Ahmad A (2003) *Brassinosteroids: Bioactivity and Crop Productivity*. Kluwer Academic Publishers, Dordrecht.

Hayat S, Hasan SA, Yusuf M, Hayat Q, Ahmad A (2010) Effect of 28-homobrassinolide on photosynthesis, fluorescence and antioxidant system in the presence or absence of salinity and temperature in *Vigna radiata*. *Environ Exp Bot* 69: 105–112.

Hayat S, Hayat Q, Alyemeni MN, Ahmad A (2013) Proline enhances antioxidative enzyme activity, photosynthesis and yield of *Cicer arietinum* L. exposed to cadmium stress. *Acta Bot Croat* 72: 323–335.

Heagle AS (1989) Ozone and crop yield. *Annu Rev Phytopathol* 27: 397–412.

Hellal FA, Abdelhameid M, Abo-Basha DM, Zewainy RM (2012) Alleviation of the adverse effects of soil salinity stress by foliar application of silicon on faba bean (*Vica faba* L.). *J Appl Sci Res* 8: 4428–4433.

Hossain MM, Miah MNA, Rahman MA, Islam MA, Islam MT (2008) Effect of salt stress on growth and yield attributes of mungbean. *Bangladesh Res Pub J* 1: 324–336.

Hossain MA, Hasanuzzaman M, Fujita M (2010) Up-regulation of antioxidant and glyoxalase systems by exogenous glycinebetaine and proline in mung bean confers tolerance to cadmium stress. *Physiol Mol Biol Plants* 16: 259–272.

Hossain MA, Hasanuzzaman M, Fujita M (2011) Coordinate induction of antioxidant defense and glyoxalase system by exogenous proline and glycinebetaine is correlated with salt tolerance in mung bean. *Front Agric China* 5: 1–14.

Howladar SM, Rady MM (2012) Effects of calcium paste as a seed coat on growth, yield and enzymatic activities in NaCl stressed-pea plants. *Afr J Biotechnol* 11: 14140–14145.

Hussein MM, El-Gereadly NHM, El-Desuki M (2006) Role of putrescine in resistance to salinity of pea plants (*Pisum sativum* L.). *J Appl Sci Res* 2: 598–604.

Hussain SS, Ali M, Ahmad M, Siddique KHM (2011) Polyamines: natural and engineered abiotic and biotic stress tolerance in plants. *Biotechnol Adv* 29: 300–311.

Ibrahim SK, Shalaby MAF, Zaki MES, Abou-Sedera FA, Abd Allah MSA (2012) Alleviation of high temperature stress on snap bean (*Phaseolus vulgaris* L.) by benzyladenine and putrescine. *J Appl Sci Res* 8: 192–199.

Imami S, Jamshidi S, Shahrokhi S (2011) Salicylic acid foliar and soil application effect on chickpea resistance to chilling stress. *IPCBEE* 24: 403–407.

Islam MR, Hamid A, Khaliq QA, Haque MM, Ahmed JU, Karim MA (2010) Effects of soil flooding on roots, photosynthesis and water relations in mungnean (*Vigna radiata* (L.) Wilczek). *Bangladesh J Bot* 39: 241–243.

Islam MT, Jahan NA, Sen AK, Pramanik MHR (2012) Effects of salinity on morpho-physiological attributes and yield of lentil genotypes. *Int J Sustain Crop Prod* 7: 12–18.

Janas KM, Zielińska-Tomaszewska J, Rybaczek D, *et al.* (2010) The impact of copper ions on growth, lipid peroxidation, and phenolic compound accumulation and localization in lentil (*Lens culinaris* Medic.) seedlings. *J Plant Physiol* 167: 270–276.

Jiménez-Bremont JF, Becerra-Flora A, Hernández-Lucero E, Rodríguez-Kessler M, Acosta-Gallegos JA, Ramírez-Pimentel JG (2006) Proline accumulation in two bean cultivars under salt stress and the effect of polyamines and ornithine. *Biol Plantarum* 50: 763–766.

Johkan M, Oda M, Maruo T, Shinohara Y (2011) Crop production and global warming. In: Casalegno S (ed.), *Global Warming Impacts – Case Studies on the Economy, Human Health, and on Urban and Natural Environments*. InTech, Rijeka, Croatia, pp. 139–152.

Juozaityte R, Ramaškeviciene A, Sliesaravicius A, *et al.* (2008) Effects of UVB radiation on photosynthesis pigment system and growth of pea (*Pisum sativum* L.). Scientific works of the Lithuanian Institute of Horticulture and Lithuanian University of Agriculture. *Sodininkyste ir Daržininkyste* 27: 179–186.

Kapoor K, Srivastava A (2010) Assessment of salinity tolerance of *Vigna mungo* var. Pu-19 using *ex vitro* and *in vitro* methods. *Asian J Biotechnol* 2: 73–85.

Kardoni F, Mosavi SJS, Parande S, Torbaghan ME (2013) Effect of salinity stress and silicon application on yield and component yield of faba bean (*Vicia faba*). *Intl J Agri Crop Sci* 6: 814–818.

Katerji N, van Hoorn JW, Hamdy A, Mastrorilli M, Oweis T (2005a) Salt tolerance analysis of chickpea, faba bean and durum wheat varieties. I. Chickpea and faba bean. *Agric Water Manage* 72: 177–194.

Katerji N, van Hoorn JW, Hamdy A, Mastrorilli M, Nachit MM, Oweis T (2005b) Salt tolerance analysis of chickpea, faba bean and durum wheat varieties. II. Durum wheat. *Agric Water Manage* 72: 195–207.

Katerji N, Mastrorilli M, Lahmer FZ, Maalouf F, Oweis T (2011) Faba bean productivity in saline–drought conditions. *Eur J Agron* 35: 2–12.

Kaur S, Gupta AK, Kaur N (1998) Gibberellic acid and kinetin partially reverse the effect of water stress on germination and seedling growth in chickpea. *Plant Growth Regul* 25: 29–33.

Kaur S, Gupta AK, Kaur N (2000) Effect of GA_3, kinetin and indole acetic acid on carbohydrate metabolism in chickpea seedlings germinating under water stress. *Plant Growth Regul* 30: 61–70.

Kaushal N, Gupta K, Bhandhari K, Kumar S, Thakur P, Nayyar H (2011) Proline induces heat tolerance in chickpea (*Cicer arietinum* L.) plants by protecting vital enzymes of carbon and antioxidant metabolism. *Physiol Mol Biol Plants* 17: 203–213.

Kesselmeier J, Staudt M (1999) Biogenic volatile organic compounds (VOC): An overview on emission, physiology and ecology. *J Atmos Chem* 33: 23–88.

Khan MS, Zaidi A, Wani PA, Oves M (2009) Role of plant growth promoting rhizobacteria in the remediation of metal contaminated soils. *Environ Chem Lett* 7: 1–19.

Khan W, Prithviraj B, Smith DL (2003) Photosynthetic responses of corn and soybean to foliar application of salicylates. *J Plant Physiol* 160: 485–492.

Khedar OP, Singh RV, Shrimali M, Singh NP (2008) *Pulses: Status and Cultivation Technology*. Aavishkar Publishers, Jaipur.

Kitano M, Saitoh K, Kuroda T (2006) Effect of high temperature on flowering and pod set in soybean. *Scientific Report of the Faculty of Agriculture, Okayama University* 95: 49–55.

Klessig DF, Malamy J (1994) The salicylic acid signal in plants. *Plant Mol Biol* 26: 1439–1458.

Ko CB, Woo YM, Lee DJ, Lee MC, Kim CS (2007) Enhanced tolerance to heat stress in transgenic plants expressing the GASA4 gene. *Plant Physiol Biochem* 45: 722–728.

Koini MA, Alvey L, Allen T, *et al.* (2009) High temperature-mediated adaptations in plant architecture require the bHLH transcription factor PIF4. *Curr Biol* 19: 408–413.

Kostca-Rick R, Manning WJ (1993) Dose response studies with the antioxidant ethylenediurea (EDU) applied as soil drench to two growth substrates on two greenhouse grown varieties of *Phaseolus vulgaris* L. *Environ Pollut* 82: 63–72.

Kozai N, Beppu K, Mochioka R, Boonprakob U, Subhadrabandhu S, Kataoka I (2004) Adverse effects of high temperature on the development of reproductive organs in 'Hakuho' peach trees. *J Hort Sci Biotech* 79: 533–537.

Kumar P, Pal M, Joshi R, Sairam RK (2013) Yield, growth and physiological responses of mung bean [*Vigna radiata* (L.) Wilczek] genotypes to waterlogging at vegetative stage. *Physiol Mol Biol Plants* 19: 209–220.

Kumar S, Kaur R, Kaur N, *et al.* (2011) Heat-stress induced inhibition in growth and chlorosis in mungbean (*Phaseolus aureus* Roxb.) is partly mitigated by ascorbic acid application and is related to reduction in oxidative stress. *Acta Physiol Plant* 33: 2091–2101.

Kumar S, Kaushal N, Nayyar H, Gaur P (2012) Abscisic acid induces heat tolerance in chickpea (*Cicer arietinum* L.) seedlings by facilitated accumulation of osmoprotectants. *Acta Physiol Plant* 34: 1651–1658.

Kumari A, Sheokand S, Swaraj K (2010) Nitric oxide induced alleviation of toxic effects of short term and long term Cd stress on growth, oxidative metabolism and Cd accumulation in chickpea. *Braz J Plant Physiol* 22: 271–284.

Kumari M, Sinhal VK, Srivastava A, Singh, VP (2011) Zinc alleviates cadmium induced toxicity in *Vigna radiata* (L.) Wilczek. *J Phytol* 3: 43–46.

Kumutha D, Ezhilmathi K, Sairam RK, Srivastava GC, Deshmukh PS, Meena, RC (2009) Waterlogging induced oxidative stress and antioxidant activity in pigeonpea genotypes. *Biol Planta* 53: 75–84.

Kusano T, Berberich T, Tateda C, Takahashi Y (2008) Polyamines: essential factors for growth and survival. *Planta* 228: 367–381.

Kusuma BLJ, Begum SS (2013) Direct *in vitro* regeneration of *Arachis hypogaea* L. (Jl-24) and study of salt tolerance. *Indian J Appl Res* 3: 52–54.

Lee EH, Upadhyay A, Agrawal M, Rowland RA (1997) Mechanism of ethylenediurea (EDU) induced ozone protection: Reexamination of free radical scavenger systems in snap bean exposed to O_3. *Environ Exp Bot* 38: 199–209.

Lee KH, LaRue TA (1992) Exogenous ethylene inhibits nodulation of *Pisum sativum* L. cv Sparkle. *Plant Physiol* 100: 1759–1763.

Lee SK, Sohn EY, Hamayun M, Yoon JY, Lee IJ (2010) Effect of silicon on growth and salinity stress of soybean plant grown under hydroponic system. *Agrofor Syst* 80: 333–340.

Lelieveld J, Crutzen PJ (1990) Influences of cloud photochemical processes on tropospheric ozone. *Nature* 343: 227–232.

Li D, Liu H, Qiao Y, *et al.* (2013) Effects of elevated CO_2 on the growth, seed yield, and water use efficiency of soybean (*Glycine max* (L.) Merr.) under drought stress. *Agric Water Manage* 129: 105–112.

Li SW, Leng Y, Feng L, Zeng ZY (2014) Involvement of abscisic acid in regulating antioxidative defense systems and IAA-oxidase activity and improving adventitious rooting in mung bean [*Vigna radiata* (L.) Wilczek] seedlings under cadmium stress. *Environ Sci Pollut Res* 21: 525–537.

Ligero F, Lluch C, Olivares J (1987) Evolution of ethylene from roots and nodulation rate of alfalfa (*Medicago sativa* L.) plants inoculated with *Rhizobium meliloti* as affected by the presence of nitrate. *J Plant Physiol* 129: 461–467.

Liso R, De Tullio MC, Ciraci S, *et al.* (2004) Localisation of ascorbic acid, ascorbic acid oxidase, and glutathione in roots of *Cucurbita maxima* L. *J Exp Bot* 55: 2589–2597.

Liu B, Liu X, Li YS, Herbert SJ (2013) Effects of enhanced UV-B radiation on seed growth characteristics and yield components in soybean. *Field Crops Res* 154: 158–163.

Loscos J, Matamoros MA, Becana M (2008) Ascorbate and homoglutathione metabolism in common bean nodules under stress conditions and during natural senescence. *Plant Physiol* 146: 1282–1292.

Lu S, Wang X, Guo Z (2013) Differential responses to chilling in *Stylosanthes guianensis* (Aublet) Sw. and its mutants. *Agron J* 105: 377–382.

Mahajan S, Tuteja N (2005) Cold, salinity and drought stresses: An overview. *Arch Biochem Biophys* 444: 139–158.

Malgorzata B, Nevena S, Zlatko Z, Daniela G (2009) Physiological response of some tomato genotypes (*Lycopersicon esculentum* L.) to high temperature stres. *J Cent Eur Agri* 9: 723–732.

Malik JA, Goel S, Kaur N, Sharma S, Singh I, Nayyar H (2012) Selenium antagonises the toxic effects of arsenic on mungbean (*Phaseolus aureus* Roxb.) plants by restricting its uptake and enhancing the antioxidative and detoxification mechanisms. *Environ Exp Bot* 77: 242–248.

Mansoor S, Naqvi FN (2012) Effect of gibberellic acid on α-amylase activity in heat stressed mung bean (*Vigna radiata* L.) seedlings. *Afr J Biotechnol* 11: 11414–11419.

Maqbool A, Shafiq S, Lake L (2010) Radiant frost tolerance in pulse crops – a review. *Euphytica* 172: 1–12.

Mathur N, Singh J, Bohra S, Bohra A, Vyas A (2006) Biomass production, productivity and physiological changes in moth bean genotypes at different salinity levels. *Am J Plant Physiol* 1: 210–213.

Mazid M, Ali B, Hayat S, Ahmad A (2010) The effect of 4-chloroindole-3-acetic acid on some growth parameters in mung bean under cadmium stress. *Turk J Biol* 34: 9–13.

Mazza CA, Giménez PI, Kantolic AG, Ballare CL (2013) Beneficial effects of solar UV-B radiation on soybean yield mediated by reduced insect herbivory under field conditions. *Physiol Plant* 147: 307–315.

McNeill J, Barrie FR, Buck WR, *et al.* (eds) (2011) *International Code of Nomenclature for Algae, Fungi and Plants (Melbourne Code). Regnum Vegetabile 154*. International Association for Plant Taxonomy. Available at: http://www.iapt-taxon.org/nomen/main.php?page=title[accessed 21 February 2014].

Mehlhorn H (1990) Ethylene-promoted ascorbate peroxidase activity protects plants against hydrogen peroxide, ozone and paraquat. *Plant Cell Environ* 13: 971–976.

Mensah JK, Ihenyen J (2009) Effects of salinity on germination, seedling establishment and yield of three genotypes of mung bean (*Vigna mungo* L. Hepper) in Edo State, Nigeria. *Nigerian Ann Nat Sci* 8: 17–24.

Mihailovic N, Drazic G (2011) Incomplete alleviation of nickel toxicity in bean by nitric oxide supplementation. *Plant Soil Environ* 57: 396–401.

Mohammed AHMA, Mohamed HI, Zaki LM, Mogazy AM (2012) Pre-exposure to gamma rays alleviates the harmful effect of salinity on cowpea plants. *J Stress Physiol Biochem* 8: 199–217.

Moran JF, Becana M, Iturbeormaetxe I, Frechilla S, Klucas RV, Apariciotejo P (1994) Drought induces oxidative stress in pea-plants. *Planta* 194: 346–352.

Moussa HR, Mohamed MAEH (2011) Role of nitric acid or H_2O_2 in antioxidant defense system of *Pisum sativum* L. under drought stress. *Nat Sci* 9: 211–216.

Mroczek-Zdyrska M, Wójcik M (2012) The influence of selenium on root growth and oxidative stress induced by lead in *Vicia faba* L. minor plants. *Biol Trace Elem Res* 147: 320–328.

Mukharjee SP, Choudhuri MA (1983) Implications of water stress-induced changes in the levels of endogenous ascorbic acid and hydrogen peroxide in *Vigna* seedlings. *Physiol Plant* 58: 166–170.

Munné-Bosch S (2005) The role of α-tocopherol in plant stress tolerance. *J Plant Physiol* 162: 743–748.

Munné-Bosch S (2007) α-Tocopherol: a multifaceted molecule in plants. *Vitam Horm* 6: 375–392.

Muscolo A, Sidari M, Anastasi U, Santonoceto C, Maggio A (2014) Effect of PEG-induced drought stress on seed germination of four lentil genotypes. *J Plant Interactions* 9: 354–363.

Nahar K, Hasanuzzaman M, Alam MM, Fujita M (2012) Exogenous glutathione alleviates short-term abiotic stress by modulating antioxidant defense and methylglyoxal detoxification system in V. radiata seedlings. Annual Meeting of the Society for Experimental Biology, June 29–July 02, 2012. Salzburg, Austria.

Nahar K, Hasanuzzaman M, Alam MM, Fujita M (2013) Exogenous glutathione-induced drought stress tolerance in Vigna radiata seedlings through enhanced antioxidant defense and methylglyoxal system. Interdrought IV Conference, September 02–September 09, 2013, Perth, Australia.

Natsheh B, Barghouthi Z, Amereih S, Salman M (2012) Effect of irrigation with sea water on germination and growth of lentil (*Lens culinaris* Medic). *J. Water. Resour. Protec.* 4,307–310.

Navalón A, Blanc R, Vilchez JL (1997) Determination of 1-naphthylacetic acid in commercial formulations and natural waters by solid-phase spectrofluorimetry. *Mikcrochim Acta* 126: 33–38.

Navarro L, Bari R, Achard P, *et al.* (2008) DELLAs control plant immune responses by modulating the balance of jasmonic acid and salicylic acid signaling. *Curr Biol* 18: 650–655.

Nayyar H, Bains TS, Kumar S, Kaur G (2005) Chilling effects during seed filling on accumulation of seed reserves and yield of chickpea. *J Sci Food Agric* 85: 1925–1930.

Nayyar H, Kaur G, Kumar S, Upadhyaya HD (2007) Low temperature effects during seed filling on chickpea genotypes (*Cicer arietinum* L.): Probing mechanisms affecting seed reserves and yield. *J Agron Crop Sci* 193: 336–344.

Nazar R, Iqbal N, Syeed S, Khan NA (2011) Salicylic acid alleviates decreases in photosynthesis under salt stress by enhancing nitrogen and sulfur assimilation and antioxidant metabolism differentially in two mungbean cultivars. *J Plant Physiol* 168: 807–815.

Noctor G, Foyer CH (1998) Ascorbate and glutathione: keeping active oxygen under control. *Annu Rev Plant Physiol Plant Mol Biol* 49: 249–279.

Pant G, Malla S, Aruna J, Chauhan UK (2012) Effect of dry heat treatments on viability and vigour of *Cassia tora* L. seeds. *Online Int J Biosol* 2: 58–64.

Papiernik SK (2005) Effects of salinity, imazethapyr, and chlorimuron application on soybean growth and yield. *Commun Soil Sci Plant Anal* 36: 951–967.

Parande S, Zamani GR, Zahan MHS, Ghaderi MG (2013) Effects of silicon application on the yield and components of yield in the common bean (*Phaseolus vulgaris*) under salinity stress. *Intl J Agron Plant Prod* 4: 1574–1579.

Pastor J, Hernández AJ, Prieto N, Fernández-Pascual M (2003) Accumulating behaviour of *Lupinus albus* L. growing in a normal and a decalcified calcicluvisol polluted with Zn. *J Plant Physiol* 160: 1457–1465.

Patel HV, Parmar SR, Chudasama CJ Mangrola AV (2013) Interactive studies of zinc with cadmium & arsenic on seed germination and antioxidant properties of *Phaseolus aureus* Roxb. *Int J Plant Anim Environ Sci* 3: 166–174.

Peng L, Xing-hua X, Qin Z, *et al.* (2011) Effect of NAA on growth and photosynthetic characteristic of soybean seedling under drought and re-watering. *Soybean Sci* 2011–01.

Piramila BHM, Prabha AL, Nandagopalan V, Stanley AL (2012) Effect of heat treatment on germination, seedling growth and some biochemical parameters of dry seeds of black gram. *Int J Pharm Phytopharm Res* 1: 194–202.

Poonam S, Kaur H, Geetika S (2013) Effect of jasmonic acid on photosynthetic pigments and stress markers in *Cajanus cajan* (L.) Millsp. seedlings under copper stress. *Am J Plant Sci* 4: 817–823.

Popelka JC, Terryn N, Higgins TJV (2004) Gene technology for grain legumes: can it contribute to the food challenge in developing countries? *Plant Sci* 167: 195–206.

Porch TG, Jahn M (2001) Effects of high temperature stress on microsporogenesis in heat-sensitive and heat-tolerant genotypes of *Phaseolus vulgaris*. *Plant Cell Environ* 24: 723–731.

Potters G, De Gara L, Asard H, Horemans N (2002) Ascorbate and glutathione: guardians of the cell cycle, partners in crime? *Plant Physiol Biochem* 40: 537–548.

Puckette MC, Weng H, Mahalingam R (2007) Physiological and biochemical responses to acute ozone-induced oxidative stress in *Medicago truncatula*. *Plant Physiol Biochem* 45: 70–79.

Qados AMSA (2011) Effect of salt stress on plant growth and metabolism of bean plant *Vicia faba* (L.). *J Saudi Soc Agric Sci* 10: 7–15.

Queiroz HM, Sodek L, Haddad CRB (2012) Effect of salt on the growth and metabolism of *Glycine max*. *Braz Arch Biol Technol* 55: 809–817.

Radhakrishnan R, Lee IJ (2013) Ameliorative effects of spermine against osmotic stress through antioxidants and abscisic acid changes in soybean pods and seeds. *Acta Physiol Plant* 35: 263–269.

Rady MM (2011) Effect of 24-epibrassinolide on growth, yield, antioxidant system and cadmium content of bean (*Phaseolus vulgaris* L.) plants under salinity and cadmium stress. *Sci Hortic* 129: 232–237.

Rady MA, Varma CB, Howladar SM (2013) Common bean (*Phaseolus vulgaris* L.) seedlings overcome NaCl stress as a result of presoaking in *Moringa oleifera* leaf extract. *Sci Hort* 162: 63–70.

Raskin I (1992) Role of salicylic acid in plants. *Annu Rev Plant Physiol Plant Mol Biol* 43: 439–463.

Rasool S, Ahmed A, Siddiqi TO, Ahmed P (2013) Changes in growth, lipid peroxidation and some key antioxidant enzymes in chickpea genotypes under salt stress. *Acta Physiol Planta* 35: 1039–1050.

Ravindran KC, Indrajith A, Pratheesh PV, Sanjiviraja K, Balakrishnan V (2010) Effect of ultraviolet-B radiation on biochemical and antioxidant defence system in *Indigofera tinctoria* L. seedlings. *Int J Eng Sci Tech* 2: 226–232.

Rezaei MA, Kaviani B, Masouleh AK (2012) The effect of exogenous glycine betaine on yield of soybean [*Glycine max* (L.) Merr.] in two contrasting cultivars Pershing and DPX under soil salinity stress. *Plant Omics J* 5: 87–93.

Rhine MD, Stevens G, Shannon G, Wrather A, Sleper D (2010) Yield and nutritional responses to waterlogging of soybean cultivars. *Irrig Sci* 28: 135–142.

Robert-Seilaniantz A, Navarro L, Bari R, Jones JD (2007) Pathological hormone imbalances. *Curr Opin Plant Biol* 10: 372–379.

Saidi I, Ayouni M, Dhieb A, Chtourou Y, Chaïbi W, Djebali W (2013) Oxidative damages induced by short-term exposure to cadmium in bean plants: Protective role of salicylic acid. *South Afr J Bot* 85: 32–38.

Sairam RK, Dharmar K, Chinnusamy V, Meena R. (2009) Waterlogging-induced increase in sugar mobilization, fermentation, and related gene expression in the roots of mung bean (*Vigna radiata*). *J Plant Physiol* 166: 602–616.

Saleh AAH (2007) Amelioration of chilling injuries in mungbean (*Vigna radiata* L.) seedlings by paclobutrazol, abscicic acid and hydrogen peroxide. *Am J Plant Physiol* 2: 318–322.

Sandorf I, Holländer-Czytko H (2002) Jasmonate is involved in the induction of tyrosine aminotransferase and tocopherol biosynthesis in *Arabidopsis thaliana*. *Planta*, 216: 173–179.

Sarma H (2011) Metal hyperaccumulation in plants: A review focusing on phytoremediation technology. *J Environ Sci Technol* 4: 118–138.

Sasse JM (2003) Physiological actions of brassinosteroids: an update. *J Plant Growth Regul* 22: 276–288.

Scebba F, Pucciareli I, Soldatini GF, Ranieri A (2003) O_3-induced changes in the antioxidant systems and their relationship to different degrees of susceptibility of two clover species. *Plant Sci* 165: 583–593.

Senaratna T, Touchell D, Bunn E, Dixon K (2000) Acetyl salicylic acid (Aspirin) and salicylic acid induce multiple stress tolerance in bean and tomato plants. *Plant Growth Regul* 30: 157–161.

Shahid MA, Pervez MA, Balal RM, *et al.* (2011) Brassinosteroid (24-epibrassinolide) enhances growth and alleviates the deleterious effects induced by salt stress in pea (*Pisum sativum* L.). *Aust J Crop Sci* 5: 500–510.

Shahzad M, Zörb C, Geilfus CM, Mühling KH (2013) Apoplastic Na^+ in *Vicia faba* leaves rises after short-term salt stress and is remedied by silicon. *J Agron Crop Sci* 199: 161–170.

Sharaf AEM, Farghal II, Sofy MR (2009) Role of gibberellic acid in abolishing the detrimental effects of Cd and Pb on broad bean and lupin plants. *Res J Agric Biol Sci* 5: 668–673.

Sharma KK, Lavanya M (2002) Recent developments in transgenics for abiotic stress in legumes of the semi-arid tropics. *JIRCAS Working Report, pp.* 61–73.

Sheokand S, Bhankar V, Sawhney V (2010) Ameliorative effect of exogenous nitric oxide on oxidative metabolism in NaCl treated chickpea plants. *Braz J Plant Physiol* 22: 81–90.

Sheokand S, Kumari A, Sawhney V (2008) Effect of nitric oxide and putrescine on antioxidative responses under NaCl stress in chickpea plants. *Physiol Mol Biol Plants* 14: 355–362.

Sibole JV, Cabot C, Poschenrieder C, Barcelo J (2003) Effcient leaf ion partitioning, an overriding condition for abscisic acid-controlled stomatal and leaf growth responses to NaCl salinization in two legumes. *J Exp Bot* 54: 2111–2119.

Sidari M, Muscolo A, Anastasi U, Pretti G, Santonoceto C (2007) Response of four genotypes of lentil to salt stress conditions. *Seed Sci Technol* 35: 497–503.

Siddiqui A, Dawar S, Zaki MJ, Hamid N (2011) Role of ultra violet (UV-C) radiation in the control of root infecting fungi on groundnut and mung bean. *Pak J Bot* 43: 2221–2224.

Smirnoff N (1996) The function and metabolism of ascorbic acid in plants. *Ann Bot* 78: 661–669.

Stern RA, Flaishman M, Applebaum S, Ben-Arie R (2007) Effect of synthetic auxins on fruit development of 'Bing' cherry (*Prunus avium* L.). *Sci Hortic* 114: 275–280.

Stevens PF (2014) "Fabaceae". Angiosperm Phylogeny Website. Version 12, July 2012. http://www.mobot.org/MOBOT/research/APweb/ [accessed 28 February 2014].

Striker GG, Manzur ME, Grimoldi AA (2011) Increasing defoliation frequency constrains regrowth of the forage legume *Lotus tenuis* under flooding. The role of crown reserves. *Plant Soil* 343: 261–272.

Subbarao GV, Johansen C, Slinkard AE, Nageswara Rao RC, Saxena NP, Chauhan YS (1995) Strategies for improving drought resistance in grain legumes. *Crit Rev Plant Sci* 14: 469–523.

Sultana R, Choudhary AK, Pal AK, Saxena KB, Prasad BD, Singh R (2014) Abiotic stresses in major pulses: Current status and strategies. In: Gaur RK, Sharma P (eds), *Approaches to Plant Stress and Their Management*. Springer, New Delhi, pp. 173–190.

Sun TP, Gubler F (2004) Molecular mechanism of gibberellin signaling in plants. *Annu Rev Plant Biol* 55: 197–223.

Sung CH, Hong JK (2010) Sodium nitroprusside mediates seedling development and attenuation of oxidative stresses in Chinese cabbage. *Plant Biotechnol Rep* 4: 243–251.

Swamy PM, Smith B (1999) Role of abscisic acid in plant stress tolerance. *Curr Sci* 76: 1220–1227.

Tacarindua CRP, Shiraiwa T, Homma K, Kumagai E, Sameshima R (2013) The effects of increased temperature on crop growth and yield of soybean grown in a temperature gradient chamber. *Field Crops Res* 154: 74–81.

Taffouo VD, Kouamou JK, Ngalangue LMT, Ndjeudji BAN, Akoa A (2009) Effects of salinity stress on growth, ions partitioning and yield of some cowpea (*Vigna unguiculata* L. Walp) cultivars. *Intl J Bot* 5: 135–143.

Taie HAA, Abdelhamid MT, Dawood MG, Nassar RMA (2013) Pre-sowing seed treatment with proline improves some physiological, biochemical and anatomical attributes of faba bean plants under sea water stress. *J Appl Sci Res* 9: 2853–2867.

Talano MA, Cejas RB, González PS, Agostini E (2013) Arsenic effect on the model crop symbiosis *Bradyrhizobium*-soybean. *Plant Physiol Biochem* 63: 8–14.

Talukdar D (2013) Comparative morpho-physiological and biochemical responses of lentil and grass pea genotypes under water stress. *J Nat Sci Biol Med* 4: 398–402.

Tambussi EA, Bartoli CG, Guiamet JJ, Beltrano J, Araus JL (2004) Oxidative stress and photodamage at low temperatures in soybean (*Glycine max* L. Merr.) leaves. *Plant Sci* 167: 19–26.

Tan W, Meng QW, Brestic M, Olsovska K, Yang X (2011) Photosynthesis is improved by exogenous calcium in heat-stressed tobacco plants. *J Plant Physiol* 168: 2063–2071.

Tavakkoli E, Paull J, Rengasamy P, McDonald GK (2012) Comparing genotypic variation in faba bean (*Vicia faba* L.) in response to salinity in hydroponic and field experiments. *Field Crops Res* 127: 99–108.

Terán H, Singh SP (2002a) Selection for drought resistance in early generations of common bean populations. *Can J Plant Sci* 82: 491–497.

Terán H, Singh SP (2002b) Comparison of sources and lines selected for drought resistance in common bean. *Crop Sci* 42: 64–70.

Thalooth AT, Tawfik MM, Mohamed HM (2006) A comparative study on the effect of foliar application of zinc, potassium and magnesium on growth, yield and some chemical constituents of mungbean plants grown under water stress conditions. *World J Agric Sci* 2: 37–46.

Turan Ö, Ekmekçi Y (2011) Activities of photosystem II and antioxidant enzymes in chickpea (*Cicer arietinum* L.) cultivars exposed to chilling temperatures. *Acta Physiol Plant* 33: 67–78.

Umebese CE, Bankole AE (2013) Impact of salicylic acid on antioxidants, biomass and osmotic adjustments in *Vigna unguiculata* L. Walp. during water deficit stress. *Afr J Biotechnol* 12: 5200–5207.

Van Assche JA, Debucquoy KLA, Rommens WAF (2003) Seasonal cycles in the germination capacity of buried seeds of some Leguminosae (Fabaceae). *New Phytol* 158: 315–323.

Vance CP, Graham PH, Allan DL (2000) Biological nitrogen fixation phosphorus: a critical future need. In: Pedrosa FO, Hungria M, Yates MG, Newton WE (eds), *Nitrogen Fixation: from Molecules to Crop Productivity*. Kluwer Academic Publishers, Dordrecht, The Netherlands, pp. 506–514.

Van Hoorn JW, Katerji N, Hamdy A, Mastrorilli M (2001) Effect of salinity on yield and nitrogen uptake of four grain legumes and on biological nitrogen contribution from the soil. *Agric Water Manage* 51: 87–98.

Wahid A, Ghani A (2008) Varietal differences in mungbean (*Vigna radiata*) for growth, yield, toxicity symptoms and cadmium accumulation. *Ann Appl Biol* 152: 59–69.

Wang F, Chen F, Cai Y, Zhang G, Wu F (2011) Modulation of exogenous glutathione in ultrastructure and photosynthetic performance against Cd stress in the two barley genotypes differing in Cd tolerance. *Biol Trace Elem Res* 144: 1275–1288.

Wang XS, Han JG (2007) Effects of NaCl and silicon on ion distribution in the roots, shoots and leaves of two alfalfa cultivars with different salt tolerance. *Soil Sci Plant Nutr* 53: 278–285.

Wasternack C, Hause B (2002) Jasmonates and octadecanoids – signals in plant stress response and development. *Prog Nucleic Acid Res* 72: 165–221.

Wigoda N, Ben-Nissan G, Granot D, Schwartz A, Weiss D (2006) The gibberellin-induced, cysteine-rich protein GIP2 from *Petunia* hybrid exhibits in planta antioxidant activity. *Plant J* 48: 796–805.

Wu JC, Sun SH, Ke YT, Xie CP, Chen FX (2011) Effects of glutathione on chloroplast membrane fluidity and the glutathione circulation system in young loquat fruits under low temperature stress. *Acta Hort* 887: 221–225.

Yakubu H, Izge AU, Hussaini MA, Jibrin JM, Bello OG, Isyaku MS (2013) Varietal response and gibberellic acid concentrations on yield and yield traits of groundnut (*Arachis hypogaea* L.) under wet and dry conditions. *Acad J Agric Res* 1: 1–8.

Yalpani N, Enyedi AJ, Leon J, Raskin I (1994) Ultraviolet light and ozone stimulate accumulation of salicylic acid, pathogenesis-related proteins and virus resistance in tobacco. *Planta* 193: 372–376.

Yan K, Chen W, He X, Zhang G, Xu S, Wang L (2010) Responses of photosynthesis, lipid peroxidation and antioxidant system in leaves of *Quercus mongolica* to elevated O_3. *Environ Exp Bot* 69: 198–204.

Yang JD, Yun JY, Zhang TH, Zhao HL (2006) Presoaking with nitric oxide donor SNP alleviates heat shock damages in mung bean leaf discs. *Bot Stud* 47: 129–136.

Yathavakilla SKV, Caruso JA (2007) A study of Se-Hg antagonism in *Glycine max* (soybean) roots by size exclusion and reversed phase HPLC-ICPMS. *Analyt Bioanalyt Chem* 389: 715–723.

Yoon JY, Hamayun M, Lee SK, Lee IJ (2009) Methyl jasmonate alleviated salinity stress in soybean. *J Crop Sci Biotechnol* 12: 63–68.

Youn JT, Van K, Lee JE, *et al.* (2008) Waterlogging effects on nitrogen accumulation and N_2 fixation of supernodulating soybean mutants. *J Crop Sci Biotech* 11: 111–118.

Zare M, Nejad MG, Bazrafshan F (2012) Influence of drought stress on some traits in five mung bean (*Vigna radiata* (L.) R. Wilczek) genotypes. *Intl J Agron Plant Prod* 3: 234–240.

Zeid IM (2004) Response of bean (*Phaseolus vulgaris*) to exogenous putrescine treatment under salinity stress. *Pak J Biol Sci* 7: 219–225.

Zengin FK, Munzuroglu O (2005) Effects of some heavy metals on content of chlorophyll, proline and some antioxidant chemicals in bean (*Phaseolus vulgaris* L.) seedlings. *Acta Biol Cracov Ser Bot* 47: 157–164.

Zhang F, Zhang H, Xia Y, Wang G, Xu L, Shen Z (2011) Exogenous application of salicylic acid alleviates cadmium toxicity and reduces hydrogen peroxide accumulation in root apoplasts of *Phaseolus aureus* and *Vicia sativa*. *Plant Cell Rep* 30: 1475–1483.

Zhou ZS, Guo K, Elbaz AA, Yang ZM (2009) Salicylic acid alleviates mercury toxicity by preventing oxidative stress in roots of *Medicago sativa*. *Environ Exp Bot* 65: 27–34.

Zuccarini P (2008) Effects of silicon on photosynthesis, water relations and nutrient uptake of *Phaseolus vulgaris* under NaCl stress. *Biol Plant* 52: 157–160.

CHAPTER 12

Genetic and molecular responses of legumes in a changing environment

Mustafeez Mujtaba Babar[1], Najam-us-Sahar Sadaf Zaidi[1], Mohamed Mahgoub Azooz[2] and Alvina Gul Kazi[1]

[1] *Atta-ur-Rahman School of Applied Biosciences, National University of Sciences and Technology (NUST), Islamabad, Pakistan*
[2] *Department of Botany, Faculty of Science, South Valley University, Qena, Egypt*

12.1 Introduction

A number of environmental and biological factors tend to adversely affect the health of a plant. These stress factors result in significant damage, especially to the quality and yield of cultivated plants. Leguminous plants are no exception in this regard. Legumes form an important part of the botanical sphere and are valuable sources of nutrients, drugs and biofuels. Moreover, they contribute significantly to the environmental and agricultural fields by establishing symbiotic relationships with nitrogen-fixing bacteria. Keeping in mind the importance of legumes, any threat to the health of these plants can present a severe challenge to agriculturalists and plant pathologists. The environmental stress factors are mainly changes in the soil's nutrient content, water levels, temperature changes and other variations in the microenvironment (Liancourt *et al.*, 2013; Lipiec *et al.*, 2013; Putten *et al.*, 2013). Plants, like all organisms, tend to respond to these stress factors by bringing about both physiological and morphological changes. A complex series of events, at the genetic and subcellular levels, is triggered to protect the plant from these harms and to induce tolerance against these stress conditions (Duque *et al.*, 2013; Reyer *et al.*, 2013). A number of genes, biomolecules, solutes, solvents, ions and regulatory elements are involved in developing resistance against an environmental stress factor (Echevarría-Zomeño *et al.*, 2013; Jacques *et al.*, 2013; Kwon *et al.*, 2013; Lindemose *et al.*, 2013). The current review focuses on the genetic, molecular and cellular changes adopted by legumes to defend themselves against environmental threats. Various stress factors that burden the growth of these plants and the observed subcellular changes are also reviewed.

12.2 Legumes: a botanical treasure

Legumes belong to the family Leguminosae (Fabaceae), commonly referred to as the bean or pea family. They are among the most widely cultivated plant species. Different varieties of these plants are found in diverse temperature and geographical zones. The family comprises around 23,000 species, providing an important source of nutrients, bioenergy, medicines and biomaterials (Tetsukazu *et al.*, 2013). These properties make them one of the most widely cultivated plant familes. Legumes are grown over an area of approximately 250 million hectares around the globe (Abate *et al.*, 2012).

Nutritionally, legumes are rich sources of proteins, carbohydrates, amino acids, vitamins and minerals. Many members of the family form part of staple diets in middle to lower income countries. The high protein content makes them an ideal nutritional choice. Legumes also contain a variety of phytochemicals including flavonoids, coumarins, anthraquinones, glycosides and lectins (Kumar *et al.*, 2012). These plant-derived secondary metabolites are used in many traditional medicinal systems to treat diseases of the cardiovascular, respiratory, genitourinary and nervous systems (Bouchenak & Lamri-Senhadji, 2013; Butt & Sultan, 2013). Apart from their direct use in human nutrition and medicinal systems, legumes play an

Legumes under Environmental Stress: Yield, Improvement and Adaptations, First Edition. Edited by Mohamed Mahgoub Azooz and Parvaiz Ahmad.

important role as sources of bioenergy. The woody legumes have been used for building and fuel (Rangan, 2013). Certain varieties have a very distinctive appearance and are utilized as ornamental plants (Ganguli & Kennedy, 2013).

Legumes also contribute to ecosystems. They form symbiotic relationships with many bacterial species and help in fixing atmospheric nitrogen. These bacteria, belonging to the rhizobacterial group, establish relationships with the roots of leguminous plants. The plant provides shelter and nutrients to the bacteria while the bacteria help in converting the nitrogen into a usable form for the plant, thereby, increasing the fertility of soil and promoting the growth of legumes (Qadri *et al.*, 2013). Another group of rhizobacteria also helps the plants to survive environmental stress conditions. These bacterial species are collectively referred to as plant growth promoting rhizobacteria (PGPRs) (Ram *et al.*, 2013).

Hence, legumes are important for the livelihood of a large human and animal population associated with the agriculture sector. Their significance as a food source, as a fuel and in medicinal systems makes them an important subject for plant biotechnology. Any threat to plant health can have significant outcomes in terms of financial losses and an increased burden on the biosphere. An understanding of the various environmental threats facing leguminous plants is, therefore, essential. Moreover, the study of the various genetic and molecular responses to these threats is also of vital importance as the manipulation of these factors can help in developing more stress-resistant legume varieties.

12.3 Environmental threats to legumes

Legumes, like other plants, are exposed to a variety of stimuli. Among these, both biotic and abiotic stress factors have been associated with adverse effects on the plants' health. Biotic stress factors include viruses, bacteria, fungi, nematodes, weeds and herbs (Ansi *et al.*, 2010; Bhatnagar-Mathur *et al.*, 2012; Boudreau, 2013). Among the physical constraints, changes in climatic conditions, and edaphic and geographical factors are of prime importance (Duque *et al.*, 2013). Moreover, changes in temperature, altered availability of water to the plant, and variation in the quantities of solutes and other nutrients in the soil directly affect plant growth (Jha *et al.*, 2014). These abiotic stress factors greatly promote attack by biological agents; for instance, waterlogging is a fundamental cause of fungal and bacterial infections in leguminous plants (Lipiec *et al.*, 2013).

Among the environmental factors, changes in soil characteristics cause alterations in the growth patterns of legumes. Increased acidity, and elevated salt and metal concentrations can all have detrimental effects on plant health. Usually, these stress factors do not work in isolation to affect the plants; rather they initiate a complex cascade of events that cause damage to the plant. For example, heat along with saline soil results in increase in the transpiration rate and absorption of more water through the roots. The absorbed water contains a higher proportion of salt, causing the plant to be salt stressed. Similarly, a diminished supply of water to the plant leads to osmotic stress in legumes. This causes a reduced growth rate of the plant. The exposure of plants to sudden changes in temperature leads to a diminished ability to carry out normal life processes. Legumes can easily tolerate slight increases and decreases in temperatures (Gupta *et al.*, 2014). However, when there is a significant rise or fall in temperature, these plants fail to tolerate these changes and are subject to either tissue or organ loss leading, sometimes, to the death of the plant. Cold season legumes like lentils, beans and peas are sensitive to temperature rises (Cannon, 2013; Kumar *et al.*, 2013a). Warm season legumes like groundnut and mung bean, if subjected to a sudden fall in temperature, also suffer considerable damage (Hamidou *et al.*, 2013; Li *et al.*, 2014). Among other environmental factors, exposure to high-intensity light, high wind speeds and natural disasters are significant causes of decreased legume output.

These physical stress factors increase the susceptibility of legumes to infestation with biotic pathological agents, resulting in decreased yield of the plant and a significant fall in the nitrogen-fixing capability of these plants (Lipiec *et al.*, 2013). These can, therefore, be classified as a part of a complex 'stress matrix' that results in a multifactorial constrained state (Mittler, 2006). Plants tend to respond to these stress factors in a variety of ways. These responses usually involve upregulation of certain genetic elements followed by the production of regulatory and effector response elements (Duque *et al.*, 2013; Libault, 2013; Lindemose *et al.*, 2013). The specific or non-specific pathways activated against a particular stress factor involve a number of chemicals, biomolecules and genetic players. These responses have been very well characterized in case of salt, water and temperature stresses and are reviewed here. Other physical constraints activate similar pathways to generate morphological and physiological changes in the legume body in order to survive the stress conditions (Hossain *et al.*, 2013).

12.4 Genetic and molecular responses to salt stress

Among the various physical stress factors, soil salinity is considered to be the most significant in terms of the resultant losses. These damages are of numerous types, varying from growth retardation to plant death. Legumes have developed a number of physiological and biochemical mechanisms to respond to this abiotic stress factor (Tuteja *et al.*, 2012). These responses are now very well characterized owing to recent developments in plant biotechnology. Modern molecular techniques have helped in yielding important information about the subcellular genetic and molecular mechanisms involved in responding to salt stress (Grieve *et al.*, 2012). Saline-stressed environments affect plant health by causing disruption of osmotic and ionic homeostasis (Shavrukov, 2013). Legumes, therefore, have to respond by evolving specialized mechanisms to develop tolerance against the salt stress.

An elevated salt content in the plant body is the consequence of uptake of the salt-rich water to carry out transpiration, photosynthesis and other metabolic processes (Aroca & Ruiz-Lozano 2012; Geissler *et al.*, 2013). These pathways are, therefore key to controlling the salt and water intake of a plant. If not regulated adequately, the salt and water intake can irreversibly damage the legume's health. At the cellular level, increased salinity results in hyperionic and hyperosmotic pressure in the plants (Q. Liu *et al.*, 2013; Rasool *et al.*, 2013). High levels of sodium and chloride ions in the soil cause retardation in the growth of legumes. Dysregulation of cell cycle, metabolic and biochemical disorganization, oxidative stress and reduced nutrient uptake are the main events characterizing the legume under saline stress (Krasensky & Jonak, 2012; Suzuki *et al.*, 2012).

Legumes might respond to salt stress in various ways. The main aim of all the strategies is to direct the salt away from the actively growing meristematic tissue. Most plants regulate the salt content of their cells by channelling these ions into the xylem tissue, resulting in the decreased provision of salt to other plant parts. Meristematic cells are not directly connected to the plant's vascular tissues and hence are not directly exposed to the hyperionic environment (Asensio *et al.*, 2012). Moreover, in certain plants the response to high salt concentrations may take the form of closure of the stomata to reduce the transpirational and ion fluxes through the plant body. This mechanism is, however, a long-term survival strategy as reducing photosynthesis and respiration can help the plant to thrive over a prolonged time period (Sabra *et al.*, 2012). Another, more commonly observed response is the accumulation of ions in the older leaves of the plants. These leaves, which ultimately detach from the plant body, serve as sinks for salt flushed from the plant body. Other morphological and physiological mechanisms adapted by legumes to survive salt stress include alterations of growth rate and transpiration rate, changes in cell structures and alteration of cell-cell interactions (Shabala *et al.*, 2012).

Legumes are glycophytes and they require less than 125 ppm of salt in the soil (Horie *et al.*, 2012). In comparison, halophytes (*Mesembryanthemum* spp., *Attalea* spp.) can grow only when larger amounts of salt are available to the plant. The mechanisms by which these extremophytes survive and propagate can provide valuable insights to generate resistance among the glycophytes to survive in high-salt growth conditions. Halophytes tend to exhibit intracellular compartmentalization to segregate the salt from other cellular contents (Y. Yang *et al.*, 2012). This results in the protection of the metabolic machinery of the cell. Moreover, certain halophytes have evolved specialized structures, known as glands, to extrude the salt from the plant (Tan *et al.*, 2013). The segmentation of the intracellular content into two parts, an organic portion comprising the cellular structures and a salt-rich inorganic part, prevents damage to cell organelles and other cellular entities. Halophytes, additionally, show a high degree of salt tolerance in comparison to the glycophytes (Krauss & Ball, 2013). The exact origin of these mechanisms, whether constitutive or adaptive, is still under investigation. At the cellular level, halophytes tend to show large vacuolization, greater quantities of ions (sodium, chloride, potassium, hydrogen) and alteration of cellular turgor pressure to thrive in saline growth conditions (Dasgupta *et al.*, 2013; Elkahoui *et al.*, 2013). These adaptive mechanisms result from the interplay of genetic and molecular events, exhibited as cellular and organ-level changes in the plant body.

Much like other plants, legumes respond to salt stress by initiating a signalling cascade that involves the perception of a stress factor as a stimulus followed by the activation of certain molecular pathways to activate and respond specifically to the physical constraint (Mantri *et al.*, 2012; Miransari 2012). Figure 12.1 represents the series of events that are involved in responding to salt stress. The response may be exhibited directly against the stressor or by the activation of certain secondary pathways that ultimately results in the development of tolerance against the particular stressor. Production of metabolites, expression of proteins and regulation of metabolic pathways are the main secondary events observed in response to salt stress in legumes (Arbona *et al.*, 2013). Regulators of the secondary pathways mainly include the transcription factors that regulate the direct effectors of the salt stress response elements. In case of a saline-stressed environment, legumes tend to produce specific response elements targeted either at decreasing salt intake or regulating salt-rich water intake. Another class of response elements, however, may not be directly involved in responding to a particular stress factor, but may help in coordination and regulation of responses to the external stimuli (Bayuelo-Jiménez *et al.*, 2012). The main changes observed – ionic regulation, metabolite production, water channelling and genetic regulation – are of prime importance in case of salt stress, and are discussed further here.

12.4.1 Ionochemical imbalance

An increase in the salt content in the soil results in a corresponding increase in the ionic levels of sodium, chlorine, calcium, potassium and hydrogen in the legume body (Li *et al.*, 2012; Lindberg *et al.*, 2012; Tahkokorpi *et al.*, 2012; Turner *et al.*, 2013). This ionic imbalance leads to irregularity in the transmembrane exchange of ions, nutrients and energy. The main regulators of intercellular transport, ATPases, pyrophosphatases and ABC transporters, are affected by the ionic overload (Shabala & Munns, 2012; Yao *et al.*, 2012; Banasiak *et al.*, 2013). Additionally, transport across the organellar membranes is also altered. The changes in H^+ ions are coupled to the alteration in the active transport mechanisms across the vacuolar tonoplast membrane. The thermodynamic and electrophoretic flux is re-established only once the salt stress conditions are relieved. Changes in the chemical and environmental factors lead to the modified expression of the transport proteins and their regulators, for instance ATPases and pyrophosphatases. Among the genetic factors responsible for the control of ion transport in salt-stressed conditions are the Salt Overly Sensitive (*SOS*) genes, primarily *SOS3*. The intracellular and transmembrane transport of Na^+, K^+, H^+ and Ca^{2+} ions are all regulated by this genetic locus (Yu *et al.*, 2012). The SOS

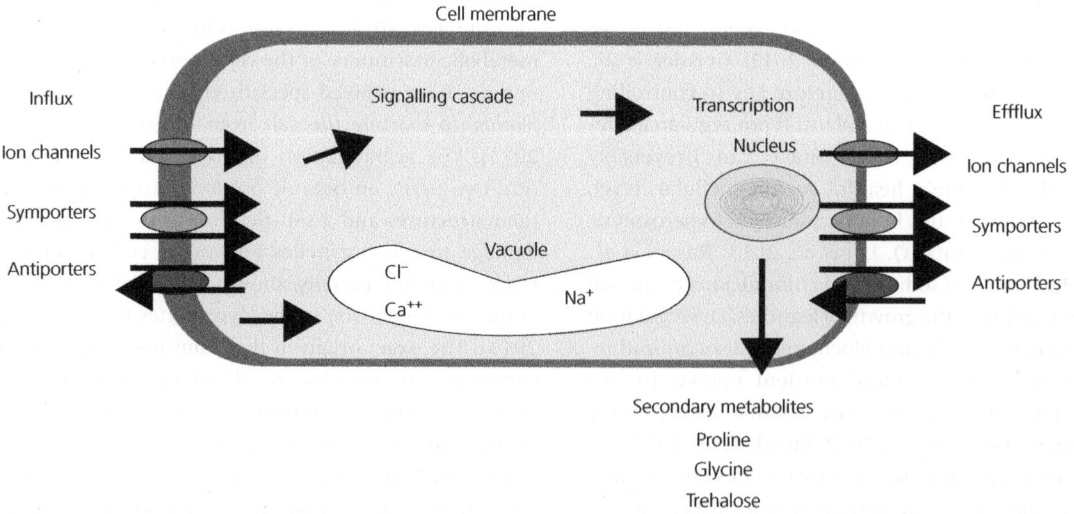

Figure 12.1 Series of cellular changes in response to saline stress.

stress compensation mechanism has been studied in a number of legumes including chickpea and *Medicago* species (Molina *et al.*, 2011; Samineni *et al.*, 2011; Arraouadi *et al.*, 2012). Other genes implicated in alteration of transmembrane ionic transport (especially Na^+ ions) include the sodium-hydrogen exchangers and superoxidase dismutase gene families (Asensio *et al.*, 2012; Hedrich *et al.*, 2012). The increased transporter activity leads to alterations in metabolic and ionic potentials, changes in protein morphology and regulation of transcription.

The genetic regulation of transport proteins results in alteration of the rate of ionic exchange across the plasma membrane. Soon after the induction of salt stress, the influx of Na^+ ions is coupled with the rapid exit of Cl^- ions from the cell. The movement is usually attributed to the use of certain anion transporters. Another hypothesis involves the symport of H^+/Cl^- ions across the plasma membrane. Cations like Na^+ and K^+ also compete for their entry into the cell, using the same cellular transporters (Nilius & Honoré, 2012). These transporters can be divided into two functional categories depending upon the degree of specificity for the cations. The non-selective transporters utilize the same proteins for the influx of more than one kind of ion. The ion-selective transporter proteins, on the other hand, like the K^+ ion transporter proteins, are involved mainly in the transmembrane exchange of a single ionic entity. Potassium ion transport proteins, for instance, include K^+ channels and transporters, Na^+/K^+ symporters and voltage-gated ion channels, among various types (Takahashi & Tateda, 2013). These transport mechanisms, either alone or in combination, contribute significantly to the development of tolerance against the salt-stressed environment in legumes. The increased demand for K^+ ions in comparison to Na^+ ions results in the intracellular accumulation of various ions even in saline-stressed environments (Thiel *et al.*, 2013).

Apart from manipulation of transmembrane ion transport mechanisms, ionic sequestration into the plant vacuoles results in the ability to tolerate salt stress more effectively. Osmotic adjustment within the cell is carried out by the regulation of different ions, chiefly Na^+ and Cl^- ions (Jiang *et al.*, 2013). Induction of cytoplasmic acidification and vacuolar alkalization follows the ionic modification of the cellular environment. Overexpression NHX genes in legumes (Na^+ and H^+ exchange) are responsible for these stress-compensation mechanisms (Bassil *et al.*, 2012; Roy & Tester, 2013). Studies on soybean have been especially helpful in deciphering the molecular mechanisms involved in NHX-mediated stress compensation (Ha *et al.*, 2013; Roy & Tester, 2013). Similarly, solute regulation, especially, in the form of alteration of calcium ion levels, leads ultimately to the generation of signal transduction and adaptation to the salt stress.

12.4.2 Metabolite production

An essential compensatory mechanism adapted by many legume species is the production of certain metabolites or response elements that can prevent, neutralize or counteract the physical constraints. The salt and drought response elements usually encountered are involved in the regulation of osmotic balance and, therefore, are referred to as osmolytes. They help in sustaining the water levels within the cell. Table 12.1 reviews the various metabolites produced in plants in response to a saline stress. Among the different metabolites, sugars (sucrose, trehalose and fructans) and sugar alcohols (inositols and glycerols) are the main players for osmoregulation in legumes (L. Yang *et al.*, 2012; Joshi *et al.*, 2013; Lyu *et al.*, 2013). Sugars and sugar alcohols complex with proteins to act as osmoprotectants by lining the cellular surfaces. Moreover, they may

Table 12.1 Metabolites produced by legumes in response to salt stress.

Molecules	Possible role(s)	References
Amino acids		
Ectoine	Osmoprotection	Saxena *et al.*, 2013
Proline	Redox control, respiration regulation	Hayat *et al.*, 2012
Carbohydrates		
Trehalose	Osmoprotection	Redillas *et al.*, 2012
Polyols	Osmoregulation, redox control	Palma *et al.*, 2012
Proteins		
Dehydrins	Dessication control, redox control	Galani *et al.*, 2013
Chaperones	Protein folding, heat-/cold-/salt-shock proteins	Nakaminami *et al.*, 2012
Amines		
Glycine betaine	Osmoprotection	Chen & Murata, 2011
Dimethyl sulphonium	Protein protection	Parvaiz & Satyawati, 2008

serve as molecular chaperones to regulate the transcription and act as inhibitors of ionic solutes preventing retardation of enzymatic activity. Hence, the response demonstrated by these players is dependent upon the type of tissue and the initial stress factor. Moreover, these metabolites are not only directly involved in osmotic regulation but also may result in the manipulation of other pathways and players. Some additional metabolites produced as a result of salt stress conditions include charged molecules like glycine betaine and proline, which are involved in the adjustment of water content in the plant body (Nounjan *et al.*, 2012; Waditee-Sirisattha *et al.*, 2012). Transcriptional activation of various genes, including pyrroline 5-carboxylate synthase, leads to an increase in the accumulation of proline. This proline, via a $NADP^+$-mediated pathway, then helps in protecting the plant against various abiotic stress conditions (Kim & Nam, 2012). Apart from protecting plants from salt stress, certain metabolites may also play a role in tolerance against reactive oxygen species. In legumes, these mechanisms are shown by adaptive mediators including superoxide dismutase (SOD), ascorbate peroxidase (APX) and phospholipid hyperoxide glutathione peroxidase (PHGPX) (Morais *et al.*, 2012; Ramana *et al.*, 2012; Zelinová *et al.*, 2013; Sultana *et al.*, 2014). Enhanced peroxidase activity in response to salt stress in *Ammopiptanthus* species has been recently studied using expressed sequence tags (ESTs) (M. Liu *et al.*, 2013). Similarly, dimethylsulphoniopropionate (DMSP) is also regulated in salt and accompanying osmotic stress in legumes and other plants (Arora *et al.*, 2012; Wdowiak-Wróbel *et al.*, 2013). The pathways involved in the production of these secondary metabolites also affect the basic metabolic pathways including amino acid and sugar synthesis and catabolic processes, as most of the secondary metabolites are either derivatives of amino acids or sugar molecules. Hence, glycine betaine, proline, inositol and pinnitol are all involved in the regulation of basic metabolic processes in salt-stressed legumes.

12.4.3 Water channelling and inter-tissue coordination

As observed in many plant species, another mechanism employed by most legumes is the ability to channel water away from meristematic tissue. This is accompanied by a decrease in osmotic penetration into cortical tissue, which is marked by decreased water uptake from the root. The physiological basis of this phenomenon is an alteration in the number and degree of water channels (aquaporins) or water/solute channels (aquaglyceroporins) that are available for osmotic exchange (Betti *et al.*, 2012; Bhardwaj *et al.*, 2013). These water channels are activated and phosphorylated by calcium-dependent protein kinases (Osakabe *et al.*, 2013). Other proteins involved in water channelling include tonoplast intrinsic protein (TIP) and plasma membrane proteins (PM proteins) (Marino *et al.*, 2012; Shavrukov, 2013).

The response of a plant to salt stress takes the form of a coordinated strategy involving different tissues of the body. Hence, communication between those different plant parts has a basic role in initiating the genetic and physiological changes of a response to saline stress. This coordination mainly involves chemical substances but in certain cases the generation and manipulation of action potentials may solely be involved in the process. Generally, the signalling molecules involved in cellular communication in legumes include enzymes (kinases, phosphatases) or plant growth hormones (auxins, cytokinin, abscisic acid, ethylene). Abscisic acid (ABA), for instance, initiates a complex genetic and signalling cascade in response to salt stress in legumes (Miransari, 2013). The biosynthetic pathways (ABA), ion channels and related catalytic pathways (abscisic acid insensitive 1, *ABI1*) and genes involved in providing resistance to plant growth hormones (e.g. auxin resistance gene 2, *AXR2*) are sequentially involved in coordination and control of stress response signals. Potassium and calcium ion channel conduction is mainly regulated by activation by these signalling pathways. The ion channels involved in this case may be K^+ or Ca^{2+} conductors. Osmolytes and a number of enzymes (oxidases, invertases and oxidoreductases) are also involved in signalling cascades (Lopez-Gomez *et al.*, 2012). These and many other chemical entities cause signal mediation and effector cell excitation, thereby yielding the responses observed to a particular stress signal.

The chemicals produced as a result of a stress factor do not necessarily act on the neighbouring tissues and cells. In response to an osmotic stress, for instance, the photosynthetic capacity of leaves was altered accompanied by a decrease in Na^+ ion uptake from the roots. The regulation of photosynthesis in response to a salt stress is mainly attributable to the production of *myo*-inositol. Moreover, indole acetic acid (IAA) assists the generation of phosphatidylinositol metabolites that, ultimately, help in neutralizing the stress conditions (Patel *et al.*, 2012). The overexpression of *myo*-inositol and IAA in response to salt stress has been studied in a number of legumes including chickpea and cacao (Fraire-Velázquez & Balderas-Hernández 2013; Kaur *et al.*, 2013; Li *et al.*, 2013). In this way, the genetically controlled cellular responses assist in responding to a physical constraint through the involvement of a wide range of tissues and mediators.

12.4.4 Regulatory elements

The stress response elements or mediators encoded by specific regulator genes help the plant to respond to a stress stimulus by altering various steps in signal transduction pathways or by affecting the effector molecules/pathways. Ultimately they are linked to providing assistance in salt and osmotic regulation in plants. They act as modulators of transcription or translation and are classified on the basis of the interaction between the transcription factors and the promoter regions of stress response genes. They either specifically or non-specifically respond to the stress conditions. Abscisic acid is one of the main chemical players in this regard. It may be activated and regulated by a variety of different elements. Dehydration responsive protein, rd22, for example is activated in the case of salt and osmotic stress conditions (Wang *et al.*, 2012). Interaction between the rd22 gene and the corresponding regulatory elements, both *cis* and *trans* regulators, results in the transcriptional activation of a number of pathways. A series of genetic interactions, including those with the Ca^{2+}-dependent protein kinases and DNA binding proteins, are responsible for the efficient response to stress factors. Apart from ABA, a number of other genetic factors might be involved in regulation of responses to salt and osmotic stress in leguminous plants. The Salt Overly Sensitive (*SOS*) genes encodes the salt response elements that are mediated by K^+ exchange across the plasma membrane (Hill *et al.*, 2013). Regulation of these genes helps in controlling the transmembrane osmotic transport in plant cells. Moreover, the dehydration response element (DRE) proteins are involved in binding to the ethylene responsive elements and are, hence, involved in controlling the response to salt and osmotic stress (Mizoi *et al.*, 2013).

Among the plant regulatory elements possessing the ability to directly control the stress responses, mitogen-activated protein kinases (MAPKs) are of prime importance. MAPKs have been identified in a number of plants, including legumes, and are considered pivotal in rendering tolerance against the abiotic stress factors including salt and water constraints. A MAPK isolated from *Pisum sativum* (PsMAPK) has been identified as possessing significant homology with yeast proteins involved in glycerol accumulation (Ahmad *et al.*, 2012). The increased levels of glycerol in the plant body can help in controlling and regulating the overall water and osmolyte levels. Moreover, plant serine/threonine kinases display a high sequence homology with DBF-2, glycogen synthase kinase (GSK) and calcineurin genes that are involved in providing tolerance against salt and osmotic imbalances in the growth environment in legumes (Akio Kido *et al.*, 2011; Wang *et al.*, 2013).

12.5 Genetic and molecular responses to drought

An important stress factor that significantly affects the yield of legumes is insufficiency of water to maintain growth and homeostasis. Legumes respond to drought in a similar manner as they respond to saline stress, and usually the same genetic and molecular players are involved. This is mainly because both physical constraints tend to induce similar changes in the plant body, i.e. increased turgor pressure, changes in ionic concentrations and alterations in cellular morphology. Abscisic acid, proline, glycine betaine and sugars are the main drought response elements (Allario *et al.*, 2012; Z. Yang *et al.*, 2012).

General mechanisms similar to those observed with other environmental restraints, are followed in case of drought stress. Figure 12.2 manifests the cascade of events occurring in leguminous plants in response to a drought stress. A number of osmosensors, which function to interpret changes in the water levels available to the plant, have been identified. Of significant importance among these are the ATHK type osmosensors (Kumar *et al.*, 2013b). Once the signal has been detected, a series of signalling molecules (proteins) are activated. MAPK, phosphatases and phospholipases are the most important enzymes in this respect. These then interact with the effector cells to produce specific changes in the plant body (Table 12.2).

Physiologically, legumes respond to a drought threat by adopting either an escape or an avoidance

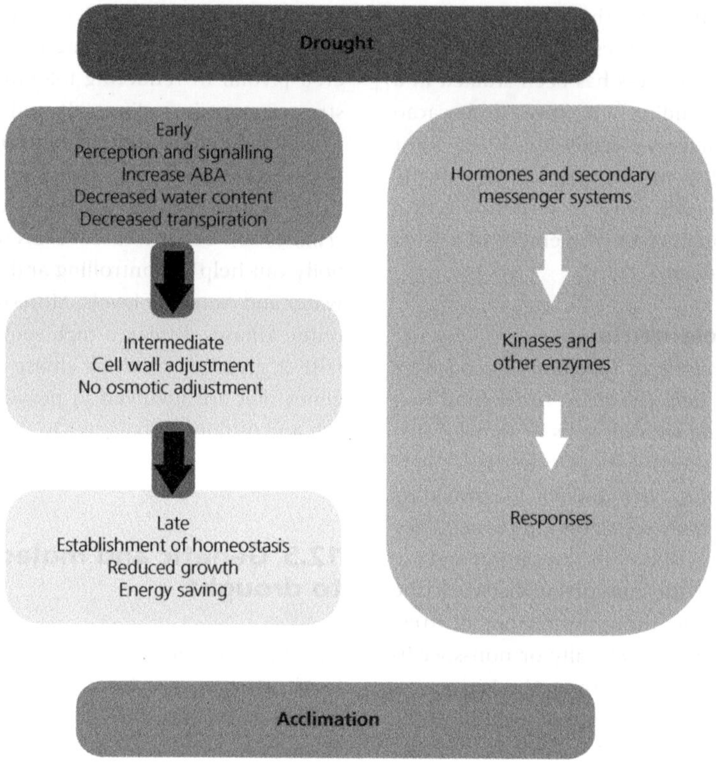

Figure 12.2 Physiological and biochemical responses to moderate drought.

Table 12.2 Signalling cascade and effector elements activated in response to environmental stress factors.

	Salt stress	Drought stress	Cold stress
Presentation	Osmotic stress/ sodium toxicity	Osmotic stress	Decreased functionality
Secondary messenger system	Phosphatidic acid, abscisic acid, calcium, hydrogen peroxide	Phosphatidic acid, abscisic acid, calcium, hydrogen peroxide	Calcium, hydrogen peroxide
Kinases	AtSK, SnRK, CBL, CDPK, MAPK, SOS	SnRK	CBL, CDPK, MAPK, SnRK
Response	Stress responsive gene expression, antiporter activation	Decreased transpiration	Membrane stabilization

Abbreviations: AtSK, Arabidopsis Shaggy/GSK3-like kinases; CBL, calcineurin B-like; CDPK, calcium-dependent protein kinases; MAPK, mitogen-activated protein kinases; SnRK, sucrose non-fermenting-1-related kinases; SOS, salt overly sensitive.

mechanism. In order to escape from the physical stress, legumes tend to bring about changes that ensure the completion of their life cycle before extreme drought conditions set in. In contrast, drought avoidance involves the retention of water within the plant body to ensure that the plant is not dehydrated. Selection of the appropriate mechanism depends on the speed at which morphological and physiological changes occur. Increased water retention is carried out by changes in root morphology to help absorb more water from the soil, by decreasing the stomatal and cuticular conductance, and by decreasing transpiration. Moreover, the expansin genes are upregulated to help in expansion of cells (Hayashi *et al.*, 2012). This improves the capacity of the plant cell to retain water. Other changes include alterations in leaf water potential, and changes in chlorophyll content and the root-to-shoot ratio, all of which synergistically contribute to increasing the water retention in legumes.

The genetic elements upregulated in response to drought influence a number of biochemical processes to mediate their effects. They may, directly, produce certain effector elements or induce the production of entities that are involved in the regulation of cellular, metabolic and biochemical responses. The genes involved in providing tolerance against drought stress include the responsive to dehydration (*rd*) and early responsive to dehydration (*erd*) series (Nakashima *et al.*, 2012; Rai *et al.*, 2012). These are stress-induced genes and can function in response to a variety of stress factors including drought, salinity and very low temperatures. The main mechanism involved in responding to drought is the presence of ABA. Whether endogenous or exogenous, ABA has been proved to induce and regulate the expression of a number of drought response elements and to help produce tolerance against this stress factor. However, certain pathways (involving *rd29A*) have been identified that do not, necessarily, depend upon ABA for the production of stress response elements (de Paiva Rolla *et al.*, 2014). In many plant species, including legumes, the endogenous levels of ABA are significantly increased in response to drought, and ABA is responsible for inducing many drought-induced response elements. These genetic elements possess specific ABA-responsive elements (ABRE) in their promoter regions that are activated on synthesis of ABA in drought conditions (Fujita *et al.*, 2013). A number of genes (*rd22, rd29*) and transcription factors (MYC, MYB) are involved in inducing ABA responses in legumes under drought (Ambawat *et al.*, 2013; Bakht *et al.*, 2013).

Dehydration Responsive Elements (DRE) are the genetic sequences involved in inducing stress responses to dehydration. DREs contains a highly conserved 9 bp sequence that is considered essential for the binding of *cis*-acting elements. A separate class of proteins known as DRE-binding proteins (DREB) has also been identified. They contain a conserved DNA binding motif (AP2 domain) that has sequence homology in the binding region but varies significantly in the region outside the protein-binding groove (Jain & Chattopadhyay, 2013). This genetic difference has formed the basis for classification of DREB into two distinct classes, DREB 1 and 2; it is DREB 2 that are mainly involved in responding to dehydration conditions. Transgenic legumes encoding DREB 2 have shown excellent capability to survive in dehydrated conditions (Acharjee & Sarmah, 2013; Jovanovic *et al.*, 2013). Hence, plant responses to drought stress rely greatly upon the genetic and molecular players involved in the signal transduction pathways (Irving & Gehring, 2013; Osakabe *et al.*, 2013)

12.6 Genetic and molecular responses to extremes in temperatures

Sudden elevations and depressions in temperature significantly affect legume output. Most, but not all, legumes prefer moderate to warm temperature conditions for their growth. However, the leguminous plants that grow in colder environments are sensitive to rises in temperature. Otherwise, a rise in temperature does not significantly affect plant health in temperate regions. In response to an increase in temperature, the plants may respond by regulating their cell cycle, metabolic rate and growth patterns. However, if the exposure is prolonged, the size and thickness of aerial parts, stomatal density and chloroplast activity are adversely affected (Theocharis *et al.*, 2012). Usually, though, if cool temperate legumes are grown in a warmer climate, they tend to survive and efficiently carry out their life processes.

In contrast, a fall in temperature is considered a major stress factor for many legumes because of the changes accompanying the temperature drop. Decreased temperature may be accompanied by frost, with its attendant dehydration and reduced nutrient supply to the plant. The plant, hence, has to develop means to survive in these harsh conditions. On perceiving a change in the growth environment, the plants tend to generate signals that, ultimately, help in the production of regulatory and effector elements. In many cases, the mechanisms and elements of resistance to drought and cold are the same (Table 12.3).

The physiological and morphological changes observed in response to a cold stress vary widely among different legume species. Figure 12.3 represents the general mechanisms that are involved in responding to a cold stress, and certain changes are observed in all plants exposed to cold temperature. Generally, alteration in cell membranes, increased ion concentration in the cell, generation of reactive oxygen species and free radicals, activation of cold-response genes and production of secondary metabolites are noted. Modification of cell membranes involves increasing their rigidity by increasing the content of unsaturated fatty acids. This helps in stabilizing the membrane and providing increased resistance of the cell to extremely low temperatures. Decreases in temperature result in a reduced ability of the plant body to carry out its metabolic processes. This is coupled to a diminished rate of photosynthesis (Theocharis *et al.*, 2012). Moreover, plant pigments like xanthophylls are produced in greater quantities to ensure the availability of antioxidants for the plant (González-Cruz & Pastenes, 2012). Other secondary metabolites like flavonoids are produced to help the plant survive cold stress (Petrussa *et al.*, 2013).

Keeping in view the diversity of responses produced, a number of genes are positively regulated in response to a cold stress. Among these the principal genes include the cold regulated (*cor*), low temperature induced (*lti*) and cold-inducible (*kin*) genes (Mantri *et al.*, 2012; Bakht *et al.*, 2013; Pang *et al.*, 2013). The cold resistance mechanisms are not always dependent upon the ABA pathway to exert their effects. ABA has been observed to be transiently increased under low temperature stress conditions. However, the levels of endogenous ABA are not significant enough to produce a stress response alone. The DRE and *rd29* act as response elements in case of a cold-induced stress (Pirzadah *et al.*, 2014). C-repeat elements (CRTs) and low temperature responsive elements have been observed to interact with the DRE in exerting their cold neutralizing properties in many plants (Humphreys & Gasior, 2013). As in case of drought, the DREB elements involved belong to the DREB 2 class (Nayak *et al.*, 2009). They are responsible for inducing specific gene expression of elements that can respond to cold. Moreover, transgenic plants containing the *cor*, *lti* and *kin* genes have been observed to develop significant tolerance to the freezing temperatures (Theocharis *et al.*, 2012; Bakht *et al.*, 2013; Castonguay *et al.*, 2013; Pang *et al.*, 2013). Another group of response elements known as sensitivity to freezing (sfr) have been implicated in interacting with the DRE and its promoter regions. Two more genes, *HOS1* and *HOS2*, are considered to be negative regulators of signals generated in response to freeze stress (Ishitani *et al.*, 1997). Cold shock domain proteins (CSDPs) are involved in the induction of freezing tolerance in leguminous plants (Theocharis *et al.*, 2012).

Table 12.3 Genetic bases for induction of tolerance to environmental stress factors.

Stress factor	Genes regulated
Salt stress	HVA, SOD, catalase, Ast, HSP, SSP, Akt, ITR, TIP
Osmotic stress	CDPK/CPK, CBL, MAPK, PKS, HOG, SPK, ABRE
Cold stress	HOS, RAV, CBF, LOS, ABRE, Rd, Cor

Abbreviations: ABRE: ABA-responsive element; Akt/KAT, K^+ selective ion channels; Ast, aspartate aminotransferase; CBF, C-repeat binding factors; CBL, calcineurin B-like; CDPK/CPK, calcium-dependent protein kinases; Cor, cold regulated genes; H/LOS, high/low expression of osmotically responsive genes; HOG, high-osmolarity glycerol response; HSP, heat-shock proteins; HVA, late embryogenesis abundant protein gene (barley derived); ITR, *myo*-inositol transporter; MAPK, mitogen-associated protein kinases; PKS, plant type III polyketide synthase; RAV, related to ABI3/VP (transcription factors); Rd, dehydration responsive elements; SOD, superoxide dismutase; SPK, calcium-dependent seed-specific protein kinase; SSP, salt shock proteins; TIP, tonoplast intrinsic protein.

An increase in intracellular calcium levels has been observed in response to a sudden drop in temperature. It has been hypothesized that Ca^{2+} plays an important role in signal transduction mechanisms. These genetic and molecular changes in the plant are responsible for inducing cold tolerance in the susceptible plants. Proline, an amino acid derivative, is involved in osmoprotection and reduction of reactive oxygen species. Many genes (*cor, lti, kin, erd*) produce proteins that act as dehydrins (Arumingtyas *et al.*, 2013). Some of these dehydrins have been positively correlated with tolerability to cold stress by increasing membrane stability and cryoprotection. Among the various osmolytes stored by legumes to survive cold

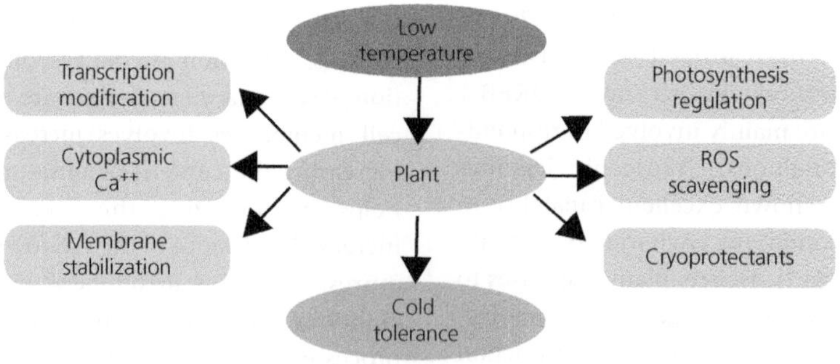

Figure 12.3 Cellular changes involved in developing cold tolerance in legumes.

stress are glutamine, betaine, sterols, glucosides and polyamines (Signorelli *et al.*, 2013).

Alteration of photosynthesis rate is one of the most common adaptations a legume makes to survive a cold-stressed environment. However, the most significant aspect is the change in carbohydrate metabolism. There is often an increase in the soluble sugar content in the plant body, which demonstrates their role in the signalling process, temperature maintenance and osmoregulation. Moreover, sugars help in osmotic regulation, by increasing the solvent retention, resulting in decreased availability of free water for freezing. Sugar molecules also interact with the lipoproteins of the plasma membrane and stabilize it. Carbohydrates are also implicated in antioxidant phenomena and signalling processes. Among the various sugars, trehalose, stachyose and raffinose are considered important in inducing cold tolerance. Hence, a number of genetic and molecular mechanisms are available for inducing tolerance in susceptible plants against low temperatures.

12.7 Plant defence mechanisms and their efficiency

Legumes respond to various stress conditions by invoking a variety of effector mechanisms. The specific ways by which plants render themselves resistant to the major physical constraints have been discussed. Apart from salinity, drought and temperature extremes, a variety of other stress factors are also responsible for causing major damage to the leguminous systems. Natural disasters, for instance, result in significant abrupt changes in the ecological parameters of the plant. The plant usually does not get enough time to develop resistance or adapt to tornadoes, floods, wild fires or high-speed winds. In contrast, exposure to high-intensity light is an ecological factor that changes very slowly and gradually. Legumes tend to adapt to this factor over a prolonged period of time, i.e. over several generations. Hence, plants can only develop tolerance to the stress factors that tend to show a constant presence of a period of several weeks to months.

Environmental stress factors do not, necessarily, have a direct effect. They may be involved in changing the overall morphology and physiology of the legume, promoting the ability of biotic stressors, such as pathogens, to infect easily and effectively. In case of a cold stress, for instance, certain bacterial and fungal species tend to infect plants with much greater ease. *Pseudomonas syringae* contributes to the ice nucleation process (Attard *et al.*, 2012), which results in failure of legumes to sense the freezing temperatures and initiate mechanisms to survive the drop in temperatures. In contrast, certain species of nitrogen-fixing bacteria, plant growth promoting rhizobacteria (PGPR), are involved in the reduction of chilling-induced damage to the plant body. These bacterial species manipulate a variety of stress-induced genes (chitinase, glucanase) and cause the production of a number of metabolites (lipids, aldehydes, sugars, ROS) to exert their effects.

It is therefore evident that there is no single, uniform survival mechanism that can be adopted by legumes to protect themselves from various environmental and physical constraints. All the pathways adopted by these plants tend to connect to one another at a certain stage. The efficiency of any process depends upon a number of factors. The intensity and time of exposure of a particular stress stimulus is the first controlling factor in regulating the response to that threat. The sensitivity with which a plant can detect a stress factor also affects the degree of the genetic and molecular changes occurring in the plant body. Additionally, the correct sequence of signal cascade components must be activated in response to a threat. All the necessary proteins, enzymes and other chemical moieties have to play their role in the cascade. Moreover, the regulatory and effector molecules produced as a result of the process need to act on specific tissue types in the plant body so as to bring about required morphological and physiological changes. These phenotypic changes are the final manifestations by which a plant responds to any environmental threat. Legumes tend to respond to various physical and environmental constraints by the concerted effort of many genes, proteins and other biomolecules, ionic species, solvent systems, and cell and tissue types. Development of resistance to various stress factors thus relies on the efficient working of all these processes and molecules within the plant body.

12.8 Conclusion and future prospects

Many of the changes exhibited by legumes in response to environmental constraints involve the interplay of a number of genetic and molecular entities. Many levels of transcriptional control are involved in controlling the genes necessary for inducing stress tolerance. The molecular pathways initiated and activated in response to activation of various genes help in regulating the responses to a particular stress factor. However, it is still unclear which chemical moieties and signals direct the stress response in a particular direction. It is imperative that we understand these mechanisms to assist in engineering stress tolerance in legumes in an effective and safe manner. Systematic analysis of gene expression, construction of molecular networks and signal transduction processes need to be studied and evaluated to achieve better biotechnological outcomes for the stress-resistant varieties.

The advent of modern genetic and protein expression tools can help in the evaluation of a variety of pathways and molecular responses in an effective and adequate manner. Transcriptomics, genomics, proteomics and metabolomics can all provide valuable insights into the genetic and subcellular molecular changes induced as a result of various stress factors. These fields have revolutionized the biological understanding of various life processes. Construction of genetic maps of the players responsible for inducing stress tolerance and functionally complementing the genetic information can help improve our overall understanding of the means by which legumes respond to various stress conditions. Once the genes involved have been identified, a variety of genetic combinations can be employed for engineering pathways that can help the plant tolerate salinity, temperature extremes and drought. Although many data are now available regarding the responses to various physical constraints, efforts need to be made to collect, integrate and interpret those data. A collaborative effort by plant physiologists, pathologists and biotechnologists is required to gain the full benefits of these research efforts.

References

Abate T, Alene AD, Bergvinson D, *et al.* (2012) Tropical grain legumes in Africa and South Asia: Knowledge and opportunities. International Crops Research Institute for the Semi Arid Tropics (Report), pp. 1–21.

Acharjee S, Sarmah BK (2013) Biotechnologically generating 'super chickpea' for food and nutritional security. *Plant Sci* 207: 108–116.

Ahmad P, Bhardwaj R, Tuteja N (2012) Plant signaling under abiotic stress environment. In: Ahmad P, Prasad MNV (eds), *Environmental Adaptations and Stress Tolerance of Plants in the Era of Climate Change.* Springer Science+Business Media, New York, pp. 297–323.

Akio Kido E, Kelle de Araujo Barbosa P, Ribamar Costa Ferreira Neto J, *et al.* (2011) Identification of plant protein kinases in response to abiotic and biotic stresses using SuperSAGE. *Curr Protein Peptide Sci* 12: 643–656.

Allario T, Brumos J, Colmenero-Flores JM, Iglesias DJ, *et al.* (2012) Tetraploid Rangpur lime rootstock increases drought tolerance via enhanced constitutive root abscisic acid production. *Plant Cell Environ* 36: 856–868.

Ambawat S, Sharma P, Yadav NR, Yadav RC (2013) MYB transcription factor genes as regulators for plant responses: an overview. *Physiol Mol Biol Plant* 19: 307–321.

Ansi A, Kumari S, Makkouk K, Muharram I, Basha R, Al-Motokel W (2010) Survey to identify virus diseases affecting food legume crops in Yemen. *Ann Agricul Sci (Cairo)* 55: 153–161.

Arbona V, Manzi M, Ollas CD, Gómez-Cadenas A (2013) Metabolomics as a tool to investigate abiotic stress tolerance in plants. *Int J Mol Sci* 14: 4885–4911.

Aroca R, Ruiz-Lozano JM (2012) Regulation of root water uptake under drought stress conditions. In: Aroca R (ed.), *Plant Responses to Drought Stress.* Springer Science+Business Media, New York, pp. 113–127.

Arora NK, Tewari S, Singh S, Lal N, Maheshwari DK (2012) PGPR for protection of plant health under saline conditions. In: Maheshwari DK (ed.), *Bacteria in Agrobiology: Stress Management.* Springer Science+Business Media, New York, pp. 239–258.

Arraouadi S, Badri M, Abdelly C, Huguet T, Aouani ME (2012) QTL mapping of physiological traits associated with salt tolerance in *Medicago truncatula* recombinant inbred lines. *Genomics* 99: 118–125.

Arumingtyas EL, Savitri ES, Purwoningrahayu RD (2013) Protein profiles and dehydrin accumulation in some soybean varieties (*Glycine max* L. Merr) in drought stress conditions. *Am J Plant Sci* 4: 134–141.

Asensio AC, Gil-Monreal M, Pires L, Gogorcena Y, Aparicio-Tejo PM, Moran JF (2012) Two Fe-superoxide dismutase families respond differently to stress and senescence in legumes. *J Plant Physiol* 169:1253–1260.

Attard E, Yang H, Delort A-M, *et al.* (2012) Effects of atmospheric conditions on ice nucleation activity of *Pseudomonas. Atmos Chem Physics* 12: 10667–10677.

Bakht J, Bano A, Shafi M, Dominy P (2013) Effect of abscisic acid applications on cold tolerance in chickpea (*Cicer arietinum* L.). *Euro J Agron* 44: 10–21.

Banasiak J, Biała W, Staszków A, *et al.* (2013) A *Medicago truncatula* ABC transporter belonging to subfamily G modulates the level of isoflavonoids. *J Exp Bot* 64: 1005–1015.

Bassil E, Coku A, Blumwald E (2012) Cellular ion homeostasis: emerging roles of intracellular NHX Na^+/H^+ antiporters in plant growth and development. *J Exp Bot* 63: 5727–5740.

Bayuelo-Jiménez J, Jasso-Plata N, Ochoa I (2012) Growth and physiological responses of *Phaseolus* species to salinity stress. *Int J Agron* 2012: e1–13.

Betti M, Pérez-Delgado C, García-Calderón M, Díaz P, Monza J, Márquez AJ (2012) Cellular stress following water deprivation in the model legume *Lotus japonicus*. *Cell* 1: 1089–1106.

Bhardwaj R, Sharma I, Kanwar M, *et al.* (2013) Aquaporins: role under salt stress in plants. In: Ahmad P, Azooz MM, Prasad MNV (eds), *Ecophysiology and Responses of Plants Under Salt Stress*. Springer Science+Business Media, New York, pp. 213–248.

Bhatnagar-Mathur P, Palit P, Kumar CS, Reddy DS, Sharma KK (2012) Grain legumes: Biotechnological interventions in crop improvement for adverse environments. In: Tuteja N, Gill SS, Tuteja R (eds), *Improving Crop Productivity in Sustainable Agriculture*. Wiley-VCH Verlag GmbH & Co., Weinheim, Germany, pp. 381–421.

Bouchenak M, Lamri-Senhadji M (2013) Nutritional quality of legumes, and their role in cardiometabolic risk prevention: a review. *J Med Food* 16: 185–198.

Boudreau MA (2013) Diseases in intercropping systems. *Annu Rev Phytopathol* 51: 499–519.

Butt MS, Sultan MT (2013) Selected functional foods for potential in disease treatment and their regulatory issues. *Int J Food Proper* 16: 397–415.

Cannon SB (2013) The model legume genomes. In: Rose RJ (ed.), *Legume Genomics*. Springer Science+Business Media, New York, pp. 1–14.

Castonguay Y, Dubé MP, Cloutier J, Bertrand A, Michaud R, Laberge S (2013) Molecular physiology and breeding at the crossroads of cold hardiness improvement. *Physiol Plant* 147: 64–74.

Chen TH, Murata N (2011) Glycinebetaine protects plants against abiotic stress: mechanisms and biotechnological applications. *Plant Cell Environ* 34: 1–20.

Dasgupta N, Nandy P, Das S (2013) Salt stress: A biochemical and physiological adaptation of some Indian halophytes of Sundarbans. In: Rout GR, Das AB (eds), *Molecular Stress Physiology of Plants*. Springer Science+Business Media, New York, pp. 155–177.

de Paiva Rolla AA, Carvalho JdFC, Fuganti-Pagliarini R, *et al.* (2014) Phenotyping soybean plants transformed with rd29A:AtDREB1A for drought tolerance in the greenhouse and field. *Transgenic Res* 23: 75–87.

Duque AS, de Almeida AM, da Silva AB, *et al.* (2013) Abiotic stress responses in plants: Unraveling the complexity of genes and networks to survive. In: Vahdati K, Leslie C (eds), *Abiotic Stress–Plant Responses and Applications in Agriculture*. InTech, pp. 49–101.

Echevarría-Zomeño S, Yángüez E, Fernández-Bautista N, Castro-Sanz AB, Ferrando A, Castellano M (2013) Regulation of translation initiation under biotic and abiotic stresses. *Int J Mol Sci* 14: 4670–4683.

Elkahoui S, Barhoumi Z, Djébali N, *et al.* (2013) Physiological and ultrastructural responses of *Catharanthus roseus* cell suspension to salt stress. *Russ J Plant Physiol* 60: 244–249.

Fraire-Velázquez S, Balderas-Hernández VE (2013) Abiotic stress in plants and metabolic responses. In: Vahdati K, Leslie C (eds), *Abiotic Stress–Plant Responses and Applications in Agriculture*. InTech, pp. 25–48.

Fujita Y, Yoshida T, Yamaguchi-Shinozaki K (2013) Pivotal role of the AREB/ABF-SnRK2 pathway in ABRE-mediated transcription in response to osmotic stress in plants. *Physiol Plant* 147: 15–27.

Galani S, Wahid A, Arshad M (2013) Tissue-specific expression and functional role of dehydrins in heat tolerance of sugarcane (*Saccharum officinarum*). *Protoplasma* 250: 577–583.

Ganguli NK, Kennedy IR (2013) Indigenous actinorhizal plants of Australia. *J Biosci* 38: 749–759.

Geissler N, Huchzermeyer B, Koyro HW (2013) Effects of salt stress on photosynthesis under ambient and elevated atmospheric CO_2 concentration. In: Ahmad P, Azooz MM, Prasad MNV (eds), *Salt Stress in Plants: Signalling, Omics and Adaptations*. Springer Science+Business Media, New York, pp. 377–413.

González-Cruz J, Pastenes C (2012) Water-stress-induced thermotolerance of photosynthesis in bean (*Phaseolus vulgaris* L.) plants: the possible involvement of lipid composition and xanthophyll cycle pigments. *Environ Exp Bot* 77: 127–140.

Grieve CM, Grattan SR, Maas EV (2012) Plant salt tolerance. Agricultural salinity assessment and management, 2nd edn. *ASCE Manual and Reports on Engineering Practice* 71: 405–459.

Gupta S, Nadarajan N, Gupta DS (2014) Legumes in omic era: retrospects and prospects. In: Gupta S, Nadarajan N, Gupta DS (eds), *Legumes in the Omic Era*. Springer Science+Business Media, New York, pp. 1–14.

Ha BK, Vuong TD, Velusamy V, Nguyen HT, Shannon JG, Lee JD (2013) Genetic mapping of quantitative trait loci conditioning salt tolerance in wild soybean (*Glycine soja*) PI 483463. *Euphytica* 193: 79–88.

Hamidou F, Halilou O, Vadez V (2013) Assessment of groundnut under combined heat and drought stress. *J Agron Crop Sci* 199: 1–11.

Hayashi S, Reid DE, Lorenc MT, *et al.* (2012) Transient Nod factor-dependent gene expression in the nodulation-competent zone of soybean (*Glycine max* [L.] Merr.) roots. *Plant Biotechno J* 10: 995–1010.

Hayat S, Hayat Q, Alyemeni MN, Wani AS, Pichtel J, Ahmad A (2012) Role of proline under changing environments: A review. *Plant Signal Behav* 7: 1456–1466.

Hedrich R, Becker D, Geiger D, Marten I, Roelfsema MRG (2012) Role of ion channels in plants. In: Okada Y (ed.), *Patch Clamp Techniques*. Springer, Science+Business Media, New York, pp. 295–322.

Hill CB, Jha D, Bacic A, Tester M, Roessner U (2013) Characterization of ion contents and metabolic responses to salt stress of different Arabidopsis AtHKT1;1 genotypes and their parental strains. *Mol Plant* 6: 350–368.

Horie T, Karahara I, Katsuhara M (2012) Salinity tolerance mechanisms in glycophytes: An overview with the central focus on rice plants. *Rice* 5: 1–18.

Hossain Z, Khatoon A, Komatsu S (2013) Soybean proteomics for unraveling abiotic stress response mechanism. *J Proteome Res* 12: 4670–4684.

Humphreys M, Gasior D (2013) Cold tolerance. In: Kole C (ed.), *Genomics and Breeding for Climate-Resilient Crops*. Springer Science+Business Media, New York, pp. 133–165.

Irving HR, Gehring C (2013) Molecular methods for the study of signal transduction in plants. In: Gehring C (ed.), *Cyclic Nucleotide Signaling in Plants. Methods in Molecular Biology* Vol. 1016. Springer/Humana Press, pp. 1–11.

Ishitani M, Xiong L, Stevenson B, Zhu J-K (1997) Genetic analysis of osmotic and cold stress signal transduction in Arabidopsis: interactions and convergence of abscisic acid-dependent and abscisic acid-independent pathways. *Plant Cell* 9: 1935–1949.

Jacques S, Ghesquière B, Van Breusegem F, Gevaert K (2013) Plant proteins under oxidative attack. *Proteomics* 13: 932–940.

Jain D, Chattopadhyay D (2013) Role of DREB-like proteins in improving stress tolerance of transgenic crops. In: Tuteja N, Sarvajeet SG (eds), *Plant Acclimation to Environmental Stress*. Springer Science+Business Media, New York, pp. 147–161.

Jha UC, Chaturvedi SK, Bohra A, Basu PS, Khan MS, Barh D (2014) Abiotic stresses, constraints and improvement strategies in chickpea. *Plant Breed* doi: 10.1111/pbr.12150.

Jiang J, Su M, Chen Y, *et al.* (2013) Correlation of drought resistance in grass pea (*Lathyrus sativus*) with reactive oxygen species scavenging and osmotic adjustment. *Biologia* 68: 231–240.

Joshi R, Ramanarao VM, Baisakh N (2013) *Arabidopsis* plants constitutively overexpressing a myo-inositol 1-phosphate synthase gene (*SaINO1*) from the halophyte smooth cord grass exhibit enhanced level of tolerance to salt stress. *Plant Physiol Biochem* 65: 61–66.

Jovanovic Z, Stanisavljevic N, Mikic A, Radovic S, Maksimovic V (2013) The expression of drought responsive element binding protein ('DREB2A') related gene from pea (*Pisum sativum* L.) as affected by water stress. *Aust J Crop Sci* 7: 1590–1596.

Kaur H, Verma P, Petla BP, Rao V, Saxena SC, Majee M (2013) Ectopic expression of the ABA-inducible dehydration-responsive chickpea l-myo-inositol 1-phosphate synthase 2 (CaMIPS2) in Arabidopsis enhances tolerance to salinity and dehydration stress. *Planta* 237: 321–335.

Kim GB, Nam YW (2012) A novel Δ1-pyrroline-5-carboxylate synthetase gene of *Medicago truncatula* plays a predominant role in stress-induced proline accumulation during symbiotic nitrogen fixation. *J Plant Physiol* 170: 291–302.

Krasensky J, Jonak C (2012) Drought, salt, and temperature stress-induced metabolic rearrangements and regulatory networks. *J Exp Bot* 63: 1593–1608.

Krauss KW, Ball MC (2013) On the halophytic nature of mangroves. *Trees* 27: 7–11.

Kumar JK, Prasad A, Richard SA (2012) In vitro antioxidant activity and preliminary phytochemical analysis of medicinal legumes. *J Pharma Res* 5: 3059–3062.

Kumar J, Srivastva E, Singh M (2013a) Genetics of early growth vigour in lentil (*Lens culinaris* Medik.). *J Genetics* 92: 323.

Kumar MN, Jane W-N, Verslues PE (2013b) Role of the putative osmosensor Arabidopsis histidine kinase1 in dehydration avoidance and low-water-potential response. *Plant Physiol* 161: 942–953.

Kwon KC, Verma D, Jin S, Singh ND, Daniell H (2013) Release of proteins from intact chloroplasts induced by reactive oxygen species during biotic and abiotic stress. *PloS One* 8(6): e67106.

Li J, Bao S, Zhang Y, *et al.* (2012) *Paxillus involutus* Strains MAJ and NAU mediate K^+/Na^+ homeostasis in ectomycorrhizal Populus × canescens under sodium chloride stress. *Plant Physiol* 159: 1771–1786.

Li Y-M, Elson M, Zhang D, *et al.* (2013) Physiological traits and metabolites of Cacao seedlings influenced by potassium in growth medium. *Am J Plant Sci* 4: 1074–1080.

Li Z, Jiang H, Zhou L, *et al.* (2014) Molecular evolution of the *HD-ZIP* I gene family in legume genomes. *Gene* 533: 218–228.

Liancourt P, Spence LA, Song DS, *et al.* (2013) Plant response to climate change varies with topography, interactions with neighbors, and ecotype. *Ecology* 94: 444–453.

Libault M (2013) Decipher plant gene regulatory networks using single plant cell types. *Int J Plant Biol Res.* 1: 1–3.

Lindberg S, Kader MA, Yemelyanov V (2012) Calcium signalling in plant cells under environmental stress. In: Ahmad P, Prasad MNV (eds), *Environmental Adaptations and Stress Tolerance of Plants in the Era of Climate Change*. Springer Science+Business Media, New York, pp. 325–360.

Lindemose S, O'Shea C, Jensen MK, Skriver K (2013) Structure, function and networks of transcription factors involved in abiotic stress responses. *Int J Mol Sci* 14: 5842–5878.

Lipiec J, Doussan C, Nosalewicz A, Kondracka K (2013) Effect of drought and heat stresses on plant growth and yield: a review. *Int Agrophys* 27: 463–477.

Liu M, Shi J, Lu C (2013) Identification of stress-responsive genes in *Ammopiptanthus mongolicus* using ESTs generated from cold- and drought-stressed seedlings. *BMC Plant Biol* 13: 88.

Liu Q, Qiao F, Ismail A, Chang X, Nick P (2013) The plant cytoskeleton controls regulatory volume increase. *Biochim Biophys Acta* 1828: 2111–2120.

Lopez-Gomez M, Tejera NA, Iribarne C, Herrera-Cervera JA, Lluch C (2012) Different strategies for salt tolerance in determined and indeterminate nodules of *Lotus japonicus* and *Medicago truncatula*. *Arch Agron Soil Sci* 58: 1061–1073.

Lyu JI, Min SR, Lee JH, *et al.* (2013) Overexpression of a trehalose-6-phosphate synthase/phosphatase fusion gene enhances tolerance and photosynthesis during drought and salt stress without growth aberrations in tomato. *Plant Cell Tissue Org Cult* 112: 257–262.

Mantri N, Patade V, Penna S, Ford R, Pang E (2012) Abiotic stress responses in plants: present and future. In: Ahmad P, Prasad MNV (eds), *Abiotic Stress Responses in Plants*. Springer Science+Business Media, New York, pp. 1–19.

Marino D, Dunand C, Puppo A, Pauly N (2012) A burst of plant NADPH oxidases. *Trends Plant Sci* 17: 9–15.

Miransari M (2012) Role of phytohormone signaling during stress. In: Ahmad P, Prasad MNV (eds), *Environmental Adaptations and Stress Tolerance of Plants in the Era of Climate Change*. Springer Science+Business Media, New York, pp. 381–393.

Miransari M (2013) Handling soybean (*Glycine max* L.) under stress. In Hakeem, KR, Ahmad P, Ozturk M (eds), *Crop Improvement: New Approaches and Modern Techniques*. Springer Science+Business Media, New York, pp. 421–439.

Mittler R (2006) Abiotic stress, the field environment and stress combination. *Trends Plant Sci* 11: 15–19.

Mizoi J, Ohori T, Moriwaki T, *et al.* (2013) GmDREB2A;2, a canonical dehydration-responsive element-binding protein2-type transcription factor in soybean, is posttranslationally regulated and mediates dehydration-responsive element-dependent gene expression. *Plant Physiol* 161: 346–361.

Molina C, Zaman-Allah M, Khan F, *et al.* (2011) The salt-responsive transcriptome of chickpea roots and nodules via deep Super SAGE. *BMC Plant Biol* 11: 31.

Morais MC, Panuccio MR, Muscolo A, Freitas H (2012) Does salt stress increase the ability of the exotic legume *Acacia longifolia* to compete with native legumes in sand dune ecosystems? *Environ Exp Bot* 82: 74–79.

Nakaminami K, Matsui A, Shinozaki K, Seki M (2012) RNA regulation in plant abiotic stress responses. *Biochim Biophys Acta (BBA) – Gene Regulatory Mechanisms (special issue)* 1819(2): 149–153.

Nakashima K, Takasaki H, Mizoi J, Shinozaki K, Yamaguchi-Shinozaki K (2012) NAC transcription factors in plant abiotic stress responses. *Biochim Biophys Acta (BBA) – Gene Regulatory Mechanisms (special issue)* 1819(2): 97–103.

Nayak SN, Balaji J, Upadhyaya HD, *et al.* (2009) Isolation and sequence analysis of DREB2A homologues in three cereal and two legume species. *Plant Sci* 177: 460–467.

Nilius B, Honoré E (2012) Sensing pressure with ion channels. *Trends Neurosci* 35: 477–486.

Nounjan N, Nghia PT, Theerakulpisut P (2012) Exogenous proline and trehalose promote recovery of rice seedlings from salt-stress and differentially modulate antioxidant enzymes and expression of related genes. *J Plant Physiol* 169: 596–604.

Osakabe Y, Yamaguchi-Shinozaki K, Shinozaki K, Tran L-SP (2013) Sensing the environment: key roles of membrane-localized kinases in plant perception and response to abiotic stress. *J Exp Bot* 64: 445–458.

Palma F, Tejera NA, Lluch C (2012) Nodule carbohydrate metabolism and polyols involvement in the response of *Medicago sativa* to salt stress. *Environ Exp Bot* 85: 43–49.

Pang T, Ye C-Y, Xia X, Yin W (2013) De novo sequencing and transcriptome analysis of the desert shrub, *Ammopiptanthus mongolicus*, during cold acclimation using Illumina/Solexa. *BMC Genomics* 14: 488.

Pervaiz A, Satyawati S (2008) Salt stress and phyto-biochemical responses of plants – a review. *Plant Soil Environ* 54: 89–99.

Patel D, Jha CK, Tank N, Saraf M (2012) Growth enhancement of chickpea in saline soils using plant growth-promoting rhizobacteria. *J Plant Growth Regul* 31: 53–62.

Petrussa E, Braidot E, Zancani M, *et al.* (2013) Plant flavonoids – biosynthesis, transport and involvement in stress responses. *Int J Mol Sci* 14: 14950–14973.

Pirzadah TB, Malik B, Rehman RU, Hakeem KR, Qureshi MI (2014) Signaling in response to cold stress. In: Hakeem KR, Rehman RU, Inayatullah T (eds), *Plant Signaling: Understanding the Molecular Crosstalk*. Springer Science+Business Media, New York, pp. 193–226.

Putten WH, Bardgett RD, Bever JD, *et al.* (2013) Plant-soil feedbacks: the past, the present and future challenges. *J Ecol* 101: 265–276.

Qadri TN, Hamid A, Qadri J, Azooz MM, Ahmad P (2013) Role of bio-fertilizers in crop improvement. In: Hakeem, KR, Ahmad P, Ozturk M (eds), *Crop Improvement: New Approaches and Modern Techniques*. Springer Science+Business Media, New York, pp. 189–208.

Rai A, Suprasanna P, D'Souza SF, Kumar V (2012) Membrane topology and predicted RNA-binding function of the 'Early Responsive to Dehydration (ERD4)' plant protein. *PloS One* 7(3): e32658.

Ram R, Maji C, Bindroo B (2013) Role of PGPR in different crops – an overview. *Ind J Sericult* 52: 1–13.

Ramana G, Padhy SP, Chaitanya K (2012) Differential responses of four soybean (*Glycine max* L.) cultivars to salinity stress. *Legume Res* 35: 185–193.

Rangan L (2013) Pongamia – A multipurpose versatile legume. *Res J Biotechnol* 8: 1–3.

Rasool S, Ahmad A, Siddiqi T, Ahmad P (2013) Changes in growth, lipid peroxidation and some key antioxidant enzymes in chickpea genotypes under salt stress. *Acta Physiol Plant* 35: 1039–1050.

Redillas MC, Park S-H, Lee JW, *et al.* (2012) Accumulation of trehalose increases soluble sugar contents in rice plants conferring tolerance to drought and salt stress. *Plant Biotechnol Rep* 6: 89–96.

Reyer CP, Leuzinger S, Rammig A, *et al.* (2013) A plant's perspective of extremes: terrestrial plant responses to changing climatic variability. *Global Change Biol* 19: 75–89.

Roy SJ, Tester M (2013) Increasing salinity tolerance crops. In: Christou P, Savin R, Costa-Pierce BA, Misztal I, Whitelaw CBA (eds), *Sustainable Food Production*. Springer, New York, pp. 986–1002.

Sabra A, Daayf F, Renault S (2012) Differential physiological and biochemical responses of three *Echinacea* species to salinity stress. *Sci Horticul* 135: 23–31.

Samineni S, Siddique KH, Gaur PM, Colmer TD (2011) Salt sensitivity of the vegetative and reproductive stages in chickpea (*Cicer arietinum* L.): Podding is a particularly sensitive stage. *Environ Exp Bot* 71: 260–268.

Saxena SC, Kaur H, Verma P, Petla BP, Andugula VR, Majee M (2013) Osmoprotectants: Potential for crop improvement under adverse conditions. In: Tuteja N, Gill SS (eds), *Plant*

Acclimation to Environmental Stress. Springer Science+Business Media, New York, pp. 197–232.

Shabala L, Mackay A, Tian Y, Jacobsen SE, Zhou D, Shabala S (2012) Oxidative stress protection and stomatal patterning as components of salinity tolerance mechanism in quinoa (*Chenopodium quinoa*). *Physiol Plant* 146: 26–38.

Shabala S, Munns R (2012) Salinity stress: physiological constraints and adaptive mechanisms In: Shabala S (ed.), *Plant Stress Physiology*. CAB International, Wallingford, UK, pp. 59–93.

Shavrukov Y (2013) Salt stress or salt shock: which genes are we studying? *J Exp Bot* 64: 119–127.

Signorelli S, Casaretto E, Sainz M, Díaz P, Monza J, Borsani O (2013) Antioxidant and photosystem II responses contribute to explain the drought-heat contrasting tolerance of two forage legumes. *Plant Physiol Biochem* 70: 195–203.

Sultana R, Choudhary A, Pal A, Saxena K, Prasad B, Singh R (2014) Abiotic stresses in major pulses: Current status and strategies. In: Gaur RK, Sharma P (eds), *Approaches to Plant Stress and Their Management*. Springer Science+Business Media, New York, pp. 173–190.

Suzuki N, Koussevitzky S, Mittler R, Miller G (2012) ROS and redox signalling in the response of plants to abiotic stress. *Plant Cell Environ* 35: 259–270.

Tahkokorpi M, Taulavuori E, Laine K, Taulavuori K (2012) Severe salt stress in *Vaccinium myrtillus* (L.) in response to Na^+ ion toxicity. *Environ Exp Bot* 76: 49–53.

Takahashi Y, Tateda C (2013) The functions of voltage-dependent anion channels in plants. *Apoptosis* 18: 917–924.

Tan WK, Lin Q, Lim TM, Kumar P, Loh CS (2013) Dynamic secretion changes in the salt glands of the mangrove tree species *Avicennia officinalis* in response to a changing saline environment. *Plant Cell Environ* 36: 1410–1422.

Tetsukazu Y, Firouzeh J, Yusuke O, *et al.* (2013) Global legume diversity assessment: Concepts, key indicators, and strategies. *Taxonomy* 62: 249–266.

Theocharis A, Clément C, Barka EA (2012) Physiological and molecular changes in plants grown at low temperatures. *Planta* 235: 1091–1105.

Thiel G, Moroni A, Blanc G, VanEtten JL (2013) Potassium ion channels: Could they have evolved from viruses? *Plant Physiol* 162: 1215–1224.

Turner NC, Colmer TD, Quealy J, *et al.* (2013) Salinity tolerance and ion accumulation in chickpea (*Cicer arietinum* L.) subjected to salt stress. *Plant Soil* 365: 347–361.

Tuteja N, Peter Singh L, Gill SS, Gill R, Tuteja R (2012) Salinity stress: a major constraint in crop production In: Tuteja N, Gill SS, Tiburcio AF, Tuteja R (eds), *Improving Crop Resistance to Abiotic Stress*, Vol. 1. John Wiley & Sons, Inc., pp. 71–96.

Waditee-Sirisattha R, Singh M, Kageyama H, Sittipol D, Rai AK, Takabe T (2012) Anabaena sp. PCC7120 transformed with glycine methylation genes from *Aphanothece halophytica* synthesized glycine betaine showing increased tolerance to salt. *Arch Microbiol* 194: 909–914.

Wang H, Zhou L, Fu Y, *et al.* (2012) Expression of an apoplast-localized BURP-domain protein from soybean (GmRD22) enhances tolerance towards abiotic stress. *Plant Cell Environ* 35: 1932–1947.

Wang TZ, Zhang JL, Tian QY, Zhao MG, Zhang WH (2013) A *Medicago truncatula* EF-H and family gene, MtCaMP1, is involved in drought and salt stress tolerance. *PloS One* 8(4): e58952.

Wdowiak-Wróbel S, Leszcz A, Małek W (2013) Salt tolerance in *Astragalus cicer* microsymbionts: The role of glycinebetaine in osmoprotection. *Curr Microbiol* 66: 428–436.

Yang L, Ma C, Wang L, Chen S, Li H (2012) Salt stress induced proteome and transcriptome changes in sugar beet monosomic addition line M14. *J Plant Physiol* 169: 839–850.

Yang Y, Zheng Q, Liu M, Long X, Liu Z, Shen Q, Guo S (2012) Difference in sodium spatial distribution in the shoot of two canola cultivars under saline stress. *Plant Cell Physiol* 53: 1083–1092.

Yang Z, Yu J, Merewitz E, Huang B (2012c) Differential effects of abscisic acid and glycinebetaine on physiological responses to drought and salinity stress for two perennial grass species. *J Am Soc Horticul Sci* 137: 96–106.

Yao M, Zeng Y, Liu L, Huang Y, Zhao E, Zhang F (2012) Overexpression of the halophyte Kalidium foliatum H^+-pyrophosphatase gene confers salt and drought tolerance in Arabidopsis thaliana. *Mol Biol Rep* 39: 7989–7996.

Yu S, Wang W, Wang B (2012) Recent progress of salinity tolerance research in plants. *Russ J Genetics* 48: 497–505.

Zelinová V, Mistrík I, Pavlovkin J, Tamás L (2013) Glutathione peroxidase expression and activity in barley root tip after short-term treatment with cadmium, hydrogen peroxide and t-butyl hydroperoxide. *Protoplasma* 250: 1057–1065.

CHAPTER 13

Omics approaches and abiotic stress tolerance in legumes

Syeda Fatma Hasan Bukhari[1], Sadia Arshad[1], Mohamed Mahgoub Azooz[2] and Alvina Gul Kazi[1]

[1] *Atta-ur-Rahman School of Applied Biosciences, National University of Sciences and Technology (NUST), Islamabad, Pakistan*

[2] *Department of Botany, Faculty of Science, South Valley University, Qena, Egypt*

13.1 Introduction

Legumes characteristically have an unusual flower structure; their fruit is podded and approximately 88% of the known species have the ability to form rhizobial nodules (Udvardi *et* & Poole, 2013), i.e. an association of the roots with symbiotic diazotrophs (rhizobia). Both organisms benefit from this association. The rhizobia are provided with energy in the form of amino acids by the plant whereas the rhizobia in turn fix nitrogen from the atmosphere, which is taken up by the plant. This association is extremely valuable as the reduction of atmospheric dinitrogen into ammonia is second only to photosynthesis as the most important biological process on earth (Sylvia *et al.*, 2005). Biological fixation of nitrogen is also the leading form of global annual nitrogen input (Russelle, 2008).

Legumes are a major part of the human diet globally and closely follow the Gramineae in their importance to humans. The main uses of legumes are as food grains/ seeds, livestock forage, silage and green manure (Jensen *et al.*, 2012). The well-known legumes include clover, alfalfa, beans, pea, lentils, mesquite, soybeans, peanuts (groundnuts), lupins, wisteria and tamarind. Legumes, especially soybean and peanut, are predominantly used for the production of vegetable oil. The global harvested area under legumes is around one tenth of the cereal crops, and a high proportion of the legumes are grown in rainfed, low-input systems as compared to cereal crops. Thus, the average global yield of pulses is only about one-quarter the average yield of cereal crops (Akibode & Maredia, 2011).

Given their great economic and commercial importance, it is unfortunate that legumes have not been given as much priority as cereals when improving crop yield. This could partly be due to the abiotic stresses under which legumes are often cultivated. The major abiotic stresses are high temperature, salinity, water stress, acidity, sodicity and nutrient deficiencies (Reddy *et al.*, 2012; Mantri *et al.*, 2013; Varshney & Kudapa, 2013). Such stresses not only greatly decrease the crop yield and growth but also negatively affect the formation and function of the symbiotic relationship between legumes and rhizobia (Chalk *et al.*, 2010).

Temperature has a marked effect on the plant's survival, growth and formation of root nodules. Very high temperatures can therefore lead to reduced crop yield. Drought is another abiotic stress factor that has dire consequences on crop yield. The growth of shoots and roots is greatly decreased in high temperature – so much so that the roots may actually die (Brisson *et al.*, 2010). Moreover, plants can suffer poor pollination if the temperature remains high. The effect of drought is largely dependent on the duration and magnitude of the stress to which the plant is exposed. Drought conditions hinder legume symbiotic performance, retard growth and lead to extremely poor crop yields (Arrese-Igor *et al.*, 2011). In a study carried out by Basu *et al.* (2007), it was shown that under conditions of drought stress, eight cultivars of peas underwent starch glucose augmentation of hexose phosphates. Also, a decrease in leaf photosynthesis was observed.

Legumes under Environmental Stress: Yield, Improvement and Adaptations, First Edition. Edited by Mohamed Mahgoub Azooz and Parvaiz Ahmad.

Salinity is also an abiotic stress factor that is detrimental to legume survival and rhizobium–legume interactions, and so is soil sodicity (Chalk *et al.*, 2010; Shabala & Munns, 2012). Sodium chloride is also known to increase levels of abscisic acid (ABA) in the transpiring portion of a growing leaf. A six-fold increase in ABA was noticed in the distal portion of the leaf elongation zone within 10 minutes of addition of salt, which later accumulated in the proximal portion. The ABA increased three-fold in the growing blade region, and stayed so for 20 minutes. This preceded a notable decrease in both transpiration and stomatal conductivity (Poltronieri *et al.*, 2011). An ATP binding cassette (ABC) transporter, AtABCG25, expressed mainly in the roots and leaf veins and located on the outer cell membrane (Kuromori *et al.*, 2010), was identified to be the main player in hypersensitivity towards ABA in germination and seedling stages in a plant mutant. Plants that were genetically engineered to upregulate the expression of ABA were shown to experience decreased water loss from isolated leaves and also had a higher leaf temperature as compared to normal plants.

Biotechnological approaches have many advantages over conventional crop breeding. A number of applications were identified by Drehr *et al.* (2003) where marker-aided techniques offered substantial cost saving as compared to conventional methods. This is because conventional methods require the breeder to rely on observed differences that give a clue to the underlying gene (Bruce, 2012). This not only is extremely time consuming, but ever since the advent of molecular diagnostic screening, is also inefficient. Biotechnology-based tools allow the quick, reliable and efficient identification of desired genes. Furthermore, given that biotechnological tools reduce the breeding time as compared to conventional techniques, it can also be expected that new varieties would be developed and introduced to the market at a much faster pace (Farre *et al.*, 2010; Anthony & Ferroni, 2012).

Conventional breeding methods and improved management strategies may help alleviate some constraints, but it is of extreme importance that biotechnological approaches be applied in order to tackle these stresses and help sustain agriculture globally.

13.2 Omics: Solutions to abiotic stress in legumes?

Biotechnological approaches like tissue culture and genetic transformation can help tackle hindrances that cannot be addressed through classical plant breeding due to sexual incompatibility or limited natural sources of resistance (Urano *et al.*, 2010; Syrenne *et al.*, 2012). The use of biotechnology may seem to be an excellent solution, but it requires deep knowledge of the target plant (legumes in this case) and the mechanisms that are stimulated or repressed on exposure to stress. The response to stress should be identifiable from the plant level all the way down to the molecular level. Three properties of the stress determine the intensity of the plant's response: type, length and severity. The plant's response to any stress is highly governed by its genome; therefore efforts are being made to understand the responses at a molecular level (Bray, 2004). Until recently research was focused on model plants, notably *Arabidopsis*, instead of legume plants due to the countless advantages of the former. Its growth, development, flowering and seed production characteristics are similar to those of higher plants, including many legumes. Furthermore, it has a very short generation time, with seeds germinating and maturing to form more seeds within only 6 weeks. The most beneficial property of this model plant is its small genome, only 125 Mb, which allows easy manipulation (Koorneef & Meinke, 2010). Since there are many similarities between the model plant and legumes, the knowledge obtained about stress responses has been applied to legume research as well. Nevertheless, the physiological differences between the two plants cannot be ignored.

Legumes have a large genome size, and some also show polyploidy (Anderson *et al.*, 2005). This hampers research; therefore, in order to understand legume biology better, model systems have been developed to study the genetics behind nodule formation and other vital processes like resistance, potential tolerance and susceptibility to various stresses. Much of the research on legumes is carried out on model plant systems, most commonly *Medicago truncatula* and *Lotus japonicus*, because of their small, diploid genome, short life cycle, autogamous nature and abundant seed production (Handberg & Stougaard, 1992; Cook, 1999). The use of these models has greatly increased the genetic as well as

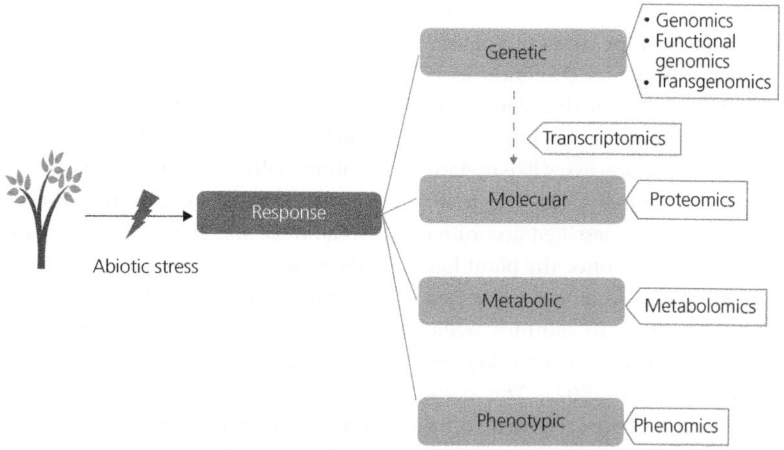

Figure 13.1

genomic data available on legumes, and the synteny (similarity) present between various legume genomes has further increased the applicability of the data available (Kalo *et al.*, 2004; Stracke *et al.*, 2004). These factors make the two models extremely valuable for the study of molecular and genetic responses of legumes to biotic as well as abiotic stresses that hinder the yield of food legume crops. Moreover, the sequencing of the soybean genome and its high similarity with the model legume systems can potentially facilitate various genetic procedures for these studies, such as positional cloning. These factors make the two models extremely valuable for the study of molecular and genetic responses of legumes to biotic as well as abiotic stresses that hinder the yield of food legume crops.

Even though genome sequence information is of great value and is a very important starting point, it is not enough to decipher gene function, and the various metabolic responses and regulatory pathways that are activated when the legume is under stress conditions. In order to circumvent these obstacles, approaches like the quantitative and qualitative analysis of the products of gene expression are very important. These need to be carried out at the transcriptomic level as well as the proteomic and metabolomic levels. Such expansive knowledge will also facilitate a more targeted use of transgenics and marker-assisted selection (Dita *et al.*, 2006). The 'omics' comprise of genomics, transcriptomics, proteomics, metabolomics, functional genomics, transgenomics and phenomics. Recent developments in these various omics have substantially contributed to greater comprehension of the molecular and genetic levels of stress responses (Langridge & Fleury, 2011), which were previously a bottleneck for transgenic breeding and genetic modification of legumes. Significant progress has been made in this regard on major legumes like soybean, peanut and chickpea, which will be discussed in the chapter (Figure 13.1).

13.3 Transcriptomics

The expression of any gene in a certain cell can only be carried out once the gene has been transcribed. The link between a gene and its translation into the protein product is messenger RNA (mRNA), which is a result of transcription. The set of all the mRNAs in a cell is known as its transcriptome. Thus the investigation of various populations of mRNAs in a cell, tissue or organism is known as transcriptomics.

A very powerful tool for the study of gene expression, transcriptomics gives an accurate representation of the qualitative and quantitative gene expression of a particular cell or tissue. Hence it can help identify candidate genes and reveal the crosstalk between different gene regulatory pathways that take part in various responses, as in a plant's response to abiotic stress (Palovaara *et al.*, 2013). Plants under abiotic stresses usually face dehydration at the cellular level, and the cellular water deficit mostly triggers genes normally involved in responses against drought or salinity. The cellular water deficit is the determining factor of a plant's response to a particular environmental condition as it triggers changes in the gene expression that define the plant's response (Kosová *et al.*, 2011). The induced or repressed genes impact many plant processes like metabolism, signalling, transport, transcription, etc. The genes that respond to abiotic stress can be classified according to the time they take to be expressed once the plant has been exposed to the abiotic stress: some are expressed immediately in a matter of seconds to minutes while others take longer to respond, from hours to days or even weeks (Ramanjulu & Bartels, 2002). The early response genes could be taken as genes responsible for initial protection and the regulation of other genes involved, including the genes encoding protein kinases and transcription factors (Opdenakker *et al.*, 2012). The genes that respond to stress, on the other hand, may play a role in facilitating the plant's adaptation to the stress, such as the genes encoding proteins like heat-shock proteins, ROS scavenger proteins, etc. (Ramanjulu & Bartels, 2002).

Several methods can potentially be used for the profiling of gene expression, including a variety of microarrays, macroarrays and differential display, which could then be validated by real-time quantitative polymerase chain reaction (qPCR) (Butte, 2002). Transcriptomic analysis can be used to screen for candidate genes that play a role in the legume's response, which can then be used in legume improvement (Garg *et al.*, 2011; Jogaiah *et al.*, 2012). Determination of large-scale transcriptome profiles can be carried out with methods like serial analysis of gene expression (SAGE), DNA microarrays or the fairly recent technique of digital gene expression (DGE), which employs next generation sequencing (NGS)-based tools like RNA-sequencing (RNA-seq) (Cramer *et al.*, 2011). The NGS approach allows the cell transcripts to be deciphered at the sequence level. This has the potential to revolutionize research on legumes, which are now undergoing sequencing of their genomes. The aforementioned approach can tackle the problems presented by the massive genomes of legumes; moreover, NGS has also circumvented issues like compromised sequence accuracy due to short sequence lengths, sequencing errors characteristic of the methods used, and also the absence of physical clones. NGS-based methods can prove to be cost-efficient platforms that can be used to predict the responses of legumes under certain abiotic stress conditions, thus allowing assessment of the legume's genome even at the functional level (Ohtsu *et al.*, 2007).

13.3.1 Chickpea

Chickpea (*Cicer arietinum*) is globally ranked third of the commercially important legume crops. Drought is a major abiotic stress affecting the yield of this legume (Jukanti *et al.*, 2012; Thudi, 2013a,b). Molina *et al.* (2008) used SuperSAGE (a modified version of SAGE) to analyse the response of chickpea roots to drought. In order to do so, 8,023,826 base pair tags were sequenced. These tags represented 17,493 unique transcripts (UniTags) from roots exposed to drought and those kept under control conditions. A total of 43% of the UniTags (7532 in number) were more than 2.7-fold differentially expressed, and the regulation of 5.0% (880) of tags increased eight-fold when exposed to stress. The large size of the tags made possible the unambiguous annotation of 22% (3858) UniTags when entered into public databases. This study comprehensively showed that multiple processes including signal transduction, transcriptional regulation, accumulation of osmolytes and also ROS scavenging experience extensive transcriptional remodelling in the roots as an early response to stress imposed by drought conditions, hence suggesting potential targets for improvement of drought tolerance in chickpea through breeding.

Another study (De Domenico *et al.*, 2012) aimed to analyse the effect of the phyto-oxylipin pathway on the chickpea response to drought conditions. In this comparative study, the key genes involved were assessed for their expression through analysis by qPCR using commercially available TaqMan assays. Samples were taken from the roots of stressed and non-stressed, drought-sensitive and drought-tolerant chickpea. It was concluded that the drought-tolerant chickpea reacted to the stress with an earlier and sustained activation of the following genes: Mt-*LOX* 1, Mt-*HPL* 1, Mt-*HPL* 2, Mt-*AOS* and Mt-*OPR*. It was also shown that the overexpression of these genes was positively correlated with levels of the major oxylipin metabolites from the AOS branch of the phyto-oxylipin pathway, ultimately leading to jasmonate synthesis. Higher levels of jasmonic acid (JA), its precursor 12-oxophytodienoic acid and JA-isoleucine were specifically found in the root tissue of the chickpea variety that showed tolerance, hence leading to the assumption that jasmonates play a vital role in the early signalling response of the legume and also are involved in tolerance mechanism in the drought-tolerant variety.

Varshney *et al.* (2009) undertook a study encompassing transcriptomics with the objective of collating as many salinity-responsive and drought-responsive ESTs of chickpea as possible. Drought stress was induced by various treatments including polyethylene glycol (PEG) induction, slow drought stress in both greenhouse and field conditions, and dehydration, whereas salt stress was instigated by the use of 80 mM NaCl. This study produced a total number of 20,162 ESTs, 4558 drought-responsive unigenes, 2595 salinity-responsive unigenes and 9569 chickpea unigenes. This was the first work to generate such a large number of ESTs in chickpea.

The effects of salt stress on the transcriptional responses involving transcription factor (TF) genes in the roots of chickpea were investigated by Gruber *et al.* (2009). Two complementary transcriptomic approaches, massive quantitative real-time PCR (qRT-PCR) and Mt16K+microarray analysis, were used to identify the salt-regulated TF genes. The PCR technique revealed 46 such genes, whereas the Mt16K+microarray analysis identified 824 salt-regulated genes, of which 84 were TF sequences. These genes showed significant changes in their expression level ($p<0.001$) in root apexes after being exposed to salinity. Analysis of stress regulation in the root apex compared with whole roots revealed several TF genes with more than 30-fold differences in expression, including certain members of the MYB, AP2/EREBP and HD-ZIP TF families.

13.3.2 Soybean

Soybean is the most intensively studied grain legume and over 1.3 million ESTs have been developed from different cDNA libraries according to Legume Information System data. This is more ESTs than any other grain legume. The discovery of new simple sequence repeat (SSR) and single nucleotide polymorphism (SNP) markers for soybeans becomes possible due to this large number of EST and bacterial artificial chromosome (BAC) sequences, which in turn facilitate the development of genetic maps with high resolution. Moreover, using a modification of the complementary DNA-amplified fragment length polymorphism (cDNA-AFLP) technique in soybean, 140 differentially expressed fragments of cDNA were observed when comparing iso-osmotic treated soybean and control soybean. Most of these genes were involved encoding transcription factors and ion transporters as well as redox enzymes (Umezawa *et al.*, 2002).

Abiotic stress can affect a legume in various ways. For instance, the symbiotic relationship between a legume and rhizobia, as well as nitrogen fixation in the root nodules, can be greatly affected by salt stress. Dong *et al.* (2013) studied the molecular mechanisms involved in regulating tolerance to salinity of functional nodules in order to achieve genetic improvement in soybean, especially in its nitrogen fixation efficiency. Deep sequencing of microRNAs (miRNAs) was carried out to decipher the miRNA expression profile in mature nodules of soybean under salt-induced stress and those under control conditions. In the study, 104 miRNAs were found to be dramatically differentially expressed during salt stress conditions. The role of eight of these miRNAs in the salt stress response was confirmed by qRT-PCR analysis as they were dynamically regulated in the functional soybean nodules in response to the salt stress. Hence the study showed that miRNAs hold great potential as players in the response of soybean nodules during salt stress.

In another study conducted by Komatsu *et al.* (2009), transcriptomic techniques were used to analyse genes induced by flooding stress. Germination of soybeans was carried out for 2 days and flooding conditions were sustained for 12 hours, following which the RNAs were extracted from the hypocotyl and the root. The transcriptomic technique, high-coverage gene expression profiling analysis, yielded 29,388 peaks. Ninety-seven of these peaks demonstrated more than a 25-fold change after 12-hour flood stress. Genes involved in ethylene biosynthesis, alcohol fermentation, cell wall loosening and pathogen defence were significantly upregulated. Also, the genes of acid phosphatase, haemoglobin and Kunitz trypsin protease inhibitor were greatly altered. This indicates that the early responses of soybean after flooding stress are potentially important in ensuring cell survival after direct damage by water.

13.4 Proteomics

Proteomics is a biotechnological approach that involves studying the proteome of an organism, i.e. the complete profile of the expressed proteins in a given cell, tissue, organ or complete organism (Memon, 2012; Lee *et al.*, 2013). It can be associated with two kinds of studies, the first being the identification of the proteome profile of a given cell, tissue, organ, organism or population, and the second being differential proteomics in which the proteome of an organism under study conditions is compared with that of the organism under control conditions. Since the proteome mirrors the actual state or response of a cell, it is the bridge between the transcriptome and the metabolome. The proteins expressed are key players in the biochemical processes, which implies that they must be closer to the phenotypic expression than DNA markers. The responses of plants to stress have mostly been studied at the DNA and RNA levels, and though this has provided great insights to improve the understanding of stress tolerance, the proteomics approach can paint a picture to evaluate the response. The mRNA levels are not always reflective of the protein expression because many proteins undergo post-translational modifications like glycosylation, phosphorylation and ubiquitinylation, which have major influences on protein function (Jorrin *et al.*, 2006). The ability of a plant to cope with a variety of abiotic stresses therefore greatly depends on the fluctuation in the number of proteins, which are either upregulated or downregulated due to alterations in gene expression. These proteins are either key players in signal transduction pathways or help in the plant's adaptation processes, like stress repair mechanisms. They thus help the plant to recover from the stress and aid in its survival (Hakeem *et al.*, 2012).

Two laboratory techniques are widely used for proteomics studies: protein electrophoresis and mass spectrometry. Differential proteomics has been extensively applied study the responses of legumes to several abiotic stresses.

13.4.1 Chickpea

A proteomics approach was used in a study to identify the dehydration-responsive proteins in the extracellular matrix (ECM) proteins of chickpea (Bhushan *et al.*, 2007). The dehydration tolerance of several commercial chickpea varieties was screened using various biochemical and physiological indexes. The temporal changes of ECM proteins after dehydration in JG-62, a tolerant chickpea variety, showed 186 proteins with a variance at a 95% significance level. Upon comparative proteomics analysis, 134 differentially expressed proteins were identified. It was demonstrated in this study that more than 100 ECM proteins, which are also involved in processes like cell wall modification, metabolism, signal transduction, cell defence and cell rescue, actively participate in the molecular mechanisms responsible for tolerance to dehydration in chickpea plants.

In another study, Pandey *et al.* (2008) observed the changes in the nuclear proteome profile in 3-week-old chickpea seedlings. These seedlings were progressively dehydrated by withdrawing water and the changes in the nuclear proteome were examined by two-dimensional gel electrophoresis. The number of differentially regulated proteins was approximately 205, and mass spectrometry analysis identified 147 differentially expressed proteins. These proteins are presumably involved in various cellular functions including gene transcription, DNA replication, cell signalling pathways and chromatin remodelling as well as acting as molecular chaperones. The chickpea nuclear proteome, as a result of the dehydration exposure, revealed a coordinated response, involving both functional and regulatory proteins. Hence this study provided an insight into the complexity of the metabolic network that operates in the nucleus during dehydration.

13.4.2 Soybean

Swigonska and Weidner (2013) studied the effects of abiotic stress on the germination of soybean seeds. Germination in soybean greatly depends on environmental factors like temperature and water availability. This also makes them highly sensitive to fluctuating or harsh environmental conditions; prolonged exposure to such abiotic stresses can not only greatly diminish the crop yield but also cause poor or delayed germination. Proteomics analysis was employed in this study to observe the effects of osmotic and cold stress on the germinated soybean seeds' roots. The seeds were germinated under one of three continuous conditions: osmotic stress (+25 °C/−0.2 MPa), cold stress (+10 °C/H_2O), and a combination of cold and osmotic stress (+10 °C/−0.2 MPa). The proteome map that was constructed for the control samples and stress-treated samples showed 1272 spots. In total, 59 proteins were present in stress-treated samples as well as in the control samples, and when divided into functional categories, the following breakdown was achieved: nine in plant defence, 10 in various metabolic pathways of carbohydrates and eight in storage. Furthermore, quite a few proteins were also involved in electron transport, protein synthesis, signal transduction, secondary metabolism, embryogenesis, development, cellular transport, cellular translocation and storage. Upon analysis of the differences in the expression

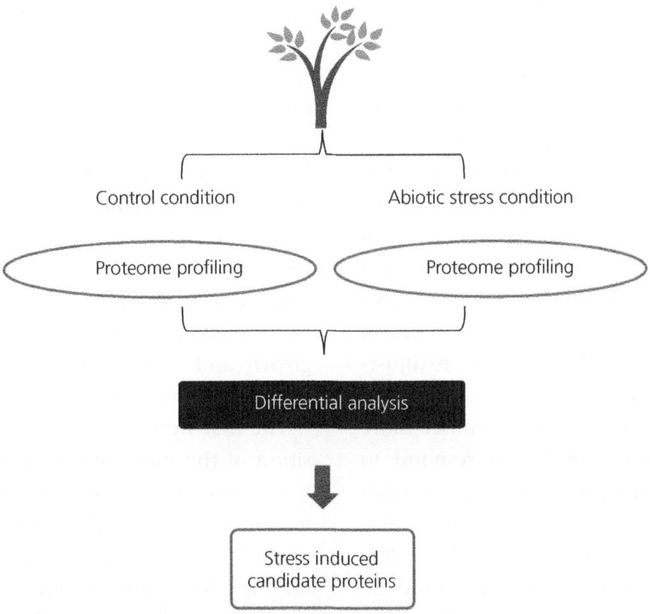

Figure 13.2

patterns, the long-term abiotic stress-induced response in soybean was traced and the similarities along with the differences between the responses to osmotic stress and cold stress were distinguished.

The effects of flooding on the cultivation of soybean have also been studied by employing proteomics techniques (Komatsu *et al.*, 2012; Ohyanagi *et al.*, 2012). Flooding is highly detrimental as it causes a marked reduction in growth and grain yield of soybean. Komatsu and colleagues used two proteomics techniques to investigate the extent of alteration in the endoplasmic reticulum (ER)-enriched fraction of soybean exposed to the stress of flood. Two-day-old soybean seedlings were exposed to flood stress for 2 days, after which the purified rough ER-enriched fraction was obtained from the root tips. The flooding-responsive proteins of this fraction were then identified using gel-free as well as 1D-gel-based proteomics techniques. Comparative proteomics analysis yielded an increase of 117 proteins and a decrease of 212 proteins in the soybean root tips exposed to flooding conditions. A total of 111 of these proteins were found to be involved in the synthesis of proteins, post-translational modification, protein folding, protein activation and protein degradation. Analysis of the mRNA expression levels of 14 of the differentially regulated proteins thought to be localized in the ER was carried out. Under 1 day of the flooding condition, 3-ketoacyl-CoA reductase 1 up-regulation was observed, whereas the downregulation of eight genes related to hormone metabolism, cell wall repair and DNA repair was observed. Specific induction of the expression of luminal-binding protein 5 also occurred in these roots, while methyltransferase PMT 2 along with arabinogalactan protein 2 were found to be downregulated. The results showed that flooding significantly affects the profile and level of protein synthesis and also glycosylation in the endoplasmic reticulum in the root tips of the soybean plants exposed to flooding stress.

These studies show that a proteomics approach can be used not only to differentially analyse the proteome of legumes under various abiotic stress conditions as well as control conditions, but also to identify candidate proteins that can help decipher the stress-responsive mechanisms of commercially important legumes (Figure 13.2).

13.5 Metabolomics

Metabolites are a reflection of the expression of genes, the interaction of proteins and various other processes responsible for gene regulation; they are closer to the phenotype of an organism than mRNA or even the proteins (Wienkoop *et al.*, 2011). Out of all the 'omics' technologies, metabolomics has proven to be the most transversal and can also be used for studying a number of different organisms without requiring any major modifications (Arbona *et al.*, 2013). The molecular phenotypes of plants, after exposure to abiotic stress, have been studied by the application of metabolomics, with the aim to discover particular patterns related to stress tolerance. The crucial role of primary metabolites like sugars, amino acids and Krebs cycle intermediates in photosynthetic dysfunction and as osmotic readjustment effectors was highlighted in these studies. It was also found that secondary metabolites are rather more specific to species and genera and thus respond to particular stresses by generating reactive oxygen species (ROS) scavengers, antioxidants, etc. They may also serve as regulatory molecules (Arbona *et al.*, 2013). It is also speculated that secondary metabolites could be induced by several abiotic stresses, thus effectively becoming a mechanism of cross-protection against abiotic stresses. Metabolomics can also be used to screen for the presence or absence of certain metabolites and also for assessing gene expression in order to provide accurate markers for the selection of stress-tolerant plants in breeding programmes.

The metabolomics approach has the potential to bridge the distance between the genotype and phenotype of an organism. Metabolic changes play a great role in plant development and responses induced by stress, hence metabolic information can accurately be used as a reflection of biological endpoints. This information can give a more precise analysis when integrated with the transcript and protein analysis. Thus the only way to fully understand a plant's response to stress at the gene expression and molecular levels would be through the integration of transcriptomics, proteomics and metabolomics (Dixon, 2001).

The metabolomics approach has also been applied to the study of legumes (Olivares *et al.*, 2011); one example is its use to determine the responses of *M. truncatula* suspension cells to various stimuli (Bell *et al.*, 2001). Although a comprehensive metabolomic approach is challenging for large-scale studies, quite a few targeted analyses have been carried out to assess the level of involvement of various subsets of metabolites in a number of stress-induced responses. Although combining the metabolomics approach with transgenetics seems to have the potential to increase the level of intrinsic resistance against various stresses and also to strengthen the role of biotechnology in the improvement of plants, emphasis should be placed on the fact that metabolic pathways never function alone and are usually part of highly complex networks (Wu & Van Etten, 2004). This means that interfering with one metabolic pathway could have a negative impact on others, resulting in concomitant detrimental traits in the modified plant. Hence large-scale analysis is extremely important to determine the metabolic networks that are responsible for the growth and development of a plant under various environmental conditions.

The first step of metabolomics experiments is acquisition of the metabolic fingerprints or profile by using a number of analytical instruments and separation techniques based on the physicochemical characteristics of each of the metabolites (Jogaiah *et al.*, 2012). Since no single technology is able to detect all the metabolites present in an organism, the use of a combination of techniques is required for studying the metabolome. These techniques include gas chromatography (GC), liquid chromatography (LC), nuclear magnetic resonance (NMR), capillary electrophoresis (CE) and mass spectrometry (MS).

13.5.1 Common bean

The common bean (*Phaseolus vulgaris* L.) is one of the most commercially important legume crops as far as human consumption is concerned. One of the abiotic stresses it usually faces is low phosphorus level in the soil (Broughton *et al.*, 2003; Singh & Schwartz, 2010). A non-biased metabolomics approach, coupled with transcriptome profiling, was used to assess the dynamics of gene expression and the overall metabolism that occurred in common bean plants kept under P-deficient conditions (Hernández *et al.*, 2007). A sum of 81 metabolites were detected, out of which 42 had been differentially expressed in the P-deficient bean plants. Stress-related metabolites like polyols were also identified in P-deficient plants' roots, thus strengthening speculation of a function for these metabolites in P stress.

13.5.2 *Lotus japonicus* (model legume)

The model legume *Lotus japonicus* has been used in studies focusing on drought stress. Through a non-targeted metabolomic approach, attempts have been made to study the response to drought conditions of model as well as forage legume species belonging to the *Lotus* genus (Pal'ove-Balang *et al.*, 2013). Increased water stress in the model legume gradually increased levels of most of the smaller molecules, indicating a progressive, probably global, reprogramming of the metabolic pathways. Unique metabolic responses to drought were also revealed among different *Lotus* species, which were highly conserved, by comparing their metabolomic profiles. It is important to note that not all pathways were conserved among all species. Hence a potent obstacle to translational approaches was highlighted – these approaches target the development of traits that are accompanied by an accumulation of compatible solute molecules. Also, a comparison between the metabolic changes induced by drought or salty conditions showed that the metabolic stress responses studied within each *Lotus* species were partially conserved. However, only a few drought- and salt-responsive metabolites were common to all of them, including sucrose synthase involved in nodule formation. The consideration of these findings can greatly aid in further understanding water stress physiology of legumes.

13.6 Genomics

Genomics is the study of the structure, function, evolution and mapping of an organism's genome. Hence it involves intensive efforts to determine the complete DNA sequence and fine-scale genetic mapping of the genome of an organism (Marko, 2013). The research carried out on a single gene cannot be classified as a genomics study unless the aim of the information gathered regarding genetic pathways and functional mechanisms is to decipher the gene's effect on the entire genome structure.

The first legume nuclear genome sequence appeared only fairly recently, and was undoubtedly fuelled by the emergence of NGS in the latter half of the 2000s. Owing to this, there is now a large trove of genomic and transcriptomic data driving a revolution in the world of biology similar in significance to the previous molecular revolution.

Genome sequencing of the model legumes *Medicago trunculata* and *Lotus japonicus* along with *Glycine max* (soybean) has been completed. Comparison of these legume genomes has revealed a key role in crop improvement and stress adaptation (Young & Bharti, 2012). Translating the genomic data from sequenced species to less researched but nonetheless important 'orphan' legumes could enhance the possibilities for tackling abiotic stress issues.

The growing amount of sequence information when combined with experimental genomics can revolutionize the way cellular processes and cells are studied. However, an important question faced by legume researchers is how advantage should be taken of this 'next generation' opportunity. The fact that a massive amount of data is now available does not mean the potential exists to solve all legume problems; 'mega data' cannot be a universal remedy (Doyle, 2013). The analysis of this data alongside its incorporation into the other omics approaches may prove to be a potent tool in tackling the abiotic stress susceptibility of legumes.

The genomics approach involves developing molecular markers so that the analysis of genetic diversity can be carried out. This also enables the manipulation of quantitative trait loci (QTL) through application of marker-assisted selection (MAS) and hence improvement of the cultivars. DNA regions, also referred to as molecular markers, can be identified that are strongly linked to genes related to stress responses and hence facilitate the development of modified breeding strategies (Yan *et al.*, 2004). These strategies can then be employed to enhance crop improvement, especially as far as stress tolerance is concerned.

13.6.1 Chickpea

The genomics approach can be applied to develop climate-resilient chickpea crops. High-throughput genotyping technologies along with modern breeding technologies (e.g. MAS and genomic selection) can aid in the cause. The efforts of the International Crops Research Institute for the Semi-Arid Tropics (ICRISAT) and partner institutes have yielded a great quantity of genomic resources for the chickpea crop, including single nucleotide polymorphism (SNP) markers, simple sequence repeats (SSRs), high-density genetic maps, diversity array technology (DArT), genome sequences and transcriptome assemblies. These resources have been used on various mapping populations to identify molecular markers that play a role in tolerance to salinity, drought and heat in chickpea. These approaches have been used to enhance stress tolerance, and in some cases, lines with up to 22% increase in yield have been developed (Varshney & Kudapa, 2013).

Marker-assisted selection has been deployed at ICRISAT in order to introgress the QTL alleles in chickpea that are associated with large root size (Saxena *et al.*, 2002). Drought has a dire effect on the yield of the crop, hence this approach could facilitate the development of a crop line having a deep root system, thus having a positive impact on the yield.

13.6.2 Soybean

In a study conducted by Li *et al.* (2005), the isolation of the DREB homologue genes, *GmDREBa*, *GmDREBb* and *GmDREBc*, from soybean was carried out. The protein deduced from each of the genes was found to contain an AP2 domain consisting of 64 amino acids. It was revealed by yeast one-hybrid assay that all three of the dehydration-responsive element binding proteins bind specifically to the dehydration-responsive element. Upon analysing these proteins with respect to their transcriptional activation abilities, it was shown that GmDREBa and GmDREBb could lead to the activation of a reporter gene, whereas GmDREBc was unable to do so. Salt, cold and drought stresses induced the transcription of *GmDREBa* and *GmDREBb* in the leaves of soybean seedlings. There was no significant change in the transcription of *GmDREBc* in leaves but its transcription in roots was probably induced by drought, salt and abscisic acid treatment. In the light of these results, it can be suggested that these three genes participate in the response of soybean to abiotic stress.

In another study, the genes for carotenoid oxygenase in soybean were identified to assess their phylogenetic relationships as well as their transcriptional response to four abiotic stresses: NaCl, PEG, and high and low temperature (Wang *et al.*, 2013). Their response to abscisic acid (ABA) treatment was also studied. All these responses were investigated to characterize the potential role they could play in a plant's resistance against these stresses. An adaptive evolution was indicated in the clades studied as the branches favoured the selection of *NCED* (9-*cis*-epoxycarotenoid oxygenase), carotenoid cleavage dioxygenase (*CCD1*) and *CCD8*. Eight genes for carotenoid oxygenase were identified in soybean. On assessment with qPCR, almost all of these genes showed significant alteration under ABA treatment and stress. The most notable changes were noticed in *CCD1* and *CCD4*. These genes are also known for the key role they play in norisoprenoid metabolism. Moreover, the transcription levels of two genes that play a major role in the strigolactone pathway (*CCD7* and *CCD8*) were notably increased when treated with ABA. This provides further proof that ABA is somehow involved in the regulation of strigolactone metabolism. Hence it was deduced that all the carotenoid oxygenase genes present in soybean are involved in its abiotic stress physiology, and ABA is presumably a core regulatory substance. These results have further provided insights into the mechanism involved in the regulation of abiotic stress response and tolerance in soybean.

13.7 Transgenomics

One of the most famous techniques employed in life sciences and medicine to develop novel phenotypes with beneficial characteristics is transgenomics. Novel phenotypes are generated using numerous genetic engineering techniques, which confer changes in the genetic code of the target organism, i.e. the transgenic organism (Correa *et al.*, 2012; Tzfira *et al.*, 2012). Non-native or foreign DNA that codes for a novel trait or an agriculturally relevant characteristic, from various sources including plants, animals, bacteria and viruses, is transferred into the target organism's germ line. In case of legumes and other plants, the transgenes are capable of increasing the levels of certain proteins and osmolytes, namely osmoprotectants (Tarczynski *et al.*, 1993), late embryogenesis proteins (Xu *et al.*, 1996), detoxification enzymes, etc., that enable defence and amplify tolerance against abiotic stresses. Two of the most efficient methods employed for the purpose of successfully transforming crops include the *Agrobacterium*-mediated gene transfer and particle bombardment technique (Pratap *et al.*, 2010).

13.7.1 Soybean

Being an important crop throughout the world, the soybean plant has been the focus for countless researches in recent times (Agarwal *et al.*, 2013; Pinto *et al.*, 2013; Wu *et al.*, 2013). Genetic engineering techniques such as *Agrobacterium*-mediated gene transfer, electric discharge particle acceleration, electroporation, etc. have been used widely to generate transgenic fertile soybean plants since the first successful attempt (Hinchee *et al.*, 1988). For instance, the *AtMYB44* gene stimulates the abscisic acid signalling pathway, which in turn triggers stomatal closure and culminates in drought tolerance as well as salt tolerance in *Arabidopsis thaliana*. Another recent attempt to transform soybean plants involved overexpression of certain transcription factors (e.g. TF DREB1A) that regulate gene expression involved in response and defence against abiotic stresses. As compared to the non-transformed plant, the transformed DREB (dehydration response element-binding protein) plants responded with better growth in terms of number of seeds and pods (Rolla *et al.*, 2013).

13.7.2 Chickpea

The major environmental or abiotic constraint on the development and productivity of the chickpea is drought, and the subsequent oxidative stress experienced by the plant reduces yield potential by 40 to 50% (Ahmad *et al.*, 2005). Therefore, intense research has been directed at mechanisms governing abiotic stress tolerance and adaptation in chickpea. The potential effect of the phyto-oxylipin pathway on drought tolerance of plants has been investigated (De Domenico *et al.*, 2012). The expression of certain genes involved in oxylipin metabolism was studied, and the early activation and overexpression of a specific lipo-oxygenase (Mt-*LOX1*), two hydroperoxide lyases (Mt-*HPL1*, Mt-*HPL2*), one allene oxide synthase (Mt-*AOS*) and oxophytodienoate reductase (Mt-*OPR*) were found to be positively correlated with significant levels of oxylipin metabolites leading to increased levels of jasmonates. These compounds, in turn, play a vital role in the tolerance mechanism of drought-tolerant chickpea varieties. The heterologous expression of the pyrroline-5-carboxylate synthetase (*P5CS*) gene through *Agrobacterium*-mediated transformation was also shown to improve salt tolerance in chickpea. As compared to the control plants, the transgenic lines were able to produce higher levels of proline, a factor that helps assuage salinity stress.

13.7.3 Peanut

Drought as well as salinity are the major environmental stresses that limit worldwide peanut production (Stansell & Pallas, 1985; Lamb *et al.*, 1997). Since the major peanut-producing countries, namely India, China and the USA, are facing critical water scarcity for peanut irrigation, research is continually being carried out to generate drought- and salinity-tolerant peanut phenotypes. Both *Agrobacterium*-mediated and biolistic bombardment techniques have been applied for the purpose of transforming peanut. Overexpression of the *Arabidopsis* vacuolar gene *AVP1* enabled greater drought and salt tolerance in transgenic peanuts simultaneously (Qin *et al.*, 2013). The vacuolar phosphatase protein encoded by the gene acts as a proton pump and generates a chemical proton gradient across vacuolar membranes. The proton pump increases the activity of other secondary transporters such as the Na^+/H^+ antiporter, which in turn helps to amplify salt tolerance in the transgenic peanut (Blumwald, 2000; Gaxiola *et al.*, 2002). Abiotic stress tolerance was enhanced in transgenic peanuts transformed with a small GTP-binding protein AhRab7 (AhRabG3f). The transcription level of the gene was positively correlated with other environmental stresses such as salt and cold, indicating its involvement in the abiotic stress tolerance mechanism (Kumar *et al.*, 2013).

13.8 Functional genomics

Over the years, understanding the genetic basis of drought tolerance in crop plants has been of prime importance for researchers. To date, many drought-responsive genes, namely *LIP9*, *OsNAC6*, *OsLEA14a*, *OsRAB16D*, *OsLEA3-1* and *Oshox24* in rice (Nakashima *et al.*, 2014) and *ADF3*, *APB*, *ASR*, *DLP*, *LTP1* and *UGE5* in pigeon pea (Deeplanaik *et al.*, 2013) have been discovered through extensive genomics projects. Nonetheless, the analysis of these drought-inducible genes and their expression remains a significant factor in developing crop varieties tolerant to abiotic stresses. Functional genomics aims at discovering and characterizing the involved genes and their corresponding proteins with respect to their functions and interactions. Mutagenesis, RNA interference (RNAi) and genetic interaction mapping are just some of the different functional genomics approaches employed for the purpose of determining gene functions and investigating the response of plants to abiotic stresses at a molecular level.

13.8.1 Gene silencing approaches

In addition to insertional mutagenesis, another potent tool for gene expression analysis and function in plants is the technique of gene silencing. To knock out or reduce the expression of genes, transfer DNA (T-DNA) insert mutations, random mutation, RNAi-based methods and virus-induced gene silencing (VIGS) can be employed (Quadrana *et al.*, 2011). A recombinant virus vector is used to introduce a fragment of plant gene into plant cells; this can trigger the plant defence mechanism, which in turn silences the plant gene and the virus. More than 30 VIGS vectors have been discovered and are being used to decipher the function of genes in cellular pathways, functions and abiotic stresses (Huang *et al.*, 2012).

13.8.2 Tilling

The basic principle of the functional genomics platform revolves around the ability to study the function of a gene or an entire cell, by introducing mutations or altering growth conditions, and determining the resulting response of the organism. One reverse-genetics technique for introducing mutations is the target-induced local lesions in genome (TILLING) technique, which utilizes traditional mutagenesis techniques to create libraries of mutagenized individuals, which are then subjected to screening for mutations and studied for determining gene function. In order to study gene function by this technique, TILLING populations have been developed in numerous legumes. For example, in case of peanut, candidate genes for allergenicity are being identified using a TILLING population of 10,000 lines (Tadege *et al.*, 2009). Similarly, the development of approximately 3500 TILLING population lines is being carried out for the identification of candidate genes for drought tolerance in chickpea (Thudi, 2013a).

In a study carried out by Betti *et al.* (2012), the mutant alleles of specific genes involved in the drought stress response of *L. japonicus* were identified using a TILLING reverse genetics tool along with a population of insertion mutants created by the LORE1 endogenous transposon. The gene that was induced most by drought conditions was orthologous to the *Oep16* gene of *Arabidopsis*. The corresponding protein of this gene belongs to a family of amino acid and pre-protein transporters that are responsible for the transport of cytosol-synthesized proteins. This study shows that the use of TILLING can provide great insight into the responses of legumes under abiotic stress and can easily help identify the proteins involved in stress-induced responses.

13.9 Phenomics

Simply put, phenomics is the study of how the genetic composition of an organism affects its appearance, function and performance. Alternatively, phenomics can be defined as the study of the growth performance as well as the composition of plants. In order to improve plant performance by plant breeding, two phenomics techniques can be employed, namely forward and reverse phenomics. Phenotyping tools are used in forward phenomics in order to 'sieve' collections of various germplasm to isolate valuable traits. Forward phenomics involves taking all of the genomes of the available varieties and passing them through a sieve, leaving behind only the germplasm of interest. One of the screens that could be used, especially for research on legumes, is abiotic stress. This approach hastens the pre-breeding phase of research. In contrast, the second approach is reverse phenomics, in which the phenotype is known and researchers investigate the mechanisms and genes responsible for the trait of interest.

Phenomics can be seen as a detailed dissection of the traits of a plant that hold potential value and hence can help understanding of the mechanisms and processes involved in stress responses and tolerance. The physiological trait isolated can be reduced to the underlying biophysical or biochemical process and even to the gene level ultimately.

Abiotic stress tolerance is a crucial trait as far as yield stability and potential are concerned. Many traits that contribute to stress tolerance can be screened at the seedling stage of legumes in a controlled environment. Two such extremely important traits are salinity and drought tolerance. Stomatal closure is one of the first effects of saline conditions on a plant, which is partly due to the osmotic effect of solutes on the root uptake of water from the soil. This leads to a reduction in photosynthesis. Hence a surrogate measure can be utilized for photosynthetic or stomatal responses under osmotic stress (Munns, 2010), which can also be extended to legumes.

Phenotyping allows the imaging of traits of interest in real time. One very potent technique is advanced phenotyping using field-portable near-infrared spectroscopy (NIRS) (Casler & van Santen, 2012). This enables direct selection of the plant breed based on its desirable traits. For instance, Casler and van Santen (2012) selected certain forage legumes based on their improved fixation of atmospheric nitrogen using this technique. The plants were screened on the basis of high tissue N concentrations, which correlated with enhanced fixation of atmospheric nitrogen. This shows how phenotypic imaging can be used for plant screening and selection. The technique can be extended to other legumes and also to other traits.

13.10 Conclusions and future prospects

Ranking third in global crop production, legumes are significant for the food security of humans and livestock globally. Therefore, it is imperative to ensure that the yield of the legume crops is adequately high. A major challenge to legume improvement programmes is that the productivity and availability of legume crops is hindered by two major limiting factors: biotic and abiotic stresses. Biotechnological approaches, including all the 'omics', have great potential to help circumvent abiotic stress-related issues. For gene discovery to step up, progress in bioinformatics and sequencing is required. Such advances will ensure the availability of the sequenced genomes of model legumes, along with the possibility of next generation sequencing, thereby increasing the potential to greatly facilitate the 'omics' approaches. Advances in these omics techniques, including those described here, will in turn help illuminate the genetic structure of legumes and thereby allow the full deployment of genetic resources for legume improvement. For instance, a number of useful tools such as sequence assembly, transformation systems and genomic DNA libraries have been developed following the release of the genome sequences of important model plants, namely *Lotus japonicus, Glycine max, Medicago truncatula, Cajanus cajan* and *Cicer arietinum*. Although soybean has been extensively studied and tolerant varieties of chickpea and peanut have been subjected to studies in response to abiotic stress conditions, proteomics studies in legumes are very limited thus demarcating a potential target for improvement (Reddy *et al.,* 2012). Comprehensive approaches, comprising the quantitative and qualitative analysis of gene expression products, are also necessary at the transcriptomic, proteomic and metabolomic levels for better elucidation of the functioning of genomes and their genes (Gupta *et al.,* 2014). In other words, in order to apply the omics techniques to legumes, information should be obtained on the abiotic stress responses at a molecular level, ranging from gene to protein, metabolic pathways and finally the phenotypic effect. The essence of the outcome of experiments with the above-mentioned techniques is the identification of key factors (genes) involved in stress responses. Following identification and validation, these candidate genes can then be used as genetic markers or employed in the production of transgenic plants. Results obtained from model legumes can be extrapolated to commercially important legume species and thus be used for guiding future research.

References

Agarwal DK, Billore SD, Sharma AN, Dupare BU, Srivastava SK (2013) Soybean: Introduction, improvement, and utilization in India – problems and prospects. *Agr Res* 2: 293–300.

Ahmad F, Gaur P, Crosser J (2005) Chickpea (*Cicer arietinum* L.). In: Singh RJ, Jauhar PP (eds), *Genetic Resources, Chromosome Engineering and Crop Improvement – Grain Legumes*. CRC Press, 1: 185–214.

Akibode S, Maredia M (2011) Global and regional trends in production, trade and consumption of food legume crops. SPIA Report, Department of Agricultural, Food and Resource Economics, Michigan State University, East Lansing, MI.

Anderson JP, Thatcher LF, Singh KB (2005) Plant defense responses: conservation between models and crops. *Funct Plant Biol* 32: 21–34.

Anthony VM, Ferroni M (2012) Agricultural biotechnology and small holder farmers in developing countries. *Curr Opin Biotech* 23: 278–285.

Arbona V, Manzi M, Ollas CD, Gómez-Cadenas A (2013) Metabolomics as a tool to investigate abiotic stress tolerance in plants. *Int J Mol Sci* 14: 4885–4911.

Arrese-Igor C, González EM, Marino D, Ladrera R, Larrainzar E, Gil-Quintana E (2011) Physiological responses of legume nodules to drought. *Plant Stress* 5: 24–31.

Basu PS, Berger JD, Turner NC, Chaturvedi SK, Ali M, Siddique KHM (2007) Osmotic adjustment of chickpea (*Cicer arietinum*) is not associated with changes in carbohydrate composition or leaf gas exchange under drought. *Ann Appl Biol* 150: 217–225.

Bell CJ, Dixon RA, Farmer AD, *et al.* (2001) The *Medicago* genome initiative: a model legume database. *Nucleic Acids Res* 29: 114–117.

Betti M, Pérez-Delgado C, García-Calderón M, Díaz P, Monza J, Márquez AJ (2012) Cellular stress following water deprivation in the model legume *Lotus japonicus*. *Cells* 1: 1089–1106.

Bhushan D, Pandey A, Choudhary MK, Datta A, Chakraborty S, Chakraborty N (2007) Comparative proteomics analysis of differentially expressed proteins in chickpea extracellular matrix during dehydration stress. *Mol Cell Proteomics* 6: 1868–1884.

Blumwald E (2000) Salt transport and salt resistance in plants and other organisms. *Curr Opin Cell Biol* 12: 431–434.

Bray EA (2004) Genes commonly regulated by water-deficit stress in *Arabidopsis thaliana*. *J Exp Bot* 55: 2331–2341.

Brisson N, Gate P, Gouache D, Charmet G, Oury FX, Huard F (2010) Why are wheat yields stagnating in Europe? A comprehensive data analysis for France. *Field Crop Res* 119: 201–212.

Broughton WJ, Hernández G, Blair M, Beebe S, Gepts P, Vanderleyden JB (2003) Beans (*Phaseolus* spp.) – model food legumes. *Plant Soil* 252: 55–128.

Bruce TJ (2012) GM as a route for delivery of sustainable crop protection. *J Exp Bot* 63: 537–541.

Butte A (2002) The use and analysis of microarray data. *Nat Rev Drug Discov* 1: 951–960.

Cannon SB (2013) The Legume Information System 2013. In: Plant and Animal Genome XXI Conference, San Diego, CA, Jan 12–16, P0952.

Carlos de Ollas, Gómez-Cadenas A, Arbona V, Manzi M (2013) Metabolomics as a tool to investigate abiotic stress tolerance in plants. *Int J Mol Sci* 14: 4885–4911.

Casler M, van Santen E (2010) Breeding objectives in forages. In: Boller B, Posselt U, Veronesi F (eds), *Handbook of Plant Breeding: Fodder Crops and Amenity Grasses*. Springer Science+Business Media, New York, pp. 115–160.

Chalk PM, Alves BJR, Boddey RM, Urquiaga S (2010) Integrated effects of abiotic stresses on inoculants performance, legume growth and symbiotic dependence estimated by 15N dilution. *Plant Soil* 328: 1–16.

Cook DR (1999) *Medicago truncatula*: a model in the making! Commentary. *Curr Opin Plant Biol* 2: 301–304.

Correa R, Stanga J, Larget B, *et al.* (2012) An assessment of transgenomics as a tool for identifying genes involved in the evolutionary differentiation of closely related plant species. *New Phytol* 193: 494–503.

Cramer GR, Urano K, Delrot S, Pezzotti M, Shinozaki K (2011). Effects of abiotic stress on plants: a systems biology perspective. *BMC Plant Biol* 11: 163.

De Domenico S, Bonsegna S, Horres R, *et al.* (2012) Transcriptomic analysis of oxylipin biosynthesis genes and chemical profiling reveal an early induction of jasmonates in chickpea roots under drought stress. *Plant Physiol Biochem* 61: 115–122.

Deeplanaik N, Kumaran RC, Venkatarangaiah K, Shivashankar SKH, Doddamani D, Telkar S (2013) Expression of drought responsive genes in pigeonpea and in silico comparison with soybean cDNA library. *J Crop Sci Biotech* 16: 243–251.

Dita MA, Rispail N, Prats E, Rubiales D, Singh KB (2006) Biotechnology approaches to overcome biotic and abiotic stress constraints in legumes. *Euphytica* 147: 1–24.

Dixon RA (2001) Natural products and plant disease resistance. *Nature* 411: 843–847.

Dong Z, Shi L, Wang Y, *et al.* (2013) Identification and dynamic regulation of microRNAs involved in salt stress responses in functional soybean nodules by high-throughput sequencing. *Int J Mol Sci* 14: 2717–2738.

Doyle JJ (2013) The promise of genomics for a "next generation" of advances in higher-level legume molecular systematics. *S Afr J Bot* 89: 10–18.

Dreher K, Khairallah M, Ribaut JM, Morris M (2003) Money matters (I): costs of field and laboratory procedures associated with conventional and marker-assisted maize breeding at CIMMYT. *Mol Breed* 11: 221–234.

Farre G, Ramessar K, Twyman RM, Capell T, Christou P (2010) The humanitarian impact of plant biotechnology: Recent breakthroughs vs bottlenecks for adoption. *Curr Opin Plant Biol* 13: 219–225.

Garg R, Patel RK, Jhanwar S, *et al.* (2011) Gene discovery and tissue-specific transcriptome analysis in chickpea with massively parallel pyrosequencing and web resource development. *Plant Physiol* 156: 1661–1678.

Gaxiola RA, Fink GR, Hirschi KD (2002) Genetic manipulation of vacuolar proton pumps and transporters. *Plant Physiol* 129: 967–973.

Gruber V, Blanchet S, Diet A, *et al.* (2009) Identification of transcription factors involved in root apex responses to salt stress in *Medicago truncatula*. *Mol Genet Genomics* 281: 55–66.

Gupta S, Nadarajan N, Gupta DS (2014) Legumes in omic era: Retrospects and prospects. In: Gupta S, Nadarajan N, Gupta DS (eds), *Legumes in the Omic Era*. Springer, New York, pp. 1–14.

Hakeem KR, Chandna R, Ahmad P, Iqbal M, Ozturk M (2012) Relevance of proteomic investigations in plant abiotic stress physiology. *OMICS: J Integr Biol* 16: 621–635.

Handberg K, Stougaard J (1992) *Lotus japonicus*: an autogamous, diploid legume species for classical and molecular genetics. *Plant J* 2: 487–496.

Hernández G, Ramírez M, Valdés-López O, *et al.* (2007) Phosphorus stress in common bean: Root transcript and metabolic responses. *Plant Physiol* 144: 752–767.

Hinchee MAW, Connor-Ward DV, Newell CA, *et al.* (1988) Production of transgenic soybean plants using agrobacterium-mediated DNA transfer. *Nat Biotechnol* 6: 915–922.

Huang CJ, Qian YJ, Li ZH, Zhou XP (2012) Virus-induced gene silencing and its application in plant functional genomics. *Sci China Life Sci* 55: 99–108.

Jensen ES, Peoples MB, Boddey RM, *et al.* (2012) Legumes for mitigation of climate change and the provision of feedstock for biofuels and biorefineries: A review. *Agron Sustain Dev* 32: 329–364.

Jogaiah S, Govind SR, Tran LSP (2012) Systems biology-based approaches toward understanding drought tolerance in food crops. *Crit Rev Biotech* 33: 23–39.

Jorrin JV, Rubiales D, Dumas-Gaudot E, *et al.* (2006) Proteomics: a promising approach to study biotic interaction in legumes: a review. *Euphytica* 147: 37–47.

Jukanti AK, Gaur PM, Gowda CLL, Chibbar RN (2012) Nutritional quality and health benefits of chickpea (*Cicer arietinum* L.): A review. *Br J Nutr* 108: 11–26.

Kalo P, Seres A, Taylor SA, *et al.* (2004) Comparative mapping between *Medicago sativa* and *Pisum sativum*. *Mol Genet Genomics* 272: 235–246.

Komatsu S, Yamamoto R, Nanjo Y, Mikami Y, Yunokawa H, Sakata K (2009) A comprehensive analysis of the soybean genes and proteins expressed under flooding stress using transcriptome and proteome techniques. *J Proteome Res* 8: 4766–4778.

Komatsu S, Kuji R, Nanjo Y, Hiraga S, Furukawa K (2012) Comprehensive analysis of endoplasmic reticulum-enriched fraction in root tips of soybean under flooding stress using proteomics techniques. *J Proteomics* 77: 531–560.

Koornneef M, Meinke D (2010) The development of Arabidopsis as a model plant. *Plant J* 61: 909–921.

Kosová K, Vítámvás P, Prášil IT, Renaut J (2011) Plant proteome changes under abiotic stress – contribution of proteomics studies to understanding plant stress response. *J Proteomics* 74: 1301–1322.

Kumar K, Kumar M, Kim SR, Ryu H, Cho YG (2013) Insights into genomics of salt stress response in rice. *Rice* 6: 27.

Kuromori T, Miyaji T, Yabuuchi H, *et al.* (2010) ABC transporter AtABCG25 is involved in abscisic acid transport and responses. *Proc Natl Acad Sci USA* 107: 2361–2366.

Lamb MC, Davidson J I, Childre JW, Martin NR (1997) Comparison of peanut yield, quality, and net returns between non irrigated and irrigated production. *Peanut Sci* 24: 97–101.

Langridge P, Fleury D (2011) Making the most of 'omics' for crop breeding. *Trends Biotechnol* 29: 33–40.

Lee J, Lei Z, Watson BS, Sumner LW (2013) Subcellular proteomics of *Medicago truncatula*. *Front Plant Sci* 4: 112–131.

Li XP, Tian AG, Luo GZ, Gong ZZ, Zhang JS, Chen SY (2005) Soybean DRE-binding transcription factors that are responsive to abiotic stresses. *Theor Appl Genet* 110: 1355–1362.

Mantri N, Basker N, Ford R, Pang E, Pardeshi V (2013) The role of micro-ribonucleic acids in legumes with a focus on abiotic stress response. *Plant Genome* 6(3); doi: 10.3835/plantgenome2013.05.0013.

Marko NF (2013) Introduction to the Genomics special edition on clinical and translational genomics. *Genomics* 102: 73–73.

Memon AR (2012) Transcriptomics and proteomics analysis of root nodules of model legume plants. In: Ashraf M, Ozturk M, Ahmed MSA, Aksoy A (eds), *Crop Production for Agricultural Improvement*. Springer, pp. 291–315.

Molina C, Rotter B, Horres R, *et al.* (2008) SuperSAGE: the drought stress-responsive transcriptome of chickpea roots. *BMC Genomics* 9: 553.

Nakashima K, Jan A, Todaka D, *et al.* (2014) Comparative functional analysis of six drought-responsive promoters in transgenic rice. *Planta* 239: 47–60.

Ohtsu K, Smith MB, Emrich SJ, *et al.* (2007) Global gene expression analysis of the shoot apical meristem of maize (*Zea mays* L.). *Plant J* 52: 391–404.

Ohyanagi H, Sakata K, Komatsu S (2012) Soybean Proteome Database 2012: update on the comprehensive data repository for soybean proteomics. *Front Plant Sci* 3: 110.

Olivares J, Soto MJ, Arrese-Igor C, *et al.* (2011) Nitrogen fixation in legumes. In: Perez de la Vega M, Torres AM, Cubero JI, Kole C (eds), *Genetics, Genomics and Breeding of Cool Season Grain Legumes*. Science Publishers, pp. 355–379.

Opdenakker K, Remans T, Vangronsveld J, Cuypers A (2012) Mitogen-activated protein (MAP) kinases in plant metal stress: Regulation and responses in comparison to other biotic and abiotic stresses. *Int J Mol Sci* 13: 7828–7853.

Palovaara J, Saiga S, Weijers D (2013) Transcriptomics approaches in the early *Arabidopsis* embryo. *Trends Plant Sci* 18: 514–521.

Pal'ove-Balang P, Betti M, Díaz P, *et al.* (2013) Abiotic stress in Lotus: aluminum and drought. In: Gaur RK, Sharma P (eds), *Molecular Approaches in Plant Abiotic Stress*. CRC Press, Boca Raton, pp. 284–303.

Pandey A, Chakraborty S, Datta A, Chakraborty N (2008) Proteomics approach to identify dehydration responsive nuclear proteins from chickpea (*Cicer arietinum* L.). *Mol Cell Proteomics* 7: 88–107.

Pinto HS, de Avila AMH, Cardoso AO (2013) Challenges to increased soybean production in Brazil. In: Board JE (ed.), *A Comprehensive Survey of Soybean Research – Genetics, Physiology, Agronomy and Nitrogen Relationships*. InTech, Croatia, pp. 199–208.

Poltronieri P, Bonsegna S, De Domenico S, Santino A (2011). Molecular mechanisms in plant abiotic stress response. *Ratar Povrt/Field Veg Crop Res* 48: 15–24.

Pratap A, Choudhary AK, Kumar J (2010) In vitro techniques towards genetic enhancement of food legumes – A review. *J Food Legume* 23: 169–185.

Qin H, Gu Q, Kuppu S, *et al.* (2013) Expression of the *Arabidopsis* vacuolar H^+-pyrophosphatase gene AVP1 in peanut to improve drought and salt tolerance. *Plant Biotechnol Rep* 7: 345–355.

Quadrana L, Rodriguez MC, López M, *et al.* (2011) Coupling virus-induced gene silencing to exogenous green fluorescence protein expression provides a highly efficient system for functional genomics in Arabidopsis and across all stages of tomato fruit development. *Plant Physiol* 156: 1278–1291.

Ramanjulu S, Bartels D (2002) Drought and desiccation-induced modulation of gene expression in plants. *Plant Cell Environ* 25: 141–151.

Reddy DS, Bhatnagar-Mathur P, Vadez V, Sharma KK (2012) Grain legumes (soybean, chickpea, and peanut): Omics approaches to enhance abiotic stress tolerance. In: Tuteja N, Gill SS, Tiburcio AF, Tuteja R (eds), *Improving Crop Resistance to Abiotic Stress*. Wiley-VCH Verlag, Singapore, pp. 995–1032.

Rolla AAP, Carvalho JFC, Fuganti-Pagliarini R, *et al.* (2013) Phenotyping soybean plants transformed with rd29A: AtDREB1A for drought tolerance in the greenhouse and field. *Transgenic Res* 23: 75–87.

Russelle MP (2008) Biological dinitrogen fixation in agriculture. In: Schepers JS, Raun WR (eds), *Nitrogen in Agricultural Systems*. American Society of Agronomy, pp. 281–359.

Sanchez DH, Shwabe F, Erban A, Udvardi MK, Kopka J (2012) Comparative metabolomics of drought acclimation in model and forage legumes. *Plant Cell Environ* 35: 136–149.

Saxena NP, Krishnamurthy L, Johansen C (2002) Genetic improvement of drought in chickpea. In: Saxena NP, Toole JCO (eds), *Field Screening for Drought Tolerance in Crop Plants with Emphasis on Rice: International Workshop on Field Screening for Drought Tolerance in Rice*. ICRISAT, Patancheru, India, pp. 128–137.

Shabala S, Munns R (2012) Salinity stress: physiological constraints and adaptive mechanisms. In: Shabala S (ed.), *Plant Stress Physiology*, CAB International, Wallingford, UK, pp. 59–93.

Singh SP, Schwartz HF (2010) Breeding common bean for resistance to diseases: a review. *Crop Sci* 50: 2199–2223.

Stansell JR, Pallas JE (1985) Yield and quality response of florunner peanut to applied drought at several growth stages. *Peanut Sci* 12: 64–70.

Stracke S, Sato S, Sandal N, *et al.* (2004) Exploitation of colinear relationships between the genomes of *Lotus japonicus*, *Pisum sativum* and *Arabidopsis thaliana*, for positional cloning of a legume symbiosis gene. *Theor Appl Genet* 108: 442–449.

Swigonska S, Weidner S (2013) Proteomic analysis of response to long-term continuous stress in roots of germinating soybean seeds. *Plant Physiol* 170: 470–479.

Sylvia DM, Fuhrmann JJ, Hartel PG, Zuberer DA (2005) Biological dinitrogen fixation: Introduction and non-symbiotic. In: Sylvia DM, Fuhrmann JJ, Hartel PG, Zuberer DA (eds), *Principles and Applications of Soil Microbiology*, 2nd edn. Prentice Hall, pp. 373–404.

Syrenne RD, Shi W, Stewart CN, Yuan JS (2012) Omics platforms: Importance of twenty-first century genome-enabled technologies in seed developmental research for improved seed quality and crop yield. In: Agarwal GK, Rakwal R (eds), *Seed Development: Omics Technologies Toward Improvement of Seed Quality and Crop Yield*. Springer, pp. 43–57.

Tadege M, Wang TL, Wen J, Ratet P, Mysore KS (2009) Mutagenesis and beyond! Tools for understanding legume biology. *Plant Physiol* 151: 978–984.

Tarczynski MC, Jensen RG, Bohnert HJ (1993) Stress protection of transgenic tobacco by production of the osmolyte mannitol. *Science* 259: 508–510.

Thudi M (2013a) Towards fine mapping of drought tolerance related QTL region in chickpea using genotyping by sequencing (GBS) approach. In: Plant and Animal Genome XXI Conference, San Diego, CA, Jan 12–16, P0362.

Thudi M (2013b) Association mapping for abiotic stress tolerance in chickpea. In: Plant and Animal Genome XXI Conference, San Diego, CA, Jan 12–16, W003.

Tzfira T, Weinthal D, Marton I, Zeevi V, Zuker A, Vainstein A (2012) Genome modifications in plant cells by custom-made restriction enzymes. *Plant Biotechnol J* 10: 373–389.

Udvardi M, Poole PS (2013) Transport and metabolism in legume–rhizobia symbioses. *Annu Rev Plant Biol* 64: 781–805.

Umezawa T, Mizuno K, Fujimura T (2002) Discrimination of genes expressed in response to the ionic or osmotic effect of salt stress in soybean with cDNA-AFLP. *Plant Cell Environ* 25: 1617–1625.

Urano K, Kurihara Y, Seki M, Shinozaki K (2010) 'Omics' analyses of regulatory networks in plant abiotic stress responses. *Curr Opin Plant Biol* 13: 132–138.

Varshney RK, Kudapa H (2013) Legume biology: the basis for crop improvement. *Funct Plant Biol* 40: v–viii.

Varshney R, Hiremath P, Lekha P, *et al.* (2009) A comprehensive resource of drought and salinity responsive ESTs for gene discovery and marker development in chickpea (*Cicer arietinum* L.). *BMC Genomics* 10: 523.

Vogel S (2013) *Prime Mover – A Natural History of Muscle*. W.W. Norton & Company, Inc.

Wang RK, Wang CE, Fei YY, Gai JY, Zhao TJ (2013) Genome wide identification and transcription analysis of soybean carotenoid oxygenase genes during abiotic stress treatments. *Mol Biol Rep* 40: 4737–4745.

Wienkoop S, Perez de la Vega M, Torres AM, Cubero JI, Kole C (2011) Proteomics and metabolomics for systems biology in legumes. In: Kole C (ed.), *Genetics, Genomics and Breeding of Cool Season Grain Legumes*. CRC Press+Science Publishers, Boca Raton, FL, pp. 303–314.

Wu JJ, Xu PF, Liu LJ, *et al.* (2013) Genetic diversity of *Phytophthora sojae* isolates in Heilongjiang province in China assessed by RAPD and EST-SSR. In: *Third International Conference on Photonics and Image in Agriculture Engineering (PIAGENG 2013)* SPIE, pp. 87620T–87620T.

Wu QD, Van Etten HD (2004) Introduction of plant and fungal genes into pea (*Pisum sativum* L) hairy roots reduces their ability to produce pisatin and affects their response to a fungal pathogen. *Mol Plant-Microbe Interact* 17: 798–804.

Xu D, Duan X, Wang B, Hong B, Ho THD, Wu R (1996) Expression of a late embryogenesis abundant protein gene, HVA1, from barley confers tolerance to water deficit and salt stress in transgenic rice. *Plant Physiol* 110: 249–257.

Yan X, Liao H, Beebe SE, Blair MW, Lynch JP (2004) QTL mapping of root hair and acid exudation traits and their relationship to phosphorus uptake in common bean. *Plant Soil* 265: 17–29.

Yang P, Zhang P, Li B, Hu T (2013) Effect of nodules on dehydration response in alfalfa (*Medicago sativa*). *Environ Expt Bot* 86: 29–34.

Young ND, Bharti AK (2012) Genome enabled insights into legume biology. *Annu Rev Plant Biol* 63: 283–305.

CHAPTER 14

MicroRNA-mediated regulatory functions under abiotic stresses in legumes

Syed Sarfraz Hussain[1], Mohammad Ali[2] and Bujun Shi[1]

[1] *Australian Centre for Plant Functional Genomics (ACPFG), University of Adelaide, Glen Osmond, Australia*

[2] *Institute of Biotechnology, Bahauddin Zakariya University, Multan, Pakistan*

14.1 Introduction

Legumes belong to the Fabaceae/Leguminosae family, with 20,000 species and representing the third largest group of angiosperms. Legumes constitute the second largest group of food and feed crops grown worldwide (Iantecheva *et al.*, 2013). Members of this family fill critical niches in most terrestrial biomes. Legumes stand third behind cereals and oilseeds in global production (Popelka *et al.*, 2004) and account for 27% of primary crop production in the world (Graham & Vance, 2003). Annually, around 250 million tonnes of grain legumes are produced worldwide, which represent major crops used for food, feed and vegetable oil. A key contribution of legumes to a sustainable environment is their ability to fix atmospheric molecular nitrogen in symbiotic association with rhizobia through a process called symbiotic nitrogen fixation (SNF) or root nodule symbiosis (RNS) in most agricultural ecosystems (Urvardi & Scheible, 2005). Legumes, known as 'green manure', fix around 200 million tons of nitrogen (Graham & Vance 2003) in soil, thus reducing the need for chemical nitrogen fertilizer. Therefore, legumes play a key role in sustainable intensified agriculture, when used in common farming practices as an intercrop in crop rotation system by dramatically improving the organic content of the field.

Certain unique features like nitrogen fixation, a protein-rich physiology and secondary metabolites make legumes stand out from other crop plants; therefore, it is extremely important to develop genomic tools in legumes to understand their unique features (Cook, 1999). Systematic research in legumes was initiated by the selection of two model systems, *Lotus japonicus* (Handberg & Stougaard, 1992) and *Medicago truncatula* (Barker *et al.*, 1990). Their genomes were sequenced (Sato *et al.*, 2008; Young *et al.*, 2011) and made available to interested scientists for further exploitation in their research. In parallel, significant progress was made in expression profiling of legume genes (Lamblin *et al.*, 2003; Oldroyd, 2005; Thompson *et al.*, 2005). Besides, draft genomes for pigeon pea (Varshney *et al.*, 2012) and chickpea (Varshney *et al.*, 2013) have also become available to the scientific community. In addition, significant scientific efforts are underway in sequencing other important legumes like peanut, pea and lentils (http://www.peanutbioscience.com/). Currently, global research efforts involving molecular approaches to gain a deeper knowledge of economically important legumes are in progress, aimed at genetic improvement. As a part of these efforts, *de novo* transcriptome assemblies were developed in different legumes (Cheung *et al.*, 2006; Deschamps & Campbell, 2010; Garg *et al.*, 2011; Hiremath *et al.*, 2011; Kupada *et al.*, 2012). These sequencing data will serve as a useful resource for research on legumes at molecular and genomic levels.

Abiotic stresses including drought, salinity, extreme temperatures and nutrient deficiency are major factors contributing to reduction or total loss of crop production (Hussain *et al.*, 2011a). Extensive efforts have been made to identify genetic elements involved in various abiotic stresses in crop plants (Hussain *et al.*, 2011a,b). It has been widely reported that microRNAs (miRNAs) are involved in the regulation of a wide variety of

Legumes under Environmental Stress: Yield, Improvement and Adaptations, First Edition. Edited by Mohamed Mahgoub Azooz and Parvaiz Ahmad.

physiological processes in plants (Sunkar *et al.*, 2007). There is also a growing body of evidence showing that miRNAs are involved in abiotic stresses. These include drought (Trindade *et al.*, 2010; T. Wang *et al.*, 2011), salinity (Ding *et al.*, 2009; Liu *et al.*, 2008), cold (Jin *et al.*, 2010), nutrient deficiency (Bari *et al.*, 2006; Pant *et al.*, 2008; Liang *et al.*, 2010), heavy metal stress (Zhou *et al.*, 2008) and oxidative stress (Sunkar & Jagadeswaran, 2008). In legumes, miRNA research is still in its infancy but has significantly progressed in the past few years, resulting in the accumulation of both conserved and species-specific miRNAs. In soybean, at least 129 miRNAs have been identified (Subramanian *et al.*, 2008; Wang *et al.*, 2009; Joshi *et al.*, 2010; Turner *et al.*, 2012). Similarly, more than 100 miRNAs were identified in the model legume *M. truncatula* (Szittya *et al.*, 2008; Jagadeeswaran *et al.*, 2009; Lelandais-Briere *et al.*, 2009). In addition, based on computational prediction and sequencing approaches, a plethora of miRNA families, including miRNA precursors and mature miRNAs, have been reported in different legume species under abiotic stresses (Valdes-Lopez *et al.*, 2008; Barrera-Figueroa *et al.*, 2011; Chen *et al.*, 2012a; Dong *et al.*, 2013; Xu *et al.*, 2013).

14.2 MicroRNAs (miRNAs): Small but significant

Currently, miRNAs have been reported in 64 plant species, and the miRNA database shows 24,521 mature miRNA sequences (Release 20.0, June 2013; http://mirbase.org/), including 6843 miRNAs from plant species. However, the rate of miRNA identification in crop plants has increased rapidly with the availability of complete genome sequences due to high-throughput sequencing methods, and improved computational and experimental protocols (Yao *et al.*, 2007; Subramanian *et al.*, 2008; Jagadeeswaran *et al.*, 2009; Lelandais-Briere *et al.*, 2009; Joshi *et al.*, 2010; Zhao *et al.*, 2010; Schreiber *et al.*,, 2011; Kim *et al.*, 2012; Li *et al.*, 2012; Wang *et al.*, 2012; Zhang *et al.*, 2012; Liang *et al.*, 2013; Lin & Lai, 2013).

Interestingly in *Arabidopsis*, over 184 miRNAs have been identified, which are predicted to regulate more than 600 genes including 225 known targets (Griffiths-Jones *et al.*, 2008; Alves *et al.*, 2009). Some of these miRNAs have been analysed at the molecular level for their roles in the regulation of target genes (Llave *et al.*, 2002; Reinhart *et al.*, 2002; Chen *et al.*, 2004; Laufs *et al.*, 2004; Duan *et al.*, 2006). The regulatory roles of miRNAs in plants have been established primarily through overexpression or by generating plants that express miRNA-resistant versions (Chen *et al.*, 2004; Park *et al.*, 2005; Schwab *et al.*, 2005; Gandikota *et al.*, 2007; Li *et al.*, 2010; Khan *et al.*, 2011; Bustos-Sanmamed *et al.*, 2013; Turner *et al.*, 2013).

Numerous studies have revealed that plant miRNAs are involved in almost all biological and metabolic processes (Comai & Zhang, 2012; Khraiwesh *et al.*, 2012; Sun 2012; Turner *et al.*, 2013) including seed germination (Reyes & Chua, 2007), flower development and sex determination (Chen, 2004; Chuck *et al.*, 2009), plant growth and development (Chen *et al.*, 2004; Chen, 2005; Willmann & Poethig, 2005; Jones-Rhoades *et al.*, 2006; Mallory & Vaucheret, 2006; Nogueira *et al.*, 2006; Yang & Xue, 2007; Lelandais-Briere *et al.*, 2010; Rubio-Somoza & Weigel, 2011), root development (Zhang *et al.*, 2005; Gutierrez *et al.*, 2009) and phytohormone signalling (Achard *et al.*, 2004; Guo *et al.*, 2005; Reyes & Chua, 2007; Meng *et al.*, 2009). In addition, miRNA involvement in biotic and abiotic stresses has been extensively investigated and reported in plants (Sunkar & Zhu, 2004; Bari *et al.*, 2006; Sunkar *et al.*, 2006; Liu *et al.*, 2008; Pant *et al.*, 2008; Zhou *et al.*, 2008; Ding *et al.*, 2009; Trindade *et al.*, 2010, 2010; Liang *et al.*, 2010; Wang *et al.*, 2011). Given the fact that miRNAs are master regulators and serve as the core of gene regulatory networks (Jones-Rhoades *et al.*, 2006), miRNA research provides great opportunities to unravel the mechanisms underlying challenging plant traits (Sun, 2012; Liang *et al.*, 2013). This chapter mainly focuses on the plethora of miRNAs and their functions in legumes.

14.3 Micro-RNA identification and functional diversity in legumes

Regulation of gene expression for proper functioning is crucial for all organisms. Plant miRNAs function as mediators to guide Argonaute (AGO) proteins in the RNA-induced silencing complex (RISC) to cleave target transcripts and/or repress translation (Bustos-Sanmamed *et al.*, 2013). Last decade has seen extensive research into miRNA functions in plant development and plant responses to abiotic and biotic stresses (Khraiwesh *et al.*, 2012). In plants, miRNAs can be identified by high-throughput sequencing like miRNA library sequencing (Ramesh *et al.*, 2013) and computer prediction (Xie *et al.*, 2010). So far, hundreds of miRNAs have been identified in many plant species, including legumes such as *M. truncatula*, *L. japonicus*, soybean, common bean, peanut, chickpea and cowpea (Subramanian *et al.*, 2008; Szittya *et al.*, 2008; Arenas-Huertero *et al.*, 2009; Lelandais-Briere *et al.*, 2009; Lu & Yang, 2010; Paul *et al.*, 2011). In miRBase (http://www.mirbase.org), an archive of miRNA sequences and annotations, a total of 1256 sequences belonging to 285 miRNA families from legumes have been deposited (Release 20: June 2013). These miRNAs have the potential to regulate species-specific biological processes in legumes (Subramanian *et al.*, 2008; Szittya *et al.*, 2008).

Conserved miRNAs are generally encoded by multigenic families (Allen *et al.*, 2004). It is noted that gene number per miRNA family was generally higher in soybean compared to *M. truncatula*. This is probably due to the large genome of soybean (1115 Mbp) and genome duplication (Schmutz *et al.*, 2010). However, for some miRNA families an opposite profile was noted. For example, miRBase reveals that miR395 and miR399 have more members in *M. truncatula* than in soybean.

On the other hand, non-conserved miRNAs that were previously described as crop-specific (soybean: Subramanian *et al.*, 2008; Wang *et al.*, 2009; Joshi *et al.*, 2010; or *M. truncatula*: Szittya *et al.*, 2008; Jagadeeswaran *et al.*, 2009, Lelandais-Briere *et al.*, 2009) have now been declared as legume-specific. Due to greater interest in legume miRNA research, hundreds of such legume-specific miRNA families have been reported in the last few years. For example, 229 and 179 miRNA families are specific to *M. truncatula* and soybean, respectively. Recently, De Luis *et al.* (2012) reported that 35 miRNA families are specifically present in *L. japonicus*. This huge diversity compared to that in rice and wheat is probably due to more interest in legumes, which resulted in a great deal of small RNA deep sequencing and analysis. High-throughput sequencing has now been employed to systematically identify stress-related miRNAs in legumes (Li *et al.*, 2010; 2011; Barrera-Figueroa *et al.*, 2011; Wang *et al.*, 2011; Chen *et al.*, 2012a,b; Zhou *et al.*, 2012a,b).

14.4 MicroRNA expression profiling under abiotic stresses in legumes

Historically, plants have evolved morphological, physiological and molecular adaptations to cope with different abiotic stresses (Hussain *et al.*, 2011a). Several lines of evidence have suggested that miRNAs play important regulatory roles in plants' responses to abiotic stresses (Sunkar & Zhu, 2004; Fuji *et al.*, 2005; Aung *et al.*, 2006; Chiou *et al.*, 2006; Okamura *et al.*, 2008). More than 40 miRNA families have been associated with abiotic stresses in plants, 13 of which have been found to be responsive to salinity and drought stresses (Barrera-Figueroa *et al.*, 2012; Nageshbabu *et al.*, 2013). It is further noted that almost all stress-related miRNAs are conserved, ultimately suggesting that miRNA-mediated regulatory roles may be evolutionarily conserved for corresponding stresses throughout the plant kingdom. However, a particular miRNA that responds to a specific abiotic stress in one species may not necessarily have the same function in another species. For example, at least 10 miRNAs involved in stress have shown opposite expression profiles in rice and *Arabidopsis* under drought stress (Barrera-Figueroa *et al.*, 2012). Different computational and experimental approaches were used in the early studies, which provided low coverage. However, with the advent of next-generation sequencing (NGS), microarrays and more advanced computational techniques, it became much easier and more cost-effective to perform genome-wide profiling for the identification of stress-responsive miRNAs. As a result, the discovery of stress-related miRNAs has expanded from a few model plants like *Arabidopsis* and rice (Zhao *et al.*, 2007; Liu *et al.*, 2008; Zhou *et al.*, 2010) to other non-model plants (Ding *et al.*, 2009; Jia *et al.*, 2009; Song *et al.*, 2011; Wang *et al.*, 2011; Barakat *et al.*, 2012; Eldem *et al.*, 2012; Li *et al.*, 2013; Ozhuner *et al.*, 2013; Shuai *et al.*, 2013; Yanik *et al.*, 2013), including orphan plants like legumes (Subramanian *et al.*, 2008; Szittya *et al.*, 2008; Arenas-Huertero *et al.*, 2009; Jagadeeswaran *et al.*, 2009, Leladais-Briere *et al.*, 2009; Pant *et al.*, 2009; Wang *et al.*, 2009; Joshi *et al.*, 2010; Lu & Yang, 2010; Trindade *et al.*, 2010; Barrera-Figueroa *et al.*, 2012; Chen *et al.*, 2012a,b; Turner *et al.*, 2012; Dong *et al.*, 2013; Xu *et al.*, 2013).

14.4.1 Drought stress

Drought is one of the main causes of yield reduction in plants. Due to the limited availability of water for agriculture in many areas of the world, it is logical to investigate the natural mechanism of drought tolerance as an integral component of understanding the biological basis of response to drought stress in plants. Legumes are extensively grown in dry or semi-arid regions of the world, usually under rain-fed agriculture. There is strong evidence that miRNAs are involved in drought stress as well as in diverse physiological processes (Sunkar & Zhu, 2004; Lu *et al.*, 2005). In one of the pioneering studies on stress-responsive miRNAs, 26 miRNAs from *Arabidopsis* were cloned and five were examined for drought regulation (Sunkar & Zhu, 2004). The result showed that three miRNAs (miR393, miR397b and miR402) were upregulated, one (miR389a) was downregulated and one (miR319c) was non-regulated by drought. Notably, all these miRNAs were also regulated by cold, NaCl and/or abscisic acid (ABA) treatments. Also in *Arabidopsis*, miR169 was found to be downregulated by drought stress in an ABA-dependent pathway. Downregulation of miR169 leads to higher expression of its target gene, *NFYA5*, which in turn enhances the drought tolerance of the plant (Li *et al.*, 2008). Several miRNAs involved in drought tolerance were discovered by global miRNA expression profiling using high-throughput techniques in legumes.

The expression pattern of miRNAs varies among different species under drought stress. For example, miR169 is downregulated in *M. truncatula* (Trindade *et al.*, 2010; Wang *et al.*, 2011) but upregulated in rice and common bean under drought (Arenas-Huertero *et al.*, 2009; Zhao *et al.*, 2009). Similarly, miR1510 is upregulated in *Glycine max* and downregulated in *M. truncatula*, while miR396 is down-regulated in *M. truncatula* and *Vigna unguiculata*, and upregulated in *G. max* (Mantri *et al.*, 2013). Some miRNAs can be differentially expressed in the same plant species under drought conditions, for example, miR398a/b in *M. truncatula*. In one study (Trindade *et al.*, 2010) it was upregulated, but in another study (Wang *et al.*, 2011) it was downregulated by drought. This difference may be attributed to experimental conditions, particularly the extent and duration of drought stress.

Wang *et al.* (2011) performed small RNA expression profiling of two libraries of *M. truncatula* under drought stress. Twenty-two miRNAs of four families and 10 miRNAs of six families were found to be upregulated and downregulated, respectively, under drought stress and this was also verified by high-throughput sequencing and quantitative real-time polymerase chain reaction (qRT-PCR). They also identified 29 new members of known miRNA families, of which eight miRNAs were found to be responsive to drought stress. Recently, a study investigated the effect of drought and salt stress on expression of conserved miRNAs in *Phaseolus vulgaris* (Nageshbabu *et al.*, 2013). Results showed that miRNAs were less sensitive to drought than salinity. However, both stresses altered the expression pattern in a dose-dependent manner. Significant changes in expression level under light drought stress were shown by miR156 and miR162, suggesting that these miRNAs may have important functions under drought stress. Furthermore, Arenas-Huertero *et al.* (2009) identified a number of novel legume-specific miRNAs in *P. vulgaris* under drought and ABA treatment. Targets of novel miRNAs were found to be involved in diverse cellular processes unique to legumes.

A complex study comprising drought and rust susceptible and resistant genotypes of soybean identified 256 miRNAs from different small RNA libraries using deep sequencing technology (Kulcheski *et al.*, 2011). Of these identified miRNAs, 71 miRNAs belong to conserved miRNA soybean families, 15 miRNAs belonging to six miRNA families were found to be conserved in other plant species, and 29 miRNAs belonging to 24 novel families were reported for the first time. The authors also identified 121 isoforms (miRNA variants) derived from 22 conserved miRNA families and four novel miRNA families. Interestingly, of the 11 novel miRNAs analysed, all were expressed differently from each other during drought stress. However, the majority were upregulated in the susceptible genotype but downregulated in the tolerant genotype under drought. This distinct miRNA behaviour under drought stress in two contrasting genotypes may potentially be involved in regulating genes associated with tolerance/intolerance. However, this needs further investigations. Similarly, Barrera-Figueroa *et al.* (2011) used deep sequencing of small RNA libraries from two cowpea genotypes (drought tolerant and susceptible) and identified 157 miRNAs belonging to 89 families by mapping sequenced small RNA reads to genomic sequences of cowpea. Forty-four drought-associated miRNAs belonging to 28 families were identified by comparing the expression levels of experimental samples with control samples. Furthermore, 30 out of 44 drought-associated miRNAs were upregulated and the rest were downregulated. The 44 drought-associated miRNAs included ones from miRNA families previously known to be associated with drought in other plant species, indicating that miRNAs are involved in conserved drought response pathways. In addition, target genes for 32 miRNAs were predicted, encoding diverse functions. Most of these targets are transcription factors. Some miRNA genes, despite high expression in libraries, were not identified mainly due to non-availability of complete genome sequences of cowpea. In conclusion, detailed investigation of the targets of drought-associated miRNAs identified so far will help in understanding the regulatory roles of miRNAs in drought tolerance of legumes.

14.4.2 Salt stress

Salt stress is another important constraint for plant growth. In particular, legume production has been greatly affected by salinity (Bayuelo-Jimenez *et al.*, 2002; Wang *et al.*, 2003; C. Chen *et al.*, 2007). MicroRNAs have a role in gene expression under different abiotic stresses and several studies have measured the differential expression of miRNAs under salt stress in different plants (Sunkar & Zhu, 2004; Liu *et al.*, 2008; Lu *et al.*, 2008; Arenas-Huertero *et al.*, 2009). To date, limited information is available on the role of miRNAs under salt stress in legumes. However, potential candidates have been identified in soybean and other legumes using *in silico* techniques (Zhang *et al.*, 2008; Chen *et al.*, 2009).

Recently, a study followed the expression levels of 12 miRNAs in *P. vulgaris* under salinity and drought stresses (Nageshbabu & Jyothi, 2013). Eleven of the 12 miRNAs had been demonstrated to be involved in either salinity or drought stress responses in previous studies in *Arabidopsis* and rice. The authors did not find any dramatic change in the expression pattern of miRNAs under salt/drought stress. This points towards the existence of fine-tuning mechanisms in plants that may be important for the regulation of gene expression without negatively impacting plant growth and development under stress. Similarly, using a comparative genomics technique, Paul *et al.* (2011) identified 18 conserved miRNAs belonging to 16 distinct miRNA families in cowpea (*Vigna unguiculata*) under salt stress. Fifteen potential targets were predicted by BLASTn searches using the miRNA sequences. Interestingly, all of the predicted targets were identified as transcription factors. Expression patterns of these miRNAs under salt stress were studied using qRT-PCR and results confirmed the presence and upregulation of seven miRNAs in roots under salt stress. The results showed the induction of miR498 under salt stress in cowpea. In *Arabidopsis*, miR498 was upregulated only during cold and drought stresses (Liu *et al.*, 2008). These results suggest that plants employ different miRNA-mediated regulatory strategies under different stresses.

Recently, Dong *et al.* (2013) studied the effect of salt on soybean nodules using deep sequencing of non-stressed nodules (NSN) and salt-stressed nodules (SSN). Computational analysis of sequence reads identified 110 known miRNAs belonging to six miRNA families with 128 novel miRNAs belonging to 64 miRNA families. Further analysis of the 128 novel miRNAs revealed that 66 miRNAs belonging to 27 miRNA families have identifiable loci in plant species, and that 12 miRNAs were conserved in legumes, of which 10 were found only in soybean. Furthermore, a total of 770 putative targets were identified for 224 miRNAs. The majority of miRNAs each have multiple targets varying from 2 to 10. Computational prediction of functions of potential targets revealed that many of them have been associated with stress responses in plants (Sunkar & Zhu, 2004; Liu *et al.*, 2008; Jia *et al.*, 2009). Interestingly, this study shed light on the potential role of miRNAs in soybean and revealed a dynamic regulation pattern of miRNAs during salt stress in functional nodules of soybean. However, further characterization of highly expressed individual miRNAs will help to elucidate the molecular mechanism of salt tolerance of functional soybean nodules.

14.4.3 Extreme temperature stress

Altered expression of miRNAs under extreme temperatures (cold and heat) has been reported in different model plant species. Heat stress disturbs cellular homeostasis and can lead to leaf etiolation, severe retardation in growth and development, and even death. Different high-throughput approaches like bioinformatics, microarrays, transcriptomics and direct cloning have been used to identify cold- and heat-responsive miRNAs in plants (Zhou *et al.*, 2008; Jian *et al.*, 2010; Xin *et al.*, 2010; Yu *et al.*, 2011). In *Arabidopsis*, the expression levels of miR393 and miR319c were upregulated by cold (Sunkar & Zhu, 2004). Similarly, several miRNAs like miR165/166, miR169, miR172, miR393, miR396, miR397 and miR408 were significantly upregulated, while other miRNAs like miR156/157, miR159/319, miR164, miR394 and miR398 showed either transient or mild regulation under cold stress (Zhou *et al.*, 2008). MicroRNA contents of *Arabidopsis* and *Populus* were also analysed using high-throughput microarray techniques (Liu *et al.*, 2008; Lu *et al.*, 2008). Deep sequencing led to the identification of 28 cold-responsive miRNAs in *Brachypodium* (Zhang *et al.*, 2009).

Wang and Long (2010) analysed four miRNAs (miR319, miR393, miR397 and miR402) from *Arabidopsis* for cold tolerance with the assumption that they might also be involved in cold tolerance in sweet pea. The results showed the presence of these miRNAs in sweet pea. RT-PCR analysis showed that the expression levels of these miRNAs increased after cold treatment in sweet pea. On the other hand, Yu *et al.* (2011) identified 35 conserved miRNA families in *Brassica rapa* using a deep sequencing approach. Among them, five miRNA families were responsive to heat tolerance. Two miRNAs (miR156h and miR156g) of the miR156 family were induced by heat stress and their putative target, *SPL2*, was downregulated following the stress. These results show that plant miRNAs can be regulated by extreme temperature stress and may function in critical defence systems for stress fitness.

14.4.4 Heavy metal stress

To date, there is a lack of detailed information on the role of miRNAs under heavy metal stress in legumes. However, a few reports have demonstrated the involvement of miRNAs in response to different heavy metal stresses – Cd (cadmium), Hg (mercury), Al (aluminium), As (arsenic) – in legumes like *M. truncatula*, *Brassica napus*, *P. vulgaris* and *Glycine max*.

Several metal stress-responsive miRNAs were identified from a small RNA library of *B. napus* seedlings exposed to 80 μM Cd (Zhou *et al.*, 2012a). Comparative analysis of four libraries (treated and control roots and shoots) revealed that Cd stress differentially regulated the expression of some miRNAs, suggesting that these miRNAs are directly or indirectly involved in different physiological processes leading to plant tolerance to Cd stress. Furthermore, the authors reported some non-conserved miRNAs were also regulated by Cd exposure. Interestingly, no species-specific miRNA was detected, probably due to the limited number of EST sequences of *B. napus* that are used for the identification of species-specific miRNAs. While a number of targets were detected for most of the conserved miRNAs, no target was found for non-conserved miRNAs possibly due to the low expression of these targets or miRNAs.

Recently Zhang *et al.* (2013) demonstrated that transgenic *B. napus* overexpressing miR395 showed improved plant growth under Cd stress (80–120 μM). The manifested transgenic plants exhibited enhanced chlorophyll accumulation and attenuation of the oxidative damage to root cells. Furthermore, the transgenic plants showed low translocation of Cd from roots to shoots compared to wild-type plants. The expression of several Cd tolerance genes confirmed the above results. Taken together, the data suggest that miR395 is involved in detoxification of Cd in *B. napus*. Previous studies showed that miR319 is also regulated under various metal stresses. MiR319 was upregulated in *M. truncatula* exposed to 80 μM Cd, 20 μM Hg and 200 μM Mn, whereas it was downregulated in *B. napus* after treatment with 80 μM Cd and 10 μM Hg (Zhou *et al.*, 2008, 2012a; Valdes-Lopez *et al.*, 2010). Similarly, miR171 and miR392 also showed induction/suppression patterns in *B. napus*, and *M. truncatula* upon treatment with different stresses like Cd, Hg and Al (Xie *et al.*, 2007; Zhou *et al.*, 2008, 2012a,2012b; Ding *et al.*, 2011; Lima *et al.*, 2011).

In conclusion, it is clear from the above reports that miRNAs constitute an integral part of regulatory circuits conferring metal stress tolerance in plants.

14.4.5 Nutrient deficiency stress

Phosphorus (P) is one of the essential macronutrients important for plant growth and development (Hackenberg *et al.*, 2013a). It is involved in many cellular processes and constitutes an integral building block of many macromolecules such as nucleic acids and biomembrane phospholipids (Vance *et al.*, 2003; Valdes-Lopez *et al.*, 2008; Yuan & Liu, 2008). Low P availability severely affects plant growth, and 30–80% of organic and insoluble forms of P are not accessible to plants (Abel *et al.*, 2002; Sha *et al.*, 2012). To react to nutrient deprivation, plants have evolved a broad spectrum of morphological, physiological and metabolic adaptations that include: decreased growth rate; establishment of mycorrhizal symbiosis for environmental P fixation and utilization; modification of root system architecture (RSA) for increased surface area; enhanced expression of phosphorus transporter genes; and adaptation of some metabolic pathways for bypassing P-requiring steps (Shulaev *et al.*, 2008; Yuan & Liu, 2008; Kehr, 2013). Over the past few years, collective studies have revealed the roles of several metabolic genes, transcription factors, ribo-regulators, plant hormones and ubiquitin-related proteins (Rubio *et al.*, 2001, 2009; Hammond *et al.*, 2003; Wu *et al.*, 2003; Franco-Zorilla *et al.*, 2004; Mission *et al.*, 2005; Z.H. Chen *et al.*, 2007; Devaiah *et al.*, 2007; Morcuende *et al.*, 2007; Mueller *et al.*, 2007; Nilsson *et al.*, 2007; Yuan & Liu, 2008).

Several miRNAs have been reported to be involved in nutrient signalling (Fuji *et al.*, 2005; Aung *et al.*, 2006; Bari *et al.*, 2006; Chiou *et al.*, 2006). Therefore, miRNAs were considered as a vital element in gene regulatory networks under nutrient deprivation conditions, particularly P-starvation stress (Schachtman & Shin, 2007; Chen *et al.*, 2008; Guo *et al.*, 2008; Valdes-Lopez & Hernandez, 2008; Yuan & Liu, 2008; Hackenberg *et al.*, 2013a). In *Arabidopsis*, P homeostasis is post-transcriptionally regulated by miR399, which forms an essential component of the phosphate starvation response 1 (PHR1) signalling pathway (Fuji *et al.*, 2005; Aung *et al.*, 2006; Bari *et al.*, 2006; Chiou *et al.*, 2006). miRNA399 was shown to be positively regulated by *AtPHR1* (Aung *et al.*, 2006; Bari *et al.*, 2006) but negatively regulated by *IPS1/At4* by means of a target mimicry mechanism (Franco-Zorilla *et al.*, 2007). A similar case was also observed in barley (Hackenberg *et al.*, 2013b). miR399 targets *UBC24*, which encodes a ubiquitin-conjugating E2 enzyme, also known as PHOSPHATE 2 (PHO2) (Fuji *et al.*, 2005; Aung *et al.*, 2006; Bari *et al.*, 2006). The *PHO2* gene negatively regulates phosphate uptake and root-to-shoot allocation. miR399 is also involved in P-deficiency signalling in rapeseed and pumpkin phloem sap and common bean roots (Pant *et al.*, 2008; Valdes-Lopez *et al.*, 2008). Overexpression of miR399 in mutant plants rescued the phenotype, resulting in high phosphate (Pi) levels in shoots and significantly reducing levels of the target transcript (Chiou *et al.*, 2006). Similarly, Pant *et al.* (2008) demonstrated that miR399 accumulated at a high level in phloem sap of rapeseed and pumpkin under P-deficiency, indicating a role in long-distance communication. This hypothesis was verified by reciprocal grafting experiments between transgenic *Arabidopsis* expressing miR399 and wild-type plants. This experiment revealed that miR399 can move from shoots to roots but not in the opposite direction. Furthermore, a reduced level of *PHO2* mRNA in the rootstock was observed. This confirmed that translocated miR399 is also functional (Lin *et al.*, 2008; Pant *et al.*, 2008; Buhtz *et al.*, 2010).

Recently, Xu *et al.* (2013) identified an additional miRNA (miR2111) that was induced under P-deficiency in soybean. This miRNA was almost undetectable under normal conditions (full nutrition) but became highly abundant under limited P. In addition, miR2111 also accumulated in phloem sap like miR399 during the P-deficient stress, indicating that miR2111 might also be involved in long-distance communication (Pant *et al.*, 2009). A Kelch repeat-containing F-box protein was identified in *Arabidopsis* and soybean as one of its targets, suggesting that miR2111 has a possible role in the control of protein abundance under P-limitation (Hsieh *et al.*, 2009). Moreover, several reports have revealed that miR156, miR778 and miR827 were upregulated whereas miR169, miR395 and miR398 were downregulated in *Arabidopsis* under P deficiency (Hsieh *et al.*, 2009; Pant *et al.*, 2009). Similarly, many miRNAs (miR157, miR160, miR165, miR166, miR169, miR393, pvu-miR2118, gma-miR1524, gma-miR1526 and gma-miR1532) showed differential expression upon P deficiency in common beans (Valdes-Lopez *et al.*, 2010). Interestingly, 167 miRNAs belonging to 35 miRNA families showed differential expression in white lupin in response to P deficiency (Zhu *et al.*, 2010). Furthermore, Zeng *et al.* (2010) demonstrated that 27 miRNA families representing at least 57 miRNA members showed significant changes in expression level under P starvation in soybean.

In regard to other nutrients, miR397, miR398, miR408 and miR857 responded to copper (Cu alteration) and have been proposed to be involved in Cu homeostasis through the regulation of Cu:Zn superoxide dismutase (CSD1 and CSD2), plastocyanin and various laccases (Abdel-Ghany & Pilon, 2008; Yamasaki *et al.*, 2008; Buhtz *et al.*, 2008; 2010; Dugas & Bartel, 2008). Similarly, miR395, which is involved in regulation of two ATP sulphurylases (APS1 and APS4) and a low-affinity sulphate transporter (SULTR2;1), has a relevant role in sulphur homeostasis during S limitation (Allen *et al.*, 2004; Jones-Rhoades *et al.*, 2006; Buhtz *et al.*, 2008, 2010; Kawashima *et al.*, 2009). miR167 is associated with lateral root growth under nitrogen (N) deficiency, miR169 and miR398a were repressed upon N limitation (Pant *et al.*, 2009; Zhao *et al.*, 2010; Wang *et al.*, 2013) and miR393 was induced by high nitrate and controls root architecture (Gifford *et al.*, 2008; Vidal *et al.*, 2010). It is further demonstrated that changes in one nutrient significantly affect the availability of other nutrients, which may result in unexpected interactions among miRNAs and consequently gene regulation (Haydon & Cobbett, 2007). A striking example is that under iron (Fe) deficiency, expression of miR399 and miR2111 was decreased in phloem and almost undetectable in roots and leaves (Buhtz *et al.*, 2008, 2010), which indicates that changes in Fe availability significantly alter P homeostasis.

14.5 MicroRNAs play important roles in nodulation and symbiosis in legumes

Nodules are specialized root lateral organs that result from symbiotic interaction between legumes and nitrogen-fixing bacteria (collectively called rhizobia); they have both agronomic and ecological importance for plants (Murray, 2011). Symbiosis begins with the exchange of certain chemical signals between the partners, which determines the host specificity. Rhizobial infection of legumes occurs primarily through root hairs. Legume roots release specific flavonoid compounds that are recognized by compatible bacterial species, which in turn release lipo-chito-oligosaccharides, called the 'Nod' signal. The bacteria colonize the root hair and gain access to the nodule cells through a specialized structure called an infection thread. This is followed by cell division in the cortex and pericycle cell layer to form a nodule primordial, thereby ultimately leading to nodule formation. A specialized structure in the nodule called a symbiosome is the functional unit of biological nitrogen fixation (Kereszt *et al.*, 2011). A number of different plant hormones are shown to play significant roles in the initiation, development and ultimately establishment of symbiosis in legumes (Suzaki *et al.*, 2013). Recent studies have shown that some legume miRNAs are functionally associated with nodulation and the establishment of symbiosis (Lelandais-Briere *et al.*, 2009; Simon *et al.*, 2009; Subramanian *et al.*, 2008; Yan *et al.*, 2013). Several miRNAs (miR156, miR160, miR166, miR167, miR169, miR319 and miR393) that are involved in auxin signalling were shown to be regulated in the early events of nodulation and symbiosis (Subramanian *et al.*, 2008; Li *et al.*, 2010).

Several researchers have identified and functionally characterized miRNAs in legume nodules using sequencing and bioinformatics approaches (Zhang *et al.*, 2008; Jagadeeswaran *et al.*, 2009; Wang *et al.*, 2009; Joshi *et al.*, 2010; Li *et al.*, 2010). Early investigations revealed the involvement of only two miRNAs, miR166 and miR169, in *Bradyrhizobium* infection and nodule development in soybean and *M. truncatula* (Combier *et al.*, 2006; Boualem *et al.*, 2008; Subramanian *et al.*, 2008). It is well known that miR166 regulates a family of HD-ZIP III transcription factors in a number of plant species including legumes (Boualem *et al.*, 2008; Subramanian *et al.*, 2008; Nogueira *et al.*, 2009; Devers *et al.*, 2011; Wang *et al.*, 2011). Overexpression of miR166 in *M. truncatula* led to a reduction in nodules and lateral root number in addition to ectopic development of vascular bundles in transgenic roots. Several lines of evidence suggest that miR166 activity during nodule development most probably relies on the conserved functions of its target in root cell differentiation. In *M. truncatula,* miR169 regulates the expression of *MtHAP2-1* (Combier *et al.*, 2006), whose expression is strongly upregulated during nodule development (El Yahyaoui *et al.*, 2004). MtHAP2-1 is a member of the NF-Y transcription factor family; it comprises three subunits – HAP2 (CBF-B, NF-YA), HAP3 (CBF-A, NF-YB) and HAP5 (CBF-C, NF-YC) – and binds the CCAAT motif present in many eukaryotic promoters (Maity & de Crombrugghe, 1998).

Overexpression of miR169 resulted in non-functional nodules and delays of 8–10 days in nodule development. Furthermore, the downregulation of *MtHAP2-1* by RNAi showed a similar arrest in nodule development. On the other hand, overexpression of miR169-resistant *MtHAP2-1* under its native promoter restored tissue differentiation. However, a downregulation of miR169 in early development of soybean nodules was noticed (Subramanian *et al.*, 2008), suggesting that miR169-dependent regulation of *MtHAP2-1* may also play a role in the early events of symbiosis.

In addition to miR166 and miR169, a number of conserved miRNAs, including miR156, miR160, miR167, miR172, miR398 and miR399, are also expressed in other nodule tissues (Lelandais-Briere *et al.*, 2009; Yan *et al.*, 2013). However, the dynamics of miRNA expression at the late stages of nodule development and symbiosis have so far not been investigated. Recently, Dong *et al.* (2013) performed deep sequencing of miRNAs from both normal and salt-stressed mature soybean nodules. They identified 110 known miRNAs belonging to 61 miRNA families, and 128 novel miRNAs belonging to 64 miRNA families. Both salt-stressed and control libraries shared nearly all the known miRNAs, and for 128 novel miRNAs only 82 were detected in both libraries. Of the remaining 34, 12 novel miRNAs were expressed only in one of the libraries, although the average abundance of each novel miRNA was much higher in the salt-stressed library.

14.6 MicroRNA-mediated approaches for functional genomics in legumes

Current molecular approaches to investigate plant biology/physiology using transcriptional profiling or large-scale DNA/protein analysis often lead to the identification of large groups of novel genes with unknown functions. One of the methods to determine the function of a new gene is to knock-down this gene using RNA interference (RNAi) technology, which can be delivered to plants by virus-induced gene silencing (VIGS) vectors. In soybean VIGS has successfully been employed to develop mutant resources (Subramanian *et al.*, 2005). However, the core of RNAi is in fact two types of small RNAs, small interfering RNAs (siRNA) and miRNAs. RNAi through siRNA-mediated gene silencing has identified several genes involved in specific developmental processes in soybean (Libault *et al.*, 2010; Takagi *et al.*, 2011). Although RNAi and VIGS have been extensively applied to develop knock-down plant mutants in the past, RNAi constructs often did not produce consistent results due to off-target effects of RNAi approaches. Therefore, a new approach based on artificial miRNAs (amiRNAs) was developed for silencing genes (Khraiwesh *et al.*, 2008; Ossowski *et al.*, 2008; Schwab *et al.*, 2010). AmiRNAs are obtained by replacing native miRNA duplexes from a natural miRNA precursor and have been shown to function as native miRNAs. Recent studies have successfully used amiRNA-based silencing constructs in *M. truncatula* (Haney & Long, 2010) and soybean (Melito *et al.*, 2010). Very recently, Verdouk and Sullivan (2013) employed an amiRNA sequence to silence the GUS Plus gene in transgenic alfalfa plants. Analysis of gene expression in the transgenic plants indicated that the amiRNA construct was effective at reducing both GUS Plus transcript levels and GUS activity. Based on these results, it is further demonstrated that amiRNA silencing constructs based on the *Arabidopsis* miR319c can be successfully utilized to specifically silence gene expression in alfalfa. Similarly, Devers *et al.* (2013) presented compelling evidence that an amiRNA-mediated gene silencing system is a useful tool to investigate functions of genes involved in root endosymbiosis in *M. truncatula*. With the amiRNAs the authors unravelled the function of a putative transcription factor gene, *MtErf1*, which showed strong induction during mycorrhizal symbiosis. Knock-down of *MtErf1* expression reduced expression levels of *Rhizophagus irregularis* genes showing reduced mycorrhizal colonization. Furthermore, knock-down of *MtErf1* by amiRNA-mediated gene silencing indicated a potential function of this gene during arbuscule development, since only truncated, non-fully branched arbuscules were observed in roots. It is further emphasized that application of this potentially efficient and effective method will likely revolutionize functional genomics in a wide variety of legumes as many reports of successful use of this approach have recently been documented for both *M. truncatula* (Haney & Long, 2010) and soybean (Melito *et al.*, 2010).

14.7 Conclusions and future prospects

Current research is directed towards exploring the essential roles in gene regulation of miRNAs, which are the largest group of gene regulatory molecules in legumes. However, these efforts provide a foundation for the evaluation of the individual roles of miRNAs in post-transcriptional regulation of developmental processes and stress responses in economically important legumes. Interestingly, miRNAs were long believed to act only in a cis- or trans- way on their targets. However, the finding that miRNAs can also transmit information over a long distance or in a signalling way regarding the nutrient status of plant organs is surprising. For example, grafting experiments provided convincing evidence that miR395 and miR399 are phloem mobile under specific nutrient deficiency (Pant *et al.*, 2009; Buhtz *et al.*, 2010). Considering the numerous regulatory functions of miRNAs, detailed follow-up studies will be invaluable to provide clear evidence of *in vivo* mobility and for exploring the possible roles of additional miRNAs in systemic regulation. Therefore, future efforts should be directed towards supplementary experimental approaches for answers to specific questions and in turn to understand the complex gene regulatory networks of miRNAs.

References

Abdel-Ghany SE, Pilon M (2008) MicroRNA-mediated systemic down-regulation of copper protein expression in response to low copper availability in Arabidopsis. *J Biol Chem* 283: 15932–15945.

Abel S, Tocconi CA, Delatorre CA (2002) Phosphate sensing in higher plants. *Physiol Plant* 115: 1–8.

Achard P, Herr A, Baulcombe DC, Harberd NP (2004) Modulation of floral development by a gibberellins-regulated microRNA. *Development* 13: 3357–3365.

Allen RS, Millgate AG, Chitty JA, *et al.* (2004) RNAi-mediated replacement of morphine with the nonnarcotic alkaloid reticuline in opium poppy. *Nat Biotechnol* 22: 1559–1566.

Alves L Jr, Niemeier S, Hauenschild A, Rehsmeier M, Merkle T (2009) Comprehensive prediction of novel microRNA targets in *Arabidopsis thaliana*. *Nucleic Acids Res* 37: 4010–4021.

Arenas-Huertero C, Pérez B, Rabanal F, *et al.* (2009) Conserved and novel miRNAs in the legume *Phaseolus vulgaris* in response to stress. *Plant Mol Biol* 70: 385–401.

Aung K, Lin SI, Wu CC, Huang YT, Su CL, Chiou TJ (2006) pho2, a phosphate over accumulator, is caused by a nonsense mutation in a microRNA399 target gene. *Plant Physiol* 141: 1000–1011.

Barakat A, Siram A, Park J, Zhebentyayeva T, Main D, Abbott A (2012) Genome-wide identification of chilling responsive microRNAs in *Prunus persica*. *BMC Genomics* 13: 481.

Bari R, Pant B, Stitt M, Scheible WR (2006) microRNA399 and *PHR1* define a phosphate-signalling pathway in plants. *Plant Physiol* 141: 988–999.

Barker D, Bianchi S, Blondon F, *et al.* (1990) *Medicago truncatula*: a model plant for studying the molecular genetics of the rhizobium–legume symbiosis. *Plant Mol Biol Rep* 8: 40–49.

Barrera-Figueroa BE, Gao L, Diop NN, *et al.* (2011) Identification and comparative analysis of drought-associated microRNAs in two cowpea genotypes. *BMC Plant Biol* 11: 127.

Barrera-Figueroa BE, Gao L, Wu Z, *et al.* (2012) High throughput sequencing reveals novel and abiotic stress-regulated microRNAs in the inflorescence of rice. *BMC Plant Biol* 12: 132.

Bayuelo-Jimenez JS, Craig R, Lynch JP (2002) Salinity tolerance of Phaseolus species during germination and early seedling growth. *Crop Sci* 42: 1584–1594.

Boualem A, Laporte P, Jovanovic M, *et al.* (2008) MicroRNA166 controls root and nodule development in *Medicago truncatula*. *Plant J* 54: 876–887.

Buhtz A, Springer F, Chappell L, Baulcombe DC, Kehr J (2008) Identification and characterization of small RNAs from the phloem of *Brassica napus*. *Plant J* 53: 739–749.

Buhtz A, Pieritz J, Springer F, Kehr J (2010) Phloem small RNAs, nutrient stress responses and systemic mobility. *BMC Plant Biol* 10: 64.

Bustos-Sanmamed P, Bazin J, Hartmann C, Crespi M, Lelasdais-Brierre C (2013) Small RNA pathways and diversity in model legumes: Lessons from genomics. *Front Plant Sci* 4: 236; doi: 10.3389/fpls.2013.00236.

Chen C, Tao C, Peng H, Ding Y (2007) Genetic analysis of salt stress responses in Asparagus bean {(*Vigna unguiculata* L.) ssp. Sesquipedalis Verds.}. *J Hered* 98: 655–665.

Chen L, Wang T, Zhao M, Tian Q, Zhang WH (2012) Identification of aluminium-responsive microRNAs in *Medicago truncatula* by genome-wide high throughput sequencing. *Planta* 235: 375–386.

Chen L, Wang T, Zhao M, Zhang W (2012) Ethylene-responsive microRNAs in roots of *Medicago truncatula* identified by high throughput sequencing at whole genome level. *Plant Sci* 184: 14–19.

Chen R, Hu Z, Zhang H (2009) Identification of microRNAs in wild soybean (*Glycine soja*). *J Integr Plant Biol* 51: 1071–1079.

Chen X (2004) A microRNA as a translational repressor of APETALA2 in Arabidopsis flower development. *Science* 303: 2022–2025.

Chen X (2005) microRNA biogenesis and function in plants. *FEBS Lett* 579: 5923–5931.

Chen Y, Lohuis D, Goldbach R, Prins M (2004) High frequency induction of RNA-mediated resistance against Cucumber mosaic virus using inverted repeat constructs. *Mol Breed* 14: 215–226.

Chen YF, Wang Y, Wu WH (2008) Membrane transporters for nitrogen, phosphate and potassium uptake in plants. *J Integr Plant Biol* 50: 835–848.

Chen ZH, Nimmo GA, Jenkins G, Nimmo HG (2007) *BHLH32* modulates several biochemical and morphological process that respond to Pi starvation in Arabidopsis. *Biochem J* 405: 191–198.

Cheung F, Hass BJ, Goldberg SMD, May GD, Xiao Y, Town CD (2006) Sequencing Medicago truncatula expressed sequenced tags using 454 Life Sciences technology. *BMC Genomics* 7: 272.

Chiou TJ, Aung K, Lin SI, Wu CC, Chiang SF, Su CL (2006) Regulation of phosphate homeostasis by microRNAs in Arabidopsis. *Plant Cell* 18: 412–421.

Chuck G, Candela H, Hake S (2009) Big impact by small RNAs in plant development. *Curr Opin Plant Biol* 12: 81–86.

Comai L, Zhang B (2012) MicroRNAs: key gene regulators with versatile functions. *Plant Mol Biol* 80: 1.

Combier JP, Frugier F, De Billy F, *et al.* (2006) *MtHAP2-1* is a key transcriptional regulator of symbiotic nodule development regulated by microRNA169 in *Medicago truncatula*. *Genes Develop* 20: 3084–3088.

Cook DR (1999) Medicago truncatula – a model in the making. *Curr Opin Plant Biol* 2: 301–304.

De Luis A, Markmann K, Cognat V, *et al.* (2012) Two microRNAs linked to nodule infection and nitrogen-fixing ability in the legume *Lotus japonicus*. *Plant Physiol* 160: 2137–2154.

Deschamps S, Campbell MA (2010) Utilization of next-generation sequencing platforms in plant genomics and genetic variant discovery. *Mol Breed* 25: 553–570.

Devaiah B, Karthikeyan AS, Ragothama KG (2007) WRKY75 transcription factor is a modulator of phosphate acquisition and root development in Arabidopsis. *Plant Physiol* 143: 1789–1801.

Devers EA, Branschield A, May P, Krajinski F (2011) Stars and symbiosis: microRNA and microRNA*-mediated transcript cleavage involved in arbuscular mycorrhizal symbiosis. *Plant Physiol* 156: 1990–2010.

Devers EA, Teply J, Reinert A, Gaude N, Krajinski F (2013) An endogenous artificial microRNA system for unravelling the function of root endosymbiosis related genes in *Medicago truncatela. BMC Plant Biol* 13: 82.

Ding D, Zhang L, Wang H, Liu Z, Zhang Z, Zheng Y (2009) Differential expression of miRNAs in response to salt stress in maize roots. *Annals Bot* 103: 29–38.

Ding Y, Chen Z, Zhu C (2011) Microarray based analysis of cadmium-responsive microRNAs in rice (*Oryza sativa*). *J Exp Bot* 62: 3563–3573.

Dong QH, Han J, Yu HP, *et al.* (2012) Computational identification of microRNAs in strawberry expressed sequence tags and validation of their precise sequences by miR-RACE. *J Hered* 103: 268–277.

Dong Z, Shi L, Wang Y, *et al.* (2013) Identification and dynamic regulation of microRNAs involved in salt stress responses in functional soybean nodules by high throughput sequencing. *Int J Mol Sci* 14: 2717–2738.

Duan CG, Wang CH, Guo HS (2006) Regulation of microRNA on plant development and viral infection. *Chinese Sci Bull* 51: 269–278.

Dugas DV, Bartel B (2008) Sucrose induction of Arabidopsis miR398 represses two Cu/Zn superoxide dismutases. *Plant Mol Biol* 67: 403–417.

El-Yahyaoui F, Kuster H, Ben Amor B, *et al.* (2004) Expression profiling in *Medicago truncatula* identifies more than 750 genes differentially expressed during nodulation, including many potential regulators of the symbiotic program. *Plant Physiol* 136: 3159–3176.

Eldem V, Celikkol-Akcay U, Ozhuner E, Bakr Y, Uranbey S, Unver T (2012) Genome-wide identification of miRNAs responsive to drought in peach (*Prunus persica*) by high throughput deep sequencing. *PLoS One* 7: e50298.

Franco-Zorrilla JM, Gonzalez E, Bustos R, Linhares F, Leyva A, Paz-Ares J (2004) The transcriptional control of plant responses to phosphate limitation. *J Exp Bot* 55: 285–293.

Franco-Zorrilla JM, Valli A, Todesco M, *et al.* (2007) Target mimicry provides a new mechanism for regulation of microRNA activity. *Nat Genet* 39: 1033–1037.

Fuji H, Chiou TJ, Lin SI, Aung K, Zhu JK (2005) A miRNA involved in phosphate starvation response in Arabidopsis. *Curr Biol* 15: 2038–2043.

Gandikota M, Birkenbihl RP, Hohmann S, Cardon GH, Saedler H, Huijser P (2007) The miRNA156/157 recognition element in the 30 UTR of the Arabidopsis SBP box gene SPL3 prevents early flowering by translational inhibition in seedlings. *Plant J* 49: 683–693.

Garg R, Patel RK, Tyagi AK, Jain M (2011) De novo assembly of chickpea transcriptome using short reads for gene discovery and marker identification. *DNA Res* 18: 53–63.

Gifford ML, Dean A, Gutierrez RA, Coruzzi GM, Birnbaum KD (2008) Cell specific nitrogen responses mediate developmental plasticity. *Proc Natl Acad Sci USA* 105: 803–808.

Graham PH, Vance CP (2003) Legumes: importance and constraints for greater use. *Plant Physiol* 131: 872–877.

Griffiths-Jones S, Saini HK, van Dongen S, Enright AJ (2008) miRBase: tools for microRNA genomics. *Nucleic Acids Res* 36: D154–D158.

Guo HS, Xie Q, Fei JF, Chua NH (2005) MicroRNA directs mRNA cleavage of the transcription factor *NAC1* to downregulate auxin signals for Arabidopsis lateral root development. *Plant Cell* 17: 1376–1386.

Guo WB, Zhang LN, Zhao J, Liao H, Zhuang CX, Yan XL (2008) Identification of temporally and spatially phosphate-starvation responsive genes in *Glycine max. Plant Sci* 175: 574–584.

Gutierrez L, Bussell JD, Pacurar DI, Schwambach J, Pacurar M, Bellini C (2009) Phenotypic plasticity of adventitious rooting in Arabidopsis is controlled by complex regulation of *AUXIN RESPONSE FACTOR* transcripts and microRNA abundance. *Plant Cell* 21: 3119–3132.

Hackenberg M, Huang PJ, Huang CY, Shi BJ, Gustafson P, Langridge P (2013a) A comprehensive expression profile of microRNAs and other classes of non-coding small RNAs in barley under phosphorus-deficient and -sufficient conditions. *DNA Res* 20: 109–125.

Hackenberg M, Shi BJ, Gustafson P, Langridge P (2013b) Characterization of phosphorus-regulated miR399 and miR827 and their isomers in barley under phosphorus-sufficient and phosphorus-deficient conditions. *BMC Plant Biol* 13: 214.

Hammond JP, Bennett MJ, Bowen C, *et al.* (2003) Change in gene expression in Arabidopsis shoots during phosphate starvation and the potential for developing smart plants. *Plant Physiol* 132: 578–596.

Handberg K, Stougaard J (1992) Lotus japonica, an autogamous, diploid legume species for classical and molecular genetics. *Plant J* 2: 487–496.

Haney CH, Long SR (2010) Plant flotillins are required for infection by nitrogen-fixing bacteria. *Proc Natl Acad Sci USA* 107: 478–483.

Haydon MJ, Cobbett CS (2007) Transporters of ligand for essential metal ions in plants. *New Phytol* 174: 499–506.

Hiremath PJ, Farmer A, Cannon SB, *et al.* (2011) Large scale transcriptome analysis in chickpea (*Cicer arietinum* L.), an orphan legume crop of the semi-arid tropics of Asia and Africa. *Plant Biotechnol J* 9: 922–931.

Hsieh LC, Lin SI, Shih ACC, *et al.* (2009) Uncovering small RNA-mediated responses to phosphate deficiency in Arabidopsis by deep sequencing. *Plant Physiol* 151: 2120–2132.

Hussain SS, Ali M, Ahmad M, Siddique KHM (2011a) Polyamines: Natural and engineered abiotic stress tolerance in plants. *Biotech Adv* 29: 300–3111.

Hussain SS, Kayani MA, Amjad M (2011b) Transcription factors as tools to engineer enhanced drought stress tolerance in plants. *Biotechnol Prog* 27: 297–306.

Iantecheva A, Mysore KS, Ratet P (2013) Transformation of leguminous plants to study symbiotic interactions. *Int J Dev Biol* 57: 577–586.

Jagadeeswaran G, Zheng Y, Li YF, *et al.* (2009) Cloning and characterization of small RNAs from *Medicago truncatula* reveals four novel legume-specific microRNA families. *New Phytol* 184: 85–89.

Jia X, Wang WX, Ren L, Chen QJ, Mendu V (2009) Differential and dynamic regulation of miR398 in response to ABA and salt stress in *Populus tremula* and *Arabidopsis thaliana*. *Plant Mol Biol* 71: 51–59.

Jian X, Zhang L, Li G, Wang X, Cao X, Fang X, Chen F (2010) identification of novel stress-regulated microRNAs from *Oryza sativa* L. *Genomics* 95: 47–55.

Jin T, Chang Q, Li W, *et al.* (2010) Stress-inducible expression of *GmDREB1* conferred salt tolerance in transgenic alfalfa. *Plant Cell Tiss Org Cult* 100: 219–227.

Jones-Rhoades MW, Bartel DP, Bartel DP (2006) MicroRNAs and their regulatory roles in plants. *Annu Rev Plant Biol* 57: 19–53.

Joshi T, Yan Z, Libault M, *et al.* (2010) Prediction of novel miRNAs and associated target genes in *Glycine max*. *BMC Bioinformatics* 1: S14.

Kawashima CG, Yoshimoto N, Maruyama-Nakashita A, *et al.* (2009) Sulphur starvation induces the expression of microRNA-395 and one of its target genes but in different cell types. *Plant J* 57: 313–321.

Kehr J (2013) Systemic regulation of mineral homeostasis by microRNAs. *Front Plant Sci* 4: 145.

Kereszt A, Mergaert P, Kondorosi E (2011) Bacteroid development in legume nodules: Evolution of mutual benefits or of sacrificial victims? *Mol Plant Microbe Interact* 24: 1300–1309.

Khan GA, Declerck M, Sorin C, Hartmann C, Crespi M, Leladais-Briere C (2011) MicroRNAs as regulators of root development and architecture. *Plant Mol Biol* 77: 47–58.

Khraiwesh B, Ossowski S, Weigel D, Reski R, Frank W (2008) Specific gene silencing by artificial microRNAs in *Physcomitrella patens*: an alternative to targeted gene knockouts. *Plant Physiol* 148: 684–693.

Khraiwesh B, Arif MA, Seumel GI, *et al.* (2010) Transcriptional control of gene expression by microRNAs. *Cell* 140: 111–122.

Khraiwesh B, Zhu JK, Zhu J (2012) Role of miRNAs and siRNAs in biotic and abiotic stress responses of plants. *Biochem Biophys Acta* 1819: 137–148.

Kim B, Yu HJ, Park SG, *et al.* (2012) Identification and profiling of novel microRNAs in the *Brassica rapa* genome based on small RNA deep sequencing. *BMC Plant Biol* 12: 218.

Kudapa H, Bharti AK, Cannon SB, *et al.* (2012) A comprehensive transcriptome assembly of pigeonpea (*Cajanus cajan* L.) using Sanger and second generation sequencing platforms. *Mol Plant* 5: 1020–1028.

Kulcheski FR, de Oliveira LF, Molina LG, *et al.* (2011) Identification of novel soybean microRNAs involved in abiotic and biotic stresses. *BMC Genomics* 12: 307.

Lamblin AF, Crow JA, Johnson JE, *et al.* (2003) MtDB: a database for personalized data mining of the model legume *Medicago truncatula* transcriptome. *Nucleic Acids Res* 31: 196–201.

Laufs P, Peaucelle A, Morin H, Traas J (2004) MicroRNA regulation of the CUC genes is required for boundary size control in Arabidopsis meristems. *Development* 131: 4311–4322.

Lelandais-Briere C, Naya L, Sallet E, *et al.* (2009) Genome-wide Medicago truncatula small RNA analysis revealed novel microRNAs and isoforms differentially regulated in roots and nodules. *Plant Cell* 21: 2780–2796.

Lelandais-Briere C, Sorin C, Declerck M, Benslimane A, Crespi M, Hartmann C (2010) Small RBA diversity in plants and its impact in development. *Curr Genomics* 11: 14–23.

Li B, Duan H, Li H, Deng XW, Yin W, Xia X (2013) Global identification of miRNAs and targets in *Populus euphratica* under salt stress. *Plant Mol Biol* 81: 525–539.

Li H, Deng Y, Wu T, Subramaniun S, Yu O (2010) Misexpression of miR482, miR1512, and miR1515 increases soybean nodulation. *Plant Physiol* 153: 1759–1770.

Li H, Dong Y, Wang Y, Yang J, Liu X (2011) Characterization of the stress associated microRNAs in Glycine max by deep sequencing. *BMC Plant Biol* 11: 170–181.

Li T, Chen J, Qiu S, *et al.* (2012) Deep sequencing and microarray hybridization identify conserved and species specific microRNAs during somatic embryogenesis in hybrid yellow poplar. *PLoS One* 7: e43451.

Li WX, Oono Y, Zhu J, *et al.* (2008) The Arabidopsis NFYA5 transcription factor is regulated transcriptionally and post-transcriptionally to promote drought resistance. *Plant Cell* 20:2238–2251.

Liang G, Yang F, Yu D (2010) MicroRNA395 mediates regulation of sulfate accumulation and allocation in *Arabidopsis thaliana*. *Plant J* 62: 1046–1057.

Liang L, Li Y, He H, Wang F, Yu D (2013) Identification of miRNAs and miRNA-mediated regulatory pathways in *Carica papaya*. *Planta* 238: 739–752.

Libault M, Farmer A, Joshi T, *et al.* (2010) An integrated transcriptome atlas of the crop model *Glycine max*, and its use in comparative analysis in plants. *Plant J* 63: 86–99.

Lima JC, Arenhart RA, Margis-Pinheiro M (2011) Aluminum triggers broad changes in microRNA expression in rice roots. *Genet Mol Res* 10: 2817–2832.

Lin SI, Chiang SF, Lin WY, *et al.* (2008) Regulatory network of microRNA399 and *PHO2* by systemic signaling. *Plant Physiol* 147: 732–746.

Lin Y, Lai ZX (2013) Comparative analysis reveals dynamic changes in miRNA and their targets and expression during somatic embryogenesis in longan (Dinocarpus longan Lour.). *PLoS One* 8: e60337.

Liu HH, Tian X, Li YJ, Wu CA, Zheng CC (2008) Microarray-based analysis of stress-regulated microRNAs in *Arabidopsis thaliana*. *RNA* 14: 836–843.

Llave C, Kasschau KD, Rector MA, Carrington JC (2002) Endogenous and silencing-associated small RNAs in plants. *Plant Cell* 14: 1605–1619.

Lu C, Tej SS, Luo S, Haudenchild CD, Meyers BC, Green PJ (2005) Elucidation of the small RNA component of the transcriptome. *Science* 309: 1567–1569.

Lu S, Sun Y, Chiang V (2008) Stress-responsive microRNAs in Populus. *Plant J* 55: 131–151.

Lu Y, Yang X (2010) Computational identification of novel microRNAs and their targets in *Vigna unguiculate*. *Comp Func Genomics* Pii: 128257; doi: 10.1155/2010/128297.

Maity SN, de Crombrugghe B (1998) Role of the CCAAT-binding protein CBF/NF-Y in transcription. *Trends Biochem Sci* 23: 174–178.

Mallory AC, Vaucheret H (2006) Functions of microRNAs and related small RNAs in plants. *Nature Genet* 38: 31–36.

Mantri N, Basker N, Ford R, Pang E, Pardeshi V (2013) The role of micro-ribonucleic acids in legumes with a focus on abiotic stress response. *Plant Genet* 6: doi: 10.3835/plantgenome2013.05.0013.

Melito S, Heuberger AL, Cook D, Diers BW, MacGuidwin AE, Bent AF (2010) A nematode demographics assay in transgenic roots reveals no significant impacts of the Rhg1 locus LRR-Kinase on soybean cyst nematode resistance. *BMC Plant Biol* 10: 104.

Meng Y, Huang F, Shi Q, *et al.* (2009) Genome-wide survey of rice microRNAs and microRNA-target pairs in the root of a novel auxin-resistant mutant. *Planta* 230: 883–898.

Misson J, Raghothama KG, Jain A, *et al.* (2005) A genome-wide transcriptional analysis using *Arabidopsis thaliana* Affimetrix gene chips determined plant responses to phosphate deprivation. *Proc Natl Acad Sci USA* 102: 11934–11939.

Morcuende R, Bari R, Gibon Y, *et al.* (2007) Genome-wide, reprogramming of metabolism and regulatory networks of Arabidopsis in response to phosphorus. *Plant Cell Environ* 30: 85–112.

Mueller R, Morant M, Jarmer H, Nilsson L, Nielsen TH (2007) Genome-wide analysis of the Arabidopsis leaf transcriptome reveals interaction of phosphate and sugar metabolism. *Plant Physiol* 143: 156–171.

Murray JD (2011) Invasion by invitation: rhizobial infection in legumes. *Mol Plant Microbe Interact* 24: 631–639.

Nageshbabu R, Jyothi MN (2013) Profile of small intergering RNAs from French bean *Physeolus vulgaris* under abiotic stress conditions. *Int J Pharm Biol Sci* 4B: 176–185.

Nageshbabu R, Jyothi MN, Sharadamma N, Sarika S, Rai DV, Devaraj VR (2013) Expression of miRNAs regulates growth and development of French bean (*Physeolus vulgaris*) under salt and drought stress conditions. *Int Res J Biol Sci* 2: 52–56.

Nilsson L, Mueller R, Nielson TH (2007) Increased expression of the MYB-related transcription factor PHR1, leads to enhanced phosphate uptake in *Arabidopsis thaliana*. *Plant Cell Environ* 30: 1499–1512.

Nogueira FT, Sarkar AK, Chitwood DH, Timmermanns MC (2006) Organ polarity in plants is specified through the opposing activity of two distinct small regulatory RNAs. *Cold Spring Harbor Symp Quant Biol* 71: 157–164.

Nogueira FT, Chitwood DH, Madi S, *et al.* (2009) Regulation of small RNA accumulation in the maize shoot apex. *PLoS Genet* 5: e1000320.

Okamura K, Phillips MD, Tyler DM, Duan H, Chou YT, Lai EC (2008) The regulatory activity of microRNA* species has substantial influence on microRNA and 3′ UTR evolution. *Nat Struct Mol Biol* 15: 354–363.

Oldroyd G (2005) Sequencing the model legume *Medicago truncatula*. *Grain Legumes* 41: 23.

Ossowski S, Schwab R, Weigel D (2008) Gene silencing in plants using artificial microRNAs and other small RNAs. *Plant J* 53: 674–690.

Ozhuner E, Eldem V, Ipek A, *et al.* (2013) Boron stress responsive microRNAs and their targets in barley. *PLoS One* 8: e59543.

Pant BD, Buhtz A, Kehr J, Scheible WR (2008) MicroRNA399 is a long-distance signal for the regulation of plant phosphate homeostasis. *Plant J* 53: 31–38.

Pant BD, Musialak-Lange M, Nuc P, *et al.* (2009) Identification of nutrient-responsive Arabidopsis and rapeseed microRNAs by comprehensive real-time polymerase chain reaction profiling and small RNA sequencing. *Plant Physiol* 150: 1541–1555.

Park MY, Wu G, Gonzalez-Sulser A, Vaucheret H, Poethig RS (2005) Nuclear processing and export of microRNAs in Arabidopsis. *Proc Natl Acad Sci USA* 102: 3691–3696.

Paul S, Kundu A, Pal A (2011) Identification and validation of conserved microRNAs along with their differential expression in roots of *Vigna unguiculata* grown under salt stress. *Plant Cell Tiss Organ Cult* 105: 233–242.

Popelka JC, Terryn N, Higgins TJV (2004) Gene technology for grain legumes: can it contribute to the food challenge in developing countries? *Plant Sci* 167: 195–206.

Ramesh V, Admane N, Hussain SM (2013) Small RNAs landscape (sRNAome) of soybean (*Glycine max* L.): Biogenesis, vital functions and potential applications. *Plant Knowledge J* 2: 24–37.

Reinhart BJ, Weinstein EG, Rhoades MW, Bartel B, Bartel, DP (2002) MicroRNAs in plants. *Genes Develop* 16: 1616–1626.

Reyes JL, Chau NH (2007) ABA induction of miR 159 controls transcript levels of two MYB factors during Arabidopsis seed germination. *Plant J* 49: 592–606.

Rubio V, Linhares F, Solano R, *et al.* (2001) A conserved MYB transcription factor involved in phosphate starvation signaling both in vascular plant and unicellular algae. *Genes Develop* 15: 2122–2133.

Rubio V, Bustos R, Irigoyen ML, Cardona-Lopez X, Rojas-Triana M, Paz-Ares J (2009) Plant hormones and nutrient signaling. *Plant Mol Biol* 69: 361–373.

Rubio-Somoza I, Weigel D (2011) MicroRNA networks and developmental plasticity in plants. *Trends Plant Sci* 16: 258–264.

Sato S, Nakamura Y, Kaneko T, *et al.* (2008) Genome structure of the legume, *Lotus japonicus*. *DNA Res* 15: 227–239.

Schachtman DP, Shin R (2007) Nutrient sensing and signaling: NPKS. *Annu Rev Plant Biol* 58: 47–69.

Schmutz J, Cannon SB, Schlueter J, *et al.* (2010) Genome sequence of the palaeopolyploid soybean. *Nature* 463: 178–183.

Schreiber AW, Shi BJ, Huang CY, Langridge P, Baumann U (2011) Discovery of barley miRNAs through deep sequencing of short reads. *BMC Genomics* 12: 129.

Schwab R, Palatnik JF, Riester M, Schommer C, Schmid M, Weigel D (2005) Specific effects of microRNAs on the plant transcriptome. *Develop Cell* 8: 517–527.

Schwab R, Ossowski S, Warthmann N, Weigel D (2010) Directed gene silencing with artificial microRNAs. *Methods Mol Biol* 592: 71–88.

Sha A, Chen Y, Ba H, *et al.* (2012) Identification of *Glycine max* microRNAs in response to phosphorus deficiency. *J Plant Biol* 55: 268–280.

Shuai P, Liang D, Zhang Z, Yin W, Xia X (2013) Identification of drought responsive and novel *Populus trichocarpa* microRNAs by high throughput sequencing and their targets using degradome analysis. *BMC Genomics* 14: 233.

Shulaev V, Cortes D, Miller G, Mittler R (2008) Metabolomics for plant stress responses. *Physiol Plant* 132: 199–208.

Simon SA, Meyers BC, Sherrier DJ (2009) MicroRNAs in rhizobia legume symbiosis. *Plant Physiol* 151: 1002–1008.

Song QX, Liu YF, Hu XY, *et al.* (2011) Identification of microRNAs and their target genes in developing soybean seeds by deep sequencing. *BMC Plant Biol* 11: 5.

Subramanian S, Graham MY, Yu O, Graham TL (2005) RNA interference of soybean isoflavone synthase genes leads to silencing in tissues distal to the transformation site and to enhanced susceptibility to *Phytophthora sojae*. *Plant Physiol* 137: 1345–1353.

Subramanian S, Fu Y, Sunkar R, Barbazuk WB, Zhu JK, Yu O (2008) Novel and nodulation regulated microRNAs in soybean roots. *BMC Genomics* 9: 160.

Sun G (2012) MicroRNAs and their diverse functions in plants. *Plant Mol Biol* 80: 17–36.

Sunkar R, Jagadeeswaran G (2008) *In silico* identification of conserved microRNAs in large number of diverse plant species. *BMC Plant Biol* 8: 37.

Sunkar R, Zhu JK (2004) Novel and stress-regulated microRNAs and other small RNAs from Arabidopsis. *Plant Cell* 16: 2001–2019.

Sunkar R, Kapoor A, Zhu JK (2006) Posttranscriptional induction of two Cu/Zn superoxide dismutase genes in Arabidopsis is mediated by down-regulation of miR398 and important for oxidative stress tolerance. *Plant Cell* 18: 2051–2065.

Sunkar R, Chinnusamy V, Zhu J, Zhu JK (2007) Small RNAs as big players in plant abiotic stress responses and nutrient deprivation. *Trends Plant Sci* 12: 301–309.

Suzaki T, Ito M, Kawaguchi M (2013) Genetic basis of cytokinin and auxin functions during root nodule development. *Front Plant Sci* 4: 42.

Szittya G, Moxon S, Santos DS, *et al.* (2008) High throughput sequencing of *Medicago truncatula* short RNAs identifies eight new miRNA families. *BMC Genomics* 9: 593.

Takagi K, Nishizawa K, Hirose A, Kita A, Ishimoto M (2011) Manipulation of saponin biosynthesis by RNA interference-mediated silencing of β-amyrin synthase gene expression in soybean. *Plant Cell Rep* 30: 1835–1846.

Thompson R, Ratet P, Kuster H (2005) Identification of gene functions by applying TILLING and insertional mutagenesis strategies on microarray based expression data. *Grain Legumes* 41: 20–22.

Trindade I, Capitao C, Dalmay T, Fevereiro MP, Santos DM (2010) miR398 and miR408 are upregulated in response to water deficit in *Medicago truncatula*. *Planta* 231: 705–716.

Turner M, Yu O, Subramanian S (2012) Genome organization and characteristics of soybean microRNAs. *BMC Genomics* 13: 169.

Turner M, Nizampatnam NR, Baron M, *et al.* (2013) Ectopic expression of miR160 results in auxin hypersensivity, cytokinin hyposensitivity, and inhibition of symbiotic nodule development in soybean. *Plant Physiol* doi: 10.1104/pp.113.220699.

Urvardi MK, Scheible WR (2005) GRAS genes and the symbiotic green revolution. *Science* 308: 1749–1750.

Valdes-Lopez O, Hernandez G (2008) Transcriptional regulation and signaling in phosphorus starvation: what about legumes? *J Integr Plant Biol* 10: 1213–1222.

Valdes-Lopez O, Arena-Huertero C, Ramirez M, *et al.* (2008) Essential role of MYB transcription factor: *PvPHR1* and microRNA: PvmiR399 in phosphorus-deficiency signaling in common bean roots. *Plant Cell Environ* 31: 1834–1843.

Valdes-Lopez O, Yang S, Aparicio-Fabre R, *et al.* (2010) MicroRNA expressionn profiles in common bean (*Phaseolus vulgaris*) under nutrient deficiency stresses and manganese toxicity. *New Phytol* 187: 805–818.

Vance CP, Udhe-Stone C, Allen DI (2003) Phosphorus acquisition and use: critical adaptation by plants for securing a non-renewable resource. *New Phytol* 157: 423–447.

Varshney RK, Chen W, Li Y, *et al.* (2012) Draft genome sequence of pigeonpea (*Cajanus cajan*), an orphan legume crop of resource-poor farmers. *Nature Biotechnol* 30: 83–89.

Varshney RK, Song C, Saxena RK, *et al.* (2013) Draft genome sequence of chickpea (*Cicer arietinum*) provides a resource for trait improvement. *Nature Biotechnol* 31: 240–246.

Verdouk JC, Sullivan ML (2013) Artificial microRNA (amiRNA) induced gene silencing in alfalfa (*Medicago sativa*). *Botany* 91: 117–122.

Vidal EA, Araus V, Lu C, *et al.* (2010) Nitrate-responsive miR393/ABF3 regulatory module controls root system architecture in *Arabidopsis thaliana*. *Proc Natl Acad Sci USA* 107: 4477–4482.

Wang C, Han J, Kibet KN, *et al.* (2012) Identification of microRNAs from Amur grapes (*Vitis amurensis* Rupr.) by deep sequencing and analysis of microRNA variations with bioinformatics. *BMC Genomics* 13: 122.

Wang T, Chen L, Zhao M, Tian Q, Zhang WH (2011) Identification of drought-responsive microRNAs in *Medicago truncatula* by genome-wide high throughput sequencing. *BMC Genomics* 12: 367.

Wang W, Vinocur B, Altman A (2003) Plant responses to drought, salinity and extreme temperature towards genetic engineering for stress tolerance. *Planta* 218: 1–14.

Wang XB, Jovel J, Udomporn P, *et al.* (2011) The 21-nucleotide, but not 22-nucleotide, viral secondary small interfering RNAs direct potent antiviral defense by two cooperative argonautes in *Arabidopsis thaliana*. *Plant Cell* 23: 1625–1638.

Wang Y, Long L (2010) Identification and isolation of the cold-resistance related miRNAs in *Pisum sativum* Linn. *Journal of Liaoning Normal University (Natural Science Edition)* 2.

Wang Y, Li P, Cao X, Wang X, Zhang A, Li X (2009) Identification and expression analysis of miRNAs from nitrogen-fixing soybean nodules. *Biochem Biophys Res Commun* 378: 799–803.

Wang Y, Zhang C, Hao Q, *et al.* (2013) Elucidation of miRNAs-mediated responses to low nitrogen stress by deep sequencing of two soybean genotypes. *PLoS One* 8: e67423.

Willmann MR, Poethig RS (2005) Time to grow up: the temporal role of small RNAs in plants. *Curr Opin Plant Biol* 8: 548–552.

Wu P, Ma L, Hou X, *et al.* (2003) Phosphate starvation triggers distinct alterations of genome expression in Arabidopsis roots and leaves. *Plant Physiol* 132: 1260–1271.

Xie F, Frazier TP, Zhang B (2010) Identification and characterization of microRNAs and their targets in the bioenergy plant switchgrass (*Panicum virgatum*). *Planta* 232: 417–434.

Xie FL, Huang SQ, Guo K, *et al.* (2007) Computational identification of novel microRNA and targets in *Brassica napus*. *FEBS Lett* 58: 1464–1474.

Xin M, Wang Y, Yao Y, *et al.* (2010) Diverse set of microRNAs are responsive to powdery mildew infection and heat stress in wheat (*Triticum aestivum* L.). *BMC Plant Biol* 10: 123.

Xu F, Liu Q, Chen L, *et al.* (2013) Genome-wide identification of soybean microRNAs and their targets reveals their organ-specificity and responses to phosphate starvation. *BMC Genomics* 14: 66

Yamasaki H, Abdel-Ghany SE, Cohu CM, Kobayashi Y, Shikanai T, Pilon M (2008) Regulation of copper homeostasis by micro-RNA in Arabidopsis. *J Biol Chem* 282: 16369–16378.

Yan Z, Hossain S, Wang J, *et al.* (2013) miR172 regulates soybean nodulation. *Mol Plant Microbe Interact* 26: 1371–1377

Yang T, Xue L (2007) Functional diversity of miRNAs in plants. *Plant Sci* 172: 423–432.

Yanik H, Turktas M, Dundar E, Hernendez P, Dorado G, Unver T (2013) Genome wide identification of alternative bearing-associated microRNAs (miRNAs) in olive (*Olea europaea* L.). *BMC Plant Biol* 13: 10.

Yao Y, Guo G, Ni Z, *et al.* (2007) Cloning and characterization of microRNAs from wheat (*Triticum aestivum* L.). *Genome Biol* 8: R96.

Young ND, Debelle F, Oldroyd GED, *et al.* (2011) The Medicago genome provides insight into the evolution of rhizobial symbioses. *Nature* 480: 520–524.

Yu H, Song C, Jia Q, *et al.* (2011) Computational identification of microRNAs in apple expressed sequence tags and validation of their precise sequences by miR-RACE. *Plant Physiol* 141: 56–70.

Yuan H, Liu D (2008) Signaling components involved in plant responses to phosphate starvation. *J Integr Plant Biol* 50: 849–859.

Zeng HQ, Zhu YY, Huang SQ, Yang ZM (2010) Analysis of phosphorus deficient responsive microRNAs and cis-elements for soybean (*Glycine max* L.). *J Plant Physiol* 167: 1289–1297.

Zhang BH, Pan XP, Wang QL, Cobb GP, Anderson TA (2005) Identification and characterization of new plant microRNAs using EST analysis. *Cell Res* 15: 336–360.

Zhang B, Pan X, Stellwag EJ (2008) Identification of soybean microRNAs and associated target genes. *Planta* 229: 161–182.

Zhang J, Subramanian S, Stacey G, Yu O (2009) Flavones and flavonols play distinct roles during nodulation of *Medicago truncatula* by *Sinorhizobium meliloti*. *Plant J* 53: 171–183.

Zhang J, Zhang S, Han S, *et al.* (2012) Genome-wide identification of microRNAs in larch and stage-specific modulation of 11 conserved microRNAs and their targets during somatic embryogenesis. *Planta* 236: 647–657.

Zhang LW, Song JB, Shu XX, Zhang Y, Yang ZM (2013) miR395 is involved in detoxification of cadmium in Brassica napus. *J Hazard Materials* 250–251: 204–211.

Zhao B, Ge L, Liang R, *et al.* (2009) Members of miR169 family are induced by high salinity and transiently inhibit the NF-YA transcription factor. *BMC Mol Biol* 10: 29.

Zhao CZ, Xia H, Frazier TP, *et al.* (2010) Seep sequencing identifies novel and conserved microRNAs in peanuts (*Arachis hypogaea* L.). *BMC Plant Biol* 10: 3

Zhao L, KimYJ, Dinh T, Chen X (2007) MiR172 regulates stem cell fate and defines the inner boundary of *APETALA3* and *PISTILLATA* expression domain in Arabidopsis floral meristem. *Plant J* 51: 840–849.

Zhou LG, Liu YH, Liu ZC, Kong DY, Duan M, Luo LJ (2010) Genome-wide identification and analysis of drought-responsive microRNAs in *Oryza sativa*. *J Exp Bot* 61: 4157–4168.

Zhou X, Wang G, Sutoh K, Zhu J, Zhang W (2008) Identification of cold-inducible microRNAs in plants by transcriptome analysis. *BBA-Gene Regul Mech* 1779: 780–788.

Zhou ZS, Song JB, Yang ZM (2012a) Genome-wide identification of *Brassica napus* microRNAs and their targets in response to cadmium. *J Exp Bot* 63: 4597–4613.

Zhou ZS, Zeng HQ, Liu ZP, Yang ZM (2012b) Genome-wide identification of *Medicago truncatula* microRNAs and their targets reveals their different regulation by heavy metal. *Plant Cell Environ* 35: 86–99.

Zhu YY, Zeng HQ, Dong CX, Yin XM, Shen QR, Yang ZM (2010) microRNA expression profiles associated with phosphorus deficiency in white lupin (*Lupinus albus* L.) *Plant Sci* 178: 23–29.

CHAPTER 15

Biotechnology approaches to overcome biotic and abiotic stress constraints in legumes

Mustafeez Mujtaba Babar[1], Najam-us-Sahar Sadaf Zaidi[1], Mohamed Mahgoub Azooz[2] and Alvina Gul Kazi[1]

[1] *Atta-ur-Rahman School of Applied Biosciences, National University of Sciences and Technology (NUST), Islamabad, Pakistan*

[2] *Department of Botany, Faculty of Science, South Valley University, Qena, Egypt*

15.1 Introduction

Legumes are one of the most important, yet diverse, groups of plants, finding applications in a broad range of fields. They possess high nutritional value, are significant sources of biofuel energy, and have medicinal importance and environmental significance. They are involved in symbiotic relationships with a number of types of soil bacteria, and thereby contribute to environmental nitrogen fixation. Hence, legumes are important components in agriculture, as they help in converting atmospheric nitrogen into a usable form. However, many stress factors, biotic and abiotic, have been related to a decrease in the overall yield of the plants. Maintenance and improvement of legume health has, therefore, become an important target for both agriculturalists and biotechnologists.

The biological stress factors mainly comprise pathogenic bacteria, viruses, fungi, nematodes and other insects, whereas the abiotic constraints include drought, salinity, temperature changes and light. The response of leguminous plants to these stress conditions has been the focus of research to effectively investigate the inherent defence mechanisms of the plants (Ahuja *et al.*, 2010). A number of cellular and molecular mechanisms are involved in the stress-tolerance process. Manipulation of these signal cascades can help in improving the overall effectiveness of the plant to resist changes in their homeostasis. Current biotechnological intervention strategies are based upon the identification and isolation of stress resistance genes in stressed plants (Bhatnagar-Mathur *et al.*, 2012). These genes are then transferred to the target legume for expression of the stress-neutralizing factors.

The current review deals with understanding the importance of leguminous plants and the various threats they face. The genetic, subcellular and molecular mechanisms involved in legume defences and their phenotypic manifestations are presented. Exploitation and manipulation of these resistance mechanisms for providing effective strategies for the control and prevention of the adverse effects associated with these conditions are then considered.

15.2 Legumes and their importance

Legumes, a collective term used for members of the family Leguminosae (Fabaceae), form the single most diverse botanical group, providing benefits to humanity across a variety of different sectors. With their exceptional contributions to the human and animal food chain, economic sector, ecology and medicines, legumes are among the most important domesticated plants. Legumes are grown on an estimated 250 million hectares of agricultural land around the world, providing financial benefit to many thousands of farmers globally (Abate *et al.*, 2012; Vance, 1997).

Their wide geographical distribution in various climatic conditions makes them useful for domestication. Nutritional data reveal that legumes are a rich source of carbohydrates, fibre, essential amino acids, vitamins and minerals (Venn *et al.*, 2012; Bouchenak & Lamri-Senhadji, 2013). They form the protein food source of choice for

Legumes under Environmental Stress: Yield, Improvement and Adaptations, First Edition. Edited by Mohamed Mahgoub Azooz and Parvaiz Ahmad.

vegetarians and those in society who are unable to afford expensive sources of protein like red meat (Day, 2013). Being gluten free, legumes are the preferred alternative to wheat and related staples (García-Manzanares & Lucendo, 2011). On the ecological front, legumes are involved in biological nitrogen fixation, which not only leads to decreased demand for nitrogen input but also helps in increasing net primary productivity, resulting thereby in the production of useful chemical energy with much less input of exogenous enrichment sources (Casler *et al.*, 2009; Jensen *et al.*, 2012). Moreover, they form an important part of permaculture systems, contributing positively to nitrogen fixation and, hence, the nutrient enrichment process (Bates & Hemenway, 2010). The medicinal effects of the phytochemicals derived from legumes are very important. These secondary metabolites produced mainly as defence and signalling chemicals, have been in human use for centuries due to their potent health benefits (Muzquiz *et al.*, 2012). Alkaloids, flavonoids, coumarins, anthraquinones, terpenes, glycosides and lectins are some of the representative phytochemicals derived from the members of this family(De Hoff *et al.*, 2009; Dixon & Pasinetti, 2010; Velazquez *et al.*, 2010; Kumar *et al.*, 2012a; Spanou *et al.*, 2012; Tiwari & Pathak, 2013; Wink, 2013). These substances are used in complementary and alternative medicine for the treatment of cardiovascular, respiratory, urinary, metabolic and nervous system disorders (Joshi *et al.*, 2011). Moreover, they formed the basis for the discovery and development of many semisynthetic and synthetic drugs, now employed in allopathic medicine (Hoareau & DaSilva, 1999) (Figure 15.1).

One of the most important aspects of leguminous plants is their ability to participate in the symbiotic relationship with the soil bacteria, rhizobia (Oldroyd, 2013). On secretion of specific chemical signals from the root nodules of the legumes, the rhizobia tend to convert atmospheric nitrogen gas to the plant-consumable forms, ammonia and ammonium compounds. This mutualism helps in the nitrogen fixation process and contributes positively to agriculture and ecosystems. These bacteria are, therefore, included in a separate class referred to as plant growth-promoting rhizobacteria (PGPR). Legumes comprise many plant species, including woody plants, ornamental plants, crops and vegetables. Woody legumes grow in habitats across the world, putting them among the preferred choices for timber, firewood and construction material (Adams *et al.*, 2010). Ornamental species, like *Lathyrus odoratus* (sweet pea), can have wide variation in flower shape, colour and other morphological characteristics, making them conspicuous for pollination (Mikić *et al.*, 2011).

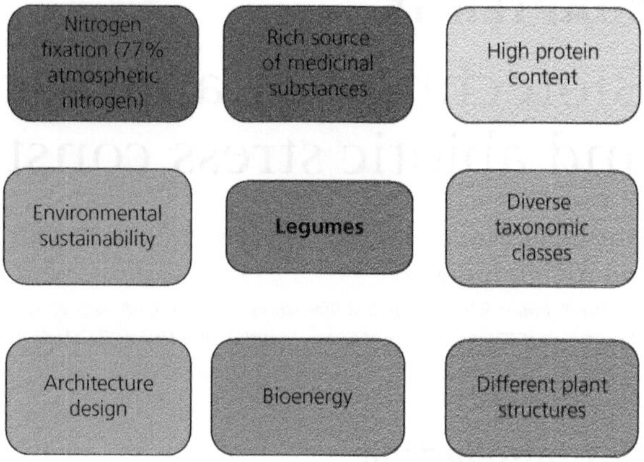

Figure 15.1 Importance of legumes in various sectors.

The family Leguminosae has over 900 genera comprising around 23,000 plant species (The Plant List, 2010). Classified into three subfamilies, Mimosoideae, Caesalpinoideae and Faboideae (Papilionoideae), based upon their morphological characteristics, these plants are distributed widely throughout the globe. Members of the Mimosoideae are woody shrubs and vines, distributed mainly in the warm and tropical regions of the Americas and Asia. Acacias (Bouchenak-Khelladi *et al.*, 2010), mimosas and members of the genera *Inga* and *Pentaclethra* are some of the most important plants of this subfamily. The other two subfamilies, Caesalpinoideae and Faboideae, follow a cosmopolitan geographical distribution and are found globally in tropical and subtropical regions. Both these subfamilies contain chiefly woody shrubs and trees; however, the Faboideae contains a number of non-woody annual herbs as well (Łotocka *et al.*, 2012). It has the largest number of genera of the three subfamilies and includes *Lathyrus* spp., *Medicago* spp., *Lotus* spp. and *Astragalus* spp. The Caesalpiniodeae is most diverse in terms of morphology and biogeographical distribution (Bruneau *et al.*, 2001). A number of leaf, flower and stem types are seen in this group, and the presence of its members in all regions of the world makes them the representative timber and ornamental species of the family, with characteristics of the other two subfamilies as well. Genera in the Caesalpinoideae include *Gleditsia*, *Amherstia*, *Gymnocladus*, *Delonix*, *Bauhinia*, *Caesalpinia* and *Senna*.

Though the importance of legumes is unmatched among all the classes of plants, the biological and physical threats they face are equally significant. To their advantage, the morphological and genetic diversity of legumes is very well documented. Variation in genes involved in floral ontogeny, seed dormancy, root mass, harvest index and other morphoagronomic traits has led to establishing the phylogenetic relationships among the various members of this family. Moreover, certain genes involved in biochemical processes, including the alcohol dehydrogenase gene and the plastid-encoded *Matk* gene, among others, have helped in simplifying the genetic, developmental and evolutionary relationships of legumes. This understanding of leguminous plants helps in identifying the targets that can potentially be exploited for controlling the response to various stress factors that can threaten the plants.

15.3 Legumes in stressed environments

Legumes, like other plants, are continuously exposed to many biological and environmental stimuli, which may have harmful effects on plant growth. These detriments, alone or in combination, generate a series of events that significantly affects plant physiology. Some of the biotic and abiotic constraints affecting legumes are discussed here.

15.3.1 Biotic stress

Legumes, like other domesticated plant varieties, when exposed to a biological threat suffer a decrease in overall agricultural output. These conditions are mainly observed in response to attack by a variety of pathogenic organisms including viruses, bacteria, fungi, insects and other infestations.

15.3.1.1 Viral diseases

Some of the major viral infections of legumes are observed in peas. More than 20 different plant viruses have been identified in peas, and the pea seedborne mosaic virus (PSbMV), pea/bean leaf roll virus (P/BLRV) and pea enation mosaic virus (PEMV) are of major economic importance (Vemulapati *et al.*, 2011; Makkouk *et al.*, 2012; Maury *et al.*, 2012). Infections can have disastrous financial impacts for the farmer. Soybean, a major economic commodity, is also affected by viral infections. The significant infections are caused by soybean mosaic virus (transmitting through the seeds) and tobacco ring spot virus (Zhang *et al.*, 2012; Sastry, 2013). Similarly, black gram and mung bean are infected by yellow mosaic virus (Vargas *et al.*, 2013). The virus, transmitted by the whitefly *Bemisia tabaci*, causes the appearance of irregular yellow and green patches on the plant, ultimately causing the total deterioration of the plant system. Other viral agents infecting leguminous plants include beet western yellow virus, subterranean clover red leaf virus and other luteoviruses, cucumber mosaic virus, bud necrosis virus and cowpea golden mosaic virus (Jones, 2012; Makkouk *et al.*, 2012).

15.3.1.2 Bacterial diseases

Bacterial infections in legumes are relatively limited compared to viral and fungal infections. *Pseudomonas syringae* and *Xanthomonas campestris* have been associated with bacterial blight and bacterial pustule in soybean plants (Baltrus *et al.*, 2012; Khare & Khare, 2012). Blight and pustule are usually observed in the cool, high altitudes and humid, tropical environments, respectively. Bacterial wilt is commonly observed in the groundnut; it is caused by the *R. solanacearum* and manifest as the damping and killing of seedlings (Van der Linden *et al.*, 2013). These bacterial infections may be spread by infected plant material, infected seeds, contaminated nutrient sources and water. The bacterial sources can be effectively controlled by appropriate handling techniques.

15.3.1.3 Fungal diseases

Due to the widespread geo-distribution of the legumes, a variety of different fungal species infect legumes according to their habitat. A number of fungal species, including *Phakopsora pachyrhizi*, *Pythium ultimum*, *Fusarium oxysporum* and *Phytophthora sojae*, are established causes of infections in soybean (Tyler, 2007; Goellner *et al.*, 2010; Wei *et al.*, 2010; Zhang *et al.*, 2010). Wet rust and rotting of seedlings and roots are the most important diseases, responsible for significant economic losses in soybean crops. Similarly, the important fungal diseases infecting lentils are dry root rot (*Macrophomina* spp.), vascular wilt (*Fusarium* spp.) and pythium root rot (*Phythium* spp.) (Singh & Azam, 2010; Mohammadi *et al.*, 2011; El-Hassan *et al.*, 2013). The fungal diseases of groundnut include root diseases, foliar diseases, seed diseases and pod diseases attributable to at least 15 different species of fungi (Ihejirika *et al.*, 2010). Most of these fungal agents are soil borne and tend to affect all the parts of the plant.

15.3.1.4 Nematode and arthropod infestations

A number of worms and insects are associated with parasitic damage to legumes. Of the many nematodes, the most significant ones infecting legumes include root-knot nematode (*Meloidogyne* spp.), cyst nematode (*Heterodera* spp.), stem nematode (*Ditylenchus* spp.) and reniform nematode (*Rotylenchulus* spp.) (Lombardo *et al.*, 2011; Vovlas *et al.*, 2011; Ahmed *et al.*, 2012; Leach *et al.*, 2012). Nematodes, in comparison to other biotic stresses, do not produce specific signs of infection; rather their infection is exhibited as a delayed appearance of normal growth characteristics, decreased ability to fix nitrogen and retardation of development. Insect pests, on the other hand, show conspicuous features of damage to all the plant body parts including leaves, stems, flowers and pods. More than 70 insect species infect soybean crops. For example, *Spodoptera litura*, *Hedylepta indicata* and *Mylabris pustulata* have been implicated in causing crop losses of up to 50% (Dialoke *et al.*, 2010; O'Neal & Johnson, 2010; Dung *et al.*, 2011). Other financially significant insects infecting legumes include aphid, hairy caterpillar, thrips, white fly (vector of the yellow mosaic virus), leaf hopper, army worm and green bug (Viteri *et al.*, 2009; Xu *et al.*, 2010; Musser *et al.*, 2011; Hodgson *et al.*, 2012; Kumar *et al.*, 2012b; Mehrkhou *et al.*, 2012).

15.3.1.5 Weeds and herbs

Unlike other biotic stress factors, weeds and herbs may not cause direct damage to the legumes. Instead they compete for the same resources that are available to the cultivated plants. They affect the growth, stature, agro-processing and overall yield of the plants. Groundnut, lentils, soybean, pigeon pea, mung pea, black gram and chickpea are all adversely affected by weeds, especially when infestation takes place during the early stages of plant life. Commonly observed weeds include various species of *Chenopodium*, *Avena*, *Sonchus*, *Convolvulus* and *Phalaris* genera (Rashid *et al.*, 2009; Döring *et al.*, 2012; Šarūnaitė *et al.*, 2012; Smitchger *et al.*, 2012). Reduction in yields of up to 90% have been noted owing to the growth of weeds in case of certain legumes (Table 15.1).

Table 15.1 Biotic and abiotic stress response factors.

Legume response to abiotic stress factors	Legume response to biotic stress factors
C-promoter binding proteins (low temperature)	Chitinase (fungal infection)
Dehydration responsive element-binding protein (low temperature and drought)	Taumatin gene (fungal infection)
Heat-shock factor (high temperature)	Nod factor receptor protein (legume–rhizobium interaction)
SQUA promoter-binding protein-like 7 (micronutrient deficiency)	Necrotic lesion response gene (viral infection)
1-pyrroline-5-carboxylate synthase (salinity)	

15.3.2 Abiotic stress

Abiotic stress factors include a list of all the environmental and physico-chemical features that affect the individual growth and development of a plant. These factors usually act in combination in form of a 'stress-matrix' resulting in heavy economic and agricultural losses (Mittler, 2006). The common abiotic constraints to plant growth include the high or diminished availability of nutrients and salts, altered levels of water in the agricultural land, deviation from the optimum growth temperatures and natural disasters, among many other edaphic, climatic and geographical factors.

15.3.2.1 Extremes in temperature

Legumes are widely distributed among the various geographical zones of the world and among the various habitats. Though the plants are resistant to slight changes in surrounding temperatures, when there is a significant increase in temperature or the elevated temperature is coupled with other stress factors, the capability of the plants to resist harsh environmental conditions is decreased. This, ultimately, results in a lack of proper growth or the death of the plant. In particular, the cold season legumes are highly sensitive to heat and increases in temperature (Thomas, 2010). These include the lentils, faba beans and pea varieties. Freezing or very low temperatures coupled with frost have been associated with yield losses in warm temperate legumes including cowpea, soybean, groundnut and mung bean (Thakur *et al.*, 2010).

15.3.2.2 Drought and waterlogging

Fluctuation in water levels in agricultural land is a major source of reduced legume yields. Drought alone does not present a major challenge to the plants. However, a combination of drought with other physical constraints like heat and salinity, prove to be a lethal combination for plant growth and development. For instance, drought in combination with high temperature results in a decrease in stomatal opening, making the overall stress-resistance mechanisms of the plant less efficient (Reynolds-Henne *et al.*, 2010). This process results in the physiological stress and, ultimately, death of the plant. Soya, alfalfa, *Medicago* spp. and a number of other legume species tend to show the greatest negative growth patterns in response to drought conditions.

At the other extreme, waterlogging results in significant physiological impact and yield loss (Palta *et al.*, 2010). Moreover, waterlogging promotes the growth of weeds and grasses in cultivable land, resulting in an increase in biotic stress for the already strained legumes (Rodenburg *et al.*, 2011). One important detriment of waterlogging is the decrease in the soil population of rhizobia, resulting in diminished nitrogen fixation capability of the plants (Lindström & Mousavi, 2010). Soybean, mungbean and grain legumes are the most important legumes affected by waterlogging.

15.3.2.3 Salt stress and nutrient depletion

Salinity, acidity and heavy metal stress belong to a group of abiotic stresses that affect plant growth by altering the soil characteristics, ultimately resulting in inappropriate growth conditions. As with other abiotic factors, the detrimental effects of these constraints are enhanced when occurring in combination with other factors. The most significant of these synergistic abiotic constraints are extremes in temperature. Heat, for instance, results in increased transpiration resulting in enhanced uptake of water loaded with salt, acid or heavy metal (Hamidou *et al.*, 2013). Salt stress, in particular, has a significant effect on a number of physiological properties of legumes, including nodulation, symbiotic nitrogen fixation and plant growth capability (López-Gómez *et al.*, 2013). Among the legumes, various varieties of peas, beans, lentils and forage legumes have been extensively studied for their responses in salt-stressed conditions. Similarly, acidic and metal-stressed soil results in inadequate growth conditions for many legumes. Conversely, the unavailability of appropriate nutrients or nutrient-depleted edaphic conditions result in lack of essential nutrients, resulting in poor growth.

15.3.2.4 Natural disasters and other abiotic constraints

The most significant of the natural disasters that affect the normal growth of legumes are tornadoes, wild fires, floods and high-speed winds (Pagano, 2013). These elements cause widespread destruction of agricultural lands. In addition, other eco-physiological agents adversely affecting legume output include variations in atmospheric carbon dioxide and the amount, duration and intensity of light (Ross *et al.*, 2013; Wang *et al.*, 2012). All these factors affect the nitrogen fixation capability, photosynthetic apparatus and anatomical features of the plants.

Biotic and abiotic factors are, hence, responsible for causing significant losses to the growth and yield of leguminous plants. The plants adapt a number of cellular and molecular pathways to neutralize the effects of these stress factors. Utilizing these mechanisms to obtain economic benefits can help in decreasing the overall disease burden on the agricultural system.

15.4 Host defence mechanisms

Legumes, like other plants, resist changes to their normal living conditions. Plants have evolved a number of mechanisms to defend themselves against harsh environmental conditions and various biological threats. Production of certain chemicals, regulation of normal physiological processes and modification of anatomical features are some of the chief strategies in this respect. Underlying all types of responses is modification of the genetic and molecular assemblage (Figure 15.2).

Legumes may present a number of physiological changes on exposure to an abiotic stress. In response to drought, for instance, a leguminous plant shows a variety of different protective mechanisms. These include changes in membrane composition and loss of turgidity, which help in the osmoprotection of the plant. Allied modifications include changes in anatomical characteristics like root thickness and soil penetrating capability of roots, alterations in photosynthesis and carbohydrate metabolism, and increased accumulation of organic acids and osmolytes. Similarly, in response to

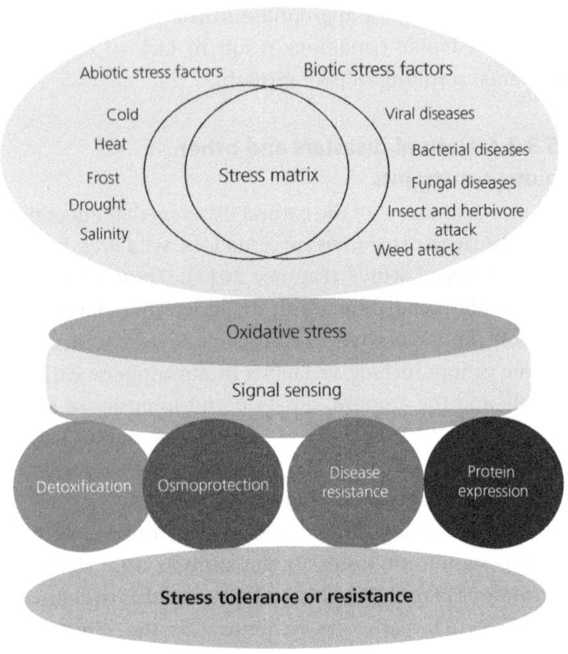

Figure 15.2 Legume response to biotic and abiotic stress factors.

high temperatures, legumes respond physiologically by decreasing the rate of photosynthesis and increasing the rate of respiration, stomatal conductance and leaf temperature. Hence, multiple responses are observed in response to a single physical constraint. Another type of response is the regulation of cellular transporters and pumps. These are mainly involved in ion exchange and regulation across membranes. An example of this osmoregulatory mechanism, in case of drought for instance, is the Na^+/H^+ antiporter system, which involves regulating the cytoplasmic pH, sodium ion levels in the cell, cell turgor and, consequently, the homeostasis of the cellular environment. Secondary to the process is the change in osmotic gradient, signalling cascade and production of certain metabolites.

The primary response, generally observed in reaction to any stress condition, is the regulation of plant metabolites, both primary and secondary. Additionally, certain metabolites are produced by the plant to detoxify or neutralize the effects of the damage caused by the stress factors. These metabolites are naturally produced but stress conditions lead to their overproduction. Principal among these is oxidative stress. A variety of neutralizing species including enzymes, non-enzyme systems and secondary metabolites, are produced as a result. Glutathione reductase, superoxide dismutase, catalase, carotenoids, ascorbate, cyanins, tocopherol, glycine and sugars are among the many ROS responsive elements (Becana *et al.*, 2010; Gill *et al.*, 2013; Liu *et al.*, 2013). All of these chemicals are considered to play important roles in counteracting the stress. Additionally, in response to salt stress and a chilling environment, certain legumes produce amines like glycine betaine that help in protection of plant. Sugars are found in increased quantities in stressed plants. Mannitol, for example, has been associated with the ability to scavenge reactive oxygen species (ROS) that are deemed harmful for legume health. Similarly, proline, an amino acid, has been found to protect the plant from salt stress in saline conditions (Kohl *et al.*, 1990). Although a similar mechanism may be deployed by legumes in case of biotic attack, the secondary metabolites are usually different compared to those produced in response to abiotic stress factors. Chief among these are the alkaloids, cyanins, terpenes, lipids and related compounds. These chemicals are potent antibacterial, antiviral, antifungal, antihelminthic and antiprotozoal entities.

Apart from the physiological response of leguminous plants to the stressed conditions, legume–rhizobia interactions are also adversely affected by these biotic and abiotic constraints. Some 70 bacterial species, classified into the genera *Sinorhizobium*, *Rhizobium*, *Mesorhizobium*, *Bradyrhizobium*, *Azorhizobium* and *Allorhizobium*, are involved in this legume–bacteria symbiotic relationship (Oldroyd *et al.*, 2011). Any physical phenomenon that adversely affects the bacterial growth causes disruption of this relationship, thus diminishing the nitrogen fixation process and the benefits associated with it. The most threatening of the various physical constraints for this symbiotic relationship are drought and high temperatures. In case of drought, for instance, shoot photosynthesis is compromised leading to a lack of carbohydrate supply to the rhizobia. This is followed by decreased biological nitrogen fixation capacity and, hence, a decrease in the overall ammonia output. Rhizobia have developed very efficient mechanisms to protect themselves and the plant nodules from drought. Firstly, they are themselves more tolerant to physical parameters compared to the legumes. Moreover, through a number of mechanisms they provide protection to the nodules of the plants. Trehalose-phosphate synthase is one such system involved in providing osmoprotection to legumes in terminal drought conditions. Similarly, in order to cope with the decreased oxygen supply, legumes contain a pigment, leghaemoglobin, which provides a constant microaerophilic environment to the affected cells (Navascués *et al.*, 2012).

Once appropriate information is available regarding the role of various chemicals in plant defence, specific biotechnological interventions can be developed and employed for improving plant yields.

15.5 Disease and stress checkpoints

Leguminous plants tend to respond to the external threats and maintain a state of homeostasis by following a generalized mechanism initially, and respond specifically to a particular type of stress factor by devising certain mechanisms to neutralize its effects. The responses generated by leguminous plants to the stress conditions require a cascade of genetic and molecular events. Initially an active signalling process is required to detect the stimulus. This is followed by the activation of genetic and molecular machinery to respond to a particular threat by regulating either the production of certain chemicals or by stimulating physiological changes in the plant body. Hence, the basis of any response is at the level of the genes. Providing resistance against biological factors is simpler compared to the abiotic constraints, as the latter affect not a single trait but a range of entities in the plant body.

In order to identify targets for providing resistance to a stress-threatened legume, the cascade of events involved in the stress response process must be understood. On initial exposure to a biotic or abiotic stress, the plant first experiences a state of shock in which the ionic balance, osmolality, membrane stability and general homeostasis are disturbed. Further, an oxidative burst may result in aggravation of the stress condition (Jaspers & Kangasjärvi, 2010; Miller *et al.*, 2010). Stress-specific cellular sensing mechanisms are activated that help in signal transduction, yielding the activation of transcription factors and genes (Hirayama & Shinozaki, 2010; Suzuki *et al.*, 2012). These players ultimately help the plant in re-establishing cellular homeostasis, aiding the plant to develop tolerance to the stress factor. Avoidance is another mechanism adapted by legumes in which the plant brings about certain morphological or physiological changes to cope with the threat. In case the stress response mechanism fails and the stress factor significantly affects the plant health, a plant may proceed to retardation and cessation of growth, ultimately leading to the death of the plant (Potters *et al.*, 2010).

For the legumes, biological stress factors present a major challenge to plant growth and development. In response to the biotic constraints, the stress response mechanism observed is, necessarily, similar to that observed in response to abiotic factors. Generation of reactive oxygen species, production of metabolites and certain physiological changes are very much evident in response to attack by herbivores or pathogenic infestations. However, as opposed to other stress factors, most of the biological factors generally tend to inhabit the plant or are exposed to the plant material for a very short time. Hence, legumes have developed a number of strategies to defend themselves. Principal among these is the production of metabolites. These chemicals, including oxylipins, flavonoids, phenylpropanoids and abscisic acid, have been isolated from stressed legumes (Pawlak-Sprada *et al.*, 2011; Samanta *et al.*, 2011; Dupont *et al.*, 2012; Reverberi *et al.*, 2012). Among the other compensatory mechanisms, enzyme activation to produce products that can help in defence against the biotic stress factor is of significant importance. Lipoxygenase activation on nematode attack and induction of glutathione transferase to induce apoptosis are established stress-induced responses in many legumes (Dalton *et al.*, 2009; Fürstenberg-Hägg *et al.*, 2013).

Responses to abiotic constraints are considered polygenic traits. A number of stress response genes regulated by multiple transcription factors are involved in generating responses against the threat. The dehydration responsive element-binding transcription factors (DREBs) comprise one such system involved in the upregulation of a number of stress-responsive genes, including *kin1*, *kin2*, *rd29A*, erd10 and *cor6.6*, all of which are involved in protecting the plant from water and cold stress conditions (Ray *et al.*, 2011; Mizoi *et al.*, 2012). Similarly, abscisic acid (ABA) and the related ABA-responsive elements (ABRE) genetic systems are involved in regulating the legume response to drought conditions (Bakht *et al.*, 2013). Heat-shock elements (HSEs) are the transcription factors involved in the activation of hsp genes (heat-shock proteins). These elements help in regulating the legume response to high temperature (Ghosh & Pal, 2012; Soares-Cavalcanti *et al.*, 2012). Hence, controlling the plant response by the regulation of transcription factors offers diverse means to protect legumes from the stress factors.

Another category of abiotic stress response elements comprises the genes encoding the machinery for synthesizing various metabolic products, including amino acids, amines, carbohydrates, sugar alcohols and various secondary metabolites (Betti *et al.*, 2012; El-Rahman *et al.*, 2012; Silvente *et al.*, 2012; Palma *et al.*, 2013; Wink, 2013). The gene products formed thereafter interact with the target sites to produce their effects. Another important class of chemicals that significantly affect the legume response to any abiotic stress is the molecular chaperones. Of these, heat-shock proteins (HSPs) serve as members of the molecular cascade involved in protection from not only heat stress but also water and salt stress at relatively lower temperatures (Al-Whaibi, 2011). Another class of proteins that assist the plant in tolerating a freezing environment are the late embryogenesis abundant (LEA)-type proteins. These molecules are involved in providing cryoprotection to the plant (Thomashow, 1999). LEA-type proteins are hydrophilic in nature and help the threatened cell by increasing membrane stability at freezing temperatures. The HSPs commonly observed in legumes include HspB1, Hsp60, Hsp70 and a variety of other molecular chaperones (Hernandez & Vierling, 1993; Brunet *et al.*, 2009; Komatsu & Ahsan, 2009; Pick *et al.*, 2012; Qu *et al.*, 2013). The LEAs mainly expressed in legumes are LEA1 (GmD19, GmPM11), LEA2 (GmPM6), LEA3 (GmPM30) and LEA4 (GmPM1, GmPM9)[1] (Liu *et al.*, 2011; Battaglia & Covarrubias, 2013).

Apart from the natural responses observed in many plants, introduction of stress-responsive foreign genes or gene cassettes can also prove to be very beneficial. Engineering of certain transcription factors like C-repeat-dehydration response elements (CRT/DRE) and CBF3 elements in legumes, like chickpea, *Medicago* spp., soybean and peanut, help in the induction of cold-regulated gene systems (Chen *et al.*, 2010; Mantri *et al.*, 2012; Reddy *et al.*, 2012). Consequently, a number of genes are regulated to elicit a response by altering metabolic and other phenotypic activities. Significant among these changes are increased levels of proline, sugars and enhanced tolerance to chilling temperatures. Apart from the transcription factors, genes encoding the production and regulation of certain metabolites are also utilized for enhancing tolerance to both biotic and abiotic stress. Proline, a potent stress-response agent, is produced by glutamic acid metabolism (Szabados & Savouré, 2010). The principal enzymes involved are pyrroline-5-carboxylate (P5C) synthase (P5CS), P5C reductase (P5CR), proline dehydrogenase (ProDH) and P5C dehydrogenase (Lehmann *et al.*, 2010). Transgenic legumes like groundnut, chickpea and clovers expressing mutated forms of P5CS have been associated with isolation of larger quantities of proline and, consequently, better salt resistance (Bhatnagar-Mathur *et al.*, 2009; Magyar-Tábori *et al.*, 2011; Vaseva *et al.*, 2012). Additionally, introduction of a trehalose overexpression system into legumes, including *Medicago* spp. and many forage varieties, likewise can help provide better osmoprotection and drought tolerance (Alam *et al.*, 2010).

An understanding of these response mechanisms can help in devising strategies to target various molecular and cellular mechanisms by molecular biology approaches. The genes identified can help in providing the targets for molecular intervention to control and respond to various stress factors, biotic and abiotic, in an effective way.

15.6 Biotechnology in legumes

Biological and physical stress conditions are responsible for significant losses in legumes. Traditional methods of control include the use of fungicides, pesticides and other hazardous chemicals for treating various biotic constraints. However, in case of any physical stress, the farmer has to bear huge financial losses due to reduced quality and agricultural yield. Molecular intervention techniques for combating stress conditions have helped in improving the yield and quality of legumes in a safer fashion. Biotechnology identifies and manipulates the potential checkpoints that can be beneficial for defence against any biotic or abiotic constraint. Moreover, the overall output of the legumes has improved, aiding the achievement of better financial returns.

15.6.1 Applications

The use of biotechnology approaches for protecting legumes from any physical or biological threat involves expertise from various fields of study. From

[1]Each LEA protein class consists of various proteins. Here, Gm refers to *Glycine max* (soybean) followed by the protein name according to nomenclatural convention.

Table 15.2 Genetically modified legumes and the genes manipulated for enhancing abiotic stress resistance.

Legume species	Genes isolated	References
Glycine max	DRE-binding protein (drought tolerance)	Chen *et al.* (2007)
Vicia faba	*PR10a* (cold tolerance)	Hanafy *et al.* (2013)
Arachis hypogaea	*PnAG3* (drought resistance)	Yu *et al.* (2012)
Phaseolus mungo	*Dn2* (aphid resistance)	Priya *et al.* (2013)

the diagnosis of an aetiological agent to the final development of a solution for the diseased condition, the dedication of a number of scientists is required. Plant pathologists, geologists, chemists, geneticists, molecular biologists, biotechnologists, microbiologists, bioinformaticians and ecologists, all are necessary for the development of a successful remedy for a disease condition. The discovery of a pathological condition, determination of aetiological factors and the players involved in mounting a response to the stress factors, and ultimately the employment of interventional strategies – all contribute to the introduction or enhancement of legumes' defences against physical and biological constraints. Molecular interventional strategies have helped in the attainment of tolerance against cold stress, herbicides, drought, salinity and high-intensity light. Moreover, these strategies have aided in the provision of resistance against viruses, bacteria, fungi and other parasites. Transgenic plants encoding genes for the enhanced production of osmoprotectants, enzymes and secondary metabolites have helped legumes in attaining stress resistance (Table 15.2).

15.6.2 Benefits obtained by biotechnology approaches

A human population explosion over the past four decades has added to the demand for both food and cash crops. Increased nutritional demand, a growing need to supply raw materials for industrial processes and a decreased land area for agricultural purposes have all contributed to the acceptance by the public of alternative methods for crop yield improvement. Biotechnology has provided innovative ideas and methods to deal with many of the severe stress conditions that would not have been targeted by any traditional means of crop improvement (Hohn & Schachermayr, 2010). These molecular techniques have helped in the improvement of legume quality, in terms of better wood, nutritional content and nitrogen fixation capability (Chandra *et al.*, 2010; Pratap *et al.*, 2010; Provorov, 2013). Incorporation of pest resistance genes into legumes helps enhance tolerance to pests, thereby decreasing the need to employ various pesticides (Keneni *et al.*, 2011). Additionally, the introduction of virus resistance genetic cassettes has helped in the control of not only viral diseases but also insect vectors that are involved in the transmission of plant viruses (Cavatorta *et al.*, 2012). However, the main focus of biotechnologists has now shifted from the control of biotic to abiotic stress factors. Biological threats cause the activation of only a single genetic pathway for the induction of disease. In comparison, physical stress factors are involved in the induction of a series of events that together result in detrimental effects on the plant. Therefore, the introduction of genes that can provide resistance against these abiotic stress factors is the main target of biotechnological intervention. Genes encoding factors involved in the defence against drought, extremes in temperatures and salinity have been successfully incorporated into a number of leguminous plants (Thao & Tran, 2012). These have resulted in the achievement of benefits, both in the quality and overall agricultural yield of legumes.

15.7 Molecular approaches to improving legume defences

The advent of modern biotechnological procedures has helped in the incorporation of better protective strategies against biological and physical stress factors. The molecular intervention strategies are based upon an understanding of various disease and stress conditions. The molecular and cellular mechanisms involved are investigated, followed by isolation of the factors involved in mounting a response to these conditions. These genetic and molecular entities are then introduced into the target plants, which are then evaluated for their ability to cope with a stress condition (Table 15.3).

15.7.1 Incorporation of disease resistance genes

A basic stage in developing ways of dealing with the biotic stress factors is gaining an understanding of the disease pathogenesis. Many bacterial and fungal species

Table 15.3 Biotechnology tools applied to leguminous plants for stress resistance.

Biotechnology tool	Purpose
Gene-phenotype relation, polymerase chain reaction	Identification of diseased plant
Culturing techniques, sampling of growth conditions	Confirmation of aetiological agent
Restriction fragment length polymorphism, amplified fragment length polymorphism, quantitative trait loci	Isolation of candidate gene
Transgenesis, site-directed mutagenesis	Introduction of a transgene into a host
Expressed sequence tags, metabolic assays, functional assays	Confirmation of expression of the gene in the target host

are dependent upon protein-degrading enzymes to gain entry into the legume body. These enzymes tend to help the prey gain easy access to the plant's internal systems, thereby causing serious harm to the plant's health; the plant responds by producing certain 'anti-stress' elements. These elements are generally classified as the pathogenesis-related (PR) proteins and antifungal proteins, on the basis of the pathogenic agents they target. Many PR proteins are activated against bacterial and fungal species in legumes, for example chitinase, thaumatin-like protein and Nod factor perception protein (Pearce *et al.*, 2010; Gough & Jacquet, 2013; Rey *et al.*, 2013). Knowledge of these PR proteins and their pathogenesis mechanisms can help in the development of agents to help tackle infection at the plant-to-plant transmission level, while not interfering with any metabolic process. Additionally, genetic cassettes can be introduced into bacteria or fungi to produce heterologous proteins that can be of biotechnological utility.

15.7.2 Isolation of disease resistance genes

A number of techniques may be employed in identifying the disease resistance genes. The more resistance genes available, the more chances there are of developing and employing effective defence mechanisms against the susceptible pathological factors. Chromosome walking is one such technique employed by plant biotechnologists to identify the resistance genes. Based upon the polymorphic nature of the genetic material of normal plants and plants under stress, this technique has been applied to identify many resistant genes. Moreover, genetic markers have been isolated by the technique known as restriction fragment length polymorphism (RFLP) mapping. It involves the comparison of enzyme-digested restriction fragments to identify the different genes observed in stress conditions. Randomly amplified polymorphic DNA analysis makes use of a similar strategy. However, in the latter case short nucleotide sequences are utilized to differentiate between the amplification patterns of genetic material in different growth conditions. The polymerase chain reaction (PCR) forms the basis of all these techniques by providing a means to rapidly amplify large genetic sequences. Once the individual genes have been identified, a walk along the genome length is carried out to identify other genes that might have taken part in providing resistance to the plant. Consequently, the whole genome may be screened from one end to the other, avoiding any chances of misreading any gene sequence. These marker genes are usually cloned in yeast artificial chromosomes (YACs), bacterial artificial chromosomes (BACs) or other cloning vectors. Moreover, complementary DNA libraries may also be developed to keep a record of the candidate gene. In certain cases, chromosome jumping may be performed to study different segments one by one, with relatively larger spanning distances. The stress resistance candidate genes once identified need to be evaluated and confirmed by observing the phenotypic presentation of the plants as well.

Another method for successful identification of a candidate gene is the gene tagging technique. It involves the establishment of a plant line that contains a molecularly characterized transposable element. Based upon the principle of gene transposition, screening is done by hybridizing the gene of interest with a particular marker. This method helps in the rapid identification of transposable elements, so that the gene involved in providing disease resistance can successfully be identified and isolated. However, high rates of transposition or the availability of transposable sites in close proximity can hinder the isolation of effective stress resistance genes. The method requires the successful inculcation of the resistance gene into the progeny. A number of techniques have been adapted to increase the frequency of transposition including the incorporation of foreign promoter elements, selection of transposable elements close to the gene of interest and decreasing the background mutation rates. Additionally, transposition-independent methods, generally relying on vector-based gene delivery, have been employed for performing gene tagging procedures with some success.

Quantitative trait loci (QTL) mapping is a similar mechanism that involves the study of genes in association with various polygenic traits that are involved in the phenotypic expression of a character. The method uses various molecular tags, like amplified fragment length polymorphisms (AFLPs) or single nucleotide polymorphisms (SNPs), to provide valuable information related to the genetic basis of any stress response.

Subtractive hybridization cloning, another method for isolating a gene of interest of unknown biochemical origin, has also been successful in leguminous plants. The procedure relies on the comparative genetic analysis between two plants, one grown under normal conditions and another subjected to the stressful environment. Biotin is linked to the DNA obtained from the plant lacking the gene of interest and then the genetic material from both the species is combined. Through repeated cycles of association and subtraction, pure genetic material containing the gene of interest can be isolated. The product can then be amplified using PCR and cloned into suitable complementary DNA (cDNA) libraries. The chief advantage of this method is the universality and comparative ease in comparison to other methods (Figure 15.3).

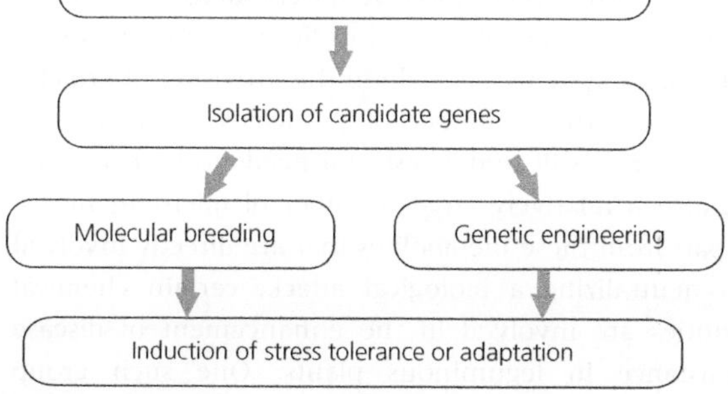

Figure 15.3 Biotechnology intervention pathway for stress resistance in legumes.

Two other methods employed for the isolation of stress resistance genes are the isolation of resistance gene proteins and the functional cloning approach. The former depends upon determining the interaction of an external stress stimulus with the plant cell's receptors. The receptor may, hence, be identified, isolated and cloned (Postel & Kemmerling, 2009). The sensory units of legumes involved in the interaction with microbes include Lysine K and M (Lys K/M) and other protein kinases (Boller & Felix, 2009; Gimenez-Ibanez *et al.*, 2009). Similarly, a number of molecular entities, like Nod factor receptor (Nfr) proteins and beta-glucan-binding protein, are involved in assisting the legume–rhizobium interaction (Choudhury & Pandey, 2013; Monaghan & Zipfel, 2012). Functional cloning, however, involves the molecular dissection of the entire genome. The genetic material is isolated, cleaved and cloned into suitable vectors to produce libraries. This DNA library is then introduced into the plant line lacking the gene and the transformants thereby produced are teffective in isolation of many stress resistance genes, one major challenge these methods present is the development of a functional assay for the final evaluation and screening purposes.

15.7.3 *In vitro* mutagenesis and transgenesis

Once the target gene has been isolated, its introduction into the target cell helps to provide resistance against a specific stress factor. The method involves the introduction of specific genes or gene segments into a target tissue to successfully carry out the protein engineering process. This requires the introduction of a synthetic primer encoding the necessary mutation. Once the primer anneals with the target genetic material, the replication process helps in the incorporation of the desired mutation or target gene at a specific site. This method has been used for neutralizing a gene or introducing a resistance gene for a particular stress factor. Another method, transgenesis, employs various biological or physicochemical methods to introduce a transgene (foreign stress-response gene) into a target cell. The gene of interest is cloned in a suitable vector, which is then introduced into a particular cell by means of vectors, liposomes, gene gun or related physicochemical methods. Once inside the cell, the gene is expressed in order to obtain the product in the form of protein or other secondary metabolites. A number of legumes have been genetically manipulated using the transgenesis process and have successfully acquired tolerance against pests, salinity, drought and chilling (Chen *et al.*, 2010; Coba de la Peña *et al.*, 2010; Hanafy *et al.*, 2013; Lüthi *et al.*, 2013).

15.7.4 Screening methods

Confirmation of the insertion of a transgene into a vector is imperative for determining the success of the genetic manipulation process. The screening methods are based either upon the analysis of a foreign gene in the vector or on the final product obtained as a result of expression. PCR provides one such easy method for confirmation of a gene in a vector. Once the gene incorporation has been confirmed, the steps leading to expression must also be considered to achieve an effective transgene. However, the cloning and expression procedures do not necessarily result in attainment of tolerance against the stress factor. Confirmation needs to be made by performing certain functional assays. In case the introduced transgene encodes for a protein, the final product needs to be screened and analysed. However, if any morphological or physiological change is desired, the progeny are evaluated for the appearance of the desired characteristics.

15.7.5 Specific induction of stress response mechanisms

Finding the means to inhibit viral growth in susceptible legumes has not been very successful. Viruses, like other biological stress factors, initiate a series of cellular and molecular mechanisms in the host plant. Moreover, many resistance genes isolated from plants as a result of virus infection have demonstrated their ability to neutralize the signs of the virus infection, but they do not inhibit the viral replication altogether. However, expressing some of the viral resistance genes in plants has been successful. This strategy may include the incorporation of genetic sequences that bear homology or complementarity to the viral genes. Utilization of satellite RNA, defective interfering DNA, antisense RNA, all or part of the viral replicon or the use of subunits like viral capsid protein, have shown to be effective in decreasing viral replication and disease severity. The use of viral subunits ensures the activation of a monogenic resistance trait, in comparison to other methods of developing resistance. The method involves the use of a chimaeric assembly incorporating the viral gene, which is introduced into a particular part of the plant. This and other techniques for suppressing viral replication are highly dependent upon the localization of the virus within a plant organ. This may be achieved by the physical barriers between the cells, inactivation of transport proteins or by the secretion of natural chemicals that might neutralize the virus and decrease cell-to-cell transfer. Once the virus has been isolated in a particular region, the chances for cell-to-cell transmission and resulting disease signs are considerably reduced. Other biotechnological techniques employed for virus control include the increased production of antiviral agents like the inhibitors of virus replication (IVRs). IVRs, for instance, have been indicated in the suppression of replication of a number of plant viruses including tobacco mosaic virus (TMV) and begomoviruses (Jeffries, 2008).

Another technique that has proved successful in defending plants against biotic stress factors is by the induction of secondary metabolites that may help in retarding the growth of the infesting agent. The method requires the isolation of disease resistance genes involved in the detoxification process, in response to a particular threat. After isolating the gene, it is cloned in a suitable vector and the plasmid is transformed through either biological or physicochemical methods into a susceptible plant variety. The plant is then subjected to evaluation for the degree of tolerance developed to a particular stress condition. The genetic variability of various plant species has provided a means to isolate the detoxification-catalysing systems.

Genetic engineering has, therefore, helped in addressing the environmental and toxicological problems associated with earlier pest control techniques. Tabtoxin and proteinase inhibitors, for instance, are involved in the neutralization of various biotic stress factors in legumes (Hirel *et al.*, 1997; Ramírez-Suero *et al.*, 2010). The introduction of small interfering RNAs (siRNAs) into plant systems can help in the provision of broader cover against a number of pesticides. This method, however, is still under test and needs to be evaluated against a relatively larger number of microorganisms. Apart from these metabolites that are directly involved in neutralizing a biological attack, certain chemical entities are involved in the enhancement of disease resistance in leguminous plants. One such group includes the surface active peptides, which help not only in signal transduction but also in complementing the antimicrobial activities of various metabolites in response to a microbial attack.

These biotechnology approaches have, thus, helped in the provision of significant benefits to farmers and the overall agriculture sector. The improvements in yield and quality of the crops have been directly related to the incorporation of the molecular sciences into the legume fields.

15.8 Integration of GM legumes in current agricultural systems

The most effective biotechnological advance is the incorporation of disease resistance genes into the plant genome. This strategy has advantages for plant biotechnologists as it is effective in providing disease control, is inexpensive, heritable and environmentally safe. The technique utilizes the isolation of genes that may be overexpressed by certain plants in response to a biotic or abiotic stress condition. Referred to as the hypersensitive reaction, the response of the plants involves the release of certain phytochemicals that work in collaboration to detoxify the threat or neutralize the agents attacking the plant. Glucanases, chitinases, phytoalexins and protease inhibitors are the chief species released by the plants to neutralize the infectious agents. At the cellular level, the response to the stress factor necessarily involves the alteration of membrane integrity and the production of reactive oxygen species. However, the overall morphological, anatomical and physiological response varies widely among the taxonomically diverse plants. The resistance to various stresses and the degree to which a plant tolerates a stress is markedly different among different families and species. Lateral transfer of resistance from one species to another is very rare. Hence, biotechnology is the only means of transmitting these characteristics. Resistance to various factors is only expressed when there is exposure to the causative agent. Moreover, there are good chances of acceptance of the foreign disease resistance gene.

However, a number of concerns have been raised regarding the incorporation of biotechnological procedures into mainstream agricultural processes. In the case of biotic stress factors, when a single disease resistance gene is introduced into a plant species, many pathogenic organisms may become resistant to the proteins produced thereby. Additionally, a number of resistance genes are not naturally available in the botanical gene pool. Hence, for the development, introduction and evaluation of the effects of the foreign/synthetic gene introduced into the plants, a longer duration and a large number of experimental trials are necessary before the gene can be introduced into the agricultural system.

The success of molecular techniques depends greatly upon the frequency of crossing, generation times and the employment of manipulative techniques for improving yields. Therefore, they fail to exhibit any beneficial effects in cases where the plants show restrained breeding habits, longer life cycles and vegetative propagation techniques are frequently employed. Physical stress factors involve responses that are polygenic traits; i.e. a number of plant defence mechanisms are activated. Hence, prior to the incorporation of resistance genes against these traits, complex screening procedures may be required to ensure the ideal response against the threat. Moreover, a number of concerns have been raised against the adoption of biotechnology procedures by environmental, regulatory and public health authorities. Use of genetically modified organisms (GMOs) for pest control purposes have been the centre of discussion for years. The introduction of GMOs to the open field setting can threaten the overall biological cycle, resulting in the disturbance of food chains and ecological cycles.

However, the recent advances in biotechnology can help in reassuring decision makers that it can serve as a safer alternative to meet the food and energy demands of the world. The threats associated with the use of GM crops are mainly based upon hoax and unsound knowledge about the actual biotechnological techniques. A simple, yet comprehensive, educational scheme should be devised to educate the masses about the benefits of employing biotechnology methods for legumes. This would involve a combined effort by the agriculture sector, academia and industry. This would not only promote the acceptance of biotech methods by the general public but also help in establishing linkages for evaluating, studying and addressing new problems in legume agriculture. One of the main limiting factors in the employment and incorporation of biotechnology procedures in agriculture is the availability of resistance genes that can serve in the generation of effective stress responses. Consequently, only the cloned genes that confer a high degree of resistance and are resilient to biotic stress factors would be incorporated into the target species. Moreover, new gene cocktails can be developed to avoid the issues observed due to the virulent strains.

15.9 Conclusion and future prospects

Biotic and abiotic stress factors, including pests, drought, extremes in temperature and salinity, are the main contributors to the decrease in legume yields worldwide. The advent of modern biotechnology procedures has helped in attaining two main goals; firstly, identification of the molecular mechanisms involved in the stress and disease process and, secondly, the incorporation of this information for improving the quality and yield of leguminous plants. However, the introduction of genetic modification procedures into agriculture is met with highly polarized attitudes around the world. Proper understanding of the biotechnological procedures, along with the advances in genomics, transcriptomics and metabolomics, can aid the acceptance of these methods by the general population. Moreover, efforts are underway to develop safer and more effective methodologies for legume improvement by the comparative analysis of protein expression profiles and functional analysis. The isolation of a sufficient number of stress resistance genes along with accurate delivery and analysis methods can, ultimately, help in developing stress-resistant legumes.

References

Abate T, Alene AD, Bergvinson D, *et al.* (2012) Tropical grain legumes in Africa and South Asia: Knowledge and opportunities. International Crops Research Institute for the Semi-Arid Tropics, Nairobi, Kenya; available at: http://oar.icrisat.org/id/eprint/5680

Adams MA, Simon J, Pfautsch S (2010) Woody legumes: a review from the South. *Tree Physiol* 30: 1072–1082

Ahmed S, Piggin C, Haddad A, Kumar S, Khalil Y, Geletu B (2012) Nematode and fungal diseases of food legumes under conservation cropping systems in northern Syria. *Soil Tillage Res* 121: 68–73.

Ahuja I, de Vos RC, Bones AM, Hall RD (2010) Plant molecular stress responses face climate change. *Trends Plant Sci* 15: 664–674.

Alam I, Kim K-H, Sharmin SA, Kim Y-G, Lee B-H (2010) Advances in the molecular breeding of forage crops for abiotic stress tolerance. *J Plant Biotechnol* 37: 425–441.

Al-Whaibi MH (2011) Plant heat-shock proteins: a mini review. *J King Saud Univ-Sci* 23: 139–150.

Bakht J, Bano A, Shafi M, Dominy P (2013) Effect of abscisic acid applications on cold tolerance in chickpea (*Cicer arietinum* L.). *Europ J Agron* 44: 10–21.

Baltrus DA, Nishimura MT, Dougherty KM, *et al.* (2012) The molecular basis of host specialization in bean pathovars of *Pseudomonas syringae*. *Mol Plant-Microbe Interact* 25: 877–888.

Bates A, Hemenway T (2010) *From Agriculture to Permaculture. State of the World 2010: Transforming Cultures from Consumerism to Sustainability*. The Worldwatch Institute, pp. 47–53.

Battaglia M, Covarrubias AA (2013) Late embryogenesis abundant (LEA) proteins in legumes. *Front Plant Sci* 4: 190.

Becana M, Matamoros MA, Udvardi M, Dalton DA (2010) Recent insights into antioxidant defenses of legume root nodules. *New Phytol* 188: 960–976.

Betti M, Pérez-Delgado C, García-Calderón M, Díaz P, Monza J, Márquez AJ (2012) Cellular stress following water deprivation in the model legume *Lotus japonicus*. *Cell* 1: 1089–1106.

Bhatnagar-Mathur P, Rao JS, Vadez V, Sharma KK (2009) Transgenic strategies for improved drought tolerance in legumes of semi-arid tropics. *J Crop Improve* 24: 92–111.

Bhatnagar-Mathur P, Palit P, Kumar CS, Reddy DS, Sharma KK (2012) Grain legumes: Biotechnological interventions in crop improvement for adverse environments. In: Tuteja N, Gill SS, Tuteja R (eds), *Improving Crop Productivity in Sustainable Agriculture*. John Wiley & Sons, Inc., pp. 381–421.

Boller T, Felix G (2009) A renaissance of elicitors: perception of microbe-associated molecular patterns and danger signals by pattern-recognition receptors. *Ann Rev Plant Biol* 60: 379–406.

Bouchenak M, Lamri-Senhadji M (2013) Nutritional quality of legumes, and their role in cardio-metabolic risk prevention: a review. *J Med Food* 16: 185–198.

Bouchenak-Khelladi Y, Maurin O, Hurter J, Van der Bank M (2010) The evolutionary history and biogeography of Mimosoideae (Leguminosae): an emphasis on African acacias. *Mol Phylogenet Evolut* 57: 495–508.

Bruneau A, Forest F, Herendeen PS, Klitgaard BB, Lewis GP (2001) Phylogenetic relationships in the Caesalpinioideae (Leguminosae) as inferred from chloroplast trnL intron sequences. *System Bot* 26: 487–514.

Brunet J, Varrault G, Zuily-Fodil Y, Repellin A (2009) Accumulation of lead in the roots of grass pea (*Lathyrus sativus* L.) plants triggers systemic variation in gene expression in the shoots. *Chemosphere* 77: 1113–1120.

Casler M, Heaton E, Shinners K, *et al.* (2009) Grasses and legumes for cellulosic bioenergy. In: Wedin WF, Fales SL (eds), *Grassland: Quietness and Strength for a New American Agriculture*. ASA-CSSS-SSSA, Madison, WI, pp. 205–219.

Cavatorta JR, Gray SM, Jahn MM (2012) Biotechnology and the control of viral diseases of crops. In: Popp JS, Jahn MM,

Matlock MD, Kemper NP (eds), *The Role of Biotechnology in a Sustainable Food Supply*. Cambridge University Press, pp. 77–89.

Chandra A, Roy AK, Kumar S (2010) Molecular techniques for improvement of forage crops. *Range Manage Agroforest* 31: 87–96.

Chen JR, Lü JJ, Liu R, *et al.* (2010) DREB1C from *Medicago truncatula* enhances freezing tolerance in transgenic *M. truncatula* and China Rose (*Rosa chinensis* Jacq.). *Plant Growth Regul* 60: 199–211.

Chen M, Wang QY, Cheng XG, *et al.* (2007) *GmDREB2*, a soybean DRE-binding transcription factor, conferred drought and high-salt tolerance in transgenic plants. *Biochem Biophys Res Commun* 353: 299–305.

Choudhury SR, Pandey S (2013) Specific subunits of heterotrimeric G proteins play important roles during nodulation in soybean. *Plant Physiol* 162: 522–533.

Coba de la Peña T, Redondo F, Manrique E, Lucas MM, Pueyo JJ (2010) Nitrogen fixation persists under conditions of salt stress in transgenic *Medicago truncatula* plants expressing a cyanobacterial flavodoxin. *Plant Biotechnol J* 8: 954–965.

Dalton DA, Boniface C, Turner Z, *et al.* (2009) Physiological roles of glutathione S-transferases in soybean root nodules. *Plant Physiol* 150: 521–530.

Day L (2013) Proteins from land plants – potential resources for human nutrition and food security. *Trends Food Sci Technol* 32: 25–42.

De Hoff PL, Brill LM, Hirsch AM (2009) Plant lectins: the ties that bind in root symbiosis and plant defense. *Mol Genetic Genom* 282: 1–15.

Dialoke S, Agu C, Ojiako F, *et al.* (2010) Survey of insect pests on pigeonpea in Nigeria. *J SAT Agric Res* 8: 8.

Divi UK, Krishna P (2010) Overexpression of the brassinosteroid biosynthetic gene AtDWF4 in Arabidopsis seeds overcomes abscisic acid-induced inhibition of germination and increases cold tolerance in transgenic seedlings. *J Plant Growth Regul* 29: 385–393.

Dixon RA, Pasinetti GM (2010) Flavonoids and isoflavonoids: from plant biology to agriculture and neuroscience. *Plant Physiol* 154: 453–457.

Döring TF, Storkey J, Baddeley J, *et al.* (2012) Legume based plant mixtures for delivery of multiple ecosystem services: Weed diversity and weed control. In: McCracken K (ed.), *SAC-SEPA Biennial Conference: Valuing Ecosystems: Policy, Economic and Management Interactions*. Edinburgh, 3rd–4th April 2012, pp. 163–168.

Dung DT, Phuong LTH, Long KD (2011) Insect parasitoid composition on soybean, some ecobiological characteristics of the parasitoid, *Xanthopimpla punctata* Fabricius on soybean leaf-folder *Omiodes indicata* (Fabricius) in Hanoi, Vietnam. *J Int Soc Southeast Asian Agricul Sci* 17: 58–69.

Dupont L, Alloing G, Pierre O, *et al.* (2012) The legume root nodule: from symbiotic nitrogen fixation to senescence. In: Nagata T (ed.), *Senescence*. InTech, p.p 137–168.

El-Hassan SA, Gowen SR, Pembroke B (2013) Use of *Trichoderma hamatum* for biocontrol of lentil vascular wilt disease: Efficacy, mechanisms of interaction and future prospects. *J Plant Protect Res* 53: 12–26.

El-Rahman SSA, Mazen MM, Mohamed HI, Mahmoud NM (2012) Induction of defence related enzymes and phenolic compounds in lupin (*Lupinus albus* L.) and their effects on host resistance against Fusarium wilt. *Eur J Plant Pathol* 134: 105–116.

Fürstenberg-Hägg J, Zagrobelny M, Bak S (2013) Plant defense against insect herbivores. *Int J Mol Sci* 14: 10242–10297.

García-Manzanares Á, Lucendo AJ (2011) Nutritional and dietary aspects of celiac disease. *Nutr Clin Pract* 26: 163–173.

Ghosh S, Pal A (2012) Identification of differential proteins of mungbean cotyledons during seed germination: a proteomic approach. *Acta Physiol Plant* 34: 2379–2391.

Gill SS, Anjum NA, Hasanuzzaman M, *et al.* (2013) Glutathione and glutathione reductase: a boon in disguise for plant abiotic stress defense operations. *Plant Physiol Biochem* 70: 204–212.

Gimenez-Ibanez S, Ntoukakis V, Rathjen JP (2009) The LysM receptor kinase CERK1 mediates bacterial perception in Arabidopsis. *Plant Signaling Behav* 4: 539–541.

Goellner K, Loehrer M, Langenbach C, Conrath U, Koch E, Schaffrath U (2010) *Phakopsora pachyrhizi*, the causal agent of Asian soybean rust. *Mol Plant Pathol* 11: 169–177.

Gough C, Jacquet C (2013) Nod factor perception protein carries weight in biotic interactions. *Trends Plant Sci* 18: 566–574.

Hamidou F, Halilou O, Vadez V (2013) Assessment of groundnut under combined heat and drought stress. *J Agron Crop Sci* 199: 1–11.

Hanafy MS, El-Banna A, Schumacher HM, Jacobsen HJ, Hassan FS (2013) Enhanced tolerance to drought and salt stresses in transgenic faba bean (*Vicia faba* L.) plants by heterologous expression of the PR10a gene from potato. *Plant Cell Rep* 32: 663–674.

Hernandez LD, Vierling E (1993) Expression of low molecular weight heat-shock proteins under field conditions. *Plant Physiol* 101: 1209–1216.

Hirayama T, Shinozaki K (2010) Research on plant abiotic stress responses in the post-genome era: past, present and future. *Plant J* 61: 1041–1052.

Hirel B, Phillipson B, Murchie E, *et al.* (1997) Manipulating the pathway of ammonia assimilation in transgenic non-legumes and legumes. *J Plant Nutr Soil Sci* 160: 283–290.

Hoareau L, DaSilva EJ (1999) Medicinal plants: a re-emerging health aid. *Elect J Biotechnol* 2: 3–4.

Hodgson EW, Kemis M, Geisinger B (2012) Assessment of Iowa soybean growers for insect pest management practices. *J Extension* 50: 4RIB6.

Hohn T, Schachermayr G (2010) Genetic modification (GM) as a new tool in the resistance toolbox. In: Strange RN, Gullino ML (eds), *The Role of Plant Pathology in Food Safety and Food Security*. Springer Science+Business Media, pp. 73–76.

Ihejirika G, Nwufo M, Obiefuna J, Ibeawuchi I (2010) Evaluation of some fungal diseases and yield of groundnut in groundnut-based cropping systems. *Arch Phytopathol Plant Protect* 43: 1044–1049.

Jaspers P, Kangasjärvi J (2010) Reactive oxygen species in abiotic stress signaling. *Physiol Plant* 138: 405–413.

Jeffries C (2008) Report of the 13th European Association for Potato Research Virology Section meeting 17–22 June 2007, Coylumbridge (Aviemore), Scotland, United Kingdom. *Potato Res* 51: 175–208.

Jensen ES, Peoples MB, Boddey RM, *et al.* (2012) Legumes for mitigation of climate change and the provision of feedstock for biofuels and biorefineries. A review. *Agron Sustain Develop* 32: 329–364.

Jones RA (2012) Virus diseases of annual pasture legumes: incidences, losses, epidemiology, and management. *Crop Pasture Sci* 63: 399–418.

Joshi B, Kumar P, Pant S (2011) Traditional use of some leguminous plants in Tarai and Bhawar regions of Kumaun Himalaya, Uttarakhand. *Environ Conservation J* 12: 65–67.

Keneni G, Bekele E, Getu E, *et al.* (2011) Breeding food legumes for resistance to storage insect pests: potential and limitations. *Sustainability* 3: 1399–1415.

Khare M, Khare U (2012) Studies on survival of *Xanthomonas campestris* pv. glycines. *Ind Phytopathol* 48: 180–181.

Kohl DH, Lin JJ, Shearer G, Schubert KR (1990) Activities of the pentose phosphate pathway and enzymes of proline metabolism in legume root nodules. *Plant Physiol* 94: 1258–1264.

Komatsu S, Ahsan N (2009) Soybean proteomics and its application to functional analysis. *J Proteom* 72: 325–336.

Kumar JK, Prasad A, Richard SA (2012a) In vitro antioxidant activity and preliminary phytochemical analysis of medicinal legumes. *J Pharma Res* 5: 3059–3062.

Kumar U, Kumar SP, Surabhi S (2012b) Spectrum of insect pest complex of soybean (*Glycine max* (L.) Merrill) at Lambapeepal Village in Kota region, India. *ISCA J Biol Sci* 1: 80–82.

Leach M, Agudelo P, Lawton-Rauh A (2012) Genetic variability of *Rotylenchulus reniformis*. *Plant Dis* 96: 30–36.

Lehmann S, Funck D, Szabados L, Rentsch D (2010) Proline metabolism and transport in plant development. *Amino Acid* 39: 949–962.

Lindström K, Mousavi SA (2010) *Rhizobium and Other N-fixing Symbioses*. Encyclopedia of Life Sciences. John Wiley & Sons, Ltd.

Liu G, Xu H, Zhang L, Zheng Y (2011) Fe binding properties of two soybean (*Glycine max* L.) LEA4 proteins associated with antioxidant activity. *Plant Cell Physiol* 52: 994–1002.

Liu J, Novero M, Charnikhova T, Ferrandino A, *et al.* (2013) Carotenoid Cleavage Dioxygenase 7 modulates plant growth, reproduction, senescence, and determinate nodulation in the model legume *Lotus japonicus*. *J Exp Bot* 64: 1967–1981.

Lombardo S, Colombo A, Rapisarda C (2011) Severe damage caused by *Meloidogyne artiellia* on cereals and leguminous in Sicily. *Redia* 94: 149–151.

López-Gómez M, Palma E, Lluch C (2013) Strategies of salt tolerance in the rhizobia–legume symbiosis. In: González-López J (ed.) *Beneficial Plant-Microbial Interactions: Ecology and Applications*. CRC Press, pp. 99–121.

Łotocka B, Kopcińska J, Skalniak M (2012) Review article: The meristem in indeterminate root nodules of Faboideae. *Symbiosis* 58: 63–72.

Lüthi C, Álvarez-Alfageme F, Ehlers JD, Higgins TJ, Romeis J (2013) Resistance of AI-1 transgenic chickpea (*Cicer arietinum*) and cowpea (*Vigna unguiculata*) dry grains to bruchid beetles (Coleoptera: Chrysomelidae). *Bull Entomol Res* 103: 373–381.

Magyar-Tábori K, Mendler-Drienyovszki N, Dobránszki J (2011) Models and tools for studying drought stress responses in pea. *Omics: J Integrat Biol* 15: 829–838.

Makkouk K, Pappu H, Kumari SG (2012) Virus diseases of peas, beans, and faba bean in the Mediterranean region. In: Loebenstein G, Lecoq H (eds), *Advances in Virus Research. Viruses and Virus Diseases of Vegetables in the Mediterranean Basin*. Academic Press, pp. 367–402.

Mantri N, Patade V, Penna S, Ford R, Pang E (2012) Abiotic stress responses in plants: present and future. In: Ahmad P, Prasad MNV (eds), *Abiotic Stress Responses in Plants: Metabolism, Productivity and Sustainability*. Springer Science+Business Media, pp. 1–19.

Maury Y, Sharma B, Tyagi A, Khetarpal R, Vishwanath S, Kumar D (2012) Detection of pea seed-borne mosaic virus in commercial seeds of pea, and germplasm of pea and lentil. *Ind Phytopathol* 44: 107–111.

Mehrkhou F, Talebi AA, Moharramipour S, Naveh VH (2012) Demographic parameters of Spodoptera exigua (Lepidoptera: Noctuidae) on different soybean cultivars. *Environ Entomol* 41: 326–332.

Mikić A, Rubiales D, Smýkal P, Stoddard FL (2011) The legume manifesto: (Net) workers on fabaceae, unite! *Ratar Povrt/Field Veg Crop Res* 48: 253–258.

Miller G, Suzuki N, Ciftci-Yilmaz S, Mittler R (2010) Reactive oxygen species homeostasis and signaling during drought and salinity stresses. *Plant Cell Environ* 33: 453–467.

Mittler R (2006) Abiotic stress, the field environment and stress combination. *Trends Plant Sci* 11: 15–19.

Mizoi J, Shinozaki K, Yamaguchi-Shinozaki K (2012) AP2/ERF family transcription factors in plant abiotic stress responses. *Biochim Biophys Acta (Special Issue) Gene Regulatory Mechanisms* 1819: 86–96.

Mohammadi N, Goltapeh EM, Kari D, Babaie A, HR P (2011) The genetic diversity of Iranian isolates causing *Fusarium* wilt of lentil. *J Agric Tech* 7: 1809–1822.

Monaghan J, Zipfel C (2012) Plant pattern recognition receptor complexes at the plasma membrane. *Curr Opin Plant Biol* 15: 349–357.

Musser FR, Catchot AL, Gibson BK, Knighten KS (2011) Economic injury levels for southern green stink bugs (Hemiptera: Pentatomidae) in R7 growth stage soybeans. *Crop Protect* 30: 63–69.

Muzquiz M, Varela A, Burbano C, Cuadrado C, Guillamón E, Pedrosa MM (2012) Bioactive compounds in legumes: pronutritive and antinutritive actions. Implications for nutrition and health. *Phytochem Rev* 11: 227–244.

Navascués J, Pérez-Rontomé C, Gay M, *et al.* (2012) Leghemoglobin green derivatives with nitrated hemes evidence production of highly reactive nitrogen species during aging of legume nodules. *Proc Nat Acad Sci USA* 109: 2660–2665.

Oldroyd GE (2013) Speak, friend, and enter: signalling systems that promote beneficial symbiotic associations in plants. *Nature Rev Microbiol* 11: 252–263.

Oldroyd GE, Murray JD, Poole PS, Downie JA (2011) The rules of engagement in the legume-rhizobial symbiosis. *Annu Rev Genet* 45: 119–144.

O'Neal ME, Johnson KD (2010) Insect pests of soybean and their management. In: Singh G (ed.), *The Soybean: Botany, Production and Uses*. CAB International, pp. 300–324.

Pagano MC (2013) Mitigating the impacts of climate change on disaster risks through soil ecosystem services. In: Filho WL (ed.), *Climate Change and Disaster Risk Management*. Springer-Verlag, Berlin, Heidelberg, pp. 205–215.

Palma F, Tejera NA, Lluch C (2013) Nodule carbohydrate metabolism and polyols involvement in the response of *Medicago sativa* to salt stress. *Environ Exp Bot* 85: 43–49.

Palta J, Ganjeali A, Turner N, Siddique K (2010) Effects of transient subsurface waterlogging on root growth, plant biomass and yield of chickpea. *Agricul Water Manag* 97: 1469–1476.

Pawlak-Sprada S, Stobiecki M, Deckert J (2011) Activation of phenylpropanoid pathway in legume plants exposed to heavy metals. Part II. Profiling of isoflavonoids and their glycoconjugates induced in roots of lupine (*Lupinus luteus*) seedlings treated with cadmium and lead. *Acta Biochim Pol* 58: 217–223.

Pearce G, Yamaguchi Y, Barona G, Ryan CA (2010) A subtilisin-like protein from soybean contains an embedded, cryptic signal that activates defense-related genes. *Proc Nat Acad Sci USA* 107: 14921–14925.

Pick T, Jaskiewicz M, Peterhänsel C, Conrath U (2012) Heat shock factor HsfB1 primes gene transcription and systemic acquired resistance in Arabidopsis. *Plant Physiol* 159: 52–55.

Postel S, Kemmerling B (2009) Plant systems for recognition of pathogen-associated molecular patterns. *Semin Cell Dev Biol* 20: 1025–1031.

Potters G, Horemans N, Jansen MA (2010) The cellular redox state in plant stress biology – a charging concept. *Plant Physiol Biochem* 48: 292–300.

Pratap A, Choudhary A, Kumar J (2010) In vitro techniques towards genetic enhancement of food legumes – a review. *J Food Legume* 23: 169–185.

Priya S, Anjana P, Major S (2013) Identification of the RAPD marker linked to powdery mildew resistant gene(s) in black gram by using bulk segregant analysis. *Res J Biotechnol* 8: 2.

Provorov N (2013) Improvement of symbiotic nitrogen fixation in plants: molecular-genetic approaches and evolutionary models. *Russ J Plant Physiol* 60: 27–32.

Qu AL, Ding YF, Jiang Q, Zhu C (2013) Molecular mechanisms of the plant heat stress response. *Biochem Biophys Res Commu* 432: 203–207.

Ramírez-Suero M, Khanshour A, Martinez Y, Rickauer M (2010) A study on the susceptibility of the model legume plant *Medicago truncatula* to the soil-borne pathogen *Fusarium oxysporum*. *Euro J Plant Pathol* 126: 517–530.

Rashid A, Khan RU, Marwat SK (2009) Importance of weed control in chickpea under rainfed condition. *Am-Eurasian J Agricul Environ Sci* 5: 456–459.

Ray S, Dansana PK, Giri J, *et al.* (2011) Modulation of transcription factor and metabolic pathway genes in response to water-deficit stress in rice. *Funct Integrat Genom* 11: 157–178.

Reddy DS, Bhatnagar-Mathur P, Vadez V, Sharma KK (2012) Grain legumes (soybean, chickpea, and peanut): Omics approaches to enhance abiotic stress tolerance. In: Tuteja N, Gill SS, Tiburico AF, Tuteja R (eds), *Improving Crop Resistance to Abiotic Stress*. Wiley-VCH Verlag, Singapore, pp. 993–1030.

Reverberi M, Fabbri AA, Fanelli C (2012) Oxidative stress and oxylipins in plant fungus interaction. In: Witzany G (ed.), *Biocommunication of Fungi*. Springer Science+Business Media, Dordrecht, pp. 273–290.

Rey T, Nars A, Bonhomme M, Bottin A, *et al.* (2013) NFP, a LysM protein controlling Nod factor perception, also intervenes in *Medicago truncatula* resistance to pathogens. *New Phytol* 198: 875–886.

Reynolds-Henne CE, Langenegger A, Mani J, Schenk N, Zumsteg A, Feller U (2010) Interactions between temperature, drought and stomatal opening in legumes. *Environ Exp Bot* 68: 37–43.

Rodenburg J, Zossou-Kouderin N, Gbèhounou G, *et al.* (2011) *Rhamphicarpa fistulosa*, a parasitic weed threatening rain-fed lowland rice production in sub-Saharan Africa – A case study from Benin. *Crop Protect* 30: 1306–1314.

Ross D, Newton P, Tate K, Luo D (2013) Impact of a low level of CO_2 enrichment on soil carbon and nitrogen pools and mineralization rates over ten years in a seasonally dry, grazed pasture. *Soil Biol Biochem* 58: 265–274.

Samanta A, Das G, Das SK (2011) Roles of flavonoids in plants. *Int J Pharm Sci Tech* 6: 12–35.

Šarūnaitė L, Arlauskienė A, Deveikytė I, Maikštėnienė S, Kadžiulienė Ž (2012) Intercropping of pea and spring cereals for weed control in an organic farming system. In: Price AJ (ed.), *Weed Control*. InTech, Croatia, pp. 3–30.

Sastry KS (2013) Mechanism of seed transmission. In: Sastry KS (ed.), *Seed-Borne Plant Virus Diseases*. Springer, India, pp. 85–100.

Silvente S, Sobolev AP, Lara M (2012) Metabolite adjustments in drought tolerant and sensitive soybean genotypes in response to water stress. *PloS One* 7(6): e38554.

Singh S, Azam T (2010) Interactive effects of *Meloidogyne incognita* and *Macrophomina phaseolina* on Lentil. *Ind Nematol* 40: 31–34.

Smitchger JA, Burke IC, Yenish JP (2012) The critical period of weed control in lentil (*Lens culinaris*) in the pacific Northwest. *Weed Sci* 60: 81–85.

Soares-Cavalcanti NM, Belarmino LC, Kido EA, *et al.* (2012) Overall picture of expressed heat shock factors in *Glycine max, Lotus japonicus* and *Medicago truncatula. Genet Mol Biol* 35(Suppl. 1): 247–259.

Spanou C, Veskoukis AS, Kerasioti T, *et al.* (2012) Flavonoid glycosides isolated from unique legume plant extracts as novel inhibitors of xanthine oxidase. *PloS One* 7(3): e32214.

Suzuki N, Koussevitzky S, Mittler R, Miller G (2012) ROS and redox signalling in the response of plants to abiotic stress. *Plant Cell Environ* 35: 259–270.

Szabados L, Savouré A (2010) Proline: a multifunctional amino acid. *Trends Plant Sci* 15: 89–97.

Thakur P, Kumar S, Malik JA, Berger JD, Nayyar H (2010) Cold stress effects on reproductive development in grain crops: an overview. *Environ Exp Bot* 67: 429–443.

Thao NP, Tran LSP (2012) Potentials toward genetic engineering of drought-tolerant soybean. *Crit Rev Biotechnol* 32: 349–362.

The Plant List (2010) Version 1.1. Available at: http://www.theplantlist.org/ [accessed 24 July 2014].

Thomas K (2010) Impact of climate change on diseases of cool season grain legume crops. In: Yadav SS, McNeil DL, Redden R, Patil SA (eds), *Climate Change and Management of Cool Season Grain Legume Crops*. Springer Science+Business Media, pp. 99–113.

Thomashow MF (1999) Plant cold acclimation: freezing tolerance genes and regulatory mechanisms. *Annu Rev Plant Biol* 50: 571–599.

Tiwari P, Pathak M (2013) Screening of phytocontent of different plant parts of a medicinal legume herb. *Asia J Biotechnol Resour* 4: 49–52.

Tyler BM (2007) Phytophthora sojae: root rot pathogen of soybean and model oomycete. *Mol Plant Pathol* 8: 1–8.

Vance CP (1997) Enhanced agricultural sustainability through biological nitrogen fixation. *Nato Asi Series G Ecolo Sci* 39: 179–186.

Van der Linden L, Bredenkamp J, Naidoo S, *et al.* (2013) Gene-for-gene tolerance to bacterial wilt in Arabidopsis. *Mol Plant-Microbe Interact* 26: 398–406.

Vargas LIR, Acosta VB, Beaver JS, Purcifull DE, Polston JE (2013) Detecting bean golden yellow mosaic virus in bean breeding lines and in the common legume weed *Macroptilium lathyroides* in Puerto Rico. *J Agricul Univ Puerto Rico* 85: 165–176.

Vaseva I, Akiscan Y, Simova-Stoilova L, *et al.* (2012) Antioxidant response to drought in red and white clover. *Acta Physiol Plant* 34: 1689–1699.

Velazquez E, R Silva L, Peix A (2010) Legumes: a healthy and ecological source of flavonoids. *Curr Nutri Food Sci* 6: 109–144.

Vemulapati B, Druffel K, Eigenbrode S, Karasev A, Pappu H (2011) Genomic characterization of pea enation mosaic virus-2 from the Pacific Northwestern USA. *Arch Virol* 156: 1897–1900.

Venn B, Thies F, O'Neil C (2012) Whole grains, legumes, and health. *J Nutr Metabol* 2012: art. ID 903767.

Viteri D, Cabrera I, Estévez De Jensen C (2009) New record of *Thrips* species associated with soybeans in Puerto Rico. *Florida Entomol* 92: 181–185.

Vovlas N, Troccoli A, Palomares-Rius J, *et al.* (2011) *Ditylenchus gigas* spp. parasitizing broad bean: a new stem nematode singled out from the *Ditylenchus dipsaci* species complex using a polyphasic approach with molecular phylogeny. *Plant Pathol* 60: 762–775.

Wang D, Heckathorn SA, Wang X, Philpott SM (2012) A meta-analysis of plant physiological and growth responses to temperature and elevated CO_2. *Oecologia* 169: 1–13.

Wei L, Xue AG, Cober ER, *et al.* (2010) Pathogenicity of *Pythium* species causing seed rot and damping-off in soybean under controlled conditions. *Phytoprotection* 91: 3–10.

Wink M (2013) Evolution of secondary metabolites in legumes (Fabaceae). *South Afr J Bot* 89: 164–175.

Xu R, Li W, Zhang LF, Lin YH, Qi B, Xing H (2010) A study on the inheritance of resistance to whitefly in soybean. *Sci Agricul Sinica* 43: 72–78.

Yu L, Li C, Zheng Y, Yan C, Zhang T, Shan S (2012) Effects of drought during peanut (Arachis hypogaea L.) seeding stage on disease resistance gene PnAG3 expression and physiological index related to drought resistance. *J Agric Biotechnol* 20: 642–648.

Zhang H, Cui X, Chen X, Zhi H, Zhang S, Zhao L (2012) Determination of the complete genomic sequence and molecular biological analysis of soybean mosaic virus. *Can J Plant Pathol* 34: 288–297.

Zhang J, Xue A, Zhang H, Nagasawa A, Tambong J (2010) Response of soybean cultivars to root rot caused by *Fusarium* species. *Can J Plant Sci* 90: 767–776 .

CHAPTER 16

Gene pyramiding and omics approaches for stress tolerance in leguminous plants

Neha Handa, Renu Bhardwaj, Ashwani Kumar Thukral, Saroj Arora, Sukhmeen Kaur Kohli, Vandana Gautam and Tajinder Kaur

Department of Botanical and Environmental Sciences, Guru Nanak Dev University, Amritsar, India

16.1 Introduction

The Leguminosae forms the third largest family of flowering plants after Orchidaceae and Asteraceae and is known for its great diversity (Doyle & Luckow, 2003). The family consists of some 670 to 750 genera and 18,000 to 19,000 species, which are usually grouped into three subfamilies, namely Mimosoideae, Caesalpinoideae and Papilionoideae, and these are widely distributed in diverse habitats ranging from forests to deserts, lowlands, alpine regions and even aquatic habitats of Africa, America and Asia (Silveira *et al.*, 2010; Olajide *et al.*, 2013). The plants of this family have wide-ranging roles – as food crops, ornamentals, forage and green manure – and these days they are also explored for their role in herbal medicines (Olajide *et al.*, 2013). Moreover, members of this family are also economically important with their roles in paper production, oil production, wood for fuel and timber, and sources of various chemicals (Lewis *et al.*, 2005). Hence, the family overall has great significance in providing a livelihood and improving socioeconomic conditions for many people (Olajide *et al.*, 2013). Of major significance is that members of this family are important food crops after cereals, thereby contributing to food security of the world (Uzoechina, 2009). Legumes are especially rich in proteins and thus play a vital role in the sustenance of large human populations of both developing and developed countries. Amongst all legumes, grain legumes are one of the major sources of low-cost dietary proteins and minerals (Apata & Ologhobo, 1997). The high content of proteins is due to the accumulation of proteins that do not possess any enzymatic or structural role during development of legume seeds (Duranti, 2006). Hence, these are usually stored in membrane-bound organelles or protein bodies in parenchyma cells of cotyledons (Duranti, 2006). These storage proteins are of three types, namely albumins, globulins and prolamins, of which globulins are most abundant (Duranti, 2006). Nutritionally, these storage proteins are rich in the essential amino acid lysine but deficient in sulphur-rich methionine and cysteine. Hence, they are usually supplemented with cereals, which are rich in the latter, along with tubers (Graham & Vance, 2003). Apart from proteins, legumes also form a source of minerals such as calcium and phosphorus. Carbohydrates, some water-soluble vitamins and dietary fibre are also important components of legumes (Liu, 1997).

In addition to providing food security to the world's population, legumes have also proven beneficial to the soil in which they are grown (Morel *et al.*, 2012). According to the US Department of Agriculture (USDA), these plants have the ability to enhance the organic matter and porosity thereby improving the structure of the soil. They also help in decreasing soil pH and recycling nutrients, and they slowly release nitrogen into the soil thereby enhancing fertility (Morel *et al.*, 2012). Hence, legumes are either grown in rotation with cereals or intercropped with them. These beneficial effects of leguminous plants are mainly due to the ability of these plants to fix atmospheric nitrogen. These plants can form symbiotic associations with nitrogen-fixing bacteria and hence become major contributors to biological nitrogen fixation – a process of enzymatic conversion of atmospheric nitrogen to ammonia thereby

Legumes under Environmental Stress: Yield, Improvement and Adaptations, First Edition. Edited by Mohamed Mahgoub Azooz and Parvaiz Ahmad.

converting nitrogen into a bioavailable form. Biological nitrogen fixation is done both by free-living and symbiotic bacteria and it has been estimated that out of all nitrogen fixed biologically, the contribution by legumes is approximately 80% (Garg & Geetanjali, 2007). Hence, this process makes legumes very important plants both agronomically and economically as they fulfil the need for food of humans and livestock and also enrich the soil thereby reducing the need for nitrogenous fertilizers.

Although legumes provide a major source of nitrogen and also have key roles in agriculture, ecology, animal/human nutrition and health, their benefits are commonly masked by various types of stress affecting the plants'growth (Graham & Vance, 2003). Multiple stresses, both biotic and abiotic, are encountered by legumes during their life cycle thereby significantly affecting crop yields. For stable and improved crop yields, it becomes imperative to develop varieties that are resistant to more than one stress, and biotechnological approaches are now key to solving this problem (Kumar *et al.*, 2011).

Fungal diseases are major problems that affect legumes, but insect pests, nematodes, viruses, bacteria and parasitic weeds are also reported to adversely affect legume production (Dita *et al.*, 2006). Powdery mildew caused by *Erysiphe pisi* and downy mildew caused by *Peronospora viciae* are fungal diseases in legumes that have been reported to badly affect the yield and quality of the crop (Sillero *et al.*, 2006). Apart from mildews, rust diseases also cause significant damage to legume crops, with species belonging to the genus *Uromyces* responsible for major diseases in numerous leguminous plants (Dita *et al.*, 2006). Many rust species such as *U. appendiculatus* on common bean, *U. ciceris-arietini* on chickpea, *U. pisi* on pea, *U. viciae-fabae* on faba bean, lentil and common vetch, *U. vignae* on cowpea, *Phakopsora pachyrhizi* (causing Asian rust) and *P. meibomiae* on soybean or *Puccinia arachidis* on groundnut severely affect the growth and yield of legume crops (Rubiales *et al.*, 2002; Carmona *et al.*, 2005). The common fungal disease of chickpea, ascochyta blight, caused by *Ascochyta rabiei*, results in maximum yield losses by mostly affecting the above-ground parts of the plants (Nene & Reddy, 1987). Most of the common soil-borne diseases of legumes attack the seedling stage of the crop. These are known as damping-off diseases; for example, damping-off due to *Rhizoctonia solani* or *Pythium* spp. causes death of the plant in most cases (Denman *et al.*, 1995; Wang *et al.*, 2003). Severe seedling loss in common bean occurs due to root rot caused by *Fusarium* spp. (Hamwieh *et al.*, 2005).

Viruses such as bean common mosaic virus (BCMV) and bean common mosaic necrotic virus (BCMNV) affecting common beans also cause significant yield losses (Dita *et al.*, 2006). Insects are also one of the biotic stresses faced by legume plants. These include aphids such as *Aphis glycine*, pod borers such as *Helicoverpa armigera* and *H. punctigera* (Yoshida *et al.*, 1997), and weevils such as *Apion godmani* and *Zabrotes subfasciatus* (Garza *et al.*, 1996) are important insect pests in grain legumes. They may act as vectors, or provide sites of infection to various pathogens, or directly damage the grain legumes by feeding on them (Dita *et al.*, 2006).

Abiotic stresses also adversely affect the growth and productivity of crops. They activate a chain of morphological, physiological, biochemical and molecular changes in plants (Mathur *et al.*, 2008). Abiotic stress includes drought, salinity, waterlogging, high temperature and chilling, which ultimately lead to crop loss thereby reducing average yields of major crop plants. Crops facing abiotic stress are more prone to weeds, insects and diseases, which further increases the chances of yield losses (Reddy *et al.*, 2004). Numerous abiotic stresses related to legume crops also directly affect the symbiotic interactions between the roots and soil bacteria, thereby affecting the overall growth of the plant. Waterlogging results in severe loss in legume yields by limiting diffusion of oxygen through the soil, and consequent replacement of nitrification by denitrification (Laanbroek, 1990). Salinity and other stresses such as alkalinity and temperature affect seed germination in legumes (Guan *et al.*, 2009). Most legumes react to moderate salinity by a reduction in their growth, which may be by inhibiting plant cell expansion as a result of osmotic stress or due to accumulation of Na^+ and Cl^- ions resulting in cell injury in transpiring leaves and other plant tissues (Manchada & Garg, 2008). Salinity affects nodule formation and nitrogen fixation (Verdoy *et al.*, 2004). Salt stress leads to a reduction in the number of rhizobia that inhabit the roots (Tu, 1981). It also affects the number and shape of root hairs and interferes with infection thread formation (Zahran & Sprent, 1986). This is mainly because nodule formation, structure, physiology and function are very sensitive to salt stress, moreso than the host legume or the rhizobia (Zahran, 1999). Another major abiotic stress affecting crop yields worldwide is drought. A study of five cultivars of mung bean (*Vigna radiata* L.) subjected to drought stress revealed that stress significantly affected flowering phase, plant height, grain yield, number of grains per pod, number of pods per plant, number of side branches, days to flowering and days to physiological maturity (Lalinia *et al.*, 2012).

Therefore, there is an important requirement to develop varieties of legumes that are tolerant to various abiotic stresses in order to increase their yield. Various biotechnological techniques provide researchers and plant breeders with an opportunity to develop such varieties, which are not only stress resistant but are also high yielding and have high nutritional value.

16.2 Approaches to incorporate stress tolerance mechanisms

16.2.1 Gene pyramiding and marker-assisted selection

The main aim of plant breeding is to select desirable traits in a particular plant species and accumulate more advantageous combinations of genes in new varieties. Compared to other crops and cereals, legumes are more prone to biotic stresses such as diseases and insect pests, while among abiotic stresses, the common types affecting them include drought, heat, cold/frost and salinity, which have impact on the overall growth and yield of the plant. Therefore, a new approach should be adopted to develop high-yielding varieties that are resistant to various biotic and abiotic stresses, by combining conventional and non-conventional techniques. With the availability of newer tools and techniques in plant breeding programmes, such as marker-assisted selection (MAS) and marker-assisted gene pyramiding, it has now become easier to develop genotypes having desirable genes.

16.2.1.1 Marker-assisted gene pyramiding

Gene pyramiding is the process of combining several genes together into a single genotype: a genotype containing all the target genes forms the end product of gene pyramiding. It is mainly used to improve qualitative traits such as disease and insect resistance, and is often performed using MAS. MAS, also called marker-assisted breeding (MAB), is a technique that can help to make traditional breeding methods more efficient. Unlike genetic engineering, which transfers isolated gene sequences into a genome, MAB does not involve gene isolation and transfer of gene sequences into the genome. Rather, it involves the use of genetic markers associated with particular traits (Figure 16.1) (Collard & Mackill, 2008). DNA markers and isozymes are commonly used to mark those gene(s) that confer resistance against various diseases. Marker-assisted pyramiding has been used widely for pyramiding disease resistance genes in plants; for example, major genes for resistance against blight (Huang *et al.*, 1997; Singh *et al.*, 2001), blast (Hittalmani *et al.*, 2000) and gall midge (Kumaravadivel *et al.*, 2006) have been pyramided in rice. In wheat, the resistance genes for powdery mildew, namely *Pm2*+*Pm4a*, *Pm2*+*Pm21*, *Pm4a*+*Pm21*, have been effectively pyramided in a single selected line (Liu *et al.*, 2000). Rust resistance genes, i.e. *Lr41*, *Lr42* and *Lr43*, have also been accumulated using the MAS gene pyramiding approach (Cox *et al.*, 1994).

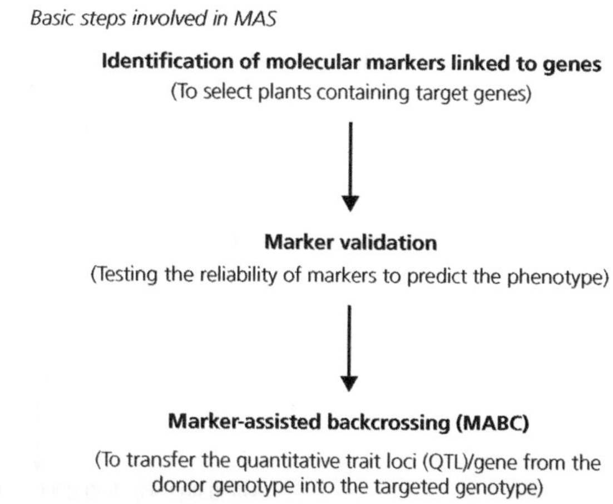

Figure 16.1 General steps of marker-assisted selection (MAS).

16.2.1.2 General scheme of marker-assisted gene pyramiding

The main objective of gene pyramiding is to obtain an ideal genotype containing all desirable traits. Servin *et al.* (2004) proposed that gene pyramiding breeding consists of two basic steps, the pedigree step and the gene fixation step (Figure 16.2). It involves accumulating target genes present in diverse populations with desired allele frequencies into one crossed population and then the selected parents are intercrossed to fix all the target genes into an ideal genotype (Servin *et al.*, 2004).

16.2.1.3 Genetic markers

The genetic differences between the species are represented by the genetic markers. They act as 'signs' or 'flags' and are located in close proximity to the genes. They neither represent the target genes nor do they affect the phenotype of the trait of interest since they are only linked to genes controlling the trait. Like genes,

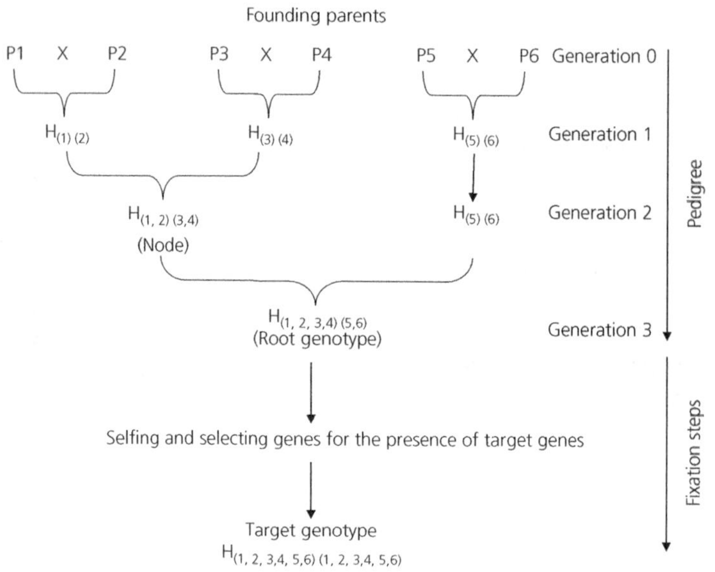

Figure 16.2 Example of gene pyramiding scheme accumulating six target genes. P1–P6: founding parents; H: hybrids. Adapted from Servin *et al.* (2004).

all genetic markers have particular genomic positions within chromosomes known as loci (Collard *et al.*, 2005).

16.2.1.4 Types of genetic markers

There are three major types of genetic markers (Lan & Chao, 2011).

1 Morphological markers, which are phenotypic traits or characters such as flower colour, seed shape or growth habits.
2 Biochemical markers; these include the variants of enzymes called isozymes, which are detected by using electrophoresis and specific staining.
3 DNA (or molecular) markers, which are most widely used and reveal variations of the DNA. Different types of DNA markers are used in several crops. Commonly used DNA markers are: simple sequence repeats (SSRs) or microsatellites; sequence characterized amplified regions (SCARs) or single nucleotide polymorphisms (SNPs); sequence tagged sites (STSs); restriction fragment length polymorphisms (RFLPs); randomly amplified polymorphic DNA (RAPD); amplified fragment length polymorphisms (AFLPs); inter simple sequence repeats (ISSRs); expressed sequence tags (ESTs), cleaved amplified polymorphic sequences (CAPSs) and diversity arrays technology (DArT) (Panigrahi *et al.*, 2013). Simple sequence repeats (SSRs) or microsatellites are the most widely used markers in major cereals and legumes (Gupta *et al.*, 1999; Gupta & Varshney, 2000).

16.2.1.5 Marker-assisted selection in legumes

The use of MAS may be advantageous for improving different crop varieties but its application in legumes is limited for combating stress mainly due to a lack of investment and the genetic complexity of traits related to stress (Dita *et al.*, 2006). But there are exceptions where successful adoption of MAS has been reported in certain legumes against important biotic stresses. Techniques including genetic engineering and marker-assisted breeding have been used for improving chickpea. A soybean variety resistant to cyst nematode was developed using MAS (Diers, 2004). Other examples of the application of MAS include pinto bean resistant to common bacterial blight (Mutlu *et al.*, 2005), and narrow-leafed lupin (*Lupinus angustifolius* L.) resistant to phomopsis stem blight (Yang *et al.*, 2002) and anthracnose (Yang *et al.*, 2004). Several other studies have also reported the introgression and pyramiding of genes of interest and quantitative trait loci (QTLs) in legumes, but gene pyramiding assisted by MAS has been reported only in few studies such as in common bean breeding for rust and anthracnose resistance (Faleiro *et al.*, 2004). Kelly and Vallejo (2004) described several molecular markers linked to the genes conferring anthracnose resistance (*Co-1* to *Co-10*) so that they can be further used to pyramid a resistant cultivar through MAS. Similarly, RAPD markers linked to the 11 genes (*Ur-1* to *Ur-11*) conferring rust resistance have been used to integrate and pyramid rust resistance into common bean cultivars. These markers have also been used to combine rust resistance with resistance to other diseases, such as bean common mosaic virus (BCMV), bean golden mosaic virus (BGMV), common bacterial blight and anthracnose (Stavely, 2000; Singh *et al.*, 2001). The genes *Rsv1*, *Rsv3* and *Rsv4* were pyramided for the trait soybean mosaic virus resistance in soybean using a marker-assisted gene pyramiding approach (Saghai Maroof *et al.*, 2008; Shi *et al.*, 2009).

16.2.1.6 Combined marker-assisted selection

To develop resistant cultivars to multiple stresses, biotechnological techniques such as transgenesis or mutagenesis and MAS can be combined to pyramid multiple resistance genes (Dita *et al.*, 2006). To increase genetic gain, combined MAS may have advantages above phenotypic screening or MAS alone (Lande & Thompson, 1990). Combined MAS may be used when additional QTLs are unknown or when manipulation of a great number of QTLs is required. This approach is more efficient than phenotypic screening alone, particularly in case of large population sizes and low trait heritability (Hospital *et al.*, 1997). MAS for improving insect resistance in tropical maize was studied by Bohn *et al.* (2001). They found that MAS alone was less efficient compared to conventional phenotypic selection, while the efficiency increased by using a combined MAS approach. One such approach was also used in soybean to manage insect resistance. This resulted in the development of resistance to corn earworm (*Helicoverpa zea*) and soybean looper (*Pseudoplusia includens*) in eight soybean lines by combining two major insect-resistance QTLs and a synthetic Bt gene (*cry1Ac*) (Walker *et al.*, 2004).

Although there are considerable efforts to improve the yield of pulses through conventional breeding methods, the rate of progress is slow. This is because the genotype of pulses is highly influenced by the environmental factors that interact with the expression of important quantitative traits (Kumar & Ali, 2006). Therefore, this prompts the need for incorporating new technologies in plant breeding programmes, such as MAS and marker-assisted gene pyramiding or growing of transgenic crops. But knowledge regarding abiotic stress QTLs in legumes is still at an early stage and requires more study for its proper application to stressed legumes.

16.2.2 Omics approaches

To address the drastic changes in plants under stress, a comprehensive study that involves both quantitative as well as qualitative analysis of genetic variation is necessary at various molecular levels such as the proteome, transcriptome and metabolome (Dita *et al.*, 2006). In recent years biotechnology has developed into a significant tool to elucidate genetic variation in plants under stress. Among the most important are the various omics approaches, which have great potential for overcoming stress in grain legumes and increasing their tolerance to stress (Reddy *et al.*, 2012). Omics approaches include genomics, proteomics, transcriptomics, metabolomics, functional genomics and genetic engineering. These can provide insights into the physiological and molecular changes in plants in response to variable stress conditions, thereby informing the genetic basis of stress adaptation.

16.2.2.1 Transcriptomics

Transcriptomics allows one to screen the expression of thousands of genes simultaneously. At any given time and set of conditions, the total set of messenger RNAs (mRNAs) in an organism is referred to as the transcriptome and their analysis is termed transcriptomics. The numerous plant genomic and transcriptomic studies of plants under stress conditions that have been done in recent years suggest that change in the plant system is due to the interaction between the environmental and internal conditions of the plant (Polesani *et al.*, 2008). Plants acclimatize to various abiotic and biotic stresses by adopting complex, broad-spectrum molecular and genetic mechanisms to counteract the effects of the stresses. Many reports suggest that changes due to stressed environment are induced at the transcriptional level (Nakashima *et al.*, 2009; Tran *et al.*, 2010; Hadiarto & Tran, 2011) (Figure 16.3).

Genome-wide sequencing (GWS) profiles the entire DNA sequence of an organism's genome in one go using oligo-probes to identify several stress-inducible genes and transcriptomic studies using cDNA and oligonucleotide microarrays (Matsui *et al.*, 2008, 2010). Another important technique used for transcriptomic studies of various legumes is quantitative real-time polymerase

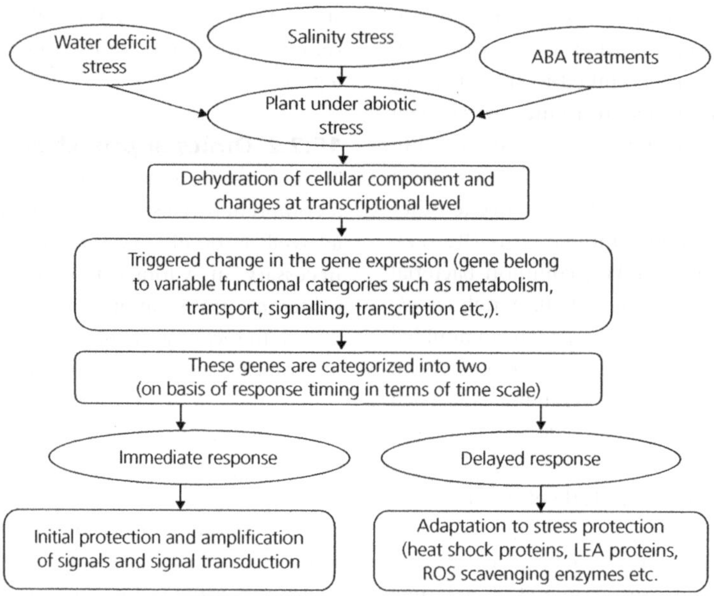

Figure 16.3 Demonstration of phenomena of stress acclimatization by plants at the transcriptional level. ABA, abscisic acid; LEA, late embryogenesis abundant; ROS, reactive oxygen species.

chain reaction (qRT-PCR) for detecting changes in gene expression under stress environments. This technique uses fluorescent dyes and probes for identification of both single-stranded DNA (ssDNA) as well as double-stranded DNA (dsDNA) followed by polymerase chain reaction (Ginzinger, 2002; Udvardi *et al.*, 2008).

Soybean is the most important food legume extensively used for molecular studies of plants under stressed environments. The discovery of SSR and SNP markers due to the availability of large numbers of BAC sequences (bacterial artificial chromosomes) and ESTs has facilitated construction of genetic maps in soybean. Another important technique used for transcriptomic studies of soybean is cDNA-AFLP (complementary DNA-amplified fragment length polymorphism). About 140 differentially expressed cDNA fragments of the soybean genome have been obtained from iso-osmotically treated plants, where the responsive genes encoded for several metabolites such as transcription factors, redox enzymes and ion transporters (Umezawa *et al.*, 2002). qRT-PCR of soybean tissues subjected to various abiotic stresses such as drought, salinity, low temperature and ABA treatments revealed upregulation of several stress-responsive genes (Le *et al.*, 2012).

Chickpea (*Cicer arietinum*) has a very high nutritional value, and of all the food grains has been most extensively studied at the transcriptional level (Udupa *et al.*, 1993; Labdi *et al.*, 1996; Singh *et al.*, 2008). Identification and insertion as well as characterization of certain stress-responsive genes in chickpea have been done recently (Shukla *et al.*, 2009; Tripathi *et al.*, 2009; Molina *et al.*, 2011). Technologies like Roche 454 and Illumina were used to sequence the transcriptome of chickpea, and various assembly programs and parameters were used to generate 43,389 transcripts from the datasets in order to perform comparative studies (Agarwal *et al.*, 2012).

Low temperature is known to cause drastic changes in the transcriptome of plants, making their transcriptional study important (Chinnusamy *et al.*, 2007). Several reports suggest that low temperature leads to changes in gene expression at transcriptional and signalling levels as well as detoxifying and degradatory metabolic phenomena (Sun *et al.*, 2007). Dinari *et al.* (2013) adopted the cDNA-AFLP technique to isolate the transcripts of chickpea under low-temperature stress. Sequencing of several differentially expressed genes was also done using RT-PCR.

Profiling mRNAs in chickpea under drought, cold and high salinity was done using cDNA microarray technology to observe changes in expression of genes in leaves, roots and flowers in susceptible and tolerant genotypes. One hundred and nine, 210 and 386 genes showed two-fold changes at the transcriptional level in these tissues, respectively, of which 2, 15 and 30 genes were differentially expressed in genotypes sensitive or tolerant to drought, low temperature and salinity , respectively (Mantri *et al.*, 2007). Another important technique used for sequencing of different cDNAs is differential display reverse transcriptase PCR (DDRT-PCR) – used in chickpea plants under water deficit conditions (Medini *et al.*, 2009).

Many reports also suggest use of the DDRT-PCR approach for transgenic peanut plants. Overexpression of the *DREB1A* gene was observed in pea plants under abiotic stress. Such reports suggest that transgenic plants can increase their tolerance to water deficit stress based on expression of the *DREB1A* gene (cold-responsive gene) isolated from *Arabidopsis*. About 51 differentially expressed transcripts were identified in peanut plants under abiotic stress conditions, of which 35 were newly expressed, 11 had increased expression and five had a decrease in their expression (Bhatnagar-Mathur *et al.*, 2007).

In recent years, six cDNA libraries have been developed from pea seeds at three different reproductive stages from tolerant and susceptible genotypes (Guo *et al.*, 2008). About 700 genes have been derived from subtractive cDNA libraries of pea plants under water stress. These studies suggest the importance of HSP70 genes and the role of certain regulators in legume plants under water deficit stress (Luo *et al.*, 2005). EST libraries have also been constructed in case of pea plant; for example, line C34-24 is derived from leaves that provide resistance to tomato spot wilt, and line A3 from its pods provides tolerance to drought as well as different doses of aflatoxin (Proite *et al.*, 2007).

16.2.2.2 Proteomics

Proteomic analysis is a powerful approach for linking gene expression to a cell's metabolism (Nouri *et al.*, 2011). A complete study of the proteins expressed in a cell or tissue at a specific time (i.e., the proteome) is one of the most important requisites for understanding biological functions at the cellular as well as the whole-organism level. Transcriptomic analysis may not always relate to protein–protein interaction, protein accumulation and metabolism under stress conditions therefore a proteomics approach is very important for determining mechanisms of stress-responsive acclimatization in plants (Gygi *et al.*, 1999). The trilogical approach of transcriptomics, proteomics and metabolomics is intensively used to study networks at the molecular level in various crop plants showing tolerance and acclimatization to environmental stresses (Sha *et al.*, 2007; Nanjo *et al.*, 2011) (Table 16.1).

Proteomics studies of various organelles and subcellular fractions are among the most informative approaches to understand the function of plant cells. However, proteomics studies of cell organelles have two major limitations: the need for a stringent purification technique and purity verification (Alberts *et al.*, 2002). A number of experimental approaches exist for the identification of protein–protein interactions and their profiling; these include yeast two-hybrid analysis and certain computational methods (Marcotte *et al.*, 1999; Salwinski & Eisenbergy, 2003; Parrish *et al.*, 2006). Proteomic analysis is performed using a number of techniques including mass spectrometry, 2D gel electrophoresis, 2D fluorescence difference gel electrophoresis (2D-FDGE) – which involves the separation of proteins on the basis of their isoelectric point – and mud PIT (multidimensional protein identification technology), which involves whole-cell lysis followed by purification of the proteins (Subramanian & Smith, 2013).

The metabolism of soybean cells is adversely affected by several abiotic stresses, which ultimately lead to deterioration in growth, productivity and proteome development. As a consequence, it leads to degradation of proteins, which ultimately leads to degradation of various phenolic compounds, metabolites, carbohydrates and lipids also present therein (Nouri *et al.*, 2011). A mathematical gene interaction model was prepared for analysis of protein–protein interactions under flooding stress conditions in soybean seedlings, and it was observed there was a change in the protein interaction patterns (Hashiguchi *et al.*, 2009).

A number of stress-induced proteins such as enzymes involved in osmolyte production are produced in response to various abiotic stresses in pea plants. (Matamoros *et al.*, 2003). About 205 protein spots were differentially expressed in pea plants adapted to water deficit stress. Profiling was done using mass spectrometry, which revealed the involvement of several functional and regulatory proteins (Pandey *et al.*, 2008).

Using nodule proteomics analysis, several stress-responsive proteins that maintain and regulate root

Table 16.1 A summary of proteomic analysis of legume plants under variable stresses.

Serial no.	Legume	Stress	Proteomics approach	Responsive proteins	References
1.	*Arachis hypogaea*	Water deficit	MALDI-TOF MS, Q-TOF MS/MS	Lipo-oxygenase, leaf proteins	Pandey *et al.* (2008)
2.	*Arachis hypogaea*	Water deficit	MALDI-TOF MS, Q-TOF MS/MS	Acetyl-CoA carboxylase, non-reductant proteins	Kottapalli *et al.* (2009)
3.	*Arachis hypogaea*	Water deficit	Simple sequence repeat analysis (SSR analysis)	Acetyl-CoA carboxylase, non-reductant proteins	Varshney *et al.* (2009)
4.	*Cicer arietinum*	Water deficit	Mass spectrometry analysis	Dehydration-responsive proteins	Bhushan *et al.* (2007)
5.	*Cicer arietinum*	Water deficit	Tentative unique sequence (TUS), ESTs, SSRs and SNPs analysis	Drought-responsive proteins	Hiremath *et al.* (2011)
6.	*Cicer arietinum*	Chilling	cDNA-AFLP approach, RT-PCR	Mitogenic activated proteins, phosphatidylinositol kinase, acetyl CoA synthetase, HSP 9	Dinari *et al.* (2013)
7.	*Glycine max*	Heavy metal (cadmium)	SDS-PAGE, Q-TOF MS	Superoxide dismutase (SOD), glutathione synthetase	Sobkowiak & Deckert (2006)
8.	*Glycine max*	Heavy metal (aluminium)	SDS-PAGE, ESI-TOF MS	Glutathione synthetase, GTP-binding proteins, ATP-binding proteins	Zhen *et al.* (2007)
9.	*Glycine max*	Water deficit	2D-gel electrophoresis, RT-PCR, SDS-PAGE	CaML (calmodulin-like proteins): CaML2 and 6b	Larrainzar *et al.* (2007)
10.	*Glycine max*	Flooding	SDS-PAGE, ESI-TOF MS	Cytosolic ascorbate peroxidase	Shi *et al.* (2008)
11.	*Glycine max*	Flooding	SDS-PAGE, MALDI-TOF MS, LC-MS/MS, protein sequencer	Cell-wall related proteins, antioxidative proteins, heat-shock proteins	Komatsu & Ahsan (2009)
12.	*Glycine max*	High temperature	SDS-PAGE, MALDI-TOF MS, LC-MS/MS, protein sequencer	Sucrose-binding protein, late embryo abundant protein, formate dehydrogenase	Ren *et al.* (2009)
13.	*Glycine max*	Flooding	SDS-PAGE, LC-MS/MS, protein sequencer	Glycolytic enzymes, ROS scavengers	Hashiguchi *et al.* (2009)
14.	*Glycine max*	Salinity	SDS-PAGE, ESI-Q-TOF MS, protein sequencer	LEA proteins, defence related proteins	Aghaei *et al.* (2009)
15.	*Glycine max*	Water deficit	SDS-PAGE, MALDI-TOF MS, LC-MS	Phosphoproteins, *in vivo* kinase activity	Mathesius (2009); Grimsrud *et al.* (2010)

16.	*Glycine max*	Salinity	SDS-PAGE, MALDI-TOF MS, protein sequencer	Glyceraldehyde-3-phosphate, dehydrogenase, fructokinase 2	Sobhanian *et al.* (2010)
17.	*Glycine max*	Chilling	SDS-PAGE, MALDI-TOF MS	LEA, dehydrogenase, phosphoenol pyruvate, carboylase	Cheng *et al.* (2010)
18.	*Glycine max*	Water deficit	SDS-PAGE, MALDI-TOF MS	Glycerol kinase, dehydratase, phloem serpin	Alam *et al.* (2010)
19.	*Glycine max*	Flooding	2D-DIGE, LC-MS/MS	Glycolytic and fermentation enzymes	Nanjo *et al.* (2010)
20.	*Glycine max*	High temperature	SDS-PAGE, MALDI-TOF MS, LC-MS/MS, protein sequencer	Low molecular weight HSPs and HSP70	Ahsan *et al.* (2010)
21.	*Glycine max*	Water deficit, chilling, ABA stress, salinity	qRT-PCR, delta CT	ABA-dependent proteins, heat-shock proteins, dehydration-responsive proteins	Le *et al.* (2012)
22.	*Medicago sativa*	Heavy metal (cadmium)	Proteomic profiling	Flavodoxin overexpression	Shvaleva *et al.* (2010)
23.	*Pisum sativa*	Wounding, pathogen infection	2D-proteome reference mapping	Ascorbic acid response proteins	Kav *et al.* (2007)

AFLP, amplified fragment length polymorphism; 2D-DIGE, 2D difference gel electrophoresis; cDNA, complementary DNA; ESI-TOF MS, electrospray ionization time-of-flight mass spectrometry; EST, expressed sequence tag; LC MS/MS, liquid chromatography tandem mass spectrometry; MALDI-TOF MS, matrix-assisted laser desorption/ionization time-of-flight mass spectrometry; qRT-PCR, quantitative real-time polymerase chain reaction; Q-TOF MS/MS, quadrupole time-of-flight tandem mass spectrometry; RT-PCR, real-time polymerase chain reaction; SDS-PAGE, sodium dodecyl sulphate polyacrylamide gel electrophoresis; SNP, single nucleotide polymorphism.

nodule metabolism in plants under abiotic stress have been identified. Root nodular proteins (nif proteins) are known to cleave ROS by making them susceptible to proteolytic cleavage (Nakano *et al.*, 2006). Proteomic analysis of nodules can be achieved using techniques such as 2D-SDS-PAGE (Siria *et al.*, 2000; Lei *et al.*, 2005; Larrainzar *et al.*, 2007; De-la-Pena *et al.*, 2008). Formation and breakdown of proteins in the nodules of plants is very important for growth, development, metabolism, etc., making proteomic analysis of the nodules crucial for an understanding of changes in protein–protein interactions and protein profiling (Feller *et al.*, 2008).

16.2.2.3 Genomics

Genomics makes a vital contribution to understand genetic diversity and underlies many modern approaches to developing modified crops and gives new opportunities for altering QTLs through marker-assisted selection. Genomics approaches include a wide range of informative techniques such as development of molecular markers, genome profiling, DNA sequencing, etc. (Reddy *et al.*, 2012). Certain genomics techniques such as recombinant profiling, genome-wide association (GWA) and QTL sequencing have provided immense amounts of information and enabled a greater understanding of the biochemical and molecular bases of genetic variation and signalling networks in numerous plant species under abiotic interaction (Cross *et al.*, 2006; Stitt *et al.*, 2010; Riedelsheimer *et al.*, 2012).

Techniques being used for genomic analysis of plants under environmental stresses include fluxomics, metabolomics, tissue-specific metabolite profiling and spatial transcriptional changes (Gifford *et al.*, 2008; Long *et al.*, 2010; Saito & Matsuda, 2010; Tohge *et al.*, 2011; Bocobza *et al.*, 2012; Ferni & Morgan, 2013).

In recent years, development of next generation sequencing (NGS) techniques has enabled the sequencing of large genomes within a few hours and allows the differential profiling of cell transcripts at the genomic level. This technique has been useful even in the genomic analysis of very large genomes including grain legumes (Oldroyd, 2005; Thompson *et al.*, 2005). Development of linkage maps based on SSRs and genetic markers for profiling and mapping of pathogen stress (i.e. ascophyta blight disease)-responsive QTLs in food legumes has also been done (Muehlbauer & Chen, 2007).

Several genetic alterations due to salt stress have been reported in soybean. Various QTLs have been identified by studying two-variant crosses in soybean, i.e. S-100 (salt tolerant) × Tokyo (salt susceptible) was evaluated using RFLP markers. An important QTL associated with salt stress acclimatization was observed to be present near the Sat_091 SSR marker on linkage group (LG) N (Lee *et al.*, 2004). Recent years have seen the development of intervarietal molecular profiling with the aid of SSRs, SNPs and gene-based molecular markers in common bean, chickpea and soybean (Hwang *et al.*, 2009; Hanai *et al.*, 2010; Nayak *et al.*, 2010) (Table 16.2).

16.2.2.4 Functional genomics

Functional genomics is a field of modern biology that aims at deciphering the functions of genes that are sequenced. Numerous genome projects generated extensive information about gene sequences of different plant species, but decoding their functions becomes the challenge that needs to be focused upon. Hence, functional genomics, which uses methodologies of reverse genetics, becomes important in recognizing the associations between an organism's genome and its phenotype (Pevsner, 2009).

This technology gains more importance when applied for the identification of genes that control stress tolerance or resistance in plants subjected to various biotic and abiotic stresses, and it can be further used to develop resistant varieties (Kudapa *et al.*, 2013). Amongst the leguminous plants, *Medicago truncatula* and *Lotus japonicus* have been used as model plants to carry out extensive molecular and genetic studies due to their small genomes and simple genetic system (Handberg & Stougaard, 1992; Cook, 1999).

The function of a gene can be identified by deactivating gene or by suppressing its expression. This can be achieved by antisense RNA suppression, targeted gene replacement, insertional mutagenesis, gene silencing through RNAi and targeted induced local lesion in genome (TILLING) approaches (Reddy *et al.*, 2012). Of these, RNAi-induced silencing – a post-transcriptional gene silencing method – has been widely adopted. In legumes, this technique has been successfully used in *Medicago*, in which the importance of PIN proteins was established by using RNAi. Reduction in PIN proteins resulted in a reduced number of nodules thereby indicating the role of these proteins in nodulation (Huo *et al.*, 2006). In soybean, RNAi was used to silence the expression of the *GmMIPS1* gene, which is responsible for encoding myoinositol-1-phosphate, which further

Table 16.2 Various molecular markers used for genomic analysis in stressed legumes.

Serial no.	Legumes	Stress	Type of marker	References
1.	*Arachis hypogaea*	Water deficit stress	SSR	Varshney *et al.* (2009)
2.	*Cicer arietinum*	Pathogen stress (*Fusarium* wilt resistance)	SSR	Iruela *et al.* (2007)
3.	*Cicer arietinum*	Pathogen stress (*Ascophyta* blight resistance)	SSR	Iruela *et al.* (2007)
4.	*Cicer arietinum*	Pathogen stress (*Ascophyta* blight resistance)	SSR	Anbessa *et al.* (2009)
5.	*Cicer arietinum*	Water deficit stress	TUS and ESTs	Hiremath *et al.* (2011)
6.	*Cicer arietinum*	Chilling stress	ESTs	Dinari *et al.* (2013)
7.	*Glycine max*	Water deficit stress	Transcriptional factor analysis, SSR, SNP	Shukla *et al.* (2006)
8.	*Glycine max*	Soybean mosaic virus resistance	SSR, AFLP	Hwang *et al.* (2006)
9.	*Glycine max*	Soybean rust resistance	SSR	Hyten *et al.* (2007)
10.	*Glycine max*	Salt and freezing stress	SSR, AFLP, SRAP and RAPD	Liao *et al.* (2008)
11.	*Glycine max*	Chilling stress	SSR, SRAP and RAPD	Cheng *et al.* (2009)
12.	*Glycine max*	Salt stress	SSR	Wei *et al.* (2009)
13.	*Glycine max*	Soybean cyst nematode	SSR	Isemura *et al.* (2010)
14.	*Glycine max*	*Phytophthora* root rot resistance	SSR	Li *et al.* (2010)
15.	*Glycine max*	Water deficit stress, salt stress, cold and ABA treatments	qRT-PCR analysis of cDNA	Le *et al.* (2012)
16.	*Lens culinaris*	Pathogen stress (*Stemphylium* blight resistance, rust resistance)	SSR, SRAP and RAPD	Saha *et al.* (2010)
17.	*Phaseolus vulgaris*	Pathogen stress (bean gold mosaic virus)	RAPD and SRAP	Blair *et al.* (2007)
18.	*Phaseolus vulgaris*	Pathogen stress (white mould disease)	AFLP	Ender *et al.* (2008)

AFLP, amplified fragment length polymorphism; cDNA, complementary DNA; EST, expressed sequence tag; qRT-PCR, quantitative real-time polymerase chain reaction; RAPD, randomly amplified polymorphic DNA; SNP, single nucleotide polymorphism; SRAP, sequence-related amplified polymorphism; SSR, simple sequence repeat; TUS, tentative unique sequence.

has an important role in growth and metabolism. Gene silencing resulted in suppression of seed formation, which suggested a relation between the *GmMIPS1* gene and seed development (Nunes *et al.*, 2006). But phenotypic instability and lack of reliable plant transformation systems are the two major shortcomings of this technique (Hannon, 2002). Hence, a technique of gene silencing using virus vectors was devised that repressed expression by harbouring the target region of the host genome (Britt & May, 2003; Baulcombe, 2004). The technique, known as virus-induced gene silencing (VIGS), was considered to be a superior approach because of its persistence in providing genotypically identical silenced plants during both vegetative and *in vitro* propagation (Kudapa *et al.*, 2013). VIGS has been reported to be applied in soybean and pea. In soybean, Kachroo and Ghabrial (2012) used a bean pod mottle virus (BPMV)-based system to identify the genes responsible for systemic immunity, resistance gene-mediated immunity and basal immunity. Apple latent spherical virus (ALSV) vector was used by Yamagishi and Yoshikawa (2009) to study the gene function during growth in the vegetative and reproductive stages of soybean plants. Similarly, in pea plants, pea early-browning virus (PEBV) was used to identify the function of genes engaged in plant-microbe symbiosis (Grønlund *et al.*, 2010). The use of VIGS, however, is limited in legumes as more research needs to be done to develop appropriate vectors and virus vector delivery methods (Kudapa *et al.*, 2013).

Another tool in functional genomics that unites chemical mutagenesis with an influential screening technique for possible mutations is targeted induced local lesion in genome (TILLING) (Henikoff *et al.*, 2004; Gilchrist & Haughn, 2005). TILLING is very appropriate for functional examination of genes as well as for applications in agriculture because phenotypic variants are produced with no introduction of foreign DNA in the plant (Dita *et al.*, 2006). Among the legumes, TILLING's capacity for assortment of mutants is suited to *L. japonicus, M. truncatula, G. max* and peanuts (Perry *et al.*, 2003). The tobacco retrotransposon Tnt1 has been used effectively for insertional mutagenesis in *M. truncatula,* showing its value as a tool for functional genomics (Tadege *et al.*, 2005). For use in reverse genetics, TILLING populations for *Medicago* (12,000 M2 plants) and *Lotus* (4904 M2 lines) were reported (Perry *et al.*, 2009; Rogers *et al.*, 2009). Apart from these model plants, TILLING populations for peanuts (10,000 lines), chickpea (~3500) and common bean (3000 M3 lines) have been reported to be produced for allergenicity, drought tolerance and root nodulation respectively (Porch *et al.*, 2009; Tadege *et al.*, 2009; Kudapa *et al.*, 2013).

16.2.2.5 Metabolomics

Proteomics and transcriptomics are significant areas in decoding a biological pathway, but these techniques are still not enough without the study of cell metabolites because nearly all biological processes are eventually controlled by cell metabolites. Therefore to understand any process in more detail, it is important to study the metabolome of the organism. The term metabolome encompasses all the metabolites in a biological cell, tissue, organ or organism, which are the end products of cellular processes (Jordan *et al.*, 2009). The metabolome can be regarded as the final stage of gene expression, one that describes the biochemical phenotype of a cell or tissue (Sumner *et al.*, 2003). Metabolomics is the study of the distinctive chemical fingerprints that are residual of definite cellular processes. Therefore, when transcriptomics and proteomics do not provide a complete picture of what might be happening in a cell, metabolic profiling can provide a snapshot of the functioning of that cell (Daviss, 2005). A full survey of the biochemical state of an organism is made possible by quantitative and qualitative study of great numbers of cellular metabolites. Thus gene function of that organism can be screened and evaluated by metabolomics (Fiehn *et al.*, 2000).

Several techniques are needed for a complete examination of the metabolite profile (Hall *et al.*, 2002). They include thin layer chromatography (Tweeddale *et al.*, 1998), high-performance liquid chromatography (HPLC) with ultraviolet and photodiode array detection (Fraser *et al.*, 2000), infrared spectroscopy (Oliver *et al.*, 1998), liquid chromatography-mass spectrometry, liquid chromatography tandem mass spectrometry, gas chromatography-mass spectrometry (Huhman & Sumner, 2002) and nuclear magnetic resonance (NMR) (Bligny & Douce, 2001).

Even though complete metabolomic studies are not easy, a number of targeted investigations have been carried out to identify the participation of subsets of metabolites in a variety of stresses in legumes (He & Dixon, 2000). Until now there have been comparatively few studies of legume seeds at the metabolomic level (Thompson *et al.*, 2009). Bell and coworkers (2001) employed metabolomics in a *M. truncatula* cell suspension to find out reactions to different stimuli. Metabolomics along with transgenomics has helped in crop improvement by increasing the intrinsic stress resistance levels in legume crops (Wu & Van Etten, 2004). Soluble metabolites of the legume *L. japonicus* were profiled by Sanchez and coworkers (2008) using gas chromatography/mass spectrometry. This examination revealed metabolic adaptations of the plants that tolerated salt stress. The modification in metabolic phenotype was without doubt concurrent with the rising salt concentration. Those metabolic alterations established a qualitative resemblance to the observed changes in patterns of gene expression of the legume. A comparison of the seed metabolome of pea lines possessing and deficient in pea albumin 2 (PA2) was carried out by Vigeolas *et al.* (2008). Pea albumin 2 is a chief reserve protein produced by introgression of a naturally occurring PA2 locus deletion into a standard genetic background. It was found that the lack of this protein was related to the dissimilarity in amino acid and polyamine contents in the seed. This was attributed in some measure to reduction in the actions of two enzymes of polyamine synthesis, i.e. spermidine synthase and arginine decarboxylase. A case study of the metabolic responses to long-term salt stress in the model legume *L. japonicus* was conducted by Sanchez *et al.* (2010). The study suggested that the major metabolic alterations of legumes in salt stress depend upon uncontrolled environmental variables. Silvente and coworkers (2012) studied the metabolite modifications in response to water stress in drought-sensitive (DM50048) and drought-tolerant (NA5009RG) genotypes of soybean. Metabolite adjustments in relation to physiological responses in the presence of short-term water stress were elucidated on physiology of both genotypes using ^{1}H NMR. These two genotypes showed major differences in physiological responses, which led to the identification of metabolites that help in overcoming stressed conditions.

Further metabolomic analysis in legumes has been mostly limited to the model species (Thompson *et al.*, 2009). This technique therefore provides an insight into the role of various metabolites that have stress-tolerance properties. But as metabolic pathways are complex and interrelated, an extensive metabolic analysis needs to be done in order to have a complete understanding of the process and its application in stress tolerance.

16.3 Conclusions

Stress, whether biotic or abiotic, causes alterations in the metabolism of the plant system. The legume family contains species that are valuable as food as well as cash crops. Therefore, such stresses can have a serious impact on the health and productivity of legumes, and consequently affect the nutrition and economy of the human population. Hence, the development of resistant legume varieties that can tolerate increasing stressful conditions in the environment has become the need of the hour.

Biotechnology is a field that provides various strategies offering great potential to counter this problem. Gene pyramiding and MAS are encouraging techniques that allow plant breeders to overcome numerous fungal and viral diseases. In addition, the omics approaches offer researchers many opportunities to develop crop varieties that can adjust and survive under extreme environmental conditions. Genome sequencing of soybean and model plants has led to identification of genes involved in stress responses. A cumulative approach consisting of all the above mentioned techniques allows deep insights at genetic and molecular levels to decode the expression and enhancement of stress responses in plants.

References

Agarwal G, Jhanwar S, Priya P, *et al.* (2012) Comparative analysis of kabuli chickpea transcriptome with desi and wild chickpea provides a rich resource for development of functional markers. *Plos One* 7: e52443.

Aghaei K, Ehsanpour AA, Shah AH, Komatsu S (2009) Proteome analysis of soybean hypocotyl and root under salt stress. *Amino Acids* 36: 91–98.

Ahsan N, Donnart T, Nouri MZ, Komatsu S (2010) Tissue-specific defense and thermo-adaptive mechanisms of soybean seedlings under heat stress revealed by proteomic approach. *J Proteome Res* 9: 4189–4204.

Alam I, Sharmin SA, Kim KH, *et al.* (2010) Proteome analysis of soybean roots subjected to short-term drought stress. *Plant Soil* 333: 491–505.

Alberts B, Johnson A, Lewis J, Raff M, Roberts K, Walter P (2002) *Molecular Biology of the Cell*, 4th edn. Garland Science, New York.

Anbessa Y, Taran B, Warkentin TD, Tullu A, Vandenberg A (2009) Genetic analyses and conservation of QTL for ascochyta blight resistance in chickpea (*Cicer arietinum* L.). *Theor Appl Genet* 119: 757–765.

Apata DF, Ologhobo AD (1997) Trypsin inhibitor and the other anti-nutritional factors in tropical legume seeds. *Trop Sci* 37: 52–59.

Baulcombe D (2004) RNA silencing in plants. *Nature* 431: 356–363.

Bell CJ, Dixon RA, Farmer AD, *et al.* (2001) The *Medicago* Genome Initiative: a model legume database. *Nucleic Acids Res* 29: 114–117.

Bhatnagar-Mathur P, Devi MJ, Srinivas Reddy D, *et al.* (2007) Stress-inducible expression of AtDREB1A in transgenic peanut (*Arachis hypogaea* L.) increases transpiration efficiency under water-limiting conditions. *Plant Cell Rep* 26: 2071–2082.

Bhushan D, Pandey A, Choudhary M, Datta A, Chakraborty S, Chakraborty N (2007) Comparative proteomics analysis of differentially expressed proteins in chickpea extracellular matrix during dehydration stress. *Mol Cell Proteomics* 6, 1868–1884.

Blair MW, Rodriguez LM, Pedraza F, Morales F, Beebe S (2007) Genetic mapping of the bean golden yellow mosaic geminivirus resistant gene bgm-1 and linkage with potyvirus resistance in common bean (*Phaseolus vulgaris* L.). *Theor Appl Genet* 114: 261–271.

Bligny R, Douce R (2001) NMR and plant metabolism. *Curr Opin Plant Biol* 4: 191–196.

Bocobza SE, Willmitzer L, Raikhel N, Aharoni A (2012) Discovery of new modules in metabolic biology using chemometabolomics. *Plant Physiol* 160: 1160–1163.

Bohn M, Groh S, Khairallah MM, Hoisington DA, Utz HF, Melchinger AE (2001) Re-evaluation of the prospects of marker-assisted selection for improving insect resistance against *Diatraea* spp. in tropical maize by cross validation and independent validation. *Theor Appl Genet* 103: 1059–1067.

Britt AB, May GD (2003) Re-engineering plant gene targeting. *Trends Plant Sci* 8: 90–95.

Carmona MA, Gally ME, Lopez SE (2005) Asian soybean rust: Incidence, severity, and morphological characterization of *Phakopsora pachyrhizi* (uredinia and telia) in Argentina. *Plant Dis* 89: 109.

Cheng L, Huan S, Sheng Y, *et al.* (2009) GMCHI, cloned from soybean [*Glycine max* (L.) Meer.] enhances survival in transgenic *Arabidopsis* under abiotic stress. *Plant Cell* 28: 145–153.

Cheng L, Gao X, Li S, *et al.* (2010) Proteomic analysis of soybean [*Glycine max* (L.) Meer.] seeds during imbibition at chilling temperature. *Mol Breed* 26: 1–17.

Chinnusamy V, Zhu J, Zhu JK (2007) Cold stress regulation of gene expression in plants. *Trends Plant Sci* 12: 444–451.

Collard BCY, Mackill DJ (2008) Marker-assisted selection: an approach for precision plant breeding in the twenty-first century. *Philos Trans Roy Soc B* 363: 557–572.

Collard BCY, Jahufer MZZ, Brouwer JB, Pang ECK (2005) An introduction to markers, quantitative trait loci (QTL) mapping and marker-assisted selection for crop improvement: the basic concepts. *Euphytica* 142: 169–196.

Cook DR (1999) *Medicago truncatula*: a model in making! Commentary. *Curr Opin Plant Biol* 2: 301–304.

Cox TS, Raupp WJ, Gill BS (1994) Leaf rust-resistance genes Lr41, Lr42 and Lr43 transferred from *Triticum tauschii* to common wheat. *Crop Sci* 34: 339–343.

Cross JM, Von Korff M, Altmann T, *et al.* (2006) Variation of enzyme activities and metabolite levels in 24 Arabidopsis accessions growing in carbon-limited conditions. *Plant Physiol* 142: 1574–1588.

Daviss B (2005) Growing pains for metabolomics. *The Scientist* 19: 25–28.

De-la-Pena C, Lei Z, Watson BS, Sumner LW, Vivanco JM (2008) Root–microbe communication through protein secretion. *J Biol Chem* 283: 25247–25255.

Denman S, Knoxdavies PS, Calitz FJ, Lamprecht SC (1995) Pathogenicity of *Pythium irregulare, Pythium sylvaticum* and *Pythium ultimum* var. *ultimum* to lucerne (*Medicago sativa*). *Australas Plant Pathol* 24: 137–143.

Diers B (2004) Soybean genetic improvement through conventional and molecular based strategies. In: *5th European Conference on Grain Legumes*, Dijon, France, pp. 147–148.

Dinari A, Niazi A, Afsharifar AR, Ramezani A (2013) Identification of upregulated genes under cold stress in cold tolerant chickpea using the cDNA-AFLP approach. *PLoS One* 8: e52757.

Dita MA, Rispail N, Elena Prats E, Rubiales D, Singh KB (2006) Biotechnology approaches to overcome biotic and abiotic stress constraints in legumes. *Euphytica* 147: 1–24.

Doyle JJ, Luckow MA (2003) The rest of the iceberg. Legume diversity and evolution in a phylogenetic context. *Plant Physiol* 131: 900–910.

Duranti M (2006) Grain legume proteins and nutraceutical properties. *Fitoterapia* 77: 67–82.

Ender M, Terpstra K, Kelly JD (2008) Marker-assisted selection for white mold resistance in common bean. *Mol Breed* 21: 149–157.

Faleiro FG, Ragagnin VA, Moreira MA, de Barros EG (2004) Use of molecular markers to accelerate the breeding of common bean lines resistant to rust and anthracnose. *Euphytica* 138: 213–218.

Feller U, Anders I, Demirevska K (2008) Degradation of rubisco and other chloroplast proteins under abiotic stress. *Gen Appl Plant Physiol* 34: 5–8.

Fernie AR, Morgan JA (2013) Analysis of metabolic flux using dynamic labelling and metabolic modelling. *Plant Cell Environ* 36: 1738–1750.

Fiehn O, Kopka J, Dormann P, Altmann T, Trethewey RN, Willmitzer L (2000) Metabolite profiling for plant functional genomics. *Nat Biotechnol* 18: 1157–1161.

Fraser PD, Pinto ME, Holloway DE, Bramley PM (2000) Application of high-performance liquid chromatography with photodiode array detection to the metabolic profiling of plant isoprenoids. *Plant J* 24: 551–558.

Garg N, Geetanjali (2007) Symbiotic nitrogen fixation in legume nodules: process and signaling. A review. *Agron Sustainable Dev* 27: 59–68.

Garza R, Cardona C, Singh SP (1996) Inheritance of resistance to the bean-pod weevil (*Apion godmani* Wagner) in common beans from Mexico. *Theor Appl Genet* 92: 357–362.

Gifford ML, Dean A, Gutierrez RA, Coruzzi GM, Birnbaum KD (2008) Cell-specific nitrogen responses mediate developmental plasticity. *Proc Nat Acad Sci USA* 105: 803–808.

Gilchrist EJ, Haughn GW (2005) TILLING without a plough: a new method with applications for reverse genetics. *Curr Opin Plant Biol* 8: 211–215.

Ginzinger DG (2002) Gene quantification using real-time quantitative PCR: An emerging technology hits the mainstream. *Exp Hematol* 30: 503–512.

Graham PH, Vance CP (2003) Legumes: Importance and constraints to greater use. *Plant Physiol* 131: 872–877.

Grimsrud PA, Os DD, Wenger CD, *et al.* (2010) Large-scale phosphoprotein analysis in *Medicago truncatula* roots provides insight into in vivo kinase activity in legumes. *Plant Physiol* 152: 19–28.

Grønlund M, Olsen A, Johansen EI, Jakobsen I (2010) Protocol: using virus induced gene silencing to study the arbuscular mycorrhizal symbiosis in *Pisum sativum*. *Plant Methods* 6: 28–35.

Guan B, Zhou D, Zhang H, Tian Y, Japhet W, Wang P (2009) Germination responses of *Medicago ruthenica* seeds to salinity, alkalinity and temperature. *J Arid Environ* 73: 135–138.

Guo B, Chen X, Dang P, *et al.* (2008) Peanut gene expression profiling in developing seeds at different reproduction stages during *Aspergillus parasiticus* infection. *BMC Dev Biol* 8: 12.

Gupta PK, Varshney RK (2000) The development and use of microsatellite markers for genetic analysis and plant breeding with emphasis on bread wheat. *Euphytica* 113: 163–185.

Gupta PK, Varshney RK, Sharma PC, Ramesh B (1999) Molecular markers and their applications in wheat breeding. *Plant Breed* 118, 369–390.

Gygi SP, Rochon Y, Franza BR, Aebersold R (1999) Correlation between protein and mRNA abundance in yeast. *Mol Cell Biol* 19: 1720–1730.

Hadiarto T, Tran LS (2011) Progress studies of drought–responsive genes in rice. *Plant Cell* 30: 297–310.

Hall R, Beale M, Fiehn O, Hardy N, Sumner L, Bino R (2002) Plant metabolomics as the missing link in functional genomics strategies. *Plant Cell* 14: 1437–1440.

Hamwieh A, Udupa SM, Choumane W, *et al.* (2005) A genetic linkage map of *Lens* sp. based on microsatellite and AFLP markers and the localization of fusarium vascular wilt resistance. *Theor Appl Genet* 110: 669–677.

Hanai LR, Santini L, Camargo LEA, *et al.* (2010) Extension of the core map of common bean with EST-SSR, RGA, AFLP, and putative functional markers. *Mol Breed* 25: 25–45.

Handberg K, Stougaard J (1992) *Lotus japonicus*: an autogamous, diploid legume species for classical and molecular genetics. *Plant J* 2: 487–496.

Hannon G (2002) Mechanisms and application of RNA interference. *Biol Reprod* 66: 70.

Hashiguchi A, Sakata Z, Komatsu S (2009) Proteome analysis of early-stage soybean seedlings under flooding stress. *J Proteome Res* 8: 2058–2069.

He XZ, Dixon RA (2000) Genetic manipulation of isoflavone 7-O-methyl transferase enhances biosynthesis of 4′-O-methylated isoflavonoid phytoalexins and disease resistance in alfalfa. *Plant Cell* 12: 1689–1702.

Henikoff S, Till BJ, Comai L (2004) TILLING traditional mutagenesis meets functional genomics. *Plant Physiol* 135: 630–636.

Hiremath PJ, Farmer A, Cannon SB, *et al.* (2011) Large-scale transcriptome analysis in chickpea (Cicer arietinum L.), an orphan legume crop of the semi-arid tropics of Asia and Africa. *Plant Biotechnol J* 9: 922–931.

Hittalmani S, Parco A, Mew TV, Zeigler RS, Huang N (2000) Fine mapping and DNA marker assisted pyramiding of the three major genes for blast resistance in rice. *Theor Appl Genet* 100: 1121–1128.

Hospital F, Moreau L, Lacoudre F, Charcosset A, Gallais A (1997) More on the efficiency of marker-assisted selection. *Theor Appl Genet* 95: 1181–1189.

Huang N, Angeles ER, Domingo J, *et al.* (1997) Pyramiding of bacterial blight resistance genes via DNA marker aided selection in rice. *Theor Appl Genet* 95: 313–320.

Huhman DV, Sumner LW (2002) Metabolic profiling of saponins in *Medicago sativa* and *Medicago truncatula* using HPLC coupled to an electrospray ion-trap mass spectrometer. *Phytochemistry* 59: 347–360.

Huo X, Schnabel E, Hughes K, Frugoli J (2006) RNAi phenotypes and the localization of a protein: GUS fusion imply a role for *Medicago truncatula* PIN genes in nodulation. *J Plant Growth Regul* 25: 156–165.

Hwang TY, Moon JK, Yu S, *et al.* (2006) Application of comparative genomics in developing molecular markers tightly linked to the virus resistance gene Rsv4 in soybean. *Genome* 49: 380–388.

Hwang TY, Sayama T, Takahashi M, *et al.* (2009) High-density integrated linkage map based on SSR markers in soybean. *DNA Res* 16: 213–225.

Hyten DL, Hartman GL, Nelson RL, *et al.* (2007) Map location of the Rpp1 locus that confers resistance to soybean rust in soybean. *Crop Sci* 47: 837–840.

Iruela M, Castro P, Rubio J, *et al.* (2007) Validation of a QTL for resistance to ascochyta blight linked to resistance to fusarium wilt race 5 in chickpea (*Cicer arietinum* L.). *Eur J Plant Pathol* 119: 29–37.

Isemura T, Kaga A, Tomooka N, Shimizu T, Vaughan DA (2010) The genetics of domestication of rice bean, *Vigna umbellata*. *Ann Bot* 106: 927–944.

Jordan KW, Nordenstam J, Lauwers GY, *et al.* (2009) Metabolomic characterization of human rectal adenocarcinoma with intact tissue magnetic resonance spectroscopy. *Dis Colon Rectum* 52: 520–525.

Kachroo A, Ghabrial S (2012) Virus-induced gene silencing in soybean. In: Watson JM, Wang MB (eds), *Antiviral Resistance in Plants*. Humana Press, New York, pp. 287–297.

Kav NNV, Srivastava S, Goonewardene L, Blade SF (2007) Proteome level changes in the roots of *Pisum sativum* in response to salinity. *Ann Appl Biol* 145: 217–230.

Kelly JD, Vallejo VA (2004) A comprehensive review of the major genes conditioning resistance to anthracnose in common bean. *Hortscience* 39: 1196–1207.

Komatsu S, Ahsan N (2009) Soybean proteomics and its application to functional analysis. *J Proteomics* 72: 325–336.

Kottapalli KR, Rakwal R, Shibato J, *et al.* (2009) Physiology and proteomics of the water deficit stress response in three contrasting peanut genotypes. *Plant Cell Environ* 32: 380–407.

Kudapa H, Ramalingam A, Nayakoti S, *et al.* (2013) Functional genomics to study stress responses in crop legumes: progress and prospects. *Funct Plant Biol* 40: 1221–1233.

Kumar J, Choudhary AK, Solanki RK, Pratap A (2011) Towards marker-assisted selection in pulses: a review *Plant Breed* 130: 297–313.

Kumar S, Ali M (2006) GE interaction and its breeding implications in pulses. *The Botanica* 56: 31–36.

Kumaravadivel N, Uma MD, Saravanan PA, Suresh H (2006) Molecular marker-assisted selection and pyramiding genes for gall midge resistance in rice suitable for Tamil Nadu Region. In: *ABSTRACTS – 2nd International Rice Congress, IRRI, Manila, Philippines*.

Laanbroek HJ (1990) Bacterial cycling of minerals that affect plant growth in waterlogged soils: A review. *Aquatic Bot* 38: 109–125.

Labdi M, Robertson LD, Singh KB, Charrier A (1996) Genetic diversity and phylogenetic relationships among the annual *Cicer* species as revealed by isozyme polymorphisms. *Euphytica* 88: 181–188.

Lalinia AA, Hoseini NM, Galostian M, Bahabadi SE, Khameneh MM (2012). Ecophysiological impact of water stress on growth and development of mungbean. *Int J Agron Plant Prod* 3: 599–607.

Lan WU, Chao W (2011) Application of molecular marker assisted selection in gene pyramiding and selection of new cultivars. *J Northeast Agric Univ* 18: 79–84.

Lande R, Thompson R (1990) Efficiency of marker assisted selection in the improvement of quantitative traits. *Genetics* 124: 743–756.

Larrainzar E, Wienkoop S, Weckwerth W, Ladrera R, Arrese-Igor C, González EM (2007). *Medicago truncatula* root nodule proteome analysis reveals differential plant and bacteroid responses to drought stress. *Plant Physiol* 144: 1495–1507.

Le DT, Aldrich DL, Valliyodan B, *et al.* (2012) Evaluation of candidate reference genes for normalization of quantitative RT-PCR in soybean tissues under various abiotic stress conditions. *PLoS One* 7: e46487.

Lee GJ, Boerma HR, Villagarcia MR, *et al.* (2004) A major QTL conditioning salt tolerance in S-100 soybean and descendant cultivars. *Theor Appl Genet* 109: 1610–1619.

Lei Z, Elmer AM, Watson BS, Dixon RA, Mendes PJ, Sumner LW (2005) A two-dimensional electrophoresis proteomic reference map and systematic identification of 1367 proteins from a cell suspension culture of the model legume *Medicago truncatula. Mol Cell Proteomics* 4: 1812–1825.

Lewis G, Schrire B, MacKinder B, Lock M (2005) *Legumes of the World.* Royal Botanical Gardens, Kew Publishing.

Li X, Han Y, Teng W, *et al.* (2010) Pyramided QTL underlying tolerance to *Phytophthora* root rot in megaenvironments from soybean cultivars 'Conrad' and 'Hefeng 25'. *Theor Appl Genet* 121: 651–658.

Liao Y, Zou HF, Wei W, *et al.* (2008) Soybean GmbZIP44, GmbZIP62 and GmbZIP78 genes function as negative regulator of ABA signaling and confer salt and freezing tolerance in transgenic *Arabidopsis. Planta* 228: 225–240.

Liu J, Liu D, Tao W, *et al.* (2000) Molecular marker-facilitated pyramiding of different genes for powdery mildew resistance in wheat. *Plant Breed* 119: 21–24.

Liu K (1997) *Soybeans, Chemistry, Technology and Utilization.* Chapman & Hall.

Long TA, Tsukagoshi H, Busch W, Lahner B, Salt DE, Benfey PN (2010) The bhlh transcription factor popeye regulates response to iron deficiency in *Arabidopsis* roots. *Plant Cell* 22: 2219–2236.

Luo M, Dang P, Guo BZ, *et al.* (2005) Generation of expressed sequence tags (ESTs) for gene discovery and marker development in cultivated peanut. *Crop Sci* 45: 346–353.

Manchada G, Garg N (2008) Salinity and its effects on the functional biology of legumes. *Acta Physiol Plant* 30: 595–618.

Mantri NL, Ford R, Coram TE, Pang EC (2007) Transcriptional profiling of chickpea genes differentially regulated in response to high-salinity, cold and drought. *BMC Genomics* 8: 303.

Marcotte EM, Pellegrini M, Ng HL, Rice DW, Yeates TO, Eisenberg D (1999) Detecting protein function and protein–protein interactions from genome sequences. *Science* 285: 751–753.

Matamoros MA, Dalton DA, Ramos J, Clemente MR, Rubio MC, Becana M (2003) Biochemistry and molecular biology of antioxidants in the rhizobia-legume symbiosis. *Plant Physiol* 133: 499–509.

Mathesius U (2009) Comparative proteomic studies of root–microbe interactions. *J Proteomics* 72: 353–366.

Mathur PB, Vadez EV, Sharma KK (2008) Transgenic approaches for abiotic stress tolerance in plants: retrospect and prospects *Plant Cell Rep* 27: 411–424.

Matsui A, Ishida J, Morosawa T, Mochizuki Y, Kaminuma E (2008) Arabidopsis transcriptome analysis under drought, cold, high-salinity and ABA treatment conditions using a tiling array. *Plant Cell Physiol* 49: 1135–1149.

Matsui A, Ishida J, Morosawa T, Okamoto M, Kim JM (2010) *Arabidopsis* tiling array analysis to identify the stress-responsive genes. *Methods Mol Biol* 639: 141–155.

Medini M, Baum M, Hamza S (2009) Transcript accumulation of putative drought responsive genes in drought-stressed chickpea seedlings. *Afr J Biotechnol* 8: 4441–4449.

Molina C, Zaman-Allah M, Khan F, Fatnassi N, Horres R (2011) The salt responsive transcriptome of chickpea roots and nodules via deep-SuperSAGE. *BMC Plant Biol* 11: 31.

Morel MA, Braña V, Castro-Sowinski S (2012) Legume crops, importance and use of bacterial inoculation to increase production. In: Goyal A (ed.), *Crop Plant.* InTech, doi: 10.5772/37413.

Muehlbauer FJ, Chen W (2007) Resistance to ascophyta blights of cool season food legumes. *Eur J Plant Pathol* 119: 135–141.

Mutlu N, Miklas P, Reiser J, Coyne D (2005) Backcross breeding for improved resistance to common bacterial blight in pinto bean (Phaseolus vulgaris L.). *Plant Breed* 124: 282–287.

Nakano M, Nobuta K, Vemaraju K, Tej SS, Skogen JW, Meyers BC (2006) Plant MPSS databases: Signature-based transcriptional resources for analyses of mRNA and small RNA. *Nucleic Acids Res* 34: 731–735.

Nakashima K, Ito Y, Yamaguchi-Shinozaki K (2009) Transcriptional regulatory networks in response to abiotic stresses in Arabidopsis and grasses. *Plant Physiol* 149: 88–95.

Nanjo Y, Skultety L, Ashraf Y, Komatsu S (2010) Comparative proteomic analysis of early-stage soybean seedlings responses to flooding by using gel and gel-free techniques. *J Proteome Res* 9: 3989–4002.

Nanjo Y, Nouri MZ, Komatsu S (2011) Quantitative proteomic analyses of crop seedlings subjected to stress conditions; a commentary. *Phytochemistry* 72: 1263–1272.

Nayak SN, Zhu H, Varghese N, *et al.* (2010) Integration of novel SSR and gene-based SNP marker loci in the chickpea genetic map and establishment of new anchor points with *Medicago truncatula* genome. *Theor Appl Genet* 120: 1415–1441.

Nene YL, Reddy MV (1987) Chickpea diseases and their control. In: Saxena MC, Singh KB (eds), *The Chickpea.* CAB International, Wallingford, Oxon, UK, pp. 233–270.

Nouri M, Toorchi M, Komatsu S (2011) Proteomics approach for identifying abiotic stress responsive proteins in soybean. In: Sudaric A (ed.), *Soybean – Molecular Aspects of Breeding.* InTech, doi: 10.5772/15518.

Nunes AC, Vianna GR, Cuneo F, *et al.* (2006) RNAi-mediated silencing of the *myo*-inositol-1-phosphate synthase gene

(*GmMIPS1*) in transgenic soybean inhibited seed development and reduced phytate content. *Planta* 224: 125–132.

Olajide OB, Fadimu OY, Osaguona PO, Saliman MI (2013) Ethnobotanical and phytochemical studies of some selected species of Leguminoseae of northern Nigeria: a study of Borgu local government area, Niger State. Nigeria. *Int J Sci Nat* 4: 546–551.

Oldroyd G (2005) Sequencing the model legume *Medicago truncatula*. *Grain Legumes* 41: 23.

Oliver SG, Winson MK, Kell DB, Baganz F (1998) Systematic functional analysis of the yeast genome. *Trends Biotechnol* 16: 373–378.

Pandey A, Chakraborty S, Datta A, Chakraborty N (2008) Proteomics approach to identify dehydration responsive nuclear proteins from chickpea (*Cicer arietinum* L). *Mol Cell Proteomics* 7: 88–107.

Panigrahi J, Mishra RR, Sahu AR, Rath SC, Seth S, Mishra SP (2013) Marker-assisted breeding for simple inherited traits conferring stress resistance in crop plants. *The Ecoscan* 3: 217–233.

Parrish JR, Gulyas KD, Finley RL (2006) Yeast two-hybrid contributions to interactome mapping. *Curr Opin Biotechnol* 17: 387–393.

Perry JA, Wang TL, Welham TJ, *et al.* (2003) A TILLING reverse genetics tool and a web-accessible collection of mutants of the legume *Lotus japonicus*. *Plant Physiol* 131: 866–871.

Perry J, Brachmann A, Welham T, *et al.* (2009) TILLING in *Lotus japonicus* identified large allelic series for symbiosis genes and revealed a bias in functionally defective ethyl methanesulfonate alleles toward glycine replacements. *Plant Physiol* 151: 1281–1291.

Pevsner J (2009) *Bioinformatics and Functional Genomics*, 2nd edn. Wiley-Blackwell, Hoboken, NJ.

Polesani M, Desario F, Ferrarini A, Zamboni A, Pezzotti M (2008) cDNA-AFLP analysis of plants and pathogen genes expressed in grapevine infected with *Plasmopara viticola*. *BMC Genomics* 9: 142–148.

Porch TG, Blair MW, Lariguet P, Galeano C, Pankhurst CE, Broughton WJ (2009) Generation of a mutant population for TILLING common bean genotype BAT 93. *J Am Soc Hort Sci* 134: 348–355.

Proite K, Leal-Bertioli SCM, Bertioli DJ, *et al.* (2007) ESTs from a wild *Arachis* species for gene discovery and marker development. *BMC Plant Biol* 7: 7.

Reddy, A.R., Chaitanya, K.V. & Vivekanandan, M. (2004) Drought induced responses of photosynthesis and antioxidant metabolism in higher plants. *J Plant Physiol* 161: 1189–1202.

Reddy SD, Mathur PB, Vadez V, Sharma KK (2012) Grain legumes (soybean, chickpea, and peanut): omics approaches to enhance abiotic stress tolerance. In: Tuteja N, Gill SS, Tiburcio AF, Tuteja R (eds), *Improving Crop Resistance to Abiotic Stress*, 1st edn. Wiley-VCH Verlag GmbH & Co.

Ren S, Bilyeu KD, Beuselinck PR (2009) Composition, vigor, and proteome of mature soybean seeds developed under high temperature. *Crop Sci* 49: 1010–1022.

Riedelsheimer C, Lisec J, Czedik-Eysenberg A, *et al.* (2012) Genome-wide association mapping of leaf metabolic profiles for dissecting complex traits in maize. *Proc Nat Acad Sci U.S.A.* 109: 8872–8877.

Rogers C, Wen J, Chen R, Oldroyd G (2009) Deletion-based reverse genetics in *Medicago truncatula*. *Plant Physiol* 151: 1077–1086.

Rubiales D, Emeran AA, Sillero JC (2002) Rusts on legumes in Europe and North Africa. *Grain Legumes* 37: 8–9.

Saghai Maroof MA, Jeong SC, Gunduz I, Tucker DM, Buss GR, Tolin SA (2008) Pyramiding of soybean mosaic virus resistance genes by marker assisted selection. *Crop Sci* 48: 517–526.

Saha GC, Sarker A, Chen W, Vandemark GJ, Muehlbauer FJ (2010) Inheritance and linkage map positions of genes conferring resistance to *Stemphylium* blight in lentil. *Crop Sci* 50: 1831–1839.

Saito K, Matsuda F (2010) Metabolomics for functional genomics, systems biology, and biotechnology. *Annu Rev Plant Biol* 61: 463–489.

Salwinski L, Eisenbergy D (2003) Computational methods of analysis of protein–protein interactions. *Curr Opin Struct Biol*, 13: 377–382.

Sanchez DH, Lippold F, Redestig H, *et al.* (2008) Integrative functional genomics of salt acclimatization in the model legume *Lotus japonicas*. *Plant J* 53: 973–987.

Sanchez DH, Szymanski J, Erban A, Udvardi MK, Kopka J (2010) Mining for robust transcriptional and metabolic responses to long-term salt stress: a case study on the model legume *Lotus japonicus*. *Plant Cell Environ* 33: 468–480.

Servin B, Martin OC, Mezard M, Hospital F (2004) Toward a theory of marker-assisted gene pyramiding. *Genetics* 168: 513–523.

Sha Valli Khan PS, Hoffmann L, Renaut J, Hausman JF (2007) Current initiatives in proteomics for the analysis of plant salt tolerance. *Curr Sci* 93: 6–11.

Shi A, Chen P, Li D, Zheng C, Zhang B, Hou A (2009) Pyramiding multiple genes for resistance to soybean mosaic virus in soybean using molecular markers. *Mol Breed* 23: 113–124.

Shi F, Yamamoto R, Shimamura S, Hiraga S, *et al.* (2008) Cytosolic ascorbate peroxidase 2 (cAPX 2) is involved in the soybean response to flooding. *Phytochemistry* 69: 1295–1303.

Shukla RK, Raha S, Tripathi V, Chattopadhyay D (2006) Expression of CAP2, an APETALA2-family transcription factor from chickpea, enhances growth and tolerance to dehydration and salt stress in transgenic tobacco. *Plant Physiol* 142: 113–123.

Shukla RK, Tripathi V, Jain D, Yadav RK, Chattopadhyay D (2009) CAP2 enhances germination of transgenic tobacco seeds at high temperature and promotes heat stress tolerance in yeast. *FEBS J* 276: 5252–5262.

Shvaleva A, de la Pena TC, Rincon A, *et al.* (2010) Flavodoxin overexpression reduces cadmium-induced damage in alfalfa root nodules. *Plant Soil* 326: 109–121.

Sillero JC, Fondevilla S, Davidson J, *et al.* (2006) Screening techniques and sources of resistance to rusts and mildews in grain legumes. *Euphytica* 147: 255–272.

Silveira JAG, Figueiredo MVB, Cavalcanti FR, Ferreira-Silva SL (2010) Legume nodule oxidative stress and N_2 fixation efficiency. In: Araujo ASF, Figueiredo MVB (eds), *Microbial Ecology of Tropical Soils.* Nova Science Publishers, Inc., pp.49–78.

Silvente S, Sobolev AV, Lara M (2012) Metabolite adjustments in drought tolerant and sensitive soybean genotypes in response to water stress. *PLoS One* 7: e388554.

Singh R, Sharma P, Varshney RK, Sharma SK, Singh NK (2008) Chickpea improvement: role of wild species and genetic markers. *Biotechnol Genet Eng Rev* 25: 267–314.

Singh S, Sidhu JS, Huang N, *et al.* (2001) Pyramiding three bacterial blight resistance genes (xa5, xa13 and Xa21) using marker-assisted selection into indica rice cultivar PR106. *Theor Appl Genet* 102: 1011–1015.

Siria HA, Natera Guerreiro N, Djordjevic MA (2000) Proteome analysis of differentially displayed proteins as a tool for the investigation of symbiosis. *Plant Microbe Interact* 13: 995–1009.

Sobhanian H, Razavizadeh R, Nanjo Y, *et al.* (2010) Proteome analysis of soybean leaves, hypocotyls and roots under salt stress. *Proteome Sci* 8: 19–34.

Sobkowiak R, Deckert J (2006) Proteins induced by cadmium in soybean cells. *J Plant Physiol* 163: 1203–1206.

Stavely JR (2000) Pyramiding rust and viral resistance genes using traditional and marker techniques in common bean. *Annual Report of the Bean Improvement Cooperative* 43: 1–4.

Stitt M, Sulpice R, Keurentjes J (2010) Metabolic networks: How to identify key components in the regulation of metabolism and growth. *Plant Physiol* 152: 428–444.

Subramanian S, Smith DL (2013) A proteomics approach to study soybean and its symbiont *Bradyrhizobium japonicum* – a review. In: Board J (ed.), *A Comprehensive Survey of International Soybean Research – Genetics, Physiology, Agronomy and Nitrogen Relationships.* Available at: http://dx.doi.org/10.5772/53728

Sumner LW, Mendes P, Dixon RA (2003) Plant metabolomics: large scale phytochemistry in the functional genomics era. *Phytochemistry* 62: 817–836.

Sun MM, Li LH, Xie H, Ma RC, He YK (2007) Differentially expressed genes under cold acclimatization in *Physcomitrella patens. Biochem Mol Biol* 40: 986–1001.

Tadege M, Ratet P, Mysore KS (2005) Insertional mutagenesis: a Swiss army knife for functional genomics of Medicago truncatula. *Trends Plant Sci* 10: 229–235.

Tadege M, Wang TL, Wen J, Ratet P, Mysore KS (2009) Mutagenesis and beyond! Tools for understanding legume biology. *Plant Physiol* 151: 978–984.

Thompson R, Burstin J, Gallardo K (2009) Post-genomics studies of developmental processes in legume seeds. *Plant Physiol* 151: 1023–1029.

Thompson R, Ratet P, Kuster H (2005) Identification of gene functions by applying TILLING and insertional mutagenesis strategies on microarray-based expression data. *Grain Legumes* 41: 20–22.

Tohge T, Ramos MS, Nunes-Nesi A, *et al.* (2011) Toward the storage metabolome: Profiling the barley vacuole. *Plant Physiol* 157: 1469–1482.

Tran LS, Nishiyama R, Yamaguchi-Shinozaki K, Shinozaki K (2010) Potential utilization of NAC transcription factors to enhance abiotic stress tolerance in plants by biotechnological approach. *GM Crops* 1: 32–39.

Tripathi V, Parasuraman B, Laxmi A, Chattopadhyay D (2009) CIPK6, a CBL-interacting protein kinase is required for development and salt tolerance in plants. *Plant J* 58: 778–790.

Tu JC (1981) Effect of salinity on Rhizobium–root hair interaction, nodulation and growth of soybean. *Can J Plant Sci* 61: 231–239.

Tweeddale H, Notley-McRobb L, Ferenci T (1998) Effect of slow growth on metabolism of *Escherichia coli,* as revealed by global metabolite pool ("Metabolome") analysis. *J Bacteriol* 180: 5109–5116.

Udupa SM, Sharma A, Sharma AP, Pai RA (1993) Narrow genetic variability in *Cicer arietinum* L. as revealed by RFLP analysis. *J Plant Biochem Biotechnol* 2: 83–86.

Udvardi MK, Czechowski T, Scheible WR (2008) Eleven golden rules of quantitative RT-PCR. *Plant Cell* 20: 1736–1737.

Umezawa T, Mizuno K, Fujimura T (2002) Discrimination of genes expressed in response to the ionic or osmotic effect of salt stress in soybean with cDNA-AFLP. *Plant Cell Environ* 25: 1617–1625.

Uzoechina OB (2009) Nutrient and anti-nutrients potentials of brown pigeon pea (*Cajanus cajan* var bicolor) seed flours. *Nigerian Food J* 27: 10–16.

Varshney RK, Bertioli DJ, Moretzsohn MC, *et al.* (2009) The first SSR-based genetic linkage map for cultivated groundnut (*Arachis hypogaea* L.). *Theor Appl Genet* 118: 729–739.

Verdoy D, Lucas MM, Manrique E, Covarrubias AA, De Felipe MR, Pueyo JJ (2004) Differential organ-specific response to salt stress and water deficit in nodulated bean (*Phaseolus vulgaris*). *Plant Cell Environ* 27: 757–767.

Vigeolas H, Chinoy C, Zuther E, Blessington B, Geigenberger P, Domoney C (2008) Combined metabolomic and genetic approaches reveal a link between the polyamine pathway and albumin 2 in developing pea seeds. *Plant Physiol* 146: 74–82.

Walker DR, Narvel JM, Boerma HR, All JN, Parrott WA (2004) A QTL that enhances and broadens Bt insect resistance in soybean. *Theor Appl Genet* 109: 1051–1057.

Wang H, Hwang SF, Chang KF, Turnbull GD, Howard RJ (2003) Suppression of important pea diseases by bacterial antagonists. *Biocontrol* 48: 447–460.

Wei W, Huang J, Hao YJ, *et al.* (2009) Soybean GmPHD-type transcription regulators improve stress tolerance in transgenic *Arabidopsis* plants. *PLoS One* 4: e7209.

Wu QD, Van Etten HD (2004) Introduction of plant and fungal genes into pea (*Pisum sativum* L.) hairy roots reduces their ability to produce pisatin and affects their response to a fungal pathogen. *Mol Plant-Microbe Interact* 17: 798–804.

Yamagishi N, Yoshikawa N (2009) Virus-induced gene silencing in soybean seeds and the emergence stage of soybean plants with apple latent spherical virus vectors. *Plant Mol Biol* 71: 15–24.

Yang H, Boersma JG, You MP, Buirchell BJ, Sweetingham MW (2004) Development and implementation of a sequence specific PCR marker linked to a gene conferring resistance to anthracnose disease in narrow-leafed lupin (*Lupinus angustifolius* L.). *Mol Breed* 14: 145–151.

Yang H, Shankar M, Buirchell BJ, Sweetingham MW, Caminero C, Smith PMC (2002) Development of molecular markers using MFLP linked to a gene conferring resistance to *Diaporthe toxica* in narrow-leafed lupin (*Lupinus angustifolius* L.). *Theor Appl Genet* 105: 265–270.

Yoshida M, Cowgill SE, Wightman JA (1997) Roles of oxalic and malic acids in chickpea trichome exudate in host-plant resistant to *Helicoverpa armigera*. *J Chem Ecol* 23: 1195–1210.

Zahran HH (1999) Rhizobium-legume symbiosis and nitrogen fixation under severe conditions and in an arid climate. *Microbiol Mol Biol Rev* 63: 968–989.

Zahran HH, Sprent JI (1986) Effects of sodium chloride and polyethylene glycol on root-hair infection and nodulation of *Vicia faba* L. plant by *Rhizobium leguminosarum*. *Planta* 167: 303–309.

Zhen Y, Qi JL, Wang SS, *et al.* (2007) Comparative proteome analysis of differentially expressed proteins induced by Al toxicity in soybean. *Physiol Plant* 131: 542–554.

CHAPTER 17

Combating phosphorus deficiency on alkaline calcareous soils by adsorption isotherm technique for legume crops in arid environments

Muhammad Waqas[1], Abdul Hannan[2] and Abdul Gaffar Sagoo[3]

[1] *Directorate General of Agricultural Extension, Office of Agricultural Officer, Kahna Nau, Lahore, Pakistan*

[2] *Office of Assistant Land Reclamation Officer – Factory area near Madni Masjid, Sargodha, Directorate of Land Reclamation, Irrigation Department, Punjab, Pakistan*

[3] *MLL Land Reclamation Research Station, Bhakkar, Punjab, Pakistan*

17.1 Introduction

The major role of mineral fertilizers is to improve crop yields, but the main constraint in achieving proven crop potential is imbalanced use of fertilizers, particularly low phosphorus (P) use as compared to nitrogen. P is an essential element classified as a macronutrient because of the relatively large amounts required by plants.

There is pervasive P deficiency for various agronomic crops grown on alkaline calcareous soils. A large proportion of applied P becomes immobile due to high levels of P fixation by $CaCO_3$ in the soil (Rahmatullah *et al.*, 1993). The fixation of P is of great importance in making soil test interpretations and fertilizer recommendations for different crops. Critical limits for P are not based on soil physical and chemical characteristics, and also are not crop specific, but are generalized and based on scattered field trials for different areas, local information and practical experiences (Nisar *et al.*, 1992).

Memon and Fox (1983) suggested that P sorption curves can be used as a basis for transferring information about P fertilizer requirements from one site to another irrespective of variations in P-sorption capacity. A P-sorption curve can be a useful tool for utilizing existing field trial data and for making predictions about fertilizer P requirements for various crops in those areas where a small amount of data are available or where detailed experimentation has not been done. The major purpose of testing soils for P is to determine the quantity of supplemental P required to prevent economic crop losses because of P deficiency; the second purpose is to monitor the quantity of available P existing over time (Fixen & Grove, 1990).

For predicting fertilizer requirements, the Freundlich model has been used to monitor nutrient adsorption in soil. This involves developing sorption isotherms because these take into account intensity, quantity and capacity factors, which are important for predicting the amounts of soil nutrients required for maximum plant growth. It can also help to predict how the nutrient status of a soil will change with cropping over time. If the critical solution level for plant growth can be identified, nutrient-sorption isotherms can be used to estimate the fertilizer needed to adjust the soil solution to the optimal level for maximum yield.

The amounts of nutrients required by a soil are affected by the clay minerals, cation exchange capacity, organic matter, $CaCO_3$ and other properties, but these need not be calculated in order to determine soil requirements using the sorption technique. The further important benefit is that a relatively small amount of time and laboratory work are needed to produce a sorption curve for site-specific fertilizer recommendations for crops (Solis & Torrent, 1989). Therefore, information about the

Legumes under Environmental Stress: Yield, Improvement and Adaptations, First Edition. Edited by Mohamed Mahgoub Azooz and Parvaiz Ahmad.

sorption capacities of soils provides an accurate estimate of the fertilizer requirements of soils. Determining fertilizer P requirements through sorption isotherms is considered more accurate than conservative soil testing procedures.

Legume crops are essential components of all agricultural systems worldwide and are important as foods for human nutrition, as feed and fodder for animal nutrition, and as soil ameliorants for soil nutrition. But they have high P requirements (Gill *et al.*, 1985). Berseem (*Trifolium alexandrinum* L.) is one of the most important winter forage legumes in Pakistan, India, Turkey, Egypt and countries of the Mediterranean region.

As an annual, berseem fits well into crop rotations and improves the soil for succeeding crops. It has wide acceptability among farmers due to its merits over other forage crops in the winter season. The crop gives five cuts during November to April and its crude protein content ranges up to 20%, with 70% digestibility, which results in significant increases in the milk production of animals. Berseem in fact is known as a 'milk multiplier'. Being a leguminous crop it fixes atmospheric nitrogen and thus is excellent for soil fertility.

Food legumes are important components of the agricultural sectors of developing countries due to their capacity to produce significant quantities of protein-rich seeds, and to improve soil physical conditions by the addition of organic matter and biological nitrogen fixation in the soil (Snapp *et al.*, 1998). Mung bean (*Vigna radiata* L.), commonly known as green gram, is an important conventional pulse crop and is widely grown in Southeast Asia, Africa, South America and Australia. It is an important short-duration grain legume crop with wide adaptability for different growing areas, low input requirements and ability to improve the soil by fixing atmospheric nitrogen. It is well matched to a large number of cropping systems and constitutes an important source of protein in the cereal-based diets of many people in Pakistan and India. It has dietary advantages over other pulses because of its high nutritive value and digestibility and its non-flatulent characteristics. Mung bean is grown extensively for use as a human food (as dry beans or fresh sprouts), but also can be used as a green manure crop and as forage for livestock.

Due to rising costs of P fertilizers and its low availability in soil to plants, it is crucial for farmers to enhance yields and P use efficiency by adjusting site-specific and crop-specific rates for different crops. The balanced use of fertilizers, their nature and method of application all play important roles in sustainable crop production. Good crop production depends upon the apposite timing and amount of fertilization (Jan & Khan, 2000). Keeping the above considerations in view, the present study was undertaken with the following objectives:

1 To assess soil levels of P in solution ('solution P') of three differently textured soils by developing their adsorption isotherms for optimum crop growth.
2 To help formulate site-specific and crop-specific fertilizer P recommendations for farmer's fields.

17.2 Methodology

These investigations were carried out to estimate the model-based phosphorus (P) fertilizer requirement of legumes in alkaline calcareous soils. The sites were selected on the basis of land forms – i.e. recent, sub-recent and old river terraces – and calcareousness. The soils were normal from a salinity point of view, alkaline in reaction, low in organic matter and deficient in available P (Table 17.1).

The sorption isotherms were examined using the Freundlich model. It is an empirical model and corresponds to a model of adsorption in which the affinity term decreases exponentially as the amount of adsorption increases. Over a limited range of concentration, the Freundlich model often describes adsorption well (Barrow, 1973). The simple form of the Freundlich model was proposed by Le Mare (1982) as follows:

$$P = aC^b$$

Table 17.1 Some basic physical and chemical properties of the three soils.

Determinants	Soil I	Soil II	Soil III
ECe (dS/m)	1.11	1.24	1.4
pHs	8.2	8.1	7.9
OM (%)	0.56	0.72	0.45
Olsen P (mg/kg)	4.9	5.8	6.5
Extractable K (mg/kg)	62	81	115
CEC ($cmol_c$/kg)	4.2	8.1	6.1
$CaCO_3$ (%)	8.6	9.2	5.5
Sand (%)	55	36	60
Silt (%)	24	38	23
Clay (%)	21	26	17
Textural class	Sandy clay loam	Clay loam	Sandy loam

CEC, cation exchange capacity; ECe, electrical conductivity of soil saturated extract; OM, organic matter; pHs, pH of soil saturated extract.

where a is the amount of P adsorbed (μg/g), C is the concentration (mg/L) and b (mL/g), is the buffer power defined by the slope of the sorption curve at the point where $P/C = 1$ (mL/g).

The parameters a and b were estimated by regression of the logarithmic form of the data obtained from adsorption isotherms. Theoretical doses of P and phosphatic fertilizers required to develop P levels in soil solutions under field conditions were calculated from this equation.

The experiment was conducted in a permanent layout that included two crops of berseem separated by one mung bean crop (three replications) with 10 treatments arranged in a randomized complete block design (RCBD) having a plot size of 5 × 3 m. All agronomic and recommended cultural practices were followed for normal growth of the crops. Berseem was sown during second week of October at a seed rate of 20 kg/ha on all three sites. A starter dose of nitrogen (20 kg/ha) along with phosphorus calculated by using a Freundlich adsorption equation was applied at sowing time. N and P were used in the form of urea and triple super phosphate, respectively. Samples at each cutting were taken and total yield of berseem was calculated. The total P, crude fibre and crude protein of each cutting of berseem were determined. Soil sampling was done immediately after harvesting of berseem. Adsorption isotherms were constructed, and P doses were calculated based on the Freundlich model for each treatment before sowing mung bean to maintain the same soil solution P level.

After berseem harvesting, mung bean was grown using seed applied at 20 kg/ha. Phosphorus doses were applied at sowing. The crop was harvested at maturity and grain yield was recorded. Post-harvest soil samples were collected for the construction of an adsorption isotherm. Again fertilizer doses were calculated and applied before sowing berseem to maintain the same soil solution P levels.

After the harvesting of mung bean berseem was again sown on all three sites. Phosphorus doses calculated from the Freundlich equation were again to achieve the same soil solution P levels at sowing time. Samples at each cutting were taken and the total yield of berseem was recorded. The total P, crude fibre and crude protein of each cutting of berseem were determined. The analytical procedures were based on the methods described by Bigham (1996) and Richards (1954) or otherwise mentioned. All the plant samples were subject to proximate analysis using standard methods. Dry matter (%), crude protein % (CP) and crude fibre % (CF) contents were determined by standard methods in the laboratory.

17.3 Results and discussion

17.3.1 Phosphorus adsorption characteristics of the selected soils

To the study the adsorption capacity of the three soils, adsorption isotherms were constructed and the parameters of the Freundlich equation for each soil are presented in Table 17.2. The buffering capacities (b) of the three soils were 58.25, 42.40 and 39.00 L/kg for sandy clay loam, clay loam and sandy loam soils, respectively. The integrity of the fit of the model was ascertained by looking at the R^2 values, which were 0.97, 0.98 and 0.96 for sandy clay loam, clay loam and sandy loam soils, respectively, indicating high conformity of the adsorption data with the Freundlich model. The relationship observed between equilibrium phosphorus concentration (EPC) and x/m (amount of nutrient adsorbed on soil colloids) was linear (Table 17.3). The values of the exponents were found to be less than unity, which relate to the characteristics of the adsorbent, i.e. soil or $CaCO_3$. Kuo

Table 17.2 P sorption parameters of the Freundlich model.

Soil	Amount adsorbed (a) (mg/kg)	Buffering capacity (b) (L/kg)	Coefficient of determination (R^2)
Sandy clay loam	92.59	58.25	0.97**
Clay loam	104.11	42.40	0.98**
Sandy loam	70.90	39.00	0.96**

**Significant at $P = 0.05$.

Table 17.3 Model and linear parameters of the Freundlich equation.

Soil	Freundlich equation	
	Model form	Linear form
Sandy clay loam	$P^a = 92.59\ C^{58.25}$	$y = 0.5825x + 1.9666$
Clay loam	$P = 104.11\ C^{42.40}$	$y = 0.3900x + 1.8507$
Sandy loam	$P = 70.90\ C^{39.00}$	$y = 0.4240x + 2.0175$

[a]P in solution.

and Lotse (1974) reported that the exponent of the Freundlich equation was independent of the time and temperature, and the values depended on P concentration in the solution. In the Freundlich equation *a* could be considered as a capacity factor, implying that a soil having a larger value of *a* has a greater adsorbing capacity than a soil having a smaller value. In the present study, clay loam soil had the highest *a* value followed by sandy clay loam and sandy loam soils, respectively. These equations showed that adsorption decreased exponentially with increasing saturation of exchange sites. Samadi (2003) reported that P adsorption in calcareous soils was significant at relatively high P concentrations only due to $CaCO_3$.

17.3.2 Computation of P doses using the Freundlich equation

Equivalent to the specific solution levels (0.05, 0.10, 0.15, 0.20, 0.25, 0.30, 0.35, 0.40 and 0.50 mg P/L), fertilizer doses were calculated by using a linear form of the Freundlich equation (Table 17.4) for the three soils. The P quantities, computed as mg P/kg, were different for sandy clay loam, clay loam and sandy loam soils. It was evident that different soils require a different P rate to maintain the same target P solution concentration.

17.3.3 Green fodder yield (Mg/ha)

17.3.3.1 Sandy clay loam soil

Data on total green fodder yield of berseem from five cuttings for three soils are presented in Table 17.5. The P doses for obtaining maximum green fodder yields were different for the three soils. In the case of sandy clay loam soil, the maximum yield in the first and second cuts was observed at the solution level of 0.30 mg P/L, which was achieved by adding 92 kg P/ha, while

Table 17.4 P doses by Freundlich adsorption model for the three soils.

Target P conc. (mg/L)	Sandy clay loam	Clay loam	Sandy loam
	P (kg/ha)		
0.00	0	0	0
0.05	32	58	44
0.10	48	78	58
0.15	61	83	68
0.20	73	105	76
0.25	83	116	83
0.30	92	125	89
0.35	100	133	94
0.40	109	141	99
0.50	124	155	108

Table 17.5 Green fodder yield in five cuttings of berseem (1st crop) on sandy clay loam soil.

Target P conc. (mg/L)	P (kg/ha)	Yield (Mg/ha)					
		1st cut	2nd cut	3rd cut	4th cut	5th cut	Total
0.00	0	13.66 e	18.01 f	14.56 f	17.74 d	10.44 e	74.53 f
0.05	32	16.33 d	21.11 e	17.78 e	21.57 c	11.56 e	88.46 e
0.10	48	17.14 d	21.42 de	20.78 d	21.66 c	14.78 d	96.44 d
0.15	61	19.52 c	22.26 cde	21.60cd	22.91 bc	15.56cd	102.11 c
0.20	73	20.08 c	23.00 bc	22.39 bc	23.00 bc	16.44 c	105.10 c
0.25	83	22.23 b	24.01 b	22.70 abc	23.14 bc	18.04 b	110.10 b
0.30	92	24.19 a	25.74 a	23.63 ab	23.83 b	19.16 ab	116.51 a
0.35	100	23.28 ab	24.08 b	24.11 a	25.81 a	20.28 a	116.91 a
0.40	109	22.23 b	23.28 bc	23.43 ab	24.16 b	19.97 a	113.12 b
0.50	124	21.89 b	22.69 bcd	22.29 bc	23.71 b	19.44 a	112.53 b
CV (%)		5.82	5.43	5.7	5.76	6.83	2.81
LSD		1.358	1.439	1.412	1.523	1.315	3.382

Means sharing same letters are statistically ≤5% level of probability.

CV, coefficient of variation; LSD, least significant difference between treatment means.

Table 17.6 Green fodder yield in five cuttings of berseem (1st crop) on clay loam soil.

Target P conc. (mg/L)	P (kg/ha)	Yield (Mg/ha)					
		1st cut	2nd cut	3rd cut	4th cut	5th cut	Total
0.00	0	13.22g	18.29g	19.42g	20.21g	16.20g	87.45g
0.05	58	14.71 f	21.36 f	22.67 f	23.36 f	19.45 f	101.54 f
0.10	78	15.18 f	23.88 e	25.33 e	26.49 e	21.52 e	112.40 e
0.15	83	17.44 e	25.36 d	27.52 d	28.23 d	22.92 d	121.47 d
0.20	105	19.03cd	27.20 c	29.20 c	29.42 c	24.31 c	129.16 c
0.25	116	20.72 a	28.48 b	30.23 bc	30.56 b	25.26 c	135.24 b
0.30	125	19.91 b	30.28 a	31.22 ab	31.09 b	27.35 b	139.84 a
0.35	133	19.43 bc	30.15 a	31.97 a	31.97 a	28.31 ab	141.83 a
0.40	141	19.03cd	29.13 b	30.24 bc	32.12 a	29.15 a	139.68 a
0.50	155	18.33 d	28.60 b	30.06 c	30.71 b	27.96 b	135.65 b
CV (%)		3.47	3.27	3.51	2.15	4.21	1.68
LSD		0.715	0.998	1.134	0.709	1.187	2.438

Means sharing same letters are statistically ≤5% level of probability.
CV, coefficient of variation; LSD, least significant difference between treatment means.

in the remaining cuts (third, fourth and fifth) the maximum yield was observed at the solution level of 0.35 mg P/L, which was achieved by adding 100 kg P/ha. The statistical analyses indicate that the maximum total green fodder yield, i.e. 116.51 Mg/ha, was obtained at a soil solution level of 0.30 mg P/L, which was achieved by adding 92 kg P/ha in sandy clay loam soil. Higher rates could not further increase the yield significantly.

17.3.3.2 Clay loam soil

In the case of clay loam soil (Table 17.6), the maximum yield from the first cut was obtained at the solution level of 0.25 mg P/L, which was achieved by adding 116 kg P/ha; the maximum yield from the second cut was obtained at the solution level of 0.30 mg P/L, which was achieved by adding 125 kg P/ha; with the third and fourth cuts the maximum yield was obtained at the solution level of 0.35 mg P/L, which was achieved by adding 133 kg P/ha, while in the fifth cut the maximum yield was observed at the solution level of 0.40 mg P/L, which was achieved by adding 141 kg P/ha. The maximum total green fodder yield of 139.84 Mg/ha was obtained at a soil solution P level of 0.30 mg/L, which was achieved with the application of 125 kg P/ha.

17.3.3.3 Sandy loam soil

In the case of sandy loam soil, the maximum yield from the first and second cuts was observed at a solution level of 0.30 mg P/L, which was achieved by adding 89 kg P/ha, whereas the maximum yield from the third cut was obtained at a solution level of 0.35 mg P/L, which was achieved by adding 94 kg P/ha, while in the remaining cuts (fourth and fifth) maximum yield was observed at the solution level of 0.40 mg P/L, which was achieved by adding 99 kg P/ha (Table 17.7). The maximum total green fodder yield of 110.34 Mg/ha was obtained at a soil solution P level of 0.30 mg/L, which was developed with the application of 89 kg P/ha; higher rates could not improve the yield significantly. Yield increased in fertilized plots up to a solution level of 0.30 mg P/L, indicating a positive response of berseem to P application. This indicates low fertilizer P recovery in alkaline and calcareous soils, and there is also general consensus that applied P is fixed in soil because of the soil's calcareousness and clay content. It was observed that the P required to adjust the same solution P levels was much higher for clay loam as compared to sandy loam soil.

The rate of fertilizer P applied was comparatively low in sandy loam soil because adsorption was less due to less clay and $CaCO_3$ contents and thus more P remained in solution form. It can be noted that green fodder yield

Table 17.7 Green fodder yield in five cuttings of berseem (1st crop) on sandy loam soil.

Target P conc. (mg/L)	P (kg/ha)	Yield (Mg/ha)					
		1st cut	2nd cut	3rd cut	4th cut	5th cut	Total
0.00	0	9.59 g	11.65 g	13.09 g	13.50 i	8.76 g	56.58 g
0.05	44	11.91 f	13.88 f	17.10 f	18.02 h	10.86 f	71.77 f
0.10	58	14.42 e	17.92 e	18.81 e	20.21 g	12.15 e	83.55 e
0.15	68	17.34 d	19.12 de	20.47 d	22.21 f	13.30 d	92.43 d
0.20	76	18.61 cd	19.95 bcd	21.88 cd	24.34 e	14.13 d	98.91 c
0.25	83	20.59 ab	21.42 abc	23.12 bc	26.12 d	15.17 c	106.42 b
0.30	89	21.19 a	21.74 a	24.59 ab	27.01 c	15.81 bc	110.34 a
0.35	94	21.42 a	21.57 ab	25.52 a	27.89 ab	16.56 ab	112.97 a
0.40	99	20.55 ab	20.69 abcd	24.16 ab	28.30 a	16.92 a	110.61 a
0.50	108	19.73 bc	19.75 cd	23.38 bc	27.48 bc	16.48 ab	106.82 b
CV (%)		6.92	8.13	6.12	2.64	5.2	2.49
LSD		1.41	1.776	1.51	0.721	0.847	2.757

Means sharing same letters are statistically ≤5% level of probability.
CV, coefficient of variation; LSD, least significant difference between treatment means.

was greatest on clay loam soil followed by sandy clay loam and sandy loam soils. Munir *et al.* (2004) and Grant *et al.* (2001) stated that P fertilizer application was needed to sustain early growth and P was critical for obtaining higher yields.

17.3.4 Dry matter and P concentration of berseem fodder

Dry matter content is an important component of a crop that is used to determine its chemical composition and nutritive value as a feed for livestock. The data indicated that there was a gradual increase in dry matter content up to the last treatment in all three soils. The statistical analysis indicated that the maximum average dry matter of 12.59% of berseem on sandy clay loam soil was obtained at a soil solution P level of 0.50 mg/L, which was achieved with the application of 124 kg P/ha (Table 17.8), followed by clay loam soil (12.49%) at the same soil solution P level, although the rate (155 kg P/ha) of fertilizer P for clay loam soil was much higher. In the case of sandy loam soil, the maximum average dry matter content, i.e. 11.90%, was obtained at the soil solution P level of 0.50 mg/L, with the lowest rate of P (108 kg/ha) compared to the other two soils. The three sites under study were deficient in P, so its application improved the status of P in the soil from deficient to optimal, which caused improvement in the dry matter (%) of berseem crops. The increase in dry matter content confirmed the necessity for P application in P-deficient soil for obtaining optimum yields.

P concentration in plants can be related to the P extraction power of roots from soil and available soil P. Normally; plants whose roots have more extensive contact with soil are better extractors of P from soil and feed well the above-ground parts of the plant. The key function of P in plants is energy storage and transfer. Energy obtained from photosynthesis is stored in phosphate compounds, which are used in subsequent growth of the plants. Data regarding P concentration (%) averages for three soils are given in Table 17.9. Each increment of P in soil solution progressively increased the P content of the berseem. Alam and Shah (2002) observed that increased rate of P application increased both dry matter and P accumulation by plants.

17.3.5 Crude protein (%) in berseem fodder

Crude protein includes true protein and non-protein nitrogen in the plant sample. Proteins are organic compounds composed of amino acids. They are the major components of vital organs and tissues, including muscles, hair and skin, and also milk and enzymes. Protein is required on a daily basis for maintenance, lactation, growth and reproduction of animals. Proteins are

Table 17.8 Dry matter (%) of berseem on the three soils.

Target P conc. (mg/L))	Sandy clay loam		Clay loam		Sandy loam	
	kg P/ha	Dry matter (%)	kg P/ha	Dry matter (%)	kg P/ha	Dry matter (%)
Control	0	8.57h	0	8.40h	0	8.49 i
0.05	32	8.99g	58	8.93g	44	8.75 hi
0.10	48	9.26g	78	9.19g	58	9.07 gh
0.15	61	9.73 f	83	9.73 f	68	9.37g
0.20	73	10.33 e	105	10.21 e	76	9.81 f
0.25	83	10.84 d	116	10.78 d	83	10.26 e
0.30	92	11.38 c	125	11.02 d	89	10.63 d
0.35	100	11.73 bc	133	11.43 c	94	11.09 c
0.40	109	12.02 b	141	11.90 b	99	11.50 b
0.50	124	12.59 a	155	12.49 a	108	11.90 a

Means sharing same letters are statistically ≤5% level of probability.

Table 17.9 P concentration (%) of berseem on the three soils.

Target P conc. (mg/L)	Sandy clay loam		Clay loam		Sandy loam	
	kg P/ha	P conc.	kg P/h	P conc.	kg P/ha	P conc.
Control	0	0.175h	0	0.185h	0	0.189g
0.05	32	0.213g	58	0.212g	44	0.204fg
0.10	48	0.227g	78	0.229 f	58	0.216 f
0.15	61	0.246 f	83	0.250 e	68	0.241 e
0.20	73	0.257 ef	105	0.263 de	76	0.254 de
0.25	83	0.269 de	116	0.273 d	83	0.268cd
0.30	92	0.279cd	125	0.288 c	89	0.279 bc
0.35	100	0.294 bc	133	0.299 bc	94	0.291 bc
0.40	109	0.307 ab	141	0.312 b	99	0.302 ab
0.50	124	0.318 a	155	0.331 a	108	0.315 a

Means sharing same letters are statistically ≤5% level of probability.

complex organic macromolecules that contain carbon, hydrogen, oxygen, nitrogen, and usually sulphur and are composed of one or more chains of amino acids. They are fundamental components of all living cells and include many substances, such as hormones and antibodies, that are necessary for the proper functioning of an organism. It is evident from the data that crude protein contents of berseem increased with an increase in solution P levels up to a certain extent: i.e. a maximum crude protein (20.61%) was obtained at a soil solution P level of 0.30 mg/L, which was developed by adding 92 kg P/ha in the case of sandy clay loam soil (Table 17.10). On clay loam soil the highest crude protein content of 19.42% was obtained at a soil solution P level of 0.35 mg/L, which was developed with the application of 133 kg P/ha. In the case of sandy loam soil, the maximum crude protein of 16.49% was also obtained at the same soil solution P level of 0.35 mg/L, which was developed with the application of 94 kg P/ha. Chaudhary and Karwasra (1984) reported that crude protein contents were increased with the application of P fertilizer.

Table 17.10 Crude protein (%) of berseem on the three soils.

Target P conc. (mg/L)	Sandy clay loam		Clay loam		Sandy loam	
	kg P/ha	Crude protein (%)	kg P/ha	Crude protein (%)	kg P/ha	Crude protein (%)
Control	0	17.34 f	0	15.75g	0	13.57 f
0.05	32	18.16 e	58	16.56 f	44	14.30 e
0.10	48	18.73 d	78	17.02 e	58	14.78 de
0.15	61	19.25 c	83	17.63 d	68	15.18cd
0.20	73	19.76 b	105	18.19 c	76	15.54 c
0.25	83	20.19 ab	116	18.67 b	83	16.15 ab
0.30	92	20.61 a	125	19.29 a	89	16.49 a
0.35	100	20.54 a	133	19.42 a	94	16.58 a
0.40	109	20.01 b	141	19.13 a	99	16.10 ab
0.50	124	19.74 b	155	18.67 b	108	15.62 bc

Means sharing same letters are statistically ≤5% level of probability.

Table 17.11 Crude fibre (%) of berseem on the three soils.

Target P conc. (mg/L)	Sandy clay loam		Clay loam		Sandy loam	
	kg P/ha	Crude fibre (%)	kg P/ha	Crude fibre (%)	kg P/ha	Crude fibre (%)
Control	0	19.57 i	0	20.31 i	0	20.05 i
0.05	32	20.38h	58	21.14h	44	20.68h
0.10	48	20.76 gh	78	21.42h	58	21.07g
0.15	61	21.26fg	83	21.93g	68	21.69 f
0.20	73	21.60 ef	105	22.34 f	76	22.23 e
0.25	83	21.92 de	116	22.79 e	83	22.62 d
0.30	92	22.35cd	125	23.28 d	89	23.19 c
0.35	100	22.86 bc	133	23.68 c	94	23.49 c
0.40	109	23.28 ab	141	24.03 b	99	23.95 b
0.50	124	23.59 a	155	24.41 a	108	24.54 a

Means sharing same letters are statistically ≤5% level of probability.

17.3.6 Crude fibre (%) in berseem fodder

Crude fibre comprises fibrous components such as cellulose, hemicellulose, pectin, gums and lignin that are found within plant cell walls. These materials are not truly digested by animals and therefore do not contribute energy when consumed. Fibre plays a role in regulating the duration of transit of food through the intestinal tract, aids in fat absorption, and affects the water-holding capacity of stools to aid in the prevention of diarrhoea and constipation. The maximum crude fibre content of 23.54% in berseem was obtained at a soil solution P level of 0.50 mg P/L in sandy clay loam soil, which was developed with the application of 124 kg P/ha (Table 17.11). The maximum crude fibre content of 24.41% in berseem was obtained at a soil solution P level of 0.50 mg P/L in clay loam soil, which was developed with the application of 155 kg P/ha. In the case of sandy loam soil, the maximum crude fiber of 24.54% was obtained at a soil solution P level of 0.50 mg/L, which was developed with the application of 108 kg P/ha. Crude fibre contents were greater in berseem grown on clay loam as compared to sandy clay loam and sandy loam soils. The increase in crude fibre contents might be due to more dry matter accumulation with the gradual

Table 17.12 Model-based P rates for mung bean for sandy clay loam soil.

Target P conc. (mg/L)	Regression equation	R^2	Rate (kg P/ha)	
			P	P_2O_5
Control	Control	–	0	0
0.05	$y = 0.4139x + 1.8255$	97**	38	87
0.10	$y = 0.4106x + 1.8362$	0.97**	53	121
0.15	$y = 0.4253x + 1.8149$	0.94**	58	133
0.20	$y = 0.4176x + 1.8188$	0.95**	67	153
0.25	$y = 0.4294x + 1.8105$	0.96**	71	163
0.30	$y = 0.4149x + 1.8274$	0.95**	82	188
0.35	$y = 0.4194x + 1.8169$	0.94**	84	192
0.40	$y = 0.4184x + 1.8232$	0.94**	91	208
0.50	$y = 0.4246x + 1.8002$	0.93**	94	215

**Significant at $P = 0.05$.

Table 17.13 Model-based P rates for mung bean for clay loam soil.

Target P conc. (mg/L)	Regression equation	R^2	Rate (kg P/ha)	
			P	P
Control	Control	–	0	0
0.05	$y = 0.501x + 1.9393$	0.95**	39	89
0.10	$y = 0.5478x + 1.9065$	0.95**	46	105
0.15	$y = 0.555x + 1.9111$	0.96**	57	131
0.20	$y = 0.5784x + 1.933$	0.96**	68	156
0.25	$y = 0.5894x + 1.9126$	0.95**	73	167
0.30	$y = 0.5253x + 1.8974$	0.97**	84	192
0.35	$y = 0.486x + 1.9088$	0.97**	97	222
0.40	$y = 0.202x + 1.9199$	0.94**	113	259
0.50	$y = 0.4183x + 1.9079$	0.94**	121	277

**Significant at $P = 0.05$.

increase of P application. Ayub *et al.* (2002) observed that crude fibre contents increased with P application along with N.

17.3.7 Response of mung bean to applied P

Adsorption isotherms were again constructed for calculation of P rates in all three soils to achieve levels of solution P as previously set. It was observed that the rates of P decreased (Tables 17.12, 17.13 and 17.14) for mung bean crops as compared to those for the first berseem crops in all three soils.

Table 17.14 Model-based P rates for mung bean for sandy loam soil.

Target P conc. (mg/L)	Regression equation	R^2	Rate (kg P/ha)	
			P	P
Control	Control	–	0	0
0.05	$y = 0.6167x + 1.9447$	0.97**	28	64
0.10	$y = 0.6534x + 1.9005$	0.98**	35	80
0.15	$y = 0.6204x + 1.9146$	0.98**	51	117
0.20	$y = 0.6411x + 1.8941$	0.98**	56	128
0.25	$y = 0.646x + 1.8897$	0.98**	63	144
0.30	$y = 0.617x + 1.9447$	0.97**	69	158
0.35	$y = 0.6565x + 1.877$	0.97**	76	174
0.40	$y = 0.657x + 1.8764$	0.98**	82	188
0.50	$y = 0.6744x + 1.8408$	0.99**	87	199

**Significant at $P = 0.05$.

17.3.8 Grain yield of mung bean (second crop)

The data regarding grain yield of mung bean as affected by different P concentrations in soil solution showed (Table 17.15) that increasing the P level significantly affected grain yield, and that the maximum grain yield of 1.138 Mg/ha was obtained at a solution P level of 0.30 mg/L, which was achieved by adding 82 kg P/ha in sandy clay loam soil. Further increases in the level of fertilizer P could not improve grain yields of mung bean. On clay loam soil chock-full grain yield of 1.191 Mg/ha was obtained at a solution P level of 0.30 mg/L, which was achieved by adding 84 kg P/ha. Whereas in case of sandy loam soil a grain yield of 0.732 Mg/ha was obtained at a solution P level of 0.35 mg/L, which was achieved by adding 76 kg P/ha, but further increases in the level of fertilizer P could not improve the grain yield of mung bean. P regulates photosynthesis and carbohydrate metabolism and is considered one of the major growth-limiting factors particularly during the reproductive stage of the growth. These findings explain the greater grain yield at higher P levels.

17.3.9 Protein content (%) of mung bean

The maximum protein content of 23.56% was obtained at a soil solution P level of 0.30 mg/L, which was developed by adding 82 kg P/ha in sandy clay loam soil (Table 17.16). The maximum protein content of 23.46% was obtained at a solution P level of 0.30 mg/L, which

Table 17.15 Effect of applied P on grain yield (Mg/ha) of mung bean on the three soils.

Target P conc. (mg/L)	Sandy clay loam		Clay loam		Sandy loam	
	kg P/ha	Grain yield	kg P/ha	Grain yield	kg P/ha	Grain yield
Control	0	0.487 i	0	0.492 h	0	0.309 h
0.05	38	0.611 h	39	0.714 g	28	0.394 g
0.10	53	0.731 g	46	0.774 f	35	0.452 f
0.15	58	0.780 f	57	0.833 e	51	0.536 e
0.20	67	0.886 e	68	0.910 d	56	0.575 d
0.25	71	1.034 b	73	1.076 b	63	0.624 c
0.30	82	1.138 a	84	1.191 a	69	0.691 b
0.35	84	1.023 bc	97	1.063 b	76	0.732 a
0.40	91	0.994 cd	113	0.995 c	82	0.650 c
0.50	94	0.981 d	121	0.985 c	87	0.619 c

Means sharing same letters are statistically ≤5% level of probability.

Table 17.16 Effect of applied P on protein content of mung bean on the three soils.

Target P conc. (mg/L)	Sandy clay loam		Clay loam		Sandy loam	
	kg P/ha	Protein content (%)	kg P/ha	Protein content (%)	kg P/ha	Protein content (%)
Control	0	21.74 e	0	21.58 e	0	20.94 d
0.05	38	21.92 de	39	21.73 de	28	21.41 d
0.10	53	22.19 cd	46	21.93 de	35	22.07 c
0.15	58	22.37 c	57	22.04 d	51	22.31 bc
0.20	67	22.74 b	68	22.56 c	56	22.46 bc
0.25	71	23.01 b	73	23.01 b	63	22.79 ab
0.30	82	23.56 a	84	23.46 a	69	22.74 ab
0.35	84	23.42 a	97	23.52 a	76	23.15 a
0.40	91	22.90 b	113	23.04 b	82	22.44 bc
0.50	94	22.77 b	121	22.89 bc	87	22.12 c

Means sharing same letters are statistically ≤5% level of probability.

was developed by adding 84 kg P/ha in clay loam soil, whereas in case of sandy loam soil the maximum protein yield of 23.15% was obtained at a solution P level of 0.35 mg/L, which was developed by adding 76 kg P/ha. Increase in grain protein contents might be due to enhanced uptake and translocation of nitrates, which provide nitrogen for amino acid synthesis. Moreover, P is involved in the synthesis of ATP, which is required in nitrogen uptake and protein synthesis.

17.3.10 Response of berseem to applied P

The number of replications was six in this study at the start of the experiment. Three replications were meant for P build-up studies and the remaining three were reserved for P depletion studies. No data were reported in this experiment for depletion studies. For P build-up, adsorption isotherms were constructed for site-specific fertilizer application to achieve the same soil solution P levels as were developed previously for first berseem and second mung bean crops. Linear forms of the Freundlich equation were used for computing P fertilizer doses (Tables 17.17, 17.18 and 17.19). It was observed that the rates of P doses decreased still further for the second berseem crop due to P build-up in the soil.

It was residual P that caused this decreased adsorption in the soils. The P doses required to develop the same soil solution levels were greatest in clay loam soil followed by sandy clay loam and sandy loam soils. The rate

Table 17.17 Model-based P rates for berseem for sandy clay loam soil.

Target P conc. (mg/L)	Regression equation	R^2	Rate (kg P/ha)	
			P	P_2O_5
Control	Control	–	0	0
0.05	$y=0.5202x+1.8042$	0.97**	27	62
0.10	$y=0.5542x+1.7775$	0.97**	33	76
0.15	$y=0.5702x+1.7579$	0.96**	39	89
0.20	$y=0.5642x+1.7658$	0.97**	47	108
0.25	$y=0.5553x+1.7693$	0.96**	54	124
0.30	$y=0.4881x+1.7408$	0.96**	61	140
0.35	$y=0.5072x+1.7205$	0.95**	62	142
0.40	$y=0.5103x+1.7211$	0.93**	66	151
0.50	$y=0.5065x+1.7162$	0.95**	73	167

**Significant at $P=0.05$.

Table 17.18 Model-based P rates for berseem for clay loam soil.

Target P conc. (mg/L)	Regression equation	R^2	Rate (kg P/ha)	
			P	P_2O_5
Control	Control	–	0	0
0.05	$y=0.4561x+1.7594$	0.95**	29	66
0.10	$y=0.4707x+1.7411$	0.95**	37	85
0.15	$y=0.4901x+1.7369$	0.94**	43	98
0.20	$y=0.4936x+1.7325$	0.95**	49	112
0.25	$y=0.4793x+1.7509$	0.95**	58	133
0.30	$y=0.4821x+1.792$	0.95**	69	158
0.35	$y=0.4906x+1.7869$	0.96**	73	167
0.40	$y=0.4329x+1.8554$	0.93**	96	220
0.50	$y=0.4465x+1.8478$	0.94**	103	236

**Significant at $P=0.05$.

of P fertilizer decreased for the second berseem crop grown after mung bean and the first berseem crop.

17.3.11 Green fodder yield (third crop)

Data on berseem total green fodder yield in five cuttings for three soils are given in Table 17.20. The fertilizer P doses for obtaining maximum green fodder yields were different in the three soils. The statistical analysis indicated that the maximum green fodder yield, i.e. 110.50 Mg/ha, was obtained at a soil solution level of 0.30 mg P/L, which was developed by adding 140 kg P_2O_5/ha in sandy clay loam soil. The maximum yield of 141.88 Mg/ha in clay loam soil was obtained at a soil solution level of 0.30 mg P/L, which was developed with the application of 158 kg P_2O_5/ha. In case of sandy loam soil, the maximum green fodder yield of 123.45 Mg/ha was obtained at a soil solution level of 0.35 mg P/L, which was developed with the application of 142 kg P_2O_5/ha.

Table 17.19 Model-based P rates for berseem for sandy loam soil.

Target P conc. (mg/L)	Regression equation	R^2	Rate (kg P/ha)	
			P	P
Control	Control	–	0	0
0.05	$y=0.6195x+1.9107$	0.98**	25	57
0.10	$y=0.6867x+1.8732$	0.98**	31	71
0.15	$y=0.6459x+1.8678$	0.99**	43	98
0.20	$y=0.6445x+1.8254$	0.98**	47	108
0.25	$y=0.6532x+1.8107$	0.95**	52	119
0.30	$y=0.661x+1.7963$	0.96**	56	128
0.35	$y=0.6658x+1.7952$	0.97**	62	142
0.40	$y=0.6672x+1.7936$	0.97**	67	153
0.50	$y=0.6776x+1.7739$	0.98**	71	162

**Significant at $P=0.05$.

17.3.12 Dry matter and P concentration of berseem fodder

The dry matter of berseem contains all the nutrients except water. Data regarding dry matter (%) of berseem in three soils are presented in Table 17.21. The statistical analysis indicated that maximum dry matter of 11.41% in the case of sandy clay loam soil was obtained at a soil solution level of 0.40 mg P/L, which was developed with the application of 66 kg P/ha. In the case of clay loam soil, the maximum dry matter of 11.21% was obtained at a soil solution level of 0.40 mg P/L, which was developed with the application of 96 kg P/ha, while in the case of sandy loam soil, the maximum dry matter of 10.51% was obtained at a soil solution level of 0.50 mg P/L, which was developed with the application of 71 kg P/ha. Data regarding the effect of applied P on P concentration (%) of berseem in three soils are presented in Table 17.22. A maximum P concentration (0.300%) was observed at a solution level of 0.50 mg P/L, which was developed by adding 73 kg P/ha in the

Table 17.20 Effect of applied P on green fodder yield (Mg/ha) of berseem on the three soils.

Target P conc. (mg/L)	Sandy clay loam		Clay loam		Sandy loam	
	kg P/ha	Yield	kg P/ha	Yield	kg P/ha	Yield
Control	0	71.26 f	0	99.05h	0	79.12h
0.05	27	80.57 e	29	109.39g	25	89.33g
0.10	33	90.20 d	37	116.45 f	31	95.69 f
0.15	39	96.56 c	43	122.91 e	43	102.86 e
0.20	47	102.29 b	49	129.43 d	47	107.70 d
0.25	54	106.33 ab	58	137.55 b	52	115.00 c
0.30	61	110.50 a	69	141.88 a	56	120.89 ab
0.35	62	110.33 a	73	140.65 a	62	123.45 a
0.40	66	106.86 ab	96	135.64 bc	67	120.18 ab
0.50	73	102.66 b	103	133.42 c	71	118.56 bc

Means sharing same letters are statistically ≤5% level of probability.

Table 17.21 Effect of applied P on dry matter (%) of berseem on the three soils.

Target P conc. (mg/L)	Sandy clay loam		Clay loam		Sandy loam	
	kg P/ha	Dry matter (%)	kg P/ha	Dry matter (%)	kg P/ha	Dry matter (%)
Control	0	9.02h	0	8.84 f	0	8.61h
0.05	27	9.37g	29	9.31 e	25	8.95 gh
0.10	33	9.72 f	37	9.59 de	31	9.30fg
0.15	39	10.09 e	43	9.92cd	43	9.55 ef
0.20	47	10.15 de	49	10.00 c	47	9.73 de
0.25	54	10.47cd	58	10.39 b	52	9.78 cde
0.30	61	10.51 c	69	10.48 b	56	10.02 bcd
0.35	62	10.86 b	73	10.68 b	62	10.08 bc
0.40	66	11.41 a	96	11.21 a	67	10.31 ab
0.50	73	11.49 a	103	11.38 a	71	10.51 a

Means sharing same letters are statistically ≤5% level of probability.

Table 17.22 Effect of applied P on P concentration (%) of berseem on the three soils.

Target P conc. (mg/L)	Sandy clay loam		Clay loam		Sandy loam	
	kg P/ha	P conc. (%)	kg P/ha	P conc. (%)	kg P/ha	P conc. (%)
Control	0	0.192 j	0	0.196 j	0	0.182 j
0.05	27	0.204 i	29	0.212 I	25	0.192 i
0.10	33	0.218h	37	0.224h	31	0.204h
0.15	39	0.228g	43	0.236g	43	0.214g
0.20	47	0.238 f	49	0.248 f	47	0.222 f
0.25	54	0.250 e	58	0.256 e	52	0.234 e
0.30	61	0.260 d	69	0.268 d	56	0.242 d
0.35	62	0.272 c	73	0.278 c	62	0.252 c
0.40	66	0.282 b	96	0.292 b	67	0.260 b
0.50	73	0.300 a	103	0.306 a	71	0.274 a

Means sharing same letters are statistically ≤5% level of probability.

Table 17.23 Effect of applied P on crude protein (%) of berseem on the three soils.

Target P conc. (mg/L)	Sandy clay loam		Clay loam		Sandy loam	
	kg P/ha	Crude protein (%)	kg P/ha	Crude protein (%)	kg P/ha	Crude protein (%)
Control	0	16.12 g	0	15.89 h	0	14.68 g
0.05	27	16.83 f	29	16.70 g	25	15.09 f
0.10	33	17.27 f	37	17.24 fg	31	15.42 ef
0.15	39	17.76 e	43	17.58 ef	43	15.74 de
0.20	47	18.25 d	49	18.13 de	47	16.08 cd
0.25	54	18.70 cd	58	18.76 bc	52	16.44 bc
0.30	61	19.21 ab	69	19.29 ab	56	16.74 ab
0.35	62	19.63 a	73	19.77 a	62	17.10 a
0.40	66	18.76 bc	96	19.06 bc	67	16.39 bc
0.50	73	18.39 cd	103	18.63 cd	71	15.90 d

Means sharing same letters are statistically ≤5% level of probability.

case of sandy clay loam soil. In the case of clay loam soil a maximum P concentration of 0.306% was observed at a solution level of 0.50 mg P/L, which was developed by adding 103 kg P/ha, whereas in the case of sandy loam soil, a maximum P concentration of 0.274% was observed at a solution level of 0.50 mg P/L, which was developed by adding 71 kg P/ha. It is clear from the observations that there is a gradual increase of P concentration from control up to the last treatment as the soil solution P level increases and consequently P doses also increase.

17.3.13 Crude protein (%) in berseem fodder

Crude protein is a key nutritional parameter that must be provided in adequate amounts in various animal diets. Crude protein, sometimes called total protein, is estimated from measuring the total nitrogen content of fodder. Data regarding effect of applied P on crude protein (%) of berseem in three soils are presented in Table 17.23.

The data indicated that crude protein (%) contents increased significantly with increase in P application rates up to certain extent. The statistical analysis indicated that the maximum crude fibre, i.e. 19.63%, was obtained at a soil solution level of 0.35 mg P/L, which was developed by adding 62 kg P/ha in the case of sandy clay loam soil. In the case of clay loam soil, the maximum crude fibre content of 19.77% was obtained at a solution level of 0.35 mg P/L, which was developed by adding 73 kg P/ha.

In the case of sandy loam soil, maximum crude fibre of 17.10% was obtained at a soil solution level of 0.35 mg P/L, which was developed by adding 62 kg P/ha; higher rates could not increase crude protein contents significantly.

17.3.14 Effect of applied P on crude fibre (%)

Crude fibre is the carbohydrate component that is hard to digest, and is made up of mostly lignin and cellulose. Crude fibre, or roughage, is 'crude' because it underestimates total fibre.

Fibre is critical for a healthy functional rumen. Fibre in the forage will improve the animal's rumen environment by stimulating increased saliva flow. This saliva is important as a 'rumen buffer', which maintains a proper pH for microbial digestion of feed. Fibre therefore affects gastro-intestinal motility and assists feed digestion and absorption. It absorbs water in the gastro-intestinal tract and provides bulk for normal faeces production, diarrhoea prevention and constipation prevention. Data regarding the effect of applied P on crude fibre (%) of berseem in three soils are presented in Table 17.24. (Table 17.24). In the case of sandy clay loam soil, the maximum crude fibre of 22.72% was obtained at a soil solution level of 0.50 mg P/L, which was developed by adding 73 kg P/ha. In the case of clay loam soil a maximum crude fibre of 22.87% was obtained at a soil solution level of 0.50 mg P/L and that solution level was developed by adding 103 kg P/ha. In the case of sandy loam soil, the maximum crude fibre of 22.54% was

Table 17.24 Effect of applied P on crude fibre (%) of berseem on the three soils.

Target P conc. (mg/L)	Sandy clay loam		Clay loam		Sandy loam	
	kg P/ha	Crude fibre (%)	kg P/ha	Crude fibre (%)	kg P/ha	Crude fibre (%)
Control	0	19.33 j	0	19.83 f	0	19.14h
0.05	27	19.65 i	29	20.20 e	25	19.52g
0.10	33	20.10h	37	20.49 e	31	19.82g
0.15	39	20.35g	43	20.87 d	43	20.15 f
0.20	47	20.76 f	49	21.13 d	47	20.44 f
0.25	54	21.12 e	58	21.59 c	52	20.83 e
0.30	61	21.43 d	69	21.89 c	56	21.23 d
0.35	62	21.92 c	73	22.28 b	62	21.56 c
0.40	66	22.34 b	96	22.46 b	67	21.99 b
0.50	73	22.72 a	103	22.87 a	71	22.54 a

Means sharing same letters are statistically ≤5% level of probability.

obtained at a soil solution level of 0.50 mg P/L, which was developed with the application of 71 kg P/ha.

17.4 Conclusions

Fertilizer rates were different for different crops on the three selected soils. Thus, fertilizer applications should be site-specific and crop-specific to avoid overuse or underuse. It was concluded that fertilizer recommendations should be based on soil tests, keeping in consideration the soil solution P level for differently textured soils adjusted on the basis of the adsorption isotherm.

References

Alam SM, Shah SA (2002) Phosphorus uptake and yield of wheat as influenced by integrated use of phosphatic fertilizer. Paper presented at the 9th International Congress of the Soil Science Society of Pakistan held at NIAB, Faisalabad, March 20, 2002.

Ayub M, Nadeem MA, Shara MS, Mahmood N (2002) Response of maize (*Zea mays* L.) fodder to different levels of nitrogen and phosphorus. *Asian J Plant Sci* 1: 352–354.

Barrow NJ (1973) Relationship between soils ability to adsorb phosphate and the residual effectiveness of superphosphate. *Aust J Soil Res* 11: 57–63.

Bigham JM (1996) *Methods of Soil Analysis. Part 3, Chemical Methods*. SSSA and ASA, Madison, WI.

Chaudhary SL, Karwasra SPS (1984) Effects of phosphorus application on yield and quality of forage sorghum. *Nepalese J Agric* 15: 27–32.

Fixen P, Grove JH (1990) Testing soils for phosphorus. In: Westerman RL (ed.), *Soil Testing and Plant Analysis*, 3rd edn. Book Series No. 3. Soil Science Society of America, Madison, WI, pp. 141–180.

Gill MA, Ali N, Nayyar MM (1985) Relative effect of phosphorus combined with potash and *Rhizobium phaseoli* on the yield of *Vigna aureus* (mung). *J Agricul Res* 23: 279–282.

Grant CA, Flaten BN, Tomasiewicz DJ, Sheppard SC (2001) The importance of early season phosphorus nutrition. *Can J Plant Sci* 81: 211–214.

Jan MT, Khan S (2000) Response of wheat yield components of N-fertilizer, their levels and application time. *Pak J Biol Sci* 3: 1227–1230.

Kuo S, Lotse EG (1974) Kinetics of phosphate adsorption and desorption by lake sediments. *Soil Sci Soc Am Proc* 38: 50–54.

Le Mare PH (1982) Sorption of isotopically exchangeable and non-exchangeable phosphate by some soils of Colombia and Brazil, and comparisons with soils of Southern Nigeria. *J Soil Sci* 33: 691–707.

Memon KS, Fox RL (1983) Utility of phosphate sorption curves in estimating the phosphorus requirements of cereal crops. Wheat (*Triticum Aestivum*). In: *Proceedings of the 3rd International Congress on Phosphorus Compounds*, Brussels, Belgium, October 4–6, 1983. The World Phosphate Institute (IMPHOS), pp. 217–230.

Munir I, Ranjha AM, Sarfraz M, Rehman O, Mehdi SM, Mahmood K (2004) Effect of residual phosphorus on sorghum fodder in two different textured soils. *Int J Agricul Biol* 6: 967–969.

Nisar A, Malik DM, Jalil A (1992). Phosphorus recommendations based on soil test crop response calibration. Tech.

Bull. 8/91. National Fertilizer Development Center, Islamabad, Pakistan, pp. 1–21.

Rahmatullah MA, Gill BZ, Sheikh, Saleem (1993) Bioavailability and distribution of P among inorganic fractions in calcareous soils. *Arid Soil Res Rehab* 8: 227–234.

Richards LA (1954) Diagnosis and improvement of saline and alkali soils. U.S. salinity lab. staff Handbook No. 60. USDA, Washington, DC.

Samadi A (2003) Predicting phosphate fertilizer requirement using sorption isotherms in selected calcareous soils of western Azerbaijan province, Iran. *Commun Soil Sci Plant Anal* 34: 2885–2899.

Snapp PL, Mafongoya S, Waddington J (1998) Organic matter technologies for integrated nutrient management in small holder cropping systems of southern Africa. *Agric Ecosyst Environ* 71: 185–200.

Solis P, Torrent J (1989) Phosphate sorption by calcareous vertisols and inceptisols of Spain. *Soil Sci Soc Am J* 53: 456–459.

Index

Note: Page numbers in *italics* refer to Figures; those in **bold** to Tables.

Legumes under Environmental Stress: Yield, Improvement and Adaptations, First Edition. Edited by Mohamed Mahgoub Azooz and Parvaiz Ahmad.

Printed and bound by CPI Group (UK) Ltd, Croydon, CR0 4YY
30/09/2026
14973386-0002